Natural Science in Archaeology

Series editors: B. Herrmann, G. A. Wagner

Andreas Hauptmann

The Archaeometallurgy of Copper

Evidence from Faynan, Jordan

With 170 Figures and 40 Tables

Springer

Volume editor

Prof. Dr. Andreas Hauptmann
Deutsches Bergbau-Museum Bochum
Forschungsstelle für Archäologie und Materialwissenschaften
Hernerstraße 45, 44787 Bochum, Germany
E-mail: andreas.hauptmann@bergbaumuseum.de

Series editors

Prof. Dr. Bernd Herrmann
Universität Göttingen
Institut für Anthropologie
Bürgerstraße 40
37073 Göttingen, Germany
E-mail: bherrmann@gwdg.de

Prof. Dr. Günther A. Wagner
Max-Planck-Institut für Kernphysik
Forschungsstelle Archäometrie
Heidelberger Akademie der Wissenschaften
Saupfercheckweg 1, 69117 Heidelberg, Germany
E-mail: g.wagner@mpi-hd.mpg.de

Inner frontispiece
Satellite view (Ikonos satellite data) of the ancient mining district of Faynan, Jordan, and its surroundings. Red dots in the image show mining and smelting sites. These are located mainly in the area of Wadi Jariye in the north, at Khirbet en-Nahas, at and near the Byzantine ruins of Faynan and at Umm el-Amad in the south. Clearly visible are the "Roman-Byzantine" terraces which might have been constructed as early as in the 4th millennium BCE.
All satellite images are shown in this volume by courtesy of European Space Imaging / © European Space Imaging GmbH.

Library of Congress Control Number: 2007925905

ISSN 1613-9712 Springer Berlin Heidelberg New York
ISBN 978-3-540-72237-3 Springer Berlin Heidelberg New York

This work is subject to copyright. All rights are reserved, whether the whole or part of the material is concerned, specifically the rights of translation, reprinting, reuse of illustrations, recitations, broadcasting, reproduction on microfilm or in any other way, and storage in data banks. Duplication of this publication or parts thereof is permitted only under the provisions of the German Copyright Law of September 9, 1965, in its current version, and permission for use must always be obtained from Springer. Violations are liable to prosecution under the German Copyright Law.

Springer is a part of Springer Science+Business Media
springer.com
German edition © Andreas Hauptmann Bochum 2000
Publication of the Deutsches Bergbau-Museum Bochum no. 155
English edition © Springer-Verlag Berlin Heidelberg 2007
All rights reserved

The use of general descriptive names, registered names, trademarks, etc. in this publication does not imply, even in the absence of a specific statement, that such names are exempt from the relevant protective laws and regulations and therefore free for general use.

Cover design: deblik, Berlin
Typesetting: Stasch · Bayreuth (stasch@stasch.com)
Production: Christine Adolph

Printed on acid-free paper   32/2132/CA – 5 4 3 2 1 0

Preface

The results presented in this book originate from field research carried out by the Deutsches Bergbau-Museum Bochum (hereafter called DBM) on the remains of early copper metallurgy in the area of Faynan, southern Jordan, between 1983 and 1993. It also contains results of later laboratory studies and fieldwork. This volume is a revised version of the German volume "Zur frühen Metallurgie des Kupfers in Faynan, Jordanien," which appeared in 2000 as supplement 11 of the journal "Der Anschnitt." The translation into English honors the wishes of many colleagues from the Near East and from English-speaking countries who have expressed the desire that a larger audience should have access to the research carried out on the ancient mining district and mining techniques. The translation was made possible by a generous grant from the foundation of the "Institute for Aegean Prehistory" (Philadelphia, USA). Prof. Dr. Susanne Kerner, Mr. Hughe Barnes, and, in parts, Dr. Hans-Dieter Kind translated the book into the English version. Much of the editorial work was done by MA Kerstin Batzel and Mr. Guillaume Ewandé.

The original research in the Faynan area was initiated within the framework of a larger project "Research into the archaeometallurgy and archaeology of mining in Faynan and the southern Arabah-valley, Jordan" funded by the VolkswagenStiftung (Hanover, Germany). The original planning and groundbreaking work was carried out by the author. Since 1997, research was continued by and in close cooperation with Prof. Dr. Thomas Levy, University of California, San Diego (USA).

Research into early metal extraction covers a number of different problems, requiring a broad, interdisciplinary approach and calling for the cooperation of various disciplines. Metal extraction from mining of the ore deposits to the metal's production, using differing smelting techniques, required craft specialization throughout the differently organized social structures of different periods of time, while the distribution of metal objects via trade was dependent on the social, cultural and political frameworks of the time.

Close cooperation with archaeologists, mainly mining archaeologists, is the indispensable precondition for the understanding of material evidence. Archaeometallurgy uses field methods (e.g., geology of ore deposits), various physical dating methods, as well as procedures for mineralogical, chemical and isotopic analyses to process a broad variety of finds. These methods and the connected problems are discussed in the first part of this volume, before the results of our own research are presented.

The initial idea for the study of early copper mining in the mining region of Faynan came from Prof. Dr. Hans-Gert Bachmann. He had worked intensively on the archaeometallurgical remains in Timna (Israel) and Sinai (Egypt), at a time when the political conditions made it impossible to investigate Timna and Faynan together. Bachmann recognized that a cultural and social interpretation of the early copper production in the Southern Levant would have been incomplete without the knowledge of the processes of metal production in Faynan (Bachmann and Hauptmann 1984). Bachmann visited Faynan for the first time in 1982, and his impression of the dimensions of the smelting activities there resulted in a declaration by Rothenberg with the significant title "Ancient Jordan City May Rival Timna's Place in Copper History" (Rothenberg 1983). It was the activities of Bachmann that led to this research project, and he continued his interest in it as far as his workload allowed him.

The first brief survey in preparation for the fieldwork was conducted in September of 1983. The first larger field season was carried out during the autumn of 1984. The next campaign took place in the spring of 1986, making it possible for Prof. Dr. Wolfgang Frey and his team (Institute for Systematic Biology and Geobotany of the Free University of Berlin) to participate. They pursued in a parallel research project, also funded by the VolkswagenStiftung, the important question of how sufficient firing material for the smelting of such huge quantities of slag had been obtained. Faynan, on the eastern side of Wadi Arabah, lies in a semi-arid region. This question became even more pressing when even the initial fieldwork showed 150 000 to 200 000 t of slag in the area of Faynan.

The third campaign was conducted in the spring of 1988, together with three Jordanian colleagues from the Department of Antiquities of Jordan, after the signing of an official cooperation agreement. Dr. Mohammed Najjar, Emsaytif Suleiman and Abd es-Samir started with excavations at an Early Bronze Age smelting site as well as the Late Neolithic settlement mound (Tell Wadi Faynan). Dr. Mohammed Najjar has been the co-director of the project since the beginning of the official cooperation agreement.

The fourth field season took place in the spring of 1990, again with the Jordanian colleagues and the botanists from Berlin. In addition, a group of archaeologists under Prof. Dr. Volkmar Fritz (then from the University of Mainz) participated. They carried out excavations in the Early Bronze Age settlement area of Barqa el-Hetiye and in Khirbet en-Nahas. A group of archaeologists from the University of Sheffield under Dr. Russell Adams started at the same time, then still independent of the Faynan-project, excavations on some Neolithic and Early Bronze Age settlements.

The fifth and last field season in the framework of this project had been planned for the spring of 1991, but had to be postponed due to the Gulf War that had started in January of that year, and finally took place in the spring of 1993. The aim of this last campaign had been to finish various field projects. That was possible for purely archaeometallurgical research, where analysis in the laboratory was the main bulk of the work. The mining archaeology project as well as the excavation of two pre-pottery settlements (Wadi Ghwair 1 and Wadi Fidan A) and the Late Chalcolithic-Early Bronze Age settlement (Wadi Fidan 4) had to be suspended.

Due to the complexity of the research design, many different scientists have been involved in this project. Firstly, my colleague Prof. Dr. Gerd Weisgerber of the DBM was the scientist who worked on the mining archaeology and with whom I carried out the entire fieldwork. Prof. Dr. Wolfgang Frey, Dr. Uli Baierle, Dr. Thomas Engel, Dr. Harald Kürschner and Christian Jagiella of the Freie Universität Berlin dealt with the botanical questions concerning the problem of early firing materials. Prof. Dr. Volkmar Fritz, Mainz, together with his team carried out the archaeological research concerning the Early Bronze Age and Iron Age settlement activities; the work was later partly continued by Matthias Flender, Tübingen/Düsseldorf. The Early Bronze Age settlement of Wadi Fidan 4 was excavated by Dr. Russell Adams, now at Ithaca College, Ithaca (New York) and Dr. Hermann Genz, now at the American University of Beirut.

In the ensuing years a number of provenance studies of several metal objects from around the Levant have been carried out. These analyses continue, and one aim is to prove whether the materials used in these objects originated from Faynan and clarify the role of this mining region in the Southern Levant and Egypt. The incorporation of objects from Israel was most commendably made possible by Dr. Miriam Tadmor, then of the Israel Museum in Jerusalem.

The isotope analysis was carried out by Prof. Dr. Friedrich Begemann and Dr. Sigrid Schmitt-Strecker of the Max-Planck-Institut für Chemie, Mainz; the neutron-activation-analysis by Prof. Dr. Ernst Pernicka, now with the University of Tübingen/ Reiss-Engelhorn-Museum in Mannheim; the thermoluminescence-dating by Dr. Irmtrud and Prof. Dr. Günther Wagner, then with the Heidelberger Academy of Science at the Max-Planck-Institut für Kernphysik. They, as well as Dr. Zofia Stos and Prof. Dr. Noel Gale, then at the Isotrace Laboratory, University of Oxford, Dr. Alberto Palmieri, Consiglio Nazionale delle Ricerche, Rome, Prof. Dr. Thomas Levy, University of San Diego, California and Dr. Russell Adams supplied a good amount of unpublished data at my disposal for this volume. For this, heartfelt thanks are expressed here.

Of all the mentioned studies, mainly those dealing directly with the ancient metallurgy have made it into this volume.

Our work has been accompanied and supported by a number of Jordanian and German individuals and institutions, to whom it is a great pleasure to give thanks. I want to mention firstly and particularly the former ambassador of Germany, the late Dr. Herwig Bartels, whose untiring interest will be always remembered. He died August 2, 2003 before this volume was finished and will be remembered by all of us. Four consecutive directors of the Department of Antiquities of Jordan have supported our research with great goodwill and often helped with unofficial dealings: Dr. Adnan Hadidi, Dr. Ghazi Bisheh, Prof. Dr. Safwan Tell, and Dr. Fawwas Al-Khreyshah.

Working in the remote and then militarily controlled area of Faynan was made possible through the assistance of the Natural Resources Authority, Amman, which at the time was represented by the acting director Mohammed Abu Ajamieh. He contributed to the logistics necessary to provide housing in the geologists' camp in Wadi Dana and supplied essential support for the research into the metal deposits. Technical help and housing was also provided by Sharif Hussein Bin Nasr, Wadi Arabah Cooperative, and his employees, who repaired the roads and tracks destroyed by floodwater at the beginning of each season and thus created the conditions that made fieldwork possible.

Contact and home in Amman was the German Protestant Institute for the Archaeology of the Holy Land. It was above all Dr. Susanne Kerner, who, if the situation required it, speedily and if necessary unconventionally assisted took care of the various members of the different campaigns.

I am most grateful to a number of team members for their work in the field, which due to the remote situation of the area was often organizationally rather difficult.

The engagement of the following people furthered the study of the archaeometallurgy of Faynan: Siegfried Averbek, Dr. Andrea Büsing-Kolbe, Dr. Andreas Brunn, Dr. Jan Cierny, Omar Daghestani, Markus Eichholz, Dr. Andreas Haasis-Berner, Jürgen Heckes, Thomas Henning, Anke Joisten-Pruschke, Werner Lieder, Christoph Roden, Dirk Rostoff, Hans Schwarz, and Angelika Weisgerber. Special thanks for many things are due to Hans-Joachim Kunkel.

Emile Masadeh, who was not only the inspector of the Department of Antiquities for our project for many years but also our friend, remains unforgotten. He died April 9, 1993.

The research project was mainly funded by the VolkswagenStiftung, Hanover. The Deutsche Forschungsgemeinschaft (DFG) supported several trips necessary to sample metal objects for provenance studies. The Vereinigung der Freunde von Kunst und Kultur im Bergbau (VFKK), Bochum, helped to balance financial deficits created through unexpected developments during the field-work.

One aim of our research into the early mining and smelting in Faynan was to pinpoint the cultural as well as economic significance of this region. For a long time archaeology neglected investigations into (pre)-historic technologies to obtain raw materials, particularly metal, most probably because of the difficulties in organizing the necessary interdisciplinary approach.

It is therefore to our delight that Faynan has become the focal point of several archaeological projects since 1994 and that a number of international institutions from the United Kingdom and the United States of America have started excavations with the aim of finding more information about the cultural and socio-political history of the area, which is today a nature reserve. It is also noticeable that the early metallurgy of the Southern Levant has been the topic of numerous studies and Ph. D. theses, which have dealt not only with the technological aspects of archaeometallurgy, but also highlighted the social and cultural repercussions.

Since 1997, extensive surveys and excavations have been conducted by Prof. Dr. Thomas Levy and his team in the greater Faynan region. The aim was at first to document *all* settlement structures in the model area of Wadi Fidan, the "gateway to the north," in order to investigate the role of early copper production particularly in the Early Bronze Age. This led to to the sensational unearthing of the copper working settlements of Khirbet Hamra Ifdan in the years 1999 and 2000 (see there), where the largest metal workshop of the Near East was discovered (Levy et al. 2002). The excavation of Khirbet en-Nahas started in 2002, shedding new light on the development of the Edomite state. And in 2003 a number of salvage excavations in Wadi Fidan were carried out in response to plans to build a dam and therefore were partly financed by the Jordan Valley Water Authority. The revised version of the Faynan-book very much profited from new satellite images which were donated by Dr. Hans-Dieter Kind. Mrs. Anette Hornschuch (DBM, Department of Information Systems) and Mike Oversberg kindly performed digital image processing and added a set of (three-dimensional) photographs of the Faynan area.

The geographical terms used in the following text are simplified transcriptions from Arabic without any diacritical signs.

Andreas Hauptmann, Bochum, May 2007

Contents

1	Introduction	1
2	**Problems and Methods of Archaeometallurgy**	7
2.1	Fieldwork	9
	– The Investigation of 'Ancient' Ore Deposits	9
	– The Documentation of Ancient Smelting Sites	13
2.2	Analytical Methods	16
	– Physical Dating	16
	– The Investigation of Ancient Slag	18
	– Chemical Analysis of Ores and Metals	27
	– Lead Isotope Analysis	31
3	**Nature and Geology in Faynan**	39
	– Toponymy	39
	– Geography	40
	– Topography	42
	– Climate	45
	– Water, Irrigation Systems and Agriculture	47
	– Wind Patterns	49
	– Vegetation and Fuel Supply for the Ancient Copper Smelters	50
4	**The Raw Material Sources**	55
4.1	Geological Overview of the Near East	55
4.2	Ancient Ore Deposits in the Near East	59
4.3	The Copper Deposit of Faynan and its Relation to Timna	63
	– Geological Framework and Genesis	63
	– The Ores	68
	– Geochemistry	73
	– Lead Isotope Analysis	79
5	**Field Evidence and Dating of Early Mining and Smelting in the Faynan District**	85
5.1	Landscape and Dating	85
5.2	Site Catalogue and Archaeometallurgical Finds	91
	I. Khirbet Faynan and Vicinity	94
	II. Find Spots in Wadi Ghwair	111

	III. The Mining District of Qalb Ratiye (JD-25, JD-GR)	112
	IV. The Mining District in Wadi Abiad (JD-26, JD-WA)	115
	V. The Mining District of Wadi Khalid (JD-3, JD-II)	116
	VI. Sites in the Lower Wadi Dana (JD-13, JD-III)	122
	VII. Ras en-Naqab (JD-5)	123
	VIII. El-Furn (Ngeib Asiemer, JD-6)	126
	IX. Khirbet en-Nahas (JD-2)	127
	X. The mining district of Umm ez-Zuhur and Madsus (JD-41)	130
	XI. Khirbet el-Jariye (JD-11)	131
	XII. Wadi el-Ghuwebe (JD-27)	132
	XIII. Sites in Wadi Fidan	133
	XIV. Barqa el-Hetiye (JD-31)	141
	XV. The Mining District of Umm el-Amad (JD-10)	144
5.3	The Development of the Mining District of Faynan: Field Evidence	145

6 Study of Archaeometallurgical Slag and Metal ... 157
6.1 Slags of the 4th Millennium BCE ... 157
- The 'Most Ancient' Slags ... 157
- The Material from the Faynan District ... 158
- Chemistry and Melting Behavior ... 159
- The Petrography of the Slag ... 164
- 'Free Silica Slags' ... 167
- The Development of Mineralogical Phases ... 169
- Sulphide Inclusions ... 178

6.2 Manganese-Rich Silicate Slag from the Early Bronze Age
to the Mameluk Period ... 180
- Manganese-Rich Slag ... 180
- The Material from the Faynan District ... 181
- Chemistry and Melting Behavior ... 182
- The Petrography of the Slag ... 186
- The Development of Mineralogical Phases ... 190

6.3 The Composition of Copper ... 199
- Trace Element Content ... 199
- The Material from the Faynan District ... 200
- Chemical Composition ... 200
- The Homogeneity of Copper ... 202
- Partitioning of Trace Elements between Metal and Slag ... 204
- Iron in Copper ... 207
- Lead Istotopy ... 211

7 Copper Smelting Technology ... 217
7.1 The Earliest Stage: Crucible Smelting ... 217
- The Archaeological Evidence ... 217
- Previous Experimental Studies and Thermodynamics of Crucible Smelting ... 219
- Reconstruction of Crucible Smelting Based on Archaeological Finds ... 223
- Modeling Other Possibilities ... 227

7.2	Early Bronze Age: Smelting in Wind-Powered Furnaces	228
	– The Beginning or Initial Early Bronze Age	228
	– Evidence for Wind-Powered and Natural Draught Furnaces, Respectively	229
	– Find-Based Reconstruction of Metallurgical Processes	232
	– The Influence of Manganese-Rich Slag on Smelting Processes	234
	– Comparisons with Smelting Experiments	236
	– Processing of Copper: Evidence from Khirbet Hamra Ifdan	239
7.3	Technologically Controllable Smelting: Iron Age to Mameluk Period	242
	– The Archaeological Evidence	242
	– Crushing Slags	245
	– Reconstruction of Smelting Processes by Iron Age Finds	246
	– Fluxing Agents	249
	– Comparison with Smelting Experiments	251
8	Export of Ore and Copper: the Importance of Faynan in Prehistoric Palestine	255
8.1	Trade of Ore from Faynan in the Neolithic	255
	– The Importance of the Colour Green in the Pre-Pottery Neolithic	255
	– Ores from Faynan in Settlements of the Southern Levant	257
8.2	Faynan As a Source for Copper in the Chalcolithic and Early Bronze Age I	261
	– Archaeological Overview	261
	– Metal Production in Settlements: Domestic Metallurgy	263
	– Chemical and Lead Isotope Analyses	264
	– Tell Abu Matar	268
	– Bir Safadi	269
	– Tuleilat Ghassul	271
	– Tell Abu Hamid	271
	– Nahal Besor/Wadi Ghazzeh	272
8.3	Metal Trade in the Early Bronze Age II–IV	272
	– Archaeological Aspects	272
	– Chemical and Lead-Isotope Analyses	275
	– Bab edh-Dhra	280
	– Numeira	283
	– Jericho	284
	– Crescent-Shaped Ingots from Faynan	285
8.4	Conclusions from the Distribution Pattern of Ore and Metal Obtained from Faynan	288
	– Ore	288
	– 'Pure' Copper	291
	– Arsenic-Antimony-Copper-Alloys	294
	– Arsenical Copper	295
	– Copper-Arsenic-Nickel-Alloys	297
	– Tin Bronzes	301
	– Faynan and Timna	302

Summary 305

References 309

Appendix 341

Index 377
Geographical Index 377
Subject Index 381

Chapter 1

Introduction

The rift valley of the Wadi Arabah divides a copper ore district, which has been mined from the proto- and pre-historic periods until modern times. In the hopes of economic reward, the area has been prospected until the very recent past. The most important source of the ore mineralizations in the western area are in Timna, roughly 30 km north of modern Eilat. The main source of copper ores on the eastern side of Wadi Arabah is Faynan, ca. 80 km south of the Dead Sea. Further south, in the Wadi Abu Khusheibah and in the Wadi Abu Qurdiyah, minor ore deposits have also been found (Fig. 1.1). One aspect of Wadi Arabah is its function as a bridge connecting five geographical as well as culturally very diverse, but nonetheless closely connected regions: Israel (the ancient Canaan), Jordan (the ancient kingdoms of Edomite

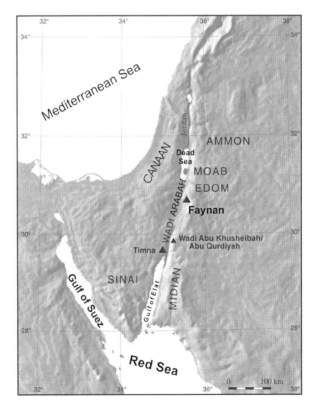

Fig. 1.1.
The course of the Wadi Arabah between the Dead Sea and the Red Sea. Barrier or interface between the surrounding historical landscapes/regions? The copper regions of Faynan, Timna, and the wadis Abu Khusheibah/Abu Qurdiyah are also shown

and Moab), Hijaz (the ancient Midian), and finally Sinai. Another aspect of the Wadi Arabah and the rift valley further north is their role as an outstanding trade route connecting the Southern Levant with Lebanon, Syria and even Anatolia. Many – altogether some hundreds – of the settlement remains on both sides of Wadi Arabah (Thompson 1975; Raikes 1980; MacDonald 1992) have direct connections with the ore deposits and bear witness to the extensive ore- and metal trade in the (pre-)historic periods of this region.

The area of Timna has been intensively researched by Beno Rothenberg since the 1950s (Conrad and Rothenberg 1980; Rothenberg 1973, 1988a, 1990 and other publications). Therefore, Timna is not only the first example of a study using archaeological methods and those methods provided by the natural sciences of a (pre-)historic region of metal production, the ancient mines and smelting sites of Timna are also very probably the most intensively researched remains denoting the early history of mining.

Faynan and the ore deposits further south were initially studied from a modern economical viewpoint, but questions concerning the ancient copper mining had already been touched upon (e.g., Bender 1968, 1974). The earliest historical description of the ancient mining and production system on the eastern side of the Arabah comes from the times of the persecution of the Christians, when the mines of Faynan were infamous as the place of confinement for the Christians condemned to work in the mines ("damnatio ad metalla", Geerlings 1983). The description comes from one of the fathers of the church, Eusebius, who was bishop of Caesarea in the 3rd/4th centuries CE (Geerlings 1985). The remains of the copper extraction were mapped for the first time in the 18th century (D'Anville 1732). No reports of earlier visits from other Europeans to that inhospitable area are recorded. Although numerous travelers going to Palestine for pilgrimage or discovery have described and sketched in the region on some of the earliest plans ever produced, the geographical details of the Wadi Arabah and particularly its still visible archaeological remains have only been recorded in the relatively recent past. From ca. 1890 onwards, Max Blanckenhorn studied the geography and geology of the Syrian-Jordanian-Palestinian area and documented among other things the Dead Sea and the Wadi Arabah (Blanckenhorn 1912). At the same time, the construction of the Hijaz railroad began to provide more thorough access to the area. The orientalist and geographer Alois Musil (1907) published geological and geographical studies of the region. The first relatively concrete descriptions of Faynan are therefore from Blanckenhorn and Musil. A few decades later, Fritz Frank (1934) and Nelson Glueck (1935, 1938, 1941) recognized the historic and archaeological importance of the mines and smelting sites. Their attempts to date these activities covered the 2nd millennium BCE, the early Iron Age and the era of the biblical kings David and Solomon to the Nabatean period and the following era of Roman occupation (Table 1.1). It was assumed that the early medieval period saw a resumption of mining activities – under Byzantine as well as under Arab rule (Bachmann and Hauptmann 1984).

During investigations of the German Geological Mission in Jordan between 1961 and 1967, the geologist Hans-Dieter Kind described and dated a number of archaeometallurgical monuments and published them in a very important publication (Kind 1965). He particularly emphasized two areas at the eastern edge of Wadi Arabah: the mines and smelting sites in Faynan and those in the area southwest of

Table 1.1. The chronological periods of Palestine with their most important aspects of cultural development from the end of the Epipaleolithic to the Hellenistic period (after Vieweger 2003)

Period	Dating	Characteristics
Natufian	11000 – 8500	Small settlements; hunter and gatherer
Pre-Pottery, Neolithic	8500 – 6000	Permanent settlements, agriculture, animal husbandry, "greenstone", lime plaster
Pottery, Neolithic	6000 – 4500	Spreading of settlements; introduction of pottery
Chalcolithic	4500 – 3500	Villages, agriculture, animal husbandry; development of copper metallurgy, earliest use of gold
Early Bronze Age I	3500 – 3100	Proto-urban settlements
EBA II, III	3100 – 2300	Development of larger, fortified towns; first use of Bronze (copper-tin-alloy), lead, silver, gold
EBA IV/Middle Bronze Age I	2300 – 2000	Transitional phase; non-urban
Middle Bronze Age II	2000 – 1550	Settlement growth; imports from Cyprus; Egyptian expansion; introduction of alphabet
Late Bronze Age I	1550 – 1400	Egyptian dominance over Palestine
Late Bronze Age II	1400 – 1250	End of city-states and Egyptian dominance; influence of Mycenaean cultures in Palestine. "Sea-people"
Iron Age I	1250 – 1000	First Iron artifacts. De-urbanization. Phoenicians
Iron Age II	1000 – 587	Founding of new cities and small territorial states (Edom, Moab). Assyrian and later Babylonian supremacy. Larger distribution of iron
Iron Age III	587? – 332	Persian supremacy
Hellenistic	332 – ≈40	Alexander the Great. Ptolemaic and Seleucid wars. Nabataeans (312 BCE–AD 106). Roman occupation since 63 BCE
Roman	≈40 BCE–AD 350	Herod; Legio III Cyrenaica; Decapolis-cities; Bishop Eusebius of Caesarea (263–339)
Late Antique/Byzantine	350 – 640	Byzantine Period
Early Islamic	640 – 750	Short period of Sassanid occupancy. Prophet Mohammed, Khalifs, Umayyad empire
Islamic Middle Ages	750 – 1291	Abbasids; Ayyubids; Crusades; Mameluks
Ottoman Empire	1250 – 1922	Mameluks, Ottomans

the ancient capital of the Nabatean kingdom Petra, in the wadis Abu Khusheibah and Abu Qurdiyah (Fig. 1.1). The latter had been, due to their relative unimportance, only superficially surveyed. Kind, who had incorporated the observations of earlier travelers in his report, underlined the necessity of precise dating. While he was only able to assume the date of the earlier mining activities, he was able to use coins to emphasize the importance of Faynan in the Hellenistic (4th–1st centuries BCE), Nabatean (until the 2nd century CE), late Roman (3rd–5th centuries CE), and Byzantine to Ottoman periods (6th–20th centuries CE).

Frank and Glueck have also briefly surveyed Wadi Fidan west of Wadi Faynan. Both described only two archaeological sites there. The first extensive surveys in this area that were so important for the earliest metal extraction were carried out by Raikes (1976, 1980) and later MacDonald (1992). Their results were an important base for the research of the last few years.

The aim of the project presented here was to use *one* example – namely that of the copper ore- and mining district of Faynan – to investigate ancient mining and the earliest methods of metal extraction as well as demonstrate the development from the early domestic mode of production to the later mass production. The hypothesis was that it would be possible to demonstrate the different stages of development in metallurgy, formulated in a model by Strahm (1994), in the metal production area of Faynan. This model includes a precursory stage of metallurgy, when ores and the native metal (mainly copper) closely associated with them were used for adornment, like other exotic goods (Schoop 1995). Then follows in an initial stage the working of the metal in a way, which started to recognize the potential metal contains as a metal. An experimental stage contained the mining and early smelting of oxide ores, characterized by trial and error methods. The use of sulphide ores in a later stage of development led to more complex metal technologies, and the final industrial stage represents a definite breakthrough in the development of metallurgy. Developing economic processes hat should be tested to a certain extent, i.e., processes of increasing production and technology, resource management, exploitation and production ("imprinting phases", Stöllner 2003) effected and were valid in this area.

It was a conscious decision to concentrate on the mining and production of metal, because it is hardly possible to come up with a meaningful cultural-anthropological interpretation covering a period of 9 000 years – as was the framework of this project – although this should be the final goal of archaeometallurgy and archaeology. Not all periods are equally represented in the metallurgical history of Faynan. Therefore, in the following text, no (historical) balanced overview is planned; rather those aspects of the work are presented, which deal directly with metal extraction.

The phases and period names given in the text follow the traditional terminology as it is used, e.g., by Weippert (1988) (Table 1.1). This terminology uses a division based on the historical metal ages, and later also political-historical events. The end of the so-called Iron Age 586 BCE is not based on the transition towards the use of another metal, but rather on the end of the Judean state and the change of the political power play connected with it. In later paragraphs it will become clear that the chronological system as it is used here should be changed. This has been proved by recent research, particularly the re-evaluation of ^{14}C-dates by new calibrations. It must therefore be mentioned again that the intention of this volume is an approximate chronological outline in order to show the course and current of the develop-

Chapter 1 · Introduction

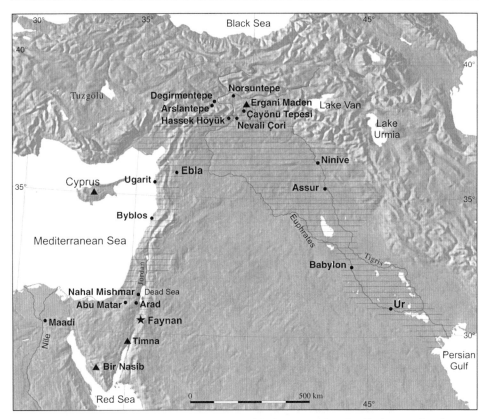

Fig. 1.2. Geographical overview of the Near East with the area of the "Fertile Crescent" and localities mentioned in the text. The area of Faynan is far south of the "Fertile Crescent" in the valley of the Wadi Arabah. In the northern and western parts of the region, sufficient rainfall allows agriculture, while Mesopotamia depended on the rivers Euphrates and Tigris with their tributaries for its water supply

ment of mining and smelting. Details of the archaeological dating for certain sites will not be given, because the studied material has only been partly published (Bachmann and Hauptmann 1984; Hauptmann al. 1985; Knauf and Lenzen 1987; Najjar et al. 1992; MacDonald 1992; Fritz 1994a,b, 1996; Levy 2002, 2004).

The metallurgy of the 4th, 3rd and 1st millennia BCE will be presented in detail, while other periods will only be mentioned in passing. The concept is to establish a scaffolding of facts to help in understanding metallurgical finds and information as well as to provide orientation for later archaeological research.

The question of economic importance and influence in the southern Levant of the ancient mining area of Faynan has great significance. Where were ores and metal exported to? During the last years a large number of metal objects has been found in Chalcolithic sites and settlements (Fig. 1.2), e.g., from Nahal Mishmar (Tadmor et al. 1995), Tell Abu Matar, Bir Safadi (Perrot 1972), Shiqmim (Levy and Shalev 1989) etc., which due to their high contents of arsenic and antimony cannot have been produced from the local ores of the Arabah region. Was the southern Levant an independent "center" of early metallurgy, where ores and metal from the Arabah region were used? If so, then where do the other kinds of metal come from and how have they been traded? These questions are also important for the metal trade of the Early Bronze Age city of Arad, where Amiran et al. (1973) assume that the city got its copper from the southern Sinai. This assumption was later supported by Ilan and Sebanne (1989) and Kempinski (1989). It is surprising that the copper deposits in Faynan played no role in these discussions. The question of Faynan's importance for the development of the state of Edom is also not fully solved. During the 2nd millennium BCE the Faynan metal sources were controlled from the west (Knauf 1992), because Cypriotic goods could no longer be imported after the breakdown of sea trade, as a result of the invasion of the sea people. But in the 7th century BCE Faynan was an important source of metal for the Assyrian army, the trade being organized by the kingdom of Edom.

This framework of questions required – through intensive surveys – a detailed examination of the geological and depositional conditions in Faynan as well as the archaeometallurgical remains. It quickly became fairly obvious that surface surveying would only identify the most intensive phases of metal extraction: the "imprinting phases" as they had been defined by Stöllner (2003). The remains of the earliest metallurgy could only be located by excavations. Consequently, one of the most important issues was to date as many mines and smelting sites as possible. Although logistical problems considerably hindered fieldwork in the 1980s, the unusually good preservation of the old mines and slag heaps greatly assisted the work.

The excellent state of preservation of the ancient remains is due to the fact that the copper deposits of Faynan are in no way comparable with the economically far more important copper mines of Chile, Zimbabwe, Zambia, USA or Russia. The Faynan region has thus not suffered any modern mining activities. The large area potentially usable for settlement activities and the wide spatial distribution of the ore outcrops resulted in a wide distribution of sites. The majority of sites in Faynan are often one-period sites or date from just a few periods – in contrast to the far more common multi-period mound sites in the Near East in general.

A large number of ore samples, slag and metal were collected from the sites. The mineralogical, chemical and isotope analyses of these samples in the laboratories have allowed a fairly good reconstruction of different steps of metal production by smelting, melting and processing and the definition of a geochemical "fingerprint" for the material in Faynan. This "fingerprint" enabled us to differentiate the ores and the metal from Faynan by their specific composition from other sources in the eastern Mediterranean area.

The important role of the ancient mining district of Timna will also become clear in the following study. The area is closely connected with Faynan both geologically and, in part, historically, and they show several common characteristics. The comparison of Faynan and Timna in several questions is therefore particularly relevant.

Chapter 2

Problems and Methods of Archaeometallurgy

The production and use of metals requires a series of different human activities. They are connected with chemical and physical transformations of materials during the change from ore into metal. These interactions can be summarized – parallel to the 'chaine operatoire' (see Stöllner 2003) – in a 'metallurgical chain' (Fig. 2.1). The activities start at the ore deposit with the mining of ore followed by technological steps used in smelting processes, when slag, raw metal and other intermediate products are produced. Different methods are used to reconstruct reaction vessels used in this process. Metal is subsequently treated in different steps – if necessary alloyed with other metals – until a final product is achieved (e.g., ingot, axe, chisel etc.).

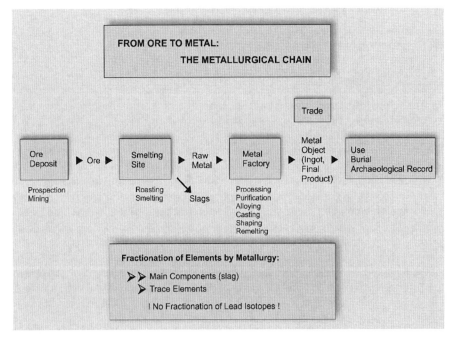

Fig. 2.1. The metallurgical chain describes single steps of human activities and the transformation of materials beginning with the extraction of ores from the deposit to the production of a finished object that was possibly traded, and which then may finally be unearthed in an archaeological context. The basis for the analyses is the fractionation of chemical elements during metallurgical processes, while the abundance ratios of lead isotopes remain constant

This might be traded and may be found later in an individual archaeological context. It is important to differentiate in this chain of interactions between mining, smelting and further processing, because each of these steps in the production line is controlled by different factors. Mining is closely bound to the ore deposit, while smelting and the sometimes very complex processing are more dependent on the framework of settlements and therefore were carried out under different social conditions. This will become clear from the survey in the mining district of Faynan in Chap. 5.

The main questions and problems of archaeometallurgy that arise from the metallurgical chain are the following:

1. Reconstructions of technological processes and craftsmanship applied to produce metal (objects);
2. Distribution of metal from a source, i.e., the reconstruction of ancient trade networks and, vice versa, provenance studies to find the source from where the raw materials of a metal object originated.

They are connected with

3. The chronology of mining and metallurgy over the millennia.

They are followed by archaeological topics such as

4. The spatial organization and social pattern of mining and metal production;
5. (Over-)regional cultural and economic impact.

Botanical and geoarchaeological investigations deal with

6. Fuel supply for metallurgy, its impact on local/regional environments, i.e., deforestation and subsequent damage to soils by erosion.

The archaeometallurgical evaluation of an ore deposit is based on the documentation of the deposit itself, archaeological studies of findings and materials used in the mining and smelting processes, and a survey of the landscape around the deposit. In order to reconstruct the organizational pattern of metal production and its socioeconomic context, the spatial distribution of mines, smelting sites, and metal workshops in a mining district in certain periods has to be analyzed, and in-depth studies (excavation of settlements, slag heaps, mines) are as necessary as special surveys. Unlike an archaeological survey, a method was developed from the 1920s on, which concentrates mainly on settlement history and distribution. The archaeometallurgical survey also calls for the application of specialized scientific and technical methods and concepts. The identification and dating of metallurgical finds and even more so of archaeological contexts as well as the sampling of those for later study in the laboratory requires specialized knowledge. A typological study of finds would not be sufficient, and a scientific analysis of the material is necessary. But even analytical studies provide only limited information. They decipher chemophysical parameters, but rarely steps of craftsmanship. This is even more the case when intermediate or final products (blooms, ingots, bars, final artifacts) are involved in the production process, or when products of metallurgical work are being recycled (recycling of slags, repeated remelting of metal).

The transfer of chemical and physical parameters of modern metal technology for the analysis in archaeological contexts is necessary, but often insufficient and has caused grave mistakes in the recent history of archaeometallurgical research. To better understand ancient techniques, it is strongly recommended to use in addition literary sources from the medieval period and especially the Renaissance in Europe. The books written by Theophilus Presbyter (12th century CE), Georg Agricola (AD 1556), Vanoccio Biringuccio (1540), and Lazarus Ercker (1574) contain a large number of observations concerning mining and metallurgical processes. There are detailed commentaries and reinterpretations, e.g., by Bartels et al. (2006).

On the other hand, successful interpretations of archaeometallurgical finds are based both on ethnographic (Anfinset 1996; Bisson 2000) and on experimental studies (Merkel 1990; Bamberger and Wincierz 1990; Herdits 1997; Shalev 1999). These two aspects allow the courses of metallurgical processes and technologies to be observed in detail from the very beginning until the making of final objects. All objects and products, controlled by individual craftsmanship and by chemo-physical laws, from slag and refractories to casting moulds, etc., can be directly assigned to specific steps in metal production. This will be shown in more detail in Chap. 7.

In this study, investigations will be presented that deal with technological studies of copper mining and metallurgy, as well as with provenance studies, typified in the copper district of Faynan. These are natural science studies mainly concerned with the geosciences, metallurgy, physics and chemistry. For readers versed in the social sciences, it might sometimes seem that these studies are made for their own sake, and that the final goal, e.g., the reconstruction of the socioeconomic organization or the kind and extent of trade connections in a larger frame have been neglected. Such a critique has been voiced, e.g., by Budd and Taylor (1995) and recently by Philip et al. (2003). The fact is that without an analysis of metallurgical finds and the evidence from fieldwork and without secure physical measurements of ores and metal (and organic materials for e.g., the dating), it would be impossible to reconstruct the specific steps of a "chaine d'operatoire" and thus to interpret ancient metal production. Knowledge about trade connections is also helped by studies of natural science. This book is therefore meant as a challenge and motivation for archaeologists to use natural science data in order to come to new interpretations in social sciences.

In the following, specific methodological problems of archaeometallurgy will be discussed and the different scientific methods and techniques applied will be explicated.

2.1 Fieldwork

The Investigation of 'Ancient' Ore Deposits

In the Old World, there are no outcrops of ore deposits existing, which have remained unworked. Usually they have been more or less completely exploited in earlier periods. These ancient mining activities near the surface are called in German 'Alter Mann' (= goaf, old workings). Possibly, the oldest description of 'ancient' mining activities comes from the Greek historian and geographer Agatharchides. In his description of the inhumane conditions of the Ptolemaic gold mining at the Red Sea he mentioned that long before his time people used copper chisels (and not iron tools!) for mining (Woelk 1966). We know also from medieval sources that old workings were already an unwelcome legacy of earlier mining, and even then made the exploitation of any ore deposit directly from the surface impossible (Weisgerber 1997).

In relation to our present problem of assessing 'ancient' ore deposits, this means that all we have remaining for further study are the residues of already exploited mineralizations. It is therefore difficult to assess the original richness of mineralizations near the surface or even the likelihood of native metals such as copper, silver, and gold being available for exploitation at the very beginning of the metal ages. Rough estimations are possible, using the literature on mining from the last two centuries and information deriving from the geochemical distribution and the production quantity of metals, a method first used by Patterson (1971) and developed later by Pernicka (1995).

The aim of studies on 'ancient' ore deposits is to come up with conclusions about the quality and quantity of ores accessible in prehistoric times. Thus, one deals with the oxidation- and secondary enrichment- or cementation-zone, because it is unlikely that ancient mining reached deeper than 100–150 m (Weisgerber and Pernicka 1995). It is important to check the availability of rich ores, even if they are locally restricted, and of oxidic ores and of native metals. Ore samples, where possible, should be taken directly from the ancient mines, in order to get material as similar as possible to that exploited by the ancient miners. In practice, this is difficult and often requires archaeological excavations in the mines, a fact that instantly limits sampling due to time and resources needed for excavation. So still today – against the opinion of Otto and Witter (1952) – one of the most difficult problems in archaeometallurgy is how to procure samples of appropriate ores from 'ancient' deposits in sufficient numbers for statistical calculations and mass estimations or in order to compare them with the analyses of artifacts (Pernicka 1995). Therefore, given the limited number of samples, it needs to be stressed that for the solution of archaeometallurgical problems, ore deposits should not only be analyzed for their isotope content but also undergo a far more thorough geochemical, mineralogical, and geological examination than has so far been the case (Moores 1985; Pernicka 1993; Stos-Gale and Gale 1994). It was possible at Faynan to deal with these problems rather satisfactorily, as the modern prospecting work had not only opened up old mines, but also created a number of fresh outcrops. It was possible to characterize the ore deposits chemically and by their isotopic composition as a result of the large number of copper inclusions in the slag: these will always produce better averages than single ore samples, as had already been pointed out by Pernicka (1992).

Modern geological textbooks only rarely contain detailed information about the composition of ore deposits near the outcrop. They normally deal with the geological origin of deposits in the cycle of plate tectonics or with economic aspects. Aspects relevant for the Bronze Age or other early periods such as, e.g., the existence of rich ores or the concentrations of those trace elements, which end up in the metal after smelting, are habitually overlooked.

Massive sulphide ore deposits in Cyprus can be taken as an example. Constantinou (1982) asserts that there is almost no malachite or azurite available, so his statement that Cyprus would have been one of the earliest copper producers of the world should be considered very critically. But Stos-Gale and Gale (1994) carried out surveys and found these oxidic ores in large enough quantities to have been of importance for the Early Bronze Age. It is more difficult to assess the role native copper played in Cyprus. It can still be found today in small, filigree dendrites, but its geochemical and isotopic composition is not compatible with that of the few Chalcolithic artifacts, which have been found on the island. Gale (1991) assumes therefore that the metal had been imported from other sources. At times, the economic importance and use of the Cu sulphate in Cyprus might have been overestimated, although metallurgically there is no problem in reducing these to copper. There is no proof yet of their processing by means of hydrometallurgical methods as has been suggested by Koucky and Steinberg (1982a).

For 'feasibility studies,' an evaluation of the metal content of an ore is usually determined by large-scale slit samples. Such analyses are of no great value for the assessment of prehistoric metal extraction, as average percentages calculated in that way do not cover the possibility of the existence of rich ores. It would be difficult, therefore, to form an opinion as to whether an ore deposit with an average 3–4% Cu content, considered quite rich nowadays, could have been exploited as a copper deposit in prehistoric times. Mining-archaeological research in Central Europe and the Near East during the last decades has shown that well-known, still economically viable ore deposits had already been exploited in ancient times (see Weisgerber and Pernicka 1995). There are even indications that such 'giants,' as they are called in geology, have reached similar importance over a larger geographical area in prehistoric times. Examples of the importance of large ore deposits were given by Hauptmann et al. (2003). They suggest that Early Bronze Age metal objects from the 'royal' tomb at Arslantepe would originate in parts from huge copper deposits at the Black Sea coast of eastern Anatolia.

On the other hand, it has also become clear that small mineralizations, which are considered unimportant today, have been worked since the earliest times (Weisgerber 1997). Examples of this phenomenon can be found everywhere in the ore-rich areas of the Old World. In the areas we are interested in, this is correct, e.g., for Sinai and again for Anatolia. Not every modern metallogenetic map, therefore, may give reliable information about the possible intensity of ancient mining.

During sampling, vertical or lateral zoning of ore deposits has to be considered. This phenomenon is observed in hydrothermal veins and massive sulphide ore deposits of volcanic origin, which occur in the fold belts of the Eastern Mediterranean mountain ranges. They often outcrop on the surface and extend vertically down to differing depths. They have achieved a certain model status in archaeometallurgy (e.g., Charles 1980; Hauptmann and Weisgerber 1985), in which the main interest has been focused on the 'secondary zoning' in the oxidation zone close to the surface. This zone is characterized by changes of the composition of ores and the metal content from the outcrop down to a larger depth.

The change is caused by atmospheric impact (rain, temperature, air) at the surface, where they alter sulphides into water soluble sulphates and carbonates, which seep down with the water. This causes depletion (but hardly a complete separation) of several elements near the surface. Iron forms a stable compound with oxygen and water, limonite (FeOOH), and forms the so-called 'iron hat' or gossan of the deposit, which was an unmistakable sign of early prospectors and still is, because of its conspicuous rusty-red color. Material from the gossan is often mixed up with roasted ore. The oxidation zone of copper deposits contains predominantly oxidized ores such as malachite, azurite, cuprite, chrysokolla, and, more often than reported in the geological literature, native copper. It should be emphasized, however, that remains of sulphides such as chalcopyrite ($CuFeS_2$), covellite (CuS), and chalcocite (Cu_2S) are present almost regularly (Locke 1926) in the 'iron hat'. Iron is in parts preserved as sulphate and forms along with lead, silver, potash and others a group of minerals/ores called jarosite. Gold also remains in the 'iron hat', and lead as well as sulphate and carbonate. In the cementation zone positioned at the level of the water table, above the actual primary ore body, sulphides become enriched with copper, cobalt, nickel, selenium, tellurium, and silver. In many mining districts, these metal-enriched layers have caused considerable (but short) economic flowering. In the case of Late Bronze Age Cyprus, the exploitation of the cementation zone inspired real technological jumps (Muhly 1980). This simplified modeling of a sulphidic ore deposit, however, does not sufficiently explain patterns of prehistoric mining and smelting techniques. Regularly, ore deposits are not formed in strictly sharp contacts to the host rock. Much more often due to hydrothermal alteration, the host rock is decomposed and mineralized in a corona around the real ore body for several tens or hundreds of meters or more. In addition, faulting and fracturing lead to transitions of mineralizations to the host rock. The pure ore ("Derberz") itself shows a transition to increasing intergrowth with the "softened" host rock which is easily mineable. A significant example of interspersed, low-grade mineralizations in heavily cracked and crumbled serpentinitic host rock are the copper ore veins exploited in prehistoric times in Oman. These are "stockwork-mineralizations" of massive sulphide deposits.

'Secondary zoning' of sulphide ore deposits has been the principal basis for models and hypotheses about phases of metallurgical development, which have sometimes been presented in the best Darwinistic evolutionary manner (Forbes 1971). Strahm (1994) has diversified this model, as has been explained in Chap. 1. In the following chapters, it will be discussed whether this 'genealogical tree of metallurgical evolution' can be shown to be valid in all cases and transferred to the situation in the Southern Levant, because the sedimentary ore deposits of the Southern Levant do not comply with the previously described type, just as, e.g., the karstic ore deposits do not conform to the type either. In the Mediterranean, Karstic ore deposits have played an important role as raw material sources for mining of iron, copper, lead, silver, and gold. They are, however, bound to the alpinotype folded carbonate rock and can be tracked in a broad belt from Spain through southern France and Italy, over the Balkan Peninsula and up to Anatolia. Well-known deposits of this type are, e.g., Laurion, Thasos, Sifnos and a number of deposits in southeast Anatolia such as the Zeytindağ near Keban. In the sedimentary copper ore deposits in the Wadi Arabah and in Sinai a 'lateral zoning', exemplified by different geochemical and mineralogical zones, can be observed. However, as ores are mainly oxidic, zoning is not of such great importance as the 'secondary zoning' in sulphide deposits. In view of all this evidence, it seems to be clear that the development of metallurgy might have depended on the individual composition of the particular ore deposit and was thus so complex that single evolutionistic steps of the model can hardly be associated as specific historical periods. The concept of metallurgical development built on 'secondary zoning', which was characterized by the change of compositions with increasing depth, has in this region hardly any validity.

The allegedly most comprehensive study of 'ancient' ore deposits from the specific viewpoint of ancient metal production had been carried out by the Max-Planck-Institut für Kernphysik (Heidelberg). The study took place in the Aegean and in Anatolia during the 1980s (Seeliger et al. 1985; Wagner et al. 1986; Pernicka 1987; Wagner et al. 1989; Wagner et al. 2000; Pernicka et al. 2003). A large number of old mines and slag heaps were visited in the framework of a project about the development of early metallurgy in the Aegean and western Anatolia. These sites were geologically, archaeologically, archaeometallurgically and mining-archaeologically investigated. Samples of ore, slag and metal were collected and analyzed in the laboratory for chemical, mineralogical and isotope compositions. The Max-Planck-Institut für Chemie (Mainz) carried out lead isotope analyses, which were not only ground breaking research but also ultimately a standard tool for archaeometallurgical research (e.g., Begemann et al. 1992, 1994, 1995; Schmitt-Strecker et al. 1992).

Independent of this project the Consiglio Nazionale delle Ricerche and the University 'La Sapienza' (Rome) conducted a survey of ore deposits in eastern Anatolia (Caneva et al. 1983; Caneva et al. 1985; Palmieri et al. 1993). The main focus of this project was the geochemical and isotopic characterization of ore deposits and metal objects for an interregional comparison. The exploitation of a smaller number of deposits could also be dated so that their actual use in prehistoric times could be proved. The strategy of that research project was therefore from the beginning on, opposite to the research project presented in this book, to include a very wide area. In this context the undertaking was very successful in that today we have proof of the early exploitation of around 200 ore deposits in eastern Anatolia.

The Documentation of Ancient Smelting Sites

For a long time, archaeometallurgy occupied itself with the study of metal artifacts, while research into ancient smelting sites received only little attention. There are only a few exceptions such as the investigation of early copper extraction at the Mitterberg near Salzburg in Austria (Zschocke and Preuschen 1932), in Timna/Israel (Rothenberg 1973; Rothenberg 1988, 1990) and research activities still in progress in Oman (Weisgerber 1981; Hauptmann 1985). Timna was one of the first sites where copper production was systematically investigated and integrated into a general culture-historical framework. This led to the impression in archaeometallurgy that mining and smelting of ores were always carried out in a close spatial connection. This model was transferred even on to the earliest stages of (extractive) metallurgy, implying the following ideas (Levy and Shalev 1989; Shalev 1994):

- The smelting of ores is always carried out in the immediate proximity of the ore source;
- Smelting is always connected with the production of large amounts of slag;
- Smelting is connected with problems of fuel supply;
- Metallurgy carried out inside a settlement is limited to metal processing (e.g., casting).

Most (pre-)historical mining and smelting sites are actually organized in this way. As examples, the gigantic slag heaps of Corta Lago in Rio Tinto (Rothenberg and Blanco-Freijeiro 1981) and on the island of Cyprus (Muhly 1980) can be given, where millions of tons of slag have accumulated from the Late Bronze Age to the Roman period. But Chalcolithic production of the metal has been proved neither in Rio Tinto nor in the huge smelting sites of Cyprus (Gale 1991). All of the larger well-known smelting sites in the Near East and in Europe date from later periods (Oman: Weisgerber 1981, Hauptmann 1985; Alps: Cierny and Weisgerber 1993, Bir Nasib in Sinai: Petrie 1906, Rothenberg 1987, eastern Anatolia: Seeliger et al. 1985; Belli 1991).

Finds of ore, small amounts of slag, remains of crucibles and copper prills from casting, which have all been found inside settlements, shed light on the fact that the first steps of metallurgy, beginning with the use of native copper and leading to the first smelting of metal in the $5^{th}/4^{th}$ millennia and its further processing have not happened in the immediate proximity of the actual ore deposits (Hauptmann et al. 1996). This pattern of development is obviously true for the entire Eastern Mediterranean. There is evidence for this to be found in numerous settlements, e.g., in Arslantepe (Hauptmann et al. 2003), Tepecik, Değirmentepe, Tülintepe (Zwicker et al. 1980; Müller-Karpe 1994; Yener 2000) and Norþuntepe at the upper Euphrates (Yalçin et al. 1992; Pernicka et al. 2002). In the Southern Levant, this is the case in Tell Abu Matar (Hauptmann 1989; Golden 1999; Shugar 2000), Shiqmim (Shalev and Northover 1987), Tell Maqass (Khalil and Riederer 1998) and nearby Tell Hujayrat al-Ghuzlan (Hauptmann et al., in prep.) and Wadi Fidan 4 (Hauptmann et al. 1993). In all these cases, the ore to be smelted was transported, seemingly in a continuation of Neolithic trade patterns, from the sources to the settlements, sometimes over a distance of more than a 100 km. No large amount of slag has been found in any of these settlements. This is not too surprising, as it is more than likely that during these early periods only high-grade ores were used. Craddock (1990) noticed the same scarcity of slag for the Bronze Age copper production in Great Britain and therefore suggested for this earliest phase of smelting a 'slagless metallurgy'. It is thus not surprising that the lengthy search for signs of the very beginning of ore smelting amongst the slag heaps close to the ore deposits proved to be in vain.

This shows that archaeological excavations of settlements are essential to provide the material necessary to gain information concerning pre-Bronze Age metal production. The smelting of ore in direct proximity of the ore source starts in the Near East only with the developed social organization of the Early Bronze Age, therefore not before the middle of the 3rd millennium BCE, when metal was produced in larger quantities (Craddock 1995). That at least is the logical conclusion to be drawn from the dating of the earliest slag heaps known so far, e.g., such as those on the island of Kythnos (Gale et al. 1985), in Timna (Rothenberg and Shaw 1990) and finally in Faynan (see below). These slag heaps have mainly accumulated at the location of their production, which means immediately adjacent to the deposits themselves and far away from any settlements.

Tons of slag has been discarded in the course of millennia, developing into larger and larger mounds. The most conspicuous (pre-)historic smelting sites can be found in areas with sparse vegetation and where various conditions have hindered the recycling of slag in more recent times. In central Europe, they are normally covered by vegetation, so that they can often only be discovered by geophysical prospection.

Proton magnetometers have been regularly employed for the study of numerous ancient smelting sites in Central Europe (Goldenberg 1990; Kruse et al. 1997). For this kind of analysis, ferromagnetic materials (magnetite, Fe_3O_4, metallic iron, maghemite, α-Fe_2O_3), which are present in the cycle of metal production, are particularly useful. Thus it seems to be a very promising method of gaining information about the existence, structure and organization of smelting sites before any archaeological excavations are carried out. Practical tests have, however, shown that the method can only be appropriately put into use for the clear-cut prospection of slag heaps (see, for instance, the geophysical prospection work at Roman iron smelting sites at Joldelund, Germany, Stümpel et al. 200.). A structural analysis of smelting sites, however, is only useful when there are already some ideas about the expected findings and knowledge concerning the natural surroundings (Presslinger et al. 1988).

Next to slag, other waste materials from smelting processes can be found at ancient smelting sites. These include ore, which can give information about the raw material, fragments of smelting furnaces (furnace walls and lining), tuyères and other refractory materials, and remains of firing material. It is noticeable that furnace fragments and tuyères have often been found in large numbers. All of those contain information on ancient technologies and on the organization of smelting sites. Table 2.1 shows an overview of the possible finds in slag heaps, their potential information content and methods of analyses. A (complete) reconstruction of prehistoric smelting furnaces has so far not been achievable, and so one is not surprised by Muhly"s (1989) statement that he had never seen a really convincing copper smelting furnace from the Bronze Age in the Eastern Mediterranean. That holds true in spite of the good conditions found in Timna 2 and 30. The conspicuous systematic and all-encompassing destruction of ancient smelting furnaces leads to the supposition that furnaces had been destroyed regularly and purposefully after the end of each smelting event. This might have possibly been done in order to extract the metal mechanically from those parts of the slag that did not become fully liquid and so remained within the furnace after the firing. This phenomenon was described also in medieval times, e.g., by Theophilus Presbyter. A similarity to the technology of the bloomery process from the 1st millennium BCE can be seen here. There, the bloom in its solid state (together with slag) had to be physically removed from the furnace by breaking up the furnace front (e.g., Straube 1986). From the mid second millennium BCE in Cyprus, fragments of a smelting furnace made of pottery have been found in Politiko Phorades (Knapp et al. 2001) and from the 1st millennium BCE shaft-like furnaces in Agia Varvara-Almyras (Fasnacht et al. 2000).

In contrast to the domestic contexts inside settlements, datable archaeological material such as, e.g., pottery is often missing on ancient slag heaps. In addition, renewed activities particularly from the recent past have often destroyed the stratigraphy of ancient remains so that it might be difficult to come up with a reliable chronological and cultural-historical assignment of a slag heap. But slag heaps are obviously rich in charcoal, which allows a dating through physical methods (see Sect. 2.2). A formal typology of different slag can be used for a reliable and quick preliminary dating. The precondition for this is an initially secure dating of single slag types.

Faynan offered exceptionally good conditions for the recording of mining and smelting activities, as the smelting sites for each chronological period are spatially divided: while during the 4th millennium BCE smelting had been carried out inside the settlements,

Chapter 2 · Problems and Methods of Archaeometallurgy

Table 2.1. Possible finds on old slag heaps and the resulting information concerning the dating and reconstruction of (s)melting processes. *AAS:* Atomic absorption spectrometry; $^{14}C:$ radiocarbon method; *EDS:* energy dispersive spectrometry (spot analysis); *ICP-OES:* inductively coupled plasma with optical emission spectrometry; *NAA:* neutron activation analysis; *SEM:* scanning electron microscopy; *XRD:* X-ray diffractometry; *XRF:* X-ray fluorescence spectrometry; *LIA:* lead isotope analysis (modified after Craddock 1995)

Material	Information	Technique	Comments
Ore; Fluxing agents; Clay	Composition of the charged ore; Metal; Provenance of ore	Chemical and mineralogical analysis (AAS, ICP-OES, XRD, microscopy etc.); LIA	Ores in ancient smelting sites are often waste. Sampling in ancient mines is more promising
Fuel	Dating; Fuel supply; Palaeoclimate; Material input for slag formation	^{14}C-dating; Botanical studies (SEM); Chemical analysis of ashes	Charcoal usually available at all smelting sites
Refractory material	Temperatures and duration of process; Composition of refractory materials; Metal composition	SEM-studies on vitrification; Petrographical studies; AAS, XRF, SEM/EDX	As precise as possible spatial localization of samples in reaction vessels. Mixing up with slagged charged ores possible
Slag	Temperatures; redox conditions; Viscosity, different steps in metal production; Kind of metal; Efficiency of smelting; Provenance of components of charge	Chemical and mineralogical analysis (AAS, ICP-OES, XRD, XRF, microscopy etc.); LIA	Ever present and important material at old smeltingsites and workshops
Metal	Efficiency of smelting process; Parameters of (s)melting process; Differentiation of raw material-alloy; Provenance of metal	Analyses of trace elements (AAS; ICP-OES, NAA) metallography; LIA	Exist as inclusions in slag. Larger pieces of metal could be waste

the later Bronze Age smelting sites were placed on the top of hills due to technological considerations. In the Iron Age and the Roman period, however, political and military control played a role in the setting of smelting locations (see Sect. 5.3). Sampling and dating were supported by profiles through slag heaps cut by erosion.

2.2 Analytical Methods

Physical Dating

Ancient mines and slag heaps have generally a low output of archaeologically datable finds. Pottery, which often allows a secure dating to a particular period, is seldom found. Occasionally, grooved hammerstones or other stone hammers are discovered in the dozens or even hundreds. They allow a rough chronological classification into the Bronze Age or earlier periods. Such a classification was possible, e.g., in the wider surroundings of the ancient mining area of Rio Tinto, where grooved hammerstones proved the existence of Chalcolithic and Early Bronze Age mining activities (Rothenberg and Blanco-Freijeiro 1981). Large amounts of grooved hammerstones from the Bronze Age have been located in the ancient mining areas of the British Isles (Crew and Crew 1990).

Attempts have been made to date the exploitation of ore deposits using the close spatial context with ancient settlements or graves. The Vogtland and the Erzgebirge in Germany provide examples for such an approach; in both areas, clear evidence for prehistoric mining is still missing, although it has been repeatedly presumed that one of the sources for Bronze Age tin might have been here. Simon (1993) drew attention to the proximity of settlements and ore outcrops. Metal analyses carried out so far show only unclear relations to nearby ore deposits. It seems to be a problematic method indeed as is shown by the considerations of Driehaus (1965): he presumes a likely connection between the 'Ducal tombs' in the Hunsrück and the utilization of the ore deposits there, which is not yet verifiable. Gale and Stos-Gale (1981), however, were able to demonstrate very convincingly using lead isotope analyses of metal artifacts from the Aegean that the deposits of Laurion had already been mined in the Early Bronze Age. Before these studies it had been assumed that the beginning of mining in this area was around the 9th century BCE.

^{14}C-dating has shown itself to be the most sensible and promising method to date ancient mining, because leftovers of pieces of kindling or torches, charcoal from fire-setting or remains of timber can often be found in the mines. Prehistoric mines in Europe and the Near East have been dated for the most part through radiocarbon. Experience has shown that most kinds of metallurgical activities can be dated with this method, because the firing and reduction material in antiquity has been, without exception, charcoal or wood. Thus in Faynan, it was easy to recover sufficient material for ^{14}C-dating. The dates presented in this volume have been determined by Dr. W. Bruns, Dr. B. Kromer and B. Münnich, Institute for Environmental Physics, University of Heidelberg. The dates have been calibrated using CALIB 3.0 from Stuiver and Reimers (1993) and are given here in real years. Although the radiocarbon method suffers from uncertainties in particular timeslots, it is still the best method to date the most important steps of metal production, because in prehistoric and historic times technological innovations not only developed relatively slowly but also diffused only gradually. To emphasize the general development was the aim of our dating; we do not assume to provide a fine-chronological framework to subdivide periods or phases. This was done, e.g., by an ample set of precise ^{14}C-dating of charcoal by accelerator mass spectrometry from secure archaeological strata, as recently done by Levy et al. (2004) at Khirbet en-Nahas. But also archaeological dating methods would result in more precise chronological data. Kuniholm (1996) has demonstrated with his work that two thirds of the period between today ('BP' = AD 1950) and 7200 BCE, so 6 000 years, can be precisely dated using dendrochronology.

'Conventional' ^{14}C-dating is based upon burning the carbon of the respective sample to CO_2 or to another gas, and the radioactivity of the sample caused by ^{14}C can be measured using a proportional counter tube. The resulting ^{14}C-concentration is dependant on the age of the sample. For this, 1–2 g of charcoal are generally sufficient. Much less material is necessary when the $^{14}C/^{12}C$-ratio is directly measured with an acceleration mass spectrometer.

Metallurgical activities can also be dated from technical ceramic by thermoluminescence (TL). This comprises fragments of furnace walls, tuyères and other parts of the furnace construction, or as in the case of Faynan, thousands of small clay-rods (see Sect. 5.2) which were regularly found in the slag heaps. TL is based on the glow, the thermoluminescence, which appears in non-conducting materials when heated up. It results from the amount of energy kept in the material from natural radioactive decay. When the material has been heated >500 °C, this energy is erased. Then the process of radioactive decay starts anew and therefore repeats the phenomenon of energy storage. If the natural radiation dose is known, which is caused mainly by the decay of ^{235}U, ^{238}U, ^{232}Th and ^{40}K, TL-intensity allows the calculation of the duration of time elapsed since the last heating.

Basically this method is also appropriate for the dating of slag, because it has also been exposed to very high temperatures during the smelting process. But complex intergrowth of silicate slag phases and metal contents in slag make it very difficult to calculate the correct dose of radiation. This results in relatively large errors of around 20% (Lorenz 1988). It therefore proved more reliable to measure quartz inclusions and fragments of technical pottery, where possible, rather than samples from the slag (Krbetscheck 2003).

TL-age measurements of copper slag and technical pottery presented in this volume have been carried out by Dr. I. and Prof. Dr. G.A. Wagner, then from the Forschungsstelle Archäometrie der Akademie der Wissenschaften (Heidelberg). The age determinations are precise enough to distinguish between slag from the Early Bronze Age, the Iron Age and the Roman period.

In principle slag could also be dated by the Fission Track method, which is based on the spontaneous nuclear fission of the uranium-isotope ^{238}U or by archaeomagnetism. Both methods have not been used here; more information can be found in the relevant literature (Wagner 1998).

The Investigation of Ancient Slag

For archaeometallurgical slag, the modern definition of that term is hardly adequate: It says that are mixtures formed during smelting from gangue and fluxes and they mainly consist of silicates and metal oxides (CaO, MnO, FeO, MgO etc.). Due to their low density they form a layer on top of the metallic phase and solidify in many cases to a glassy mass. Slags are waste products. This is in principal also true for archaeometallurgical slag, but exceptions exist that are mainly due to limited firing control in ancient smelting furnaces or crucibles. It will be proven in the following that archaeometallurgical slag is rich in metal or metal-bearing phases (e.g., copper, copper sulphide, delafossite) and often did not reach the fully liquid state. The processes producing such slag cannot always be clearly defined. It is necessary to understand the term slag, as it is used here, sometimes in the way it was defined in the mining regulations at Goslar in AD 1360 (Article 161, 162), where it describes slag as materials which had been melted once, including the actual waste product as well as reusable material, consisting of intermediate products and still usable waste products which are regularly produced during smelting (Bartels et al. 2007). This will be discussed further in Chap. 7 in the example of the finds from Khirbet Hamra Ifdan.

2.2 · Analytical Methods

After slags were extracted from a secure stratigraphical context from slag heaps or workshops, and after a first classification, a meaningful interpretation calls for texture analyses, for mineralogical phase analysis and chemical composition.

Due to rapid cooling, archaeometallurgical slags consist in general of a very fine-grained mixture of different phases. They have a complex composition, depending on the bulk chemistry. It is often difficult to determine mineralogical phases by light microscopy, with the exception of main components such as olivine, spinel, and pyroxene. Sometimes, as is the case at Faynan, slag has even solidified in a glassy state, although such 'glasses' show transitions from the microcrystalline, via the cryptocristalline up to the hyalocrystalline state formed in residual melts. In these cases, even X-ray-diffractometry cannot help to identify all phases. Scanning electron microscopy (SEM) and electron microprobe analysis (EMS) has gained increasing importance, since they provide fast and reliable (quantitative) spot analyses for phase identification.

But microscopy with polarizing light is an indispensable method particularly for texture analysis. Until very recent times it was usual to concentrate petrographic slag analysis on the description of shape and size of single grains. This was mainly carried out with the purpose of obtaining information about the cooling history of the sample.

The macroscopic analysis of the fabric of slags, meaning the orientation and distribution of identical phases or other identical structural constituents in their spatial association, has rarely been carried out before. The reason has been simply that the available finds had dimensions that were too large for thin sectioning or mounted sample preparation. Slag produced in small-scale operations, which often lead to inhomogeneous compositions, can however be easily studied for fabric analysis. In iron metallurgy this is true for fist-sized, dome-shaped slag, produced throughout the Old World, as a by-product during iron processing (so-called plano-convex bottoms, PCB, or smithing slag, see Pleiner 2000). Their chemical and mineralogical composition is very similar to bloomery slag, produced in iron smelting; it is therefore impossible using these criteria to ascertain which process produced the slag. This has very often led to quite fundamental misinterpretations of iron production and iron processing. It was only Keesmann and his team, who were able to define working criteria in using texture and fabric analyses and detailed mineralogical analyses (e.g., Keesmann 1985; Keesmann and Hilgart 1992) to safely distinguish between these types of slag.

In copper metallurgy, slags with a size comparable to smithing slags also occur. They have only recently become available for research but have since been intensively studied (Yalçin et al. 1992; Hauptmann et al. 1993; Hauptmann 2003; Shugar 2000; Schreiner 2002; Saez 2003). Without exception, the slag originates from the Chalcolithic period and the beginning of the Early Bronze Age and has been smelted in small reaction vessels, i.e., in crucibles. They differ from the smelting slag of later periods in their still very high content of metal. Studying this aspect of the earliest slag from Wadi Faynan has therefore been of particular interest.

The investigation of archaeometallurgical slag comprises questions concerning technological aspects and the provenance of the ore charge. It has therefore proved advisable to carry out as many bulk analyses as possible by conventional analytical methods, including trace elements. A precondition for such analyses is a sufficient quantity of material. This varies between a few grams and a kilo depending on the original size of the sample. If the archaeological context allowed for it, it proved prudent to analyze a whole series of samples to obtain good statistical results.

The interpretation of chemical bulk analysis can be impaired by the heterogeneous composition of a slag, particularly when the slag still contains un-decomposed constituents. But if the information required does not only involve petrological aspects, the value of the information produced by the analysis is not lost. Hence, the criticism to use chemical bulk analysis for slag investigation as expressed by Keesmann and Hilgart (1992) is only understandable in connection with their mineralogical study of non-homogenous smithing slag. In these cases, it was very helpful to analyze single sections on the micro-scale. This can be carried out by energy dispersive X-ray analysis in a scanning electron microscope. This method has the added advantage that the composition of inclusions and the surrounding molten slag can be studied separately.

'Slag – indicators of archaeometallurgical processes' was the title of a synopsis concerning the study of ancient slag by Bachmann (1978a). It was the summary and continuation of work which had been carried out since the beginning of the 1970s on ancient smelting slags: Morton and Wingrove (1969, 1972) have worked on Roman and Medieval iron slag, while Lupu and Rothenberg (1970) as well as Milton et al. (1976) have analyzed copper slag from Timna/Israel since the start of the excavations there. Steinberg and Koucky (1974) have undertaken the study of the (pre-)historic copper slag of Cyprus.

The emphasis of Bachmann''s work was a mineralogical and chemical analysis of slag in order to form the basis for the reconstruction of the technological parameters of ancient smelting processes. In accordance with modern metallurgical techniques, he suggested determining the parameters for slag basicity and viscosity, and ascertaining the melting temperatures by using phase diagrams and thus assessing the efficiency of the initial charging and the amount of control over the smelting process. Bachmann presented a calculation program for the determination of these data and added in a second publication (Bachmann 1980) another program, which calculates the viscosity. His programs were based on copper slag produced during a relatively advanced state of metallurgy. The samples investigated came from the western margin of the Wadi Arabah and the Sinai, and dated for the most part from the Late Bronze Age (14th–11th centuries BCE) and even later periods. He interpreted older material, which had been dated preliminarily into the Bronze Age or Chalcolithic period (Bachmann 1978a,b; Rothenberg 1978), very cautiously and assigned it to a "trial-and-error"-phase, because the composition of the material deviated considerably from the bulk of ancient slag.

As summarized by Hauptmann (1999), ancient smelting processes are controlled by four parameters, all of which are detectable in the composition of the slag. They are (1) the composition of the charge, (2) the firing temperature (3) the gas atmosphere during smelting redox-conditions and (4) the reaction cinetics between components of the charged material during the duration of the reaction, i.e., the length of the smelting process. Until the recent past, it was traditionally assumed that the tight clustering of ancient iron-rich silica slag compositions, similar to that produced in the making of ancient glasses (Rehren 1998), would reflect exclusively the technical skill of ancient metallurgists who, having found a suitable charge composition to produce a useful product on one occasion, were then able to reproduce that charge again to produce a similar result. In view of the limited control that ancient metallurgists could have had over the essential components of the raw materials as well as the smelting processes and the lack of significant differences among most of the slag found throughout wide geographical areas and produced over centuries, this appears to be a highly unlikely assumption.

In reality, given the interdependence of all these procedural parameters, there are only a very few ways in which all the low melting slag could have been produced throughout the Old World. Most of the studied slags, mainly from the later periods of early metallurgy, fulfill these technical conditions at least in so far as they have compositions which are close to cotectic troughs or eutectics in the low melting parts of the quaternary system $CaO–SiO_2–FeO–Al_2O_3$. In accordance with the character of the raw material, usually limonite + quartz + clay minerals + ore, if necessary with the addition of fluxing agents, a concentration towards iron-rich silicate slag is noticeable. Their simplified composition can be discussed on one of the side planes of the tetrahedron either in the system $CaO–FeO–SiO_2$ or $FeO–Al_2O_3–SiO_2$. In addition, furnace walls made of clay or rock act as a source material for the final tuning of low melting compositions (as a function of oxygen pressure and temperature).

The main elements in the composition of copper slag are shown in Fig. 2.2 and in comparison also the bloomery slag from iron extraction. They can hardly be distinguished from each other by chemical analysis alone.

Bachmann points out that the plot of slag analyses in ternary or quaternary systems would only lead to meaningful interpretations, if there is a correspondence between the theoretical and actual phase content. Taken literally, this would require a formation under conditions of equilibrium, which never happens in reality. But research so far has shown that slag, if solidified from homogenous liquids, forms parageneses during cooling, which also appear in comparable natural rock. He therefore emphasizes the necessity of mineralogical phase analysis and includes the X-ray diffraction of all major components of ancient slag (fayalite, Fe_2SiO_4, spinel, e.g., Fe_3O_4 or $MgAl_2O_4$, and pyroxene, e.g., $CaFeSi_2O_8$) in his studies.

Fig. 2.2. Average composition of ancient slag from copper smelting (medium shaded area) and iron smelting (dark shaded area) shown in the ternary system $FeO–Al_2O_3–SiO_2$ (wt.-%) (Osborn and Muan 1964). The compositions match mainly the eutectic area of the system. Following data from Bachmann (1978, 1980), Hauptmann (1985), Keesmann (1989) and unpublished data from the DBM

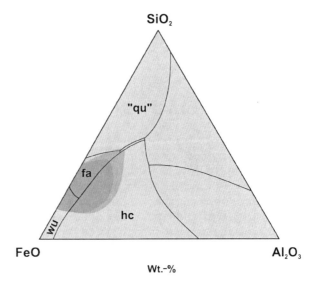

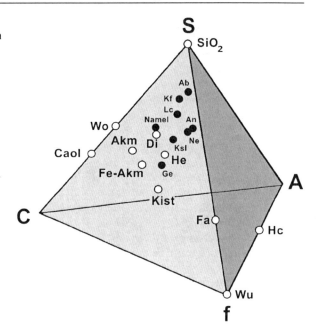

Fig. 2.3.
Important phases in iron-rich silicate slag, shown in the system CaO–FeO–Al$_2$O$_3$–SiO$_2$ (C-f-A-S). *Ab, Kf, An:* feldspars (albite, kalifeldspar, anorthite); *Lc, Ksl, Ne:* feldspathoids (leucite, kalsilite/kaliophilite, nepheline); *Namel, Ge:* melelites (Namelelite, gehlenite); *Akm, Fe-Akm:* åkermanites, ferroåkermanite; *Di, He:* clinopyroxenes (diopside, hedenbergite); *Wo, Caol:* calcium silicates (wollastonite, shannonite); *Kist, Fa:* olivines (kirschsteinite, fayalite); *Wu:* wuestite; *Hc:* hercynite; *SiO$_2$:* cristobalite, tridymite (after Keesmann 1989)

These studies have been considerably advanced by the mineralogical-petrographic studies of Keesmann. He and his team broadened the knowledge on the mineralogy of archaeometallurgical slag as well as differentiation processes during solidification in a number of detailed studies (Keesmann 1989; 1993). The choice of suitable systems to discuss slag in his work follows a 'critical paragenesis', which is usually included in the quaternary system CaO–SiO$_2$–FeO–Al$_2$O$_3$ (Fig. 2.3). He differentiates, for example, between the paragenesis fayalite – wuestite – hercynite in iron slag with the projection into the system SiO$_2$–FeO–Al$_2$O$_3$ and clinopyroxene-bearing parageneses in the system CaO–FeO–SiO$_2$. A mathematically based correction of the chemical analysis is necessary for a more accurate estimate of the crystallization temperature. Keesmann accounts for this with the varying amount of 'glass' within the complexly composed residual melt in the slag, where Al and Ca are enriched and thus produce too high values.

Keesmann also considers the previously discussed topic of the reduction-oxidation conditions. Studies on this subject carried out by Milton et al. (1976), Hauptmann (1985) and Moesta (1989a) have greatly helped in understanding ancient smelting processes. During its formation, iron-rich slag reacts very sensitively to the oxygen content or better the CO/CO$_2$-ratio in the smelting furnace. This leads to the crystallization of phases with differing Fe$^{2+/3+}$-ratios, identifiable directly from the mineralogical phase content in the slag. For the identification of the redox-conditions, Lutz (1990) and Metten (2003) also used Mößbauer-spectroscopy. The buffer equilibria of the system Fe–Si–O are powerful tools for the estimation of the oxygen partial pressure (p_{O2}) as a function of temperature. The principle, known as the Bowen-Fenner-Trend in petrology, can be defined in a very simple but important reaction, which is particularly applicable to slag due to the nearly ubiquitous appearance of the main element fayalite:

$$3\,Fe_2SiO_4 + O_2 \rightleftharpoons 2\,FeO \cdot Fe_2O_3 + 3\,SiO_2$$

2.2 · Analytical Methods

This is the QFM buffer equilibrium, which depicts the following: if p_{O_2} is sufficiently high, magnetite and a silica-rich compound crystallize first, so that Fe is mainly bound as an oxide. If p_{O_2} is low, no magnetite will be formed, but an iron-rich silicate (fayalite), or even metallic iron precipitates. In combination with the equilibria of the systems Cu–O and Mn–Si–O, it was possible to come up with important information about chronologically different smelting processes in Faynan, which will be treated extensively in Chap. 7.

These reactions are good examples of the different reducibilities of a number of oxides and other compounds, which play an important role in archaeometallurgy. The unit for reducibility is the free enthalpy (energy) of formation ($\Delta G°$), as it is illustrated in Fig. 2.4 for some oxides as a function of temperature. The figure also shows the limits of ancient smelting processes when using charcoal as the reduction- and energy source. It can be recognized that the stability of each oxide decreases with increasing temperature. It is also noticeable that the reduction of the oxides is helped along by increasing $\Delta G°$. The only line running in the opposite direction to the lower right represents the oxidation of carbon to CO. Only the oxides with an enthalpy of formation above this line can be successfully reduced. Further-

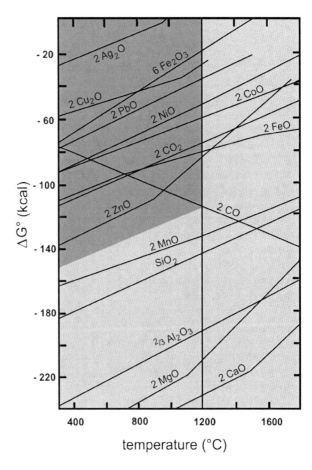

Fig. 2.4. Free enthalpy (energy) – temperature chart ($\Delta G°$) for oxide formation. The limits of ancient smelting processes are set by the temperature of 1 200 °C, which can generally be reached as an average maximum and the line of the reaction $2C + O_2 \rightarrow 2CO$. Only those oxides can be reduced to metal (darker field), whose enthalpies of fusion are above the crossing-point of these two lines; otherwise, they will become parts of slag compounds, e.g., MnO. Modified after data from Kubaschewski and Alcock (1979)

more, another limiting parameter is that ancient smelting furnaces could hardly ever achieve temperatures much above 1 200 °C. The oxides that are reducible under these conditions are therefore in the upper left-hand corner of the diagram.

It can be seen that all elements below the crossing-point of the two lines cannot be reduced to metal. They remain in the oxide state and form slag constituents. These are, e.g., SiO_2, Al_2O_3, MnO, MgO and CaO. Other elements shown here (Au, Ag, Cu, As, Sb, Ni, Co, Sn, Zn, Fe) are reducible, thus turn into metal and can be separated from the slag. The enrichment of these elements during the smelting of oxide ores increases in the sequence from below to above, as shown in Fig. 2.4. The enthalpy of formation is only valid for pure oxides, so the smelting of natural rock or ore leads to a fractionation of different elements into slag and metal, which can be calculated using the partition coefficient.

Slag not only enables us to reconstruct ancient smelting processes, in the course of provenance studies it very advantageously offers the possibility of studying raw material sources chemically as well as with lead isotope analysis. Despite the fractionation of elements during the change of ore to metal by slag formation, possibly through the addition of fluxing agents and the actual removal of the metal itself, it is feasible to obtain information about the original ore by analyzing the metal inclusions left in the slag (see Sect. 2.1).

Additionally the main, minor and trace elements of the silicate slag matrix can all give information about the origin and nature of the ore. This is of particular interest when smelting was not carried out in the immediate vicinity of the ore deposit but only after the ore had been transported over larger distances away from the mines. This problem is especially important for the initial stages of metallurgy in the 4th and 3rd millennia BCE, not only in the Near East but throughout the Old World – when smelting of ores was normally carried out inside settlements.

It has been suggested that, paralleling modern technologies, ancient metal extraction exclusively was based on a liquid-liquid-separation of slag, metal and/or matte by the specific weight. This concept, presented by Tylecote (1976) in an idealized form, holds that even in prehistoric times the *entire* charge melted completely in the furnace and was fully liquefied. The idea was that a plano-convex copper regulus formed underneath the slag in a depression at the bottom of the furnace, which could then be tapped in its liquid state. This model remained not without critics. Only a short time later it became clear that plano-convex copper reguli of the Bronze Age, so frequently found in the Near East, did not originate from a smelting furnace. They were instead deliberately customized ingots for trading, produced by several castings (Weisgerber 1981).

The concept of viewing archaeometallurgical slag *a priori* as material solidified from a *completely* liquefied melt has to be reconsidered and examined – particularly for the early stage of metallurgy. Slag from these periods often contains un-decomposed inclusions, and their correct interpretation is of fundamental importance. They can be either consciously added fluxing agents or the remains of the host rock. It is first of all necessary to prove that a specific slag had actually reached the fully liquid state before any further conclusions can be drawn. It will be shown in Chap. 6 that it is not at all self-evident that slag from the 4th millennium BCE at Faynan was completely liquefied. To put this in a wider framework, it is the contrast between 'low-tech' and 'high-tech' technologies.

It is far more informative to study the formation of slag during gradual heating of the ore charge from the first liquefying of the components at lower temperatures to a complete liquefaction of the charge. A charge composed of SiO_2, CaO, Al_2O_3, Na_2O, P_2O_5, Cu_2O and Fe oxide, chlorides and sulphates in various mineralogical components, as might have been available in Faynan, already starts melting at 600–800 °C. Firstly, clay minerals react with phosphorites, salt and Fe oxides. Quartz does not take part in the reaction until a sufficiently high temperature is reached and a sufficient duration of the reaction is achieved. Hence, partial melting takes place controlled by the geometry of eutectic troughs. Complete liquefaction is achieved only when the original charge possesses an ideal 'eutectic' composition. Such ideal compositions occur seldom in nature as can be demonstrated in a simplified way: The liquidus-solidus relations of the most important slag forming oxides SiO_2 and Fe oxide are shown in Fig. 2.5. It is noticeable that only a small range of compositions is completely liquefied at around 1 200 °C. Other compositions, particularly those high in SiO_2, are not fully molten. In such melts, one finds inclusions of undecomposed quartz, cristobalite and tridymite as is often observed in ancient slag (see also Sect. 6.1, 'Free Silica Slag'). Similarly, other components with a high melting point such as barite or chromite were found, which can be used to identify the provenance of the ore (Hauptmann et al. 1993; Metten 2003).

The notion that different melting points might have been important for the separation of specific components during the gradual liquefaction of ores with rising temperatures and the lengthening duration of the reaction has been underestimated. Metten (2003) has mentioned this possibility for matte smelting in the Alps during the Bronze Age. She describes the liquation of Cu-Fe sulphides, which start to quickly melt at ca. 900 °C, while the partly liquefied slag still contains quartz inclusions.

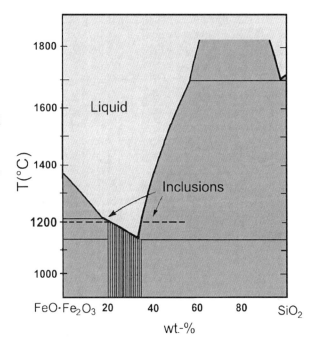

Fig. 2.5.
Simplified cross-section through the system Fe oxide–SiO_2 that shows the melting behavior of an iron-rich ore charge. The maximal area of variation (*hatched*) is shown for a charge that can be liquefied completely at 1 200 °C. Other compositions lead to undecomposed remains of the charge embedded in the melt

Iron production in the bloomery process, on the other hand, is based upon the separation of liquid slag from the solid iron bloom. The principle of liquation might also be valid for other early smelting processes, e.g., the smelting of Cu-As-ores, which also form a liquid alloy at around 900 °C, or during the formation of metallic copper and Cu sulphide just above 1 000 °C. A copper-rich sulphide such as covellite (CuS), a rather common ore intergrown with oxidic copper ores, melts at a temperature as low as 813 °C. This has been observed in Early Bronze Age slag from Shahr-i Sokhta (Iran), which acted as a solvent and collection accumulator for copper metal during smelting, as well as a lubricant that considerably enhanced the liquid-liquid separation of the copper from the slag as it solidified (Hauptmann et al. 2003).

It is argued whether slag may be utilized to define single steps of metal production. Bachmann (1978, 1980) and Sperl (1980) see important significance in the macroscopic typology of slag. This typology is meant to be a base for distinguishing separate steps of metallurgical processes from each other and to reconstruct their technical details in the Alps. This is generally correct where differences between slag from a smelting process and that from further metal processing is concerned. Otherwise, it has to be considered with care. In the Alps with their Bronze Age smelting processes of Cu-Fe sulphides, a reconstruction of those different steps cannot be securely established, neither on the basis of differing types of slag nor with ensuing chemical and mineralogical analyses. Piel et al. (1992) and Metten (2003) argue that 'plate slag' and 'tap slag' ("Plattenschlacke" and "Laufschlacke") occurring there could *also* be the results of one single smelting process. They interpret both kinds of slag as *(a)* eutectic molten material, which got separated from *(b)* the remaining charge that remained as a 'restite' or resister in the furnace. As has been described above, this phenomenon has been observed in slag produced in the Chalcolithic and the beginning of the Early Bronze Age, albeit in an undeveloped state. Archaeometallurgical plate slags are not only to be found in the archaeological context of Faynan and Timna but seem to be distributed all over the world – e.g., in the Alps (Presslinger 2004), the Harz mining district (Klappauf 2000), Nepal (Anfinset 1996) and the Ukraine (unpubl. results Bochum). We also noticed plate slag as a recurrent result of experimental archaeology (Bamberger 1990; Merkel 1990). It is reasonable to assume that the wide geographic distribution of this type of slag results much less from special local craftsmanship but is rather the product of given chemo-physical conditions during smelting, which are similar everywhere.

There seem to be indications on the other hand that even in the very early stages of metallurgy a 'highly developed technology' was known, e.g., in Los Millares in southwest Spain (Keesmann et al. 1994). Shalev and Northover (1987) analyzed slag from the Chalcolithic sites of Shiqmim/Israel and suggested that they were able to define not only smelting slag but also refining slag. Particularly for this latter case, it can be demonstrated that they have merely noticed samples of slag forming reactions indicative of different stages of the smelting process. This will be commented upon extensively later.

It can be summarized that scientific analyses of slag enable us to understand the principles of early metal extraction, meaning the thermo-chemical or thermo-dynamic processes. But the present stage of research is not sufficient to reconstruct the technical details of the artisans' craft. The wide variation of technical processes known to be used in traditional iron smelting in Africa gives a very good example of the diversity of possible techniques. But a meticulous analysis of all the facts definitely allows us to outline ancient technical processes, and that should certainly be the aim of all archaeometallurgical research. Espelund's (1991) statement: "we know the principles but not the practice, the ancients knew the practice but not the principle" is too pessimistic about our abilities.

Chemical Analysis of Ores and Metals

Since the beginning of archaeometallurgical research in the 19th century, it has been a point of discussion as to whether metallic trace elements in copper, lead, and bronze artifacts could be used to identify specific ores or ore deposits from which the metal was extracted. Efforts have been made to compile the chemical characteristics of ore deposits and the metals extracted from them and to define a chemical 'fingerprint'. The primary goal was on one side to find the sources of the raw material from which the respective metal objects might have been produced, a study that allows the reconstruction of ancient trading routes and cultural interactions. On the other side, metallurgical-technological criteria were used to define 'workshop-recipes'. Such studies have been carried out in large-scale research projects. Optical emission-spectral analysis (OES) was employed to analyze minor and trace elements in metal objects and on a smaller scale also in ores (Pittioni 1957; Junghans et al. 1960, 1968, 1974; Otto and Witter 1952).

One should bear in mind two factors, which complicate a direct comparison between ore and artifact and have repeatedly led to critical assessments concerning the value of chemical analyses in provenance studies (Merkel 1990; Chernyk 1992; Budd et al. 1995). These are, firstly, mineralogical and chemical variations of ore deposits, mainly in near-surface parts, and secondly, a fractioning of the trace element pattern due to metallurgical processes.

Although it was already in the 1950s that the two teams around Otto and Witter (1952) in Halle and around Pittioni (1957) in Vienna had realized that an assignment of a metal object to a specific ore deposit is not achievable "without the cooperation of ore geology and mineralogy and metallurgy," today there are still considerable gaps in geochemical characterizations of ancient ore deposits.

This problem is partly due to the strategy of many research projects, where normally archaeological metal artifacts were preferentially investigated. Junghans and coworkers have, from the beginning on, consciously decided to exclude ore analyses in their Europe-wide analytical project "Studien zu den Anfängen der Metallurgie (SAM)". They justified this with the large geographic region, where it would have been impossible to deal with an investigation of ore deposits in any case. In order not to lose sight of the main goal of the project, the study of the beginnings and distribution of copper metallurgy in Europe, they concentrated on the analyses of some 22 000 metal objects. The team was *a priori* not interested in provenance studies, but much more in a (technological) definition of 'workshop-recipes' and of 'metal provinces' which might have corresponded to smelting or processing centers and ore districts. Generally, they were well aware of the need for ore analyses.

It is still not possible, even with today''s knowledge, to categorically state if the chemical composition of a particular metal artifact results from natural impurities in the ore or from a conscious decision to produce an alloy. The alloy would have been made by adding an alloy agent to strengthen a particular property of the material, such as hardness or castability or to change the color of the metal. The difference between accident and planned additive is of greatest importance, because the conscious production of an alloy is an intentional manipulation of material. And that is just as much a technical advance as the intentional addition of a fluxing agent to optimize a smelting process.

The analyses of metal artifacts carried out by Junghans and his team and those by Pernicka demonstrate that copper objects often contain (natural) impurities of As, Pb and Ni of up to 10 (wt.-%), sometimes even higher. The possible pivotal point between natural impurity and conscious production of an alloy for lead-bronzes varies from 4 to 6% (Stos-Gale et al. 1986; Pernicka et al. 1990). The limit is not easily set even for tin, which is geochemically separated more from copper than any of the other mentioned metals. Also, the lithophile element definitely has chalcophile tendencies and is able to build sulphides such as stannite (Cu_2FeSnS_4). This ore appears in large amounts in several tin deposits of the Old and New World. An example of this is the ore deposit at Mušiston in Tajikistan, where it occurs together with mushistonite ($CuSn(OH)_6$) and consequently forms a 'natural bronze ore' (Alimov et al. 1998). But one should follow the suggestion from Pernicka et al. (1990), who set the limit for tin at ca. 1%, following the statistical analyses by Spindler (1971). Early Bronze Age bronzes have a hiatus around the 1% value for tin in their composition.

Even now, arsenical copper (occasionally named 'arsenic bronze') plays a key role in the discussions and publications about the question as to when alloys have been first consciously produced. This alloy appears in the 4th millennium BCE, nearly contemporaneously in the Near East and in several different regions of Central Europe (Junghans et al. 1971; Schubert 1981), and it is, in fact, the alloy of the Early Bronze Age. It is, between numerous other chemical compositions, the dominant copper alloy before tin-bronze took over. There are large variations in the As-percentage, which sometimes reaches over 15%, though it is more commonly well under 5%. Pernicka et al. (1990) suggest 2% as the limit between As-copper and purposefully produced As-bronze, even though they are fully aware of the problematic nature of this margin.

Studies in material science by Buchwald and Leisner (1990) and Budd (1991) confirm this as a reasonable demarcation, as As-concentrations >2% measurably increase ductility and hardness. The interpretation of arsenic containing copper has focused so far almost exclusively on the advantageous physical properties compared to metallic copper. It is still a point of discussion as to how arsenic copper, if at all, has been produced. There are primarily two possibilities, which had not changed for over 40 years (Otto and Witter 1952; Charles 1967; Eaton and McKerrell 1976; Moorey and Schweizer 1972; Strahm 1994). One group suggests a production of As-copper by consciously adding As-ores such as, e.g., realgar, orpiment, scherbenkobalt (concentrically banded or 'shelly' arsenic), or even an arsenic-rich mother alloy to copper (Tylecote 1985). This argument was supported by an observation of Tylecote et al. (1976), that copper objects with high As-contents could not have originated from the utilization of appropriate ores, because As would change during the smelting process through oxidation into the easily volatile As_2O_3 and therefore evaporate.

The suggestion that arsenical copper had to be intentionally produced survived for a surprisingly long time. This was the case even in regions such as the Caucasus (Selimkhanov 1972), where numerous fahlore deposits with signs of prehistoric mining are suggested (see Chernykh 1992) – but have never been investigated for a detailed mineralogical composition of ores. Such ore deposits could have easily been the place of origin of the raw material for the production of those artifacts with a high As-component. In addition, ore deposits with massive mineralizations of realgar and orpiment outcropping at the surface are known from the Caucasus (Twaltschrelidze 2001). They are located close to several copper prospects with evidence of prehistoric mining.

Intentional arsenic alloying is suggested also for EBA Crete. Reference is made to EBA I contexts at Poros Katsambas, Greece, Doonan et al., forthcoming), which contain heat-altered fragments of loellingite ($FeAs_2$) along with heavily slagged crucible fragments. The slag commonly contains Cu-As prills with up to 40% As. Ingot fragments from associated contexts contain no detectable As (by EDS). This evidence needs to be considered in light of Cu-As alloys forming the mainstay of the EBA metallurgical tradition. At the EBA II smelting site of Chrysokamino, copper prills embedded in slag show As- and Ni-concentrations as well in the lower percentage level and above. As five tiny pieces of ores analyzed from the sites are considerably lower in these two elements, a deliberate adding of As-, Ni-minerals is suggested (Bassiakos and Catapotis 2006). However, as the source(s) which supplied Chrysokamino with ore are not known (Betancourt 2006; Gale and Stos-Gale 2006), clear evidence is difficult. At EBA I Tell Abu Matar, Israel, the highly elevated level of As in copper prills embedded in slag compared with copper ores utilized for smelting (from Faynan) is explained by the process of intentionally alloying arsenical copper (Shugar 2000). It is difficult to understand that As-rich materials should have been transported from ore deposits in central Anatolia. This hypothesis is based upon an evaluation of lead isotope data.

However, most explanations try to account for the origin of arsenic-copper from ores with natural impurities of arsenic (Pernicka 1995; Lechtman 1996). Due to its geochemical behavior, arsenic forms compounds with Cu^{2+} also in the oxidation zone of ore deposits. There are a large number of Cu-arsenates, which show a greenish color similar to many secondary copper ores. Arsenic can become inadvertently part of the smelting charge via decomposed scorodite ($FeAsO_4 \cdot 2H_2O$), a mineral frequently occurring in ore deposits close to the surface (Smirnov 1954). There are regrettably only a few examples where metal analyses can be compared with the appropriate original ores. Under these circumstances, the research by Gale et al. (1985) and Stos-Gale et al. (1986) on Early Bronze Age slag from the Cycladic island of Kythnos is even more important. They have analyzed copper inclusion in slag with widely varying As-concentrations between 0.2 and 6.6%, which they can assign with compatible ores. They thus provide confirmation that arsenical copper was smelted unintentionally, but after the crushing of the slag, a deliberate choice of the copper prills rich in arsenic followed, in order to produce a suitable alloy. The arsenic-rich, silver-colored prills could easily be distinguished from the reddish pure copper prills by its color. Results of lead isotope analyses made it plausible that As-containing copper artifacts of the hoard (from Kythnos) were also made from the same metal. Tadmor et al. (1995) demonstrated, after the investigation of the hoard from Nahal Mishmar/Israel, that there is a considerable aesthetic aspect in the use of alloys. The color of metal objects changes depending on whether As-Sb-alloys or pure copper are used. In the Nahal Mishmar hoard, the so-called prestige objects are the ones made from alloys, while all the tools had been crafted from pure copper.

It is more than likely that the spread of arsenical copper marks the beginning of extractive metallurgy. This is strongly supported by the studies of Begemann et al. (1994) on metal artifacts dating to the 4th millennium BCE from Ilipinar in western Anatolia. They demonstrated that these artifacts differ from metal objects of Neolithic origin, which have exceptionally low concentrations of trace elements. Neolithic objects had been made from native copper, while the ones from Ilipinar were made of copper smelted from ores.

It should be noticed that slag from the middle of the 4th millennium BCE already has As-contents of up to 0.08%. This is true for the slag from Kythnos but also for that from Noršuntepe in Turkey (Yalçin et al. 1992). Sometimes the As-content can even rise as high as one percent as in the slag from Tülintepe (Çukur and Kunc 1989). It might be assumed that this is caused by an intra-metallic compound of As-containing speiss. The material might be interpreted as the result of early experiments with arsenic-rich minerals (Schmitt-Strecker et al. 1992).

In metallurgical processes, certain elements become fractionated. The simplest division and the one that is intended in this case is the separation of copper from components of the host rock. Therefore those elements, which follow the copper throughout all stages of a smelting process to the precipitation of the metal, need analyzing. For a long time there was no agreement concerning these elements. In the above-mentioned 'SAM-project,' eleven elements had been analyzed but only five of these were used for classification purposes. Pittioni (1957) himself used nine elements (Co, Ni, Zn, As, Sb, Ag, Sn, Pb, Bi) for his "relation ore deposit – finished object" study. Pernicka (1987) had very intensively studied the elements' behavior during smelting processes, and concluded that As, Sb, Co, Ni, Pb, Ag, Au, Bi, Ir, Se, Zn, Fe as well as the Pt-metals can be used as the principal indicators for determining the provenance of ore. Important for slag formation are, e.g., SiO_2, Al_2O_3, FeO, MnO, MgO, CaO, BaO, K_2O, Na_2O, P_2O_5, and TiO_2. The likelihood of elements to act as a slag constituent or impurity in metal is primarily dependent on their reducibility, as has been explained above (Fig. 2.4). All elements with oxides, which can be more readily reduced than the oxide of divalent iron, are enriched in the metal, while all other elements gather in the slag. Some elements such as P, Zn and Sn have a transitional character. An overview of the different chemical elements with regard to their importance for the technology of the metallurgical process and for provenance studies is shown in Table 2.2.

Normally, only a few milligrams of material are available for the analysis of metal artifacts, in contrast to that of ores and slag. For the former, a higher sensitivity to the measurement of trace elements is necessary, which can be improved by concentration methods. There are a number of analytical methods, which can be employed in archaeometallurgy. They all have the necessary sensitivity to measure elements even in the $\mu g\ g^{-1}$-range. In the study presented here, the largest part of ores, metal samples and slag have been measured by atomic absorption analysis (AAS; see chapter on analytical details and measurements in the Appendix). Inductively coupled plasma with optical emission spectrometry (ICP-OES) (Prange 1998)

Table 2.2. Grouping of selected chemical elements with regard to their importance in defining metallurgical techniques and provenance studies. The table is mainly based on different reducibility of the elements from their oxides. Values for Sn, Pb and Zn are the approximate limit for deliberately produced alloys. Slightly modified (P!) following data from Pernicka (1999)

Technology	Technology and/or provenance	Provenance
Al, Ba, Ca, Cr, Fe, Ge,	As, Cd, Co, Hg, In,	Ag, Au, Bi, Co, Ir, Ni,
K, Mg, Mn, Mo, Na, Si, Sr, Ti, W	P, Sb, Se, Te, Tl	Os, Pd, Rh, Ru, Sb
Sn ≥ ca. 1%	Sn ≤ ca. 1%	
Zn > ca. 5%	Zn < ca. 5%	
Pb > ca. 5%	Pb < ca. 5%	

and instrumental neutron activation analysis (NAA) were also used. The samples have to be dissolved for the first two methods, in which intra-element- or matrix-effect may occur, particularly in AAS. The ICP-OES does not reach the sensitivity of the AAS, but a large number of samples can be analyzed simultaneously. A major advantage of NAA is that samples can be measured in the solid state. This is particularly important when only a few milligrams of material are available, as the same sample can be used afterwards for isotope analysis. On the other hand, it is very difficult to analyze lead and bismuth using NAA. Ores, metals and slag can also be analyzed by X-ray fluorescence spectrometry, which is a common analytical method in geoscience and a routine method for pottery analysis. However, there are hardly any suitable standards available for materials in archaeometallurgy, so they need to be assessed in the laboratory.

We refrained from using proton-induced X-ray emission spectrometry (PIXE). As the high energy beam produced by PIXE is focused in a similar way to that of an electron microprobe, it is suitable for microanalyses, but less for bulk analyses. Nevertheless, the method has been repeatedly applied for analyzing bulk compositions of metals and alloys (Golden 1998; Maddin 2003) and of slags (Shugar 2000). Ancient metals and especially alloys and slags are heterogeneously composed and contain a variety of different phases, and PIXE analyses may not always produce representative results.

Lead Isotope Analysis

Explanations in the chapters above have clarified the two main problems in using chemical analyses to assign metal objects with a particular ore deposit. These are the geochemical variability of ore deposits and the fractioning of trace elements during metallurgical processes from ore to metal.

Isotopic composition of lead is one geochemical phenomenon that does not show these disadvantages. Lead isotope analysis (LIA) has been used before in geosciences to understand the geological history of ore formation (plumbotectonic model of Doe and Zartman (1979)). The LIA was rapidly taken up by archaeology, initially to assign lead and silver artifacts to a particular mining region or deposit. A history of the application of LIA in archaeology has been compiled, e.g., by Pernicka et al. (1984), Pernicka (1987) and Stos-Gale (1993). The scholarly acceptance of the LIA was similar to that given to Libby's radiocarbon dating, which became very commonly used in archaeology in general during the 1950s. The LIA was also greeted enthusiastically as a new, 'secure' method for provenance studies of metal artifacts (Muhly 1995).

In the last years, it has been mainly the Max-Planck-Institute for Chemistry (Mainz) that has carried out lead isotopes analyses in archaeological contexts, though also the Isotrace Laboratory of the University of Oxford, the Department of Archaeological Science of the University of Bradford and the Conservation Analytical Laboratory, Smithsonian Institution, in Washington have also used this method. LIA results presented in this volume have been taken using a thermal ionization mass spectrometer. The analyses of the finds from Faynan and of several metal artifacts from different sites in the Levant have been carried out by Prof. Begemann and Dr. Schmitt-Strecker (Mainz), while those from Timna have kindly been provided by Dr. Stos-Gale and Prof. Dr. Gale (then Oxford).

The LIA is based on the formation of four stable isotopes ^{204}Pb, ^{206}Pb, ^{207}Pb and ^{208}Pb, of which the last three are daughter-products of uranium and thorium (Table 2.3). ^{204}Pb is not radiogenic; its concentration corresponds to the original amount in the lead and is constant. Since the formation of the earth around 4.6 billion years ago, the initial content of U and Th has changed due to their decay. This has led to a permanently increasing quantity of radioactive isotopes ^{206}Pb, ^{207}Pb and ^{208}Pb in the earth's crust. The relative frequency of these isotopes is not constant throughout the world, because U and Th are unevenly dispersed in the earth's crust. The isotope abundance ratios depend also on the length of time during which radiogenic isotopes were able to form. Generally, lead is geochemically separated from U and Th during the formation of lead ore deposits. So at this point in time the isotopic composition of lead is frozen and remains constant.

LIA is a sensitive analysis of masses, and the isotope ratio is measured with a mass spectrometer. Their high sensitivity has also resulted in the utilization of mass spectrometry in microanalyses. For a routine measurement sample, one needs ca. 10^{-7} g lead; if necessary and with appropriate care even 10^{-9} g would be adequate for the isotope analysis (Pernicka 1987). The lead isotope standards NBS SRM 981 and 982 of the National Bureau of Standards are used for the LIA in all laboratories. In geoscience, mass spectrometers and software are usually calibrated for ^{204}Pb, because this isotope contains no radiogenic particles. But the isotopes ^{206}Pb, ^{207}Pb und ^{208}Pb can be measured more accurately due to their higher concentration. It turned

Table 2.3.
The formation of the three radiogenic lead isotopes through radioactive decay of uranium and thorium

Decay	Half-life ($\times 10^9$ a)
^{238}U → ^{206}Pb	4.468
^{235}U → ^{207}Pb	0.704
^{232}Th → ^{208}Pb	14.01

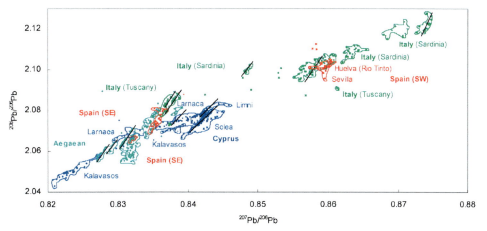

Fig. 2.6. Lead isotope abundance ratios of copper and lead ore deposits from various mining areas used in Roman times in the Mediterranean. The ages of the ore deposits range from Cambrian (high ^{207}Pb/^{206}Pb values) to Tertiary (low ^{207}Pb/^{206}Pb values). The apparent gap of LIA between 0.847 and 0.856 corresponds to an age gap between the Variscan and the precursory of the Alpine orogeny in Europe (from Klein et al. 2004)

out that the best results for provenance studies in archaeometallurgy were achieved when the three following isotope ratios were measured: ^{208}Pb/^{206}Pb, ^{207}Pb/^{206}Pb and ^{204}Pb/^{206}Pb. Today the relative error of measurement is ±0.1% and less. In principle all geological measurements normalized to ^{204}Pb can be compared with data published from archaeological studies, if the appropriate precision has been achieved.

The classification with only three isotope ratios is simple when compared to the ca. 30 different concentrations of elements usually quantified in provenance studies of pottery. But the small number, on the other hand, raises the probability that two different deposits might have matching isotope proportions and thus are overlapping and make provenance studies problematic. Lead isotope abundance ratios are given numerically and conventionally presented in two-dimensional fields. In the beginning, ellipses were used, particularly by the team from Oxford, to form the borders of single regions. This form of illustration is a compromise. In reality, each analysis has to be plotted in a three-dimensional space, so the ellipses can only be seen as sections through an ellipsoid.

The different half-lives of ^{235}U, ^{238}U, and of ^{232}Th which finally lead to the formation of the three lead isotopes ^{206}Pb, ^{207}Pb, and ^{208}Pb (Table 2.3) are suitable for determining the earth's age and geological developments, which have led, over millions of years, to the formation of ore deposits. There is a direct relation between the isotope ratios and the age of a deposit. LIA in the upper right hand corner of the diagram in Fig. 2.6 represent old ore deposits, and the data which plot into the lower left hand corner are compositions from geologically young mineralizations. The age determination mainly of geologically very old sources, using LIA, can be calculated by the 'Holmes-Houtermans'-Model (Houtermans 1960). The logical precondition for this model is based on the assumption that lead, which is to be dated, has a 'one-stage-history,' meaning that the Pb/U-ratio has not changed through time. But this model has often resulted in abnormally ancient dates, which have been caused by

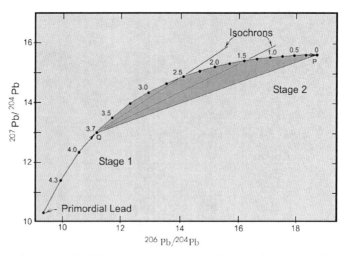

Fig. 2.7. Development of lead isotope abundance ratios from the decay of uranium dependent on the duration of time, depicted in the curved lines as a two-stage lead development model (after Stacey and Kramers 1975). Lead develops from a composition of the primordial lead of 4.57 to 3.70 Ba in a reservoir with ^{238}U/^{204}Pb $(\mu) = 7.19$ (stage 1). At point Q (3.7 Ba), μ changes via geochemical differentiation to 9.74. After that, lead develops undisturbed towards point P (present). The *straight lines* from point Q to varying points along the development-line are isochrones

high levels of U. These could be explained by, e.g., the proximity of a uranium deposit. Stacey and Kramers (1975) came therefore to the conclusion that a development, which happens in two chronological stages, would be far more reliable in describing the percentage of the isotopes (Fig. 2.7).

The characterization of a deposit by lead isotopes depends on the geological age, the percentage of ^{204}Pb in the ores and the concentration of mother-isotopes U and Th, but also on a possible remobilization of the ore. Such a remobilization appears particularly in epigenetic ores. It is therefore usual that the expected variations of the isotopic composition of the deposit are given as between 0.4% to 0.5% (Schmitt-Strecker et al. 1992; Stos-Gale 1993). This is clearly above possible experimental errors. When artifacts are compared with ores, the threshold of tolerance of the latter should always be given.

The confidence with which an exclusive statement can be made, such as where the metal of a finished object certainly does *not* come from, is with this small span of deviation much greater than for the results of chemical composition studies. In the case of a multistage ore genesis, even larger variations are possible, e.g., ca. 1.5% for the deposit of Iglesiente in Sardinia (Boni and Koeppel 1985) or even 15% for the lead deposit in the Mississippi Valley (Houtermans 1960). It is important to take a sufficiently large number of samples to capture these variations. It is even more effective to measure metal inclusions in slag, because they provide a broader based average than single ore samples (see Sect. 2.1.). No isotope fractionation due to erosion seems to happen in ore deposits, as was observed by Gulson (1986). One can therefore assume that in an ore deposit, lead in primary ores has the same composition as in the weathered zones of the 'iron hat' or of the cementation zone. This

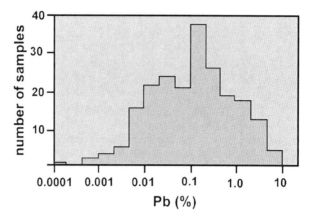

Fig. 2.8.
The histogram shows lead contents in Chalcolithic and Bronze Age copper and bronze artifacts and proves that lead isotope analysis is very appropriate for these objects (after Wagner et al. 1989)

is an important fact as it has been assumed in archaeometallurgy that fluxing agents from e.g., the 'iron hat' might have been used for smelting primary ores.

LIA is not limited to lead ores or lead artifacts. Lead is to be found in nearly all oxide and sulphide ores, which play a role in archaeometallurgy. It exists, as a natural impurity in, e.g., Ag-, Cu-, and Fe-ores. As a result, most copper, bronze, or silver and iron artifacts contain lead at a level of 10–1 000 $\mu g\ g^{-1}$ (Fig. 2.8). It is generally sufficient to take 20–50 mg of material for sampling. The method is destructive, but due to the small amount necessary it can be used for many valuable objects (glass, pottery, pigments etc.), and it still leaves enough material for a chemical analysis.

The decisive point that makes the LIA so valuable for provenance analyses is the fact that lead isotope ratios do not change during metallurgical processes. The isotope pattern remains constant, independent of what temperature the ore is roasted at or under what reducing-oxidizing conditions the metal is smelted or how often it is remelted. The pattern is therefore characteristic of a particular deposit and allows a secure assignment of the finished product to the initial raw material (Pernicka et al. 1984). A fractioning of lead isotopes during the course of metallurgical processes can so far not be proven, although it has been postulated by Budd et al. (1995). Age determinations by lead isotopes allow an estimate of the approximate age of the ore mineralization from which an artifact was made. And conversely, the information from the isotopes is the basis on which to reject other deposits as possible sources for a particular object, because their geneses had been dependent on the geological development of the region. One precondition for the viability of these results lies in the necessity of excluding with certainty that any other metal, including lead, had been alloyed during the production of this artifact. Otherwise, one would only describe the isotope characteristics of this added metal or those of the mixture between the lead of the initial and the added material.

At the beginning of the use of isotope analysis in archaeology, the results of LIA had already led to a number of new discoveries. In the last 30 years, a series of research projects have taken place to use LIA for the characterization of ore deposits and metal objects and to thus determine the provenance of an object. This has all immensely assisted in the reconstruction of the beginning of metallurgy in the Old World. The Aegean area has in particular been studied intensively, as have the ore deposits of Laurion and on the Cyclads (summarized, e.g., in Pernicka 1987; Stos-

Gale and Macdonald 1991). The Heidelberg-Mainz team produced a comprehensive overview of 'ancient' ore deposits of copper, lead and silver in Anatolia (Pernicka et al. 1984; Seeliger et al. 1985; Wagner et al. 1986, Wagner et al. 2000, Pernicka et al. 2003). Yener et al. (1991) complemented it with information from the region around the Taurus Mountains. The well-known Bronze Age trade of oxhide ingots has been for decades the research object of the Oxford group, who have also studied the origins of oxhide ingots from Cyprus and/or Sardinia (summarized, e.g., in Gale 1991a; Stos-Gale and Gale 1994).

But the growing number of analyses and the increasing knowledge in archaeometallurgy has revealed problems that complicate the apparently unproblematic interpretation of the LIA of archaeological artifacts. These are

1. Ore deposits can have identical or overlapping isotope compositions, even when they are geographically far apart. Since their age and the conditions of their formation had been so similar in several cases, their lead isotopy does not show sufficient dissimilarities for a distinctive assignment (Pernicka et al. 1984). The question of which deposit was the ore source has to remain doubtful in those cases where no archaeological evidence could help. This means that the particular strength of the LIA lies in the refutation of a certain source when lead isotopes are not fitting and not in the unambiguous assignment of an artifact to a particular deposit.

   In the beginning, the overlapping of isotope compositions was not too big a problem, because only a limited number of ore deposits and metal objects had been studied. But in the meantime, so much data has been collected, that it has become rather difficult to sort out the overlapping of certain ore deposits around the Mediterranean (Stos-Gale 1991, 1993; but see Klein et al. 2004, Fig. 8). In Oxford alone, more than 2 000 metal objects and some 1 500 ore samples coming from deposits in several European countries have been analyzed (Stos-Gale 1993, Fig. 2.9). Then the variations in the isotope ratios of ores from Wadi Arabah (Gale et al. 1990; Hauptmann et al. 1992; Stos-Gale 1993) and from Sardinia (Gale 1991a; Sayre et al. 1992) were published. It is, however, not futile to hope for any reliable differentiation of the ore deposits around the Mediterranean Sea as a whole (Pernicka 1995). But it shows that provenance studies are still successful if considered in a regional archaeological context.

   Particularly the members of American teams studied the problem of overlapping isotope patterns in different ore deposits using statistical methods (Sayre et al. 1992). Again, it can be summarized that any statement about the provenance of metal artifacts is far more reliable if the area of possible origin is regionally limited. If on the contrary, geographically undefined spheres are discussed, the answers cannot be expected to be very comprehensive;

2. Recycling of scrap metal has also to be taken into account. The isotope pattern resulting from such processes cannot be compared with the original ore source. Pernicka (1987) judged the possibility of such a secondary use of metal as relatively low, at least for the Early Bronze Age. That can be substantiated with the example of the metal artifacts from Hassek Höyük (Schmitt-Strecker et al. 1992). On the other hand, investigations into the archaeometallurgical craftsmanship of Early Bronze Age III/IV Khirbet Hamra Ifdan (see Chap. 5) clearly proved recy-

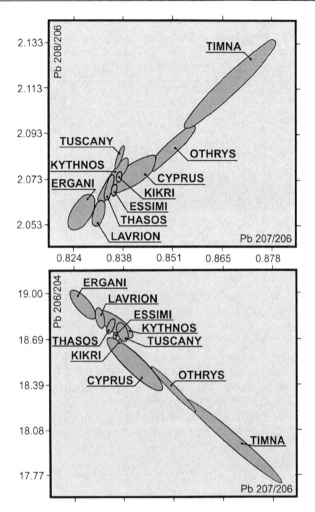

Fig. 2.9.
Lead isotope compositions of copper deposits in the Aegean, Anatolia and Wadi Arabah, displayed in isotope-diagrams normalized to ^{206}Pb. The ores indicate very substantial overlapping (after Stos-Gale 1993)

cling and remelting of copper objects. The issue becomes more problematic with the growing 'industrialization' of metal production in the later Bronze Age and the Iron Age. The discussion of the trade routes of the Cypriot oxhide ingots in the Mediterranean revolves around this problem (Muhly 1991; Gale 1991; Stos-Gale and Gale 1994). The question of whether these ingots might be made of metal from several different ore sources in Cyprus or even melted together from recycled second-hand metal has developed into a sore point in the discussion (Budd et al. 1995; Gale and Stos-Gale 1995; Hauptmann et al. 2001). What has not been taken into account so far is a possible homogenization of raw metal. The metal from one mining district such as Faynan could have been smelted at various smelting sites, but then gathered at one collection point, there processed further and remelted into larger entities. In particular, a remelting of small copper prills extracted from metal-rich slag seems to have been a common practice in archaeometallurgy throughout the ages;

3. The assignment of metal artifacts to a raw material source could be hindered further if ores from different sources had been smelted together. This problem seems to be particularly relevant for the period before the full maturing of the Bronze Age. In the eastern Mediterranean during this early period, ores were transported to the settlements not only from the nearest sources but also from quite distant deposits, and the following smelting in the settlements was rather small-scale (Shalev and Northover 1987; Khalil 1988; Ilan and Sebanne 1989; Hauptmann et al. 1993; Shalev 1994). The examples of Abu Matar (Hauptmann 1989), Arslantepe (Palmieri et al. 1993; Hauptmann et al. 2003) and Hassek Höyük (Schmitt-Strecker et al. 1992) showed that ores from different deposits had been imported and processed together.

The current problems of LIA are discussed by different scientists in a number of recently published articles in the following journals: Archaeometry (vol. 34, 1992 and vol. 35, 1993), Journal of Mediterranean Archaeology (vol. 8, 1995) and European Journal of Archaeology (vol. 3, 1, 2000 and 4, 1, 2001) They debate the chances and limits of using LIA in consideration of the mentioned problems and its acceptance in archaeological research after 25 years of application. Muhly (1995) very aptly compares the resulting critical assessment of LIA and the cautious utilization in archaeology with the initial problems that the ^{14}C-method had. He therefore cautions against calling the entire method into question on the basis of still existing problems. Although the discussion is still controversial, it becomes increasingly clear that it is necessary to avoid using LIA as the only method in provenance studies. Rather, LIA has to be incorporated into a combination of methods, such as trace element analyses, additional surveys of ore deposits, geochemical and mineralogical analyses and last but not least the archaeological context, all of which need to be considered carefully. Archaeometallurgical field work is gaining more importance than ever in the study of evidence of mining and smelting processes in the periods discussed.

Chapter 3

Nature and Geology in Faynan

Locations suitable for settlement are those areas where several natural phenomena overlap. Topography, climate, vegetation, availability of water, arable land, raw materials and building materials result in an environment that provides necessary preconditions for successful human survival. The combination of these elements in the 'basic natural kit' (Schröder and Yalçin 1991) have been so favorable in the settlement area of Faynan that the region has been almost continuously occupied from the Palaeolithic to modern times. The history of settlement in Faynan is not only a remarkable story of the utilization of an arid region, but also of the exploitation of its indigenous raw material (copper ores) and vegetation. Not surprisingly, all of this activity also affected and shaped the landscape. These processes of change are very interesting from a geoarchaeological standpoint, but they can only be touched upon here. They have been studied in detail in a research project undertaken by the CBRL (Centre of British Research in the Levant, former BIAAH) (Barker et al. 1997, 1998).

Toponymy

Knauf elucidates in Hauptmann et al. (1985) that the Arabic name Faynan (Hebrew: Punon, Greek: φινω, φαινων, Latin: Phunon; Geerlings 1985) is already mentioned in ancient Near Eastern texts. A site mentioned in a place-name list of Ramses II (1219–1213 BCE, Görg 1982) might be identified as the spot. The Arabic word 'Fainan', which means 'beautiful, long hair' in Arabic or a man with such hair, could be the etymon of the topographic name, stemming from the specific landscape and its vegetation. Faynan is situated at the foot of the southern mountains. The old name for this steep mountain rise from the Wadi Arabah up to the Jordanian plateau is 'seir' (Harding 1971) a word also meaning 'the hairy one'. The word 'Pinon' appears on the list of Edomite 'chiefs' or 'district rulers' in the Book of Genesis 36, 40–43. The designation 'Phunon' finally turns up in Numbers 33:42f. This is a list of place names, which attempts to describe the stations along the route to the Holy Land that the Israelites took under Moses' leadership. The quote in Numbers is actually an addition to the Pentateuch expanding the list of locations, which had been given in the older rendition, 'Israel in the desert' (Noth 1966). Most probably the author described the that was route taken, at the time he was writing, when traveling from the Gulf of Aqaba, through the Wadi Arabah to Moab and finally to Jericho. This historical account confirms the existence of Faynan as a town in the 5th century BCE.

Geography

The area of Faynan is located southwest of the Hashemite kingdom of Jordan between the 30th and the 31st degree of northern latitude and the 35th and 36th degree of eastern longitude. Wadi Faynan is part of the Wadi Arabah rift-system (Figs. 1.1, 1.2). It is around 50 km away from the southernmost end of the western limit of the so-called 'Fertile Crescent.' The 'Fertile Crescent' describes a region of the Near East that is in the shape of a crescent moon standing on its tip, which lies in a semi-circle around the arid and semi-arid areas north of the Arabian Peninsula. The southern limit of the 'Fertile Crescent' is roughly identical with the 300 mm isohyete and stretches from the Eastern Mediterranean coast in a wide arch first north then eastwards, north of Euphrates and Tigris and down along the mountain ranges of the Zagros mountains, southward nearly all the way to the Persian Gulf. As the name indicates, rainfall is sufficient there to allow successful farming (of crops) and animal husbandry. In the 'Fertile Crescent,' the earliest sedentary cultures in the Neolithic developed, when agriculture and animal husbandry were first practiced. Not until the 6th millennium BCE did the first settlers move into areas outside the "Fertile Crescent" towards Mesopotamia, where the foundations for the ancient Sumerian civilization were laid. The later empires of Babylon and the kingdoms east of the Jordan were completely, and the kingdoms of Assyria and Israel partly, outside the "Fertile Crescent."

In the region discussed here, which is south of Bilad el-Sham, geography and climate vary considerably. In the west, the arid desert-like area of the Negev merges into northern Sinai, while to the east of Faynan there is a steep rise towards the Jordanian plateau. These arid regions are normally not sand deserts but limestone deserts, which are covered in parts by humus. This meager soil can nevertheless support some hardy but undemanding vegetation. The Negev and the southern part of the eastern Jordanian desert are separated by the Wadi Arabah. The approximately 360 km long depression starting at Lake Tiberias and ending at the Gulf of Aqaba divides the eastern and western half of this region. The depression is part of the great Rift Valley that stretches from East Africa to southern Turkey.

The deeply carved north-south depression of the Wadi Arabah (Fig. 3.1) is a geologically, morphologically, and climatically different landscape, set apart from the above-described larger background. It needs to be treated as an independent regional unit. The depression stretches from the Dead Sea to the Gulf of Aqaba/Eilat over a distance of 175 km. The width changes between 10 km in the south and 25 km in the center and north. The elevation of the Wadi Arabah at the southern end of the Dead Sea is 406 m below sea level (measured 1994). From here it rises gradually over a distance of 50 km up to sea level. The watershed between the Dead Sea and the Red Sea is situated 100 km south of the former at a height of 200 m. The gradient towards the Red Sea is so shallow that no significant erosion channels have been created. Water running into the Wadi Arabah from the east evaporates on salt pans.

In the west, Wadi Arabah is fringed by a steep rise leading to the Negev desert, while in the east it is bordered by the higher Jordanian plateau reached by ascending a very steep incline. The plateau runs over a distance of ca. 120 km and is on average 1 200 m high. Some peaks reach 1 700 m. To the east the Jordanian plateau descends towards the central Jordanian desert. The desert consists of an escarpment, which

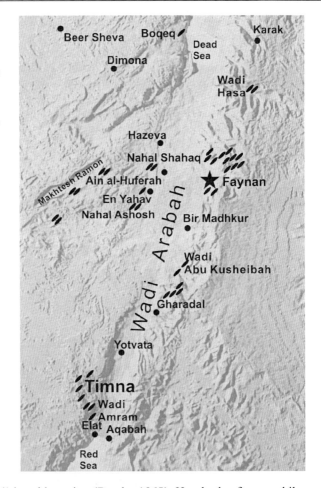

Fig. 3.1.
The Wadi Arabah running between the Dead Sea and the Red Sea is a part of the Rift Valley, which stretches from East Africa to Syria. It separates the Negev from the Jordanian Plateau. Faynan is at the eastern side of the Arabah. Modern and ancient sites mentioned in the text are shown on the map. Copper mineralizations are *hatched*

has partly been 'drowned' in arid erosion (Bender 1968). Hundreds of square kilometers are covered by sharp-edged, eroded hornstone, the so-called Hamada.

The mountain range east of Wadi Arabah consists of crystalline rocks in the Precambrian basement, covered with a huge sequence of Cambrian to Cretaceous sediments. The tectonics of the Rift Valley are active up until today (Horowitz 1979). These activities and the change from a more humid climate to the arid climate that exists today (see below) led to extensive linear erosion at the eastern side of the Arabah where gorge-like, deep wadis were formed. Only the south of the range is less steep. Until a few years ago, asphalted roads connected the Wadi Arabah with the plateau only in the north at Kerak and in the south at Aqaba. The former tracks through the Wadi Dana towards Petra and to Gharandal, which are asphalted today, were then only usable in good weather conditions. These routes might already have existed in (pre-)historic times along with the roads through the Wadi Ghuwebe up to the Edomite towns.

The east-west connections crossing the Arabah would always have been rather difficult. Therefore from the earliest times on, the Wadi Arabah must have played a very important role as a north-south trade route. The main trading route between the Nile delta and southern Palestine was further west, parallel to the Mediterranean coast, but the Arabah held a strategic importance. An old caravan road extended from the Hijaz via Maan to Gaza and crossed the Arabah between Faynan and Petra. Also, other historic caravan routes coming from the south went through the Wadi Arabah. In particular, the Nabatean capital, Petra, was well served by this north-south route (Zayadine 1985). Roughly 30 km south of the Dead Sea, the road turned to the west, where it left the valley and moved through the northern Negev towards the area of the modern town of Beersheba. The Wadi Arabah was rather densely settled, intriguingly even in prehistoric times, which might have been due to the formerly more humid climate (see Fig. 3.7). A number of settlements existed even in the Pre-Pottery Neolithic (Raikes 1980). Thompson's (1975) survey showed that the Arabah was also densely settled in the Bronze Age.

Topography

The ancient copper district of Faynan is shaped like a pointed 'niche'. The shape has been caused by a few east-west running faults that break the north-south running tectonic pattern of Wadi Arabah (see also Figs. 4.2 and 4.3). These faults are visible in the Wadi Ghuwebe in the north (at Khirbet en-Nahas, see also Figs. 5.1 and 5.3) and in Faynan itself, where the wadis Dana, Sheger and Ghwair meet.

Fig. 3.2. View towards the plain around the Byzantine ruins (Khirbet Faynan, *centre, arrow*). Behind it, the mountains steeply rising up to the Jordanian Plateau border the area. To the *left of the center* a dark horst consisting of andesite is visible. *Below it, left* of the ruins is one of the Quaternary gravel terraces

Fig. 3.3.
Quaternary gravel terraces at the confluence of Wadi Sheger into Wadi Faynan next to Faynan 1 at the foot of the Jordanian Plateau. Both wadis have cut deeply into the ground, leaving steep sides. The main activity must have happened in post-Roman times, because the wadis cut through Roman slag heaps and have left a standing section of 4–7 m height (*arrows*). As is also visible in Fig. 3.2, the wide plains of this region are nearly bare of vegetation

Their confluence is close to Khirbet Faynan and the dominant drainage continuing to the west is then first called Wadi Faynan and later Wadi Fidan. The 'niche' equals approximately the extent of the research area, where most of the ancient mines and smelting sites are located. The Byzantine settlement of Khirbet Faynan is at a height of 250–300 m above sea level. In comparison to the wide, relatively monotonous plain of the Negev and the broad Jordanian plateau to the east, which surround the area, the region of Faynan is diversely structured. The landscape shows geologically, morphologically and climatically a different pattern from the majority of the region.

To the east, the district of Faynan is bordered by a steeply rising mountain range, which even today is very rough and with few paths (Fig. 3.2). The foot of the cliffs is build up by a sequence of five Quaternary gravel terraces. The oldest terraces start at steeply plummeting mountainous outcrops and exist only in relics as pediments or glacis. Towards the west, the terraces become flatter and merge. Eventually they are covered with widely fanning mud flats deposited at the exits to the wadis. In the center of the wadis, watercourses have often cut a narrow passage down to the present wadi bottom, leaving high vertical profiles (Fig. 3.3).

Faynan and Timna have compatible morphological structures. Hauptmann and Horowitz (1980) found in Timna a succession of five gravel terraces, as is also found in the central Negev (Goldberg 1976). Horowitz (1979) suggested that the genesis of the older terraces is due to the more humid, pluvial periods of the Quaternary. His reasoning is based on the softly descending slopes of the older terraces, which show a sharp contrast to the canyon-like cuts in the younger sediments, as can be seen everywhere in the Arabah, the Negev and on the Jordanian plateau. These clefts have been brought about by torrential rains of the current arid climate. In Timna it was possible to demonstrate that the youngest wadi cuts were caused by tectonic movements in the Wadi Arabah. A most interesting research topic is the change of flow regimes in the Faynan area during the 3rd millennium BCE, which probably lead to an extensive exposure of copper mineralizations in the Dolomite-Limestone-Shale Unit and subsequently to an intensification of mining activities in the developed Early Bronze Age.

Further to the west, the wadis fan out, turning into broad, alluvial flats and end in clay pans without drainage. There are single sand dunes in this morphologically sedentary, flat area. In the wide plains at Barqa el-Hetiye, (Baierle et al. 1989, Fig. 3.4) densely covered with salt trees (Haloxylon persicum), loess sediments occur. These are typical sediments of Wadi Arabah and the Negev further to the west. Ginzbourg (1963), who has studied these sediments in the region of Beersheba, suggests that they are of aeolian origin accumulated by means of southwestern and western winds from the central and northern Sinai. In between are only a few solitary hilltops, mostly antithetic monadnocks, which developed during the forming of the rift system (Fig. 3.5).

Fig. 3.4. Area with dunes and sand- as well as loess deposits in the surroundings of Barqa el-Hetiye in the western part of the survey area. The vegetation consists of Haloxylon persicum. In the center, the excavations discussed in detail in Sect. 5.2. In the background, the screening mountain range of Jebel Hamra Ifdan

Fig. 3.5.
Cuesta landscape in the central area of the Faynan district. In the middle, the ancient site (Khirbet Faynan) of Faynan with Iron Age slag heaps (see Sect. 5.2.). In front, remains of *Ziziphus spina-christi*, probably of anthropogenic origin

Faynan is bordered in the west by the north-south running Jabal Hamra Ifdan, a low mountain range of Precambrian granititic rocks, which is a barrier towards the Wadi Arabah. The Faynan district is therefore enclosed on all sides and stands as an isolated geographic unit. The narrow outlet of the Wadi Fidan in the north of the region is consequently seen as the 'Gateway to the West' (Levy et al. 2002). The dense settlement activities in that wadi particularly during the Neolithic, Chalcolithic and Early Bronze Age conspicuously indicate the strategic importance of this narrow outlet for the cultural and economic links between Faynan and the settlements in the Negev.

Climate

The northwestern part of Palestine and the eastern bank of the Jordan River are situated in a transitional area between the Mediterranean climate to the west and the arid climate to the east and south. The rainy season starts in the northwest in October, reaches its climax in January/February, and ends with late rains in April. As the precipitation comes exclusively from the evaporation pan of the Mediterranean Sea

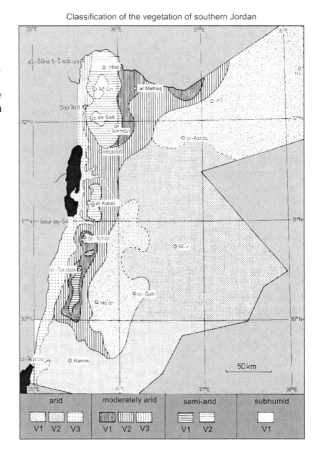

Fig. 3.6. Bioclimatic zones in Jordan. The western boundary of the Jordanian plateau with its relatively high precipitation belongs to a moderate or semi-arid climate. Directly to the west, the Wadi Arabah has an arid climate. The symbols *V1* to *V3* represent areas with increasing temperatures (after Kürschner 1986)

in the west, the rain falls as the clouds rise up onto the Jordanian Plateau, condense and thus mainly rain on the foothills and eastern side of the Wadi Arabah. The precipitation is highest, with 250 mm, locally even up to 500 mm, along the ridge of the lower mountains to the east (collected by Baierle (1993) partially for the area of Faynan). The western edge of the high plateau is therefore valuable for intensive arable and pastoral agriculture. Wet winters and usually very dry summers with a hot and arid climate in the south are characteristic. Jordan can be divided into four bioclimatic zones (Fig. 3.6).

The arid desert-like areas with precipitation below 50 mm are situated east and south of a semi-arid transitional zone with the Wadi Arabah as a spur extending to the north. The arid land also marks the boundary for settlements and agriculture at the western edge of the high plateau. Rainwater floods the watercourses, which incline towards the Wadi Arabah, e.g., the wadis Ghuwebe, Ghwair/Faynan/Fidan, and Dana (Fig. 5.1) in the lower parts of which a perennial water table close to the surface occurs (Bender 1968). Other perennial water-bearing wadis are the wadis Hasa, Khuneizira and Musa, which show surface water in the higher reaches of the mountains (Baierle 1993).

Probably conditions for human settlement in this region had been much better during the Neolithic through to the Bronze Age than today. Research has shown that the climate changed in the 10th millennium BCE. During this period at the end of the Mesolithic and the beginning of the Neolithic, the dry climate underwent a transition into a more humid phase (Fig. 3.7). In the Neolithic, a higher precipitation than today is suggested (Bintliff 1982). This change is recognizable in the dense settle-

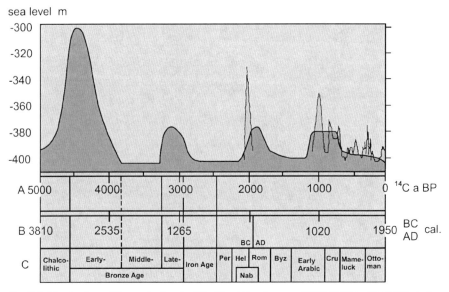

Fig. 3.7. Fluctuation of the water level of the Dead Sea from ca. 3800 BCE to AD 1950. A rise of the water level means a generally more humid phase in the region, e.g., at the beginning of the Early Bronze Age, in the Late Bronze Age, in the Roman period and in the Early Islamic period. Underneath are three different time scales: ^{14}C-date BP (*A*), ^{14}C-date calibrated BCE and AD (*B*), historical periods (*C*) (after Bruins 1994)

ment activities of this period within the climatic 'marginal zones' (Gebel 1984). The humid conditions remained stable in the Eastern Mediterranean until the beginning of the 5th millennium and were followed in around 3500 BCE by a clearly more arid phase (Barker et al. 1998). Figure 3.7 illustrates that likewise in the first half of the 3rd millennium BCE more humid conditions prevailed. They supported a growing urbanization in the Early Bronze Age II-III (Amiran 1991; Bruins 1994). Subsequent climatic changes might have been one of the deciding factors for the end of the cultural flowering of the cities at the end of the 3rd millennium BCE in the Early Bronze Age III-IV. Other periods with more humid conditions occur in the Late Bronze Age, in the Roman period and in the Early Islamic period.

Water, Irrigation Systems and Agriculture

Faynan is characterized by plentiful rainfall and spring water flowing from the mountain range. These water sources played a crucial role in the likelihood of settlement in this otherwise very deprived and desert-like region. The climatic conditions in this area between the eastern side of the Arabah and the peaks of the Jordanian plateau are characterized by unpredictable and heavy precipitation. Torrential rains lead to dramatic flooding (Baierle 1993). The ancient farmers of the Faynan area were confronted with climatic conditions, which were not particularly suited for either settling down or agricultural use of the land. Rain fell once or twice a year in amounts too large to be usefully utilized and at the wrong time of the year. Floods occurred suddenly and led to tremendous erosion damage. The water from the floods poured down the wadi beds leaving the slopes relatively untouched.

A benevolent environment, such as one that would have allowed development in different central regions in Palestine, was not the case in Faynan. But since the Neolithic it has been possible to cultivate the land and develop an infrastructure for a continuous exploitation of the ore deposits. In order to be independent of recurrent floods, agricultural activities were concentrated in their beginnings in the 9th–8th millennium BCE to the narrow catchment area of the Wadi Ghwair. This wadi is located at the foot of the ascent to the Jordanian plateau. Simmons and Najjar (1996) excavated a Pre-Pottery Neolithic settlement here (Wadi Ghwair 1, see Chap. 5). The inhabitants of the site used perennial water sources for irrigation as it was characteristic for the beginning of settled life in the Pre-Pottery Neolithic (Gebel 1984), in a period when food production became common from farming and animal husbandry. This can be concluded from other sites in the Arabah (Raikes 1985), especially from the situation in Tell as-Sultan (Jericho). Here, field irrigation has been postulated for the Neolithic (Kenyon 1960), although no archaeological evidence is available. The change of settlement patterns in the 6th, 5th, and 3rd millennia BCE should be explained in terms of changing water management techniques, at least this is suggested by Barker et al. (1997, 1998).

Settlements such as Tell Wadi Faynan or Wadi Faynan 100 (see Chap. 5) quite obviously used the perennial water supply of Wadi Dana and Wadi Faynan (see Fig. 5.1). Habitation sites in the Wadi Fidan used the water that outcrops due to the Jebel Hamra Ifdan, which effectively dams the wadi by creating a natural waterhole. They augmented this source in order to carry out agriculture by collecting surface

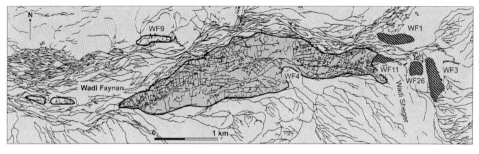
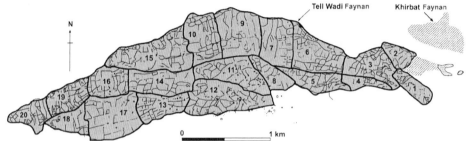

Fig. 3.8. The well-planned Roman-Nabatean irrigation system, which stretches from Khirbet Faynan (WF1) over several kilometers to the west. It is directly south of Wadi Faynan (upper drawing). The horseshoe-shaped structure on the right-hand side of the lower figure is *the* Roman slag heap. See also Figs. 5.1 and 5.3 (from Barker et al. 1997)

water. Circular depressions nearby filled with silt and clay are still indications of this water collection. Similar structures have been found at the Chalcolithic settlements in the Beersheba valley. We suggest that the well-planned irrigation system still in existence today and generally dated to the Nabatean-Roman period, which stretches today between Khirbet Faynan (see Fig. 5.1) nearly all the way to Al-Qurayqira (Fig. 3.8), might have its origin in the 4th millennium BCE. This hypothesis is indicated by two observations. The first is that pottery from the 4th millennium BCE can be found next to Roman-Nabatean pottery scattered over the terraces. The second observation comes from several remains of irrigation systems close to the settlements of Tell Hujayrat al-Ghuzlan and Tell Magass near Aqaba, in the gravel fans of the Wadi Yitim (Brückner 2002; unpublished results German Archaeological Institute, Berlin). Here, remains of terraces and channels were found, which were constructed for irrigating the water. They show parallels to Faynan.

Living conditions must also have been acceptable in times of lower precipitation such as the late antique times and the post-Byzantine centuries. This becomes clear from the account of Eusebius describing how Faynan was a bishop's see in the 3rd/4th century CE, during the Roman/Byzantine period (Geerlings 1985). The most prominent remains of the above-mentioned irrigation system are to be found in the direct vicinity of the ruins of Khirbet Faynan. Musil (1907/1908) had already described the pier of an aqueduct at the mouth of the Wadi Sheger. The channel runs all the way to a large water reservoir and then to the remains of a Byzantine water mill (see also Fig. 5.3). Only two of the three arches, which spanned the wadi at the beginning of the last century, are still standing. Other remains of the water channel can be followed in the Wadi Ghwair 1.5 km above the aqueduct until they disappear in thick vegetation. This channel irrigated an area of around five to six km². Until 1997, the abandoned fields and field boundaries from the Roman-Nabatean-period were remarkably well preserved (Barker et al. 1998).

Today irrigated land farmed by hose-pipe irrigation systems is of little significance compared to traditional farming regions east and south of the Dead Sea in the Ghor. Most of this irrigated land at Faynan is at the village of Al Qurayqira (Ain el-Fidan), partly organized through the 'Jordan Valley Authority'. The farms mainly grow tomatoes and melons. Single fields can also be found in the Wadi Ghwair and Wadi Ghuwebe.

Wind Patterns

Palestine and the eastern side of the Wadi Arabah experience westerly winds coming from the Mediterranean. In the summer the winds reach the mountains west of the Jordan at noon and the east Jordanian mountains in the afternoon. The winds can be strong, particularly on top of exposed mountain heights, and they are responsible for the high precipitation of dew. They pass over low-lying regions such as the Arabah at a great height, but warm fall winds appear, too. The air in the Jordan valley and in the Wadi Arabah stands still in summer until the early afternoon. At around 2 p.m. a strong northern wind starts that carries dust and builds vortexes (dust devils). In wintertime these winds generally start earlier and can continue for weeks (Horowitz 1979). At Aqaba, strong and continuous northern winds prevail during spring and were thought to have been one of the reasons for metallurgical activities in the southern Arabah such as Tell el-Kheleifeh (Glueck 1938). When seasons are changing at the beginning and end of the summer, northern and southern hot winds arrive from the desert areas. They are called el-hamsin or es-serqiye (adapted to 'sirocco' in Italian) and can cause temperatures of 35–40 °C, even 50 °C. They are connected with the development of areas of low pressure in the region and are very uncomfortable to experience.

Large-scale wind directions are amplified by small-scale circular currents set off by local topographical features. They can cause fast changes of wind directions. At Faynan these are seasonal upslope currents and upslope and downslope valley winds. These have been the deciding factors in the positioning of Bronze Age smelting sites and therefore of crucial importance for the development of copper production. The positioning of the smelting site at Ras en-Naqab was not only chosen because of the northern winds typical for the Arabah, but also because of the upslope winds triggered off through the warming of air layers closer to the valley floor. The position of the smelting sites in the direct vicinity of Faynan and Wadi Ghwair was mainly influenced by east-west blowing upslope and downslope valley winds. These can reach quite considerable velocity due to the gorge-like narrowing of the wadi at the foot of the range and the enormous difference in height up to the Jordanian plateau.

During several campaign seasons the wind speed was measured at the position of the furnace array of Faynan 9 (see Chap. 5). In March the speed was a constant 5–6 m s^{-1}, often coming from the west, and reaching for hours even hurricane strength (wind-force 10). Figure 3.9 shows the wind conditions, which were measured on the

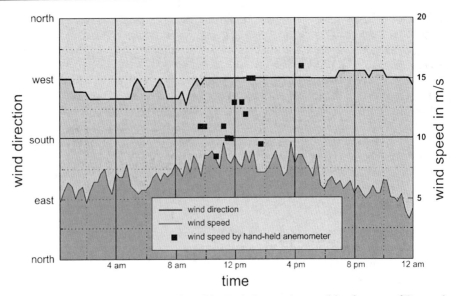

Fig. 3.9. Wind conditions at the position of the Early Bronze Age smelting furnaces of Faynan 9 on March 18, 1998. The direction of the wind constantly coming from the west between 10.00 a.m. and 0.00 and the high wind velocity between 8.00 a.m. and 4.00 p.m. should be noticed (from Kölschbach 1999)

18 March, 1998 with an anemometer with hemispherical wind catchers as an indicator of the average conditions. The wind measuring experiment happened during smelting experiments while testing the functioning of Early Bronze Age wind-powered furnaces. The success of these smelting experiments (see Sect. 7.2.) refutes the often repeated misgivings (e.g., Rothenberg 1990), that the changing wind conditions in the Arabah would make the smelting of ore in wind-powered furnaces impossible.

Vegetation and Fuel Supply for the Ancient Copper Smelters

Human occupation of Faynan necessitates a substantial use of the natural vegetation. Since the Palaeolithic, humans have burned wood. The human impact on the geo-biosphere starts with the domestication of animals in the Pre-Pottery-Neolithic (Uerpmann, quoted in Fall 1990) and has left traces in Faynan long before metal production (Barker et al. 1998). These early remains are small at Faynan and in the entire Levant, compared to those of later periods. Even in the Chalcolithic and Early Bronze Age I, human activities have not affected the quantity of the region's vegetation to any significant degree (Gophna et al. 1986/87). This is mirrored by pollen analyses in the Golan Heights, too (Schwab et al. 2004).

Environmental impacts are to be observed with cultural developments in the Early Bronze Age from the middle of the 3rd millennium BCE. From the 27th century on, the area started to become more arid (Amiran 1991). The fossilized excretions of the rock hyrax (*Procavia capensis*, a close relative of the elephant, although it looks more like a badger) from caves in the immediate vicinity of Petra have been analyzed for pollens. The analyses show that tree density of southern Jordan had degenerated due to anthropogenic impact before the 2nd millennium BCE (Fall 1990). The date points to an era when a significant increase of population, as Gophna's et al. (1986–1987) is convinced, caused a wide-ranging deterioration of vegetation, at least in the coastal plains of Israel. In the Middle Bronze Age, vegetation recovered again. It will be explained in Chap. 5 that it was also during the Early Bronze Age when the first 'industrial development' in metal production took place at Faynan. Possibly it was this development, which led to the first substantial damage to local vegetation. Recent pollen analyses in lacustrine sediments from the Birkat Ram crater lake in the Golan Heights further proved anthropogenic environmental damages mainly in the Hellenistic-Roman-Byzantine and in the periods of the Crusaders (Schwab et al. 2004).

The quantity of fuel necessary to produce the amount of slags over the millennia observed at Faynan was considerable and there are many steps in mining and metallurgical activities where supply of fuel is needed. In general (and not in every case represented at Faynan), these are mining (fire-setting), timber to support mine workings, roasting and smelting of ore, (re-)melting, casting, and smithing of metal. Fuel used in ancient times comprises wood, charcoal, dung, and in later times coal and coke (Rehren 1997). Charcoal was the main fuel used at Faynan, but archaeological evidence of charcoal production was not found.

An initial study and assessment of the vegetation at Faynan was carried out collaterally with a research project of the Institut für Systematische Botanik und Pflanzengeographie from the FU Berlin (Baierle et al. 1989; Baierle 1993). This study was meant to assemble basic information necessary for answering one of the most important questions of archaeometallurgy: to evaluate resources of firing material, which were one of the basic requirements of copper production (Engel 1992, 1993, 1995; Engel and Frey 1996). Analyses of some 9 000 pieces of charcoal, which had been collected from well-dated smelting sites, built the core of the palaeobotanical studies. The ancient wood burnt for (s)melting has been compared with the today's vegetation to trace the location of woods used and to judge amounts of wood used.

Caused by its geographical setting at the crossing point of four different flora regions and as the result of a number of varying elevations, a complex geological structure, and a rapidly changing climatic gradient, the vegetation of the area of Faynan consists of different units (Baierle et al. 1989; Baierle 1993). A simplified categorization follows:

On the higher levels of the mountains (1 100 m and higher) *Juniperus phoenicea*-und *Quercus callipronis*-open forests are the main vegetation. They are relatively resistant against cold, evergreen, and form an open deciduous forest. Today's stock can only be considered as the heavily degenerated remains of an earlier, much more widely spread flora. The building and operation of the Hijaz railroad at the beginning of the last century certainly resulted in a very severe felling of trees. This becomes very clear from Musil's (1907) accounts, who described a 'dense oak forest' above the area of Faynan. It is interesting to notice that the wood of exactly this kind of tree has been identified in the Early Bronze Age charcoal, which had been obtained from smelting sites dated to that period (Engel and Frey 1996, Engel et al., unpublished results Berlin/Bochum). Conspicuously high percentages of *Juniperus phoenicea* (31.5 wt.-%) have been found, as well as *Olea europaea* (12 wt.-%), *Pistacia* cf. *atlantica* (16.9 wt.-%) and *Quercus callipronis* (8.1 wt.-%). The likelihood that the decimation of this rich vegetation was a result of copper production cannot be excluded. The total production listed in a cursory form and estimated from the volume of the slag heaps from these periods (5 000 t, see Table 5.3) would point towards a required volume of wood of about 45 000 t during a time span of 1 000 years. But it has to be assumed that the tree population of the 3rd millennium BCE was also to be found at much lower altitudes than today, when the trees are limited to the Jordanian plateau. This would certainly have been the result of much more humid weather conditions then, and it eliminates the arguments summarized under the recent slogan of 'charcoal to the ore' concerning a longer transportation route. But this is not true for the medieval period and modern times. Charcoal of the Mameluk period also contains high percentages of *Juniperus phoenicea*, *Olea europaea* and *Quercus callipronis*. But as there is no evidence of a decisive climatic change in historical times, which could have caused the wood line to extend down the slope again (Baierle et al. 1989), it has to be assumed that this wood must have been transported from the Jordanian Plateau, from much of the same area as where it is to be found today.

The second major fuel resource is to be found in the hydrophytic vegetation of damp environments, which appears in places that are damp all year round or close to water in some wadis. The vegetation is limited to areas where a sufficient amount of water and a narrow passage guarantee that run off water from springs and rainfall appear on or close to the surface all year round, such as in parts of Wadi Dana, Fidan/Faynan/Ghwair or Ghuwebe (Fig. 3.10, see also Fig. 5.1).

Fig. 3.10. Hydrophytic vegetation at Wadi Fidan near Ain el-Fidan with *Tamarix jordanis*, *Nerium oleander* and *Retama raetam*. Such bushes make up an important part of the fuel supply for Late Bronze and Iron Age copper smelting

The major part of this vegetation includes *Tamarix jordanis*, *Nerium oleander* and *Retama raetam*. These bushes and shrubs, up to 5 m high, can be found in wadis up to 750 m altitude or around Petra up to 1 050 m altitude. These coppices have a rapid growth rate. The statistical evaluation of the charcoal composition from Iron Age smelting sites shows a preference for the use of Tamarix bushes as the firing material for copper smelting (52.7 wt.-%), followed by *Retama raetam* (20 wt.-%) and *Phoenix dactylifera* (8.4 wt.-%) (Baierle et al. 1989; Engel 1993; unpubl. results Berlin/Bochum). The complete weight of the slag from the three largest smelting sites (Khirbet en-Nahas, Khirbet el-Jariye, Faynan) was found to be about 100 000–130 000 t (see Table 5.3) and therefore suggests a wood consumption of over 800 000 t for the duration of some 800 years.

Vegetation on the stony landscape made up of limestone, dolomite, sandstone and granites consists in lower altitudes of open dwarf shrubs. Most of them are either not attractive for pasture animals or have a high growth rate.

Haloxylon persicum and *Retama raetam* grow on sprawling, sandy graveled mud fans, e.g., in the central section of Wadi Faynan, at the mouth of Wadi Fidan into the Wadi Arabah, or in the surroundings of Khirbet en-Nahas. The shrubs can also be found on the loess deposits at Barqa el-Hetiye. This is the flora that covers the majority of the Wadi Arabah (Baierle 1993). During Roman times these semi-arid shrubs made up an important proportion of the fuel supply with 22.6 wt.-% respectively 23.7 wt.-% (Baierle et al. 1989; Engel et al., unpubl. results Berlin/Bochum). A variety of other woods were also used in this period, such as acacias (*Acacia* spec.), *Chenopodiaceae* spec., the wood of the slender version of the moringa trees (*Moringa peregrina*) and tamarisks. As a result of the estimated amount of Roman slag on the slag heaps of Faynan 1 (40 000–70 000 t), careful inference points to the use of 320 000–560 000 t of wood over a period of 400 years for the smelting processes.

Vegetation changes significantly where smaller wadis join the Wadi Arabah. The slope gradient decreases, sedimentation of sand and other fine grained material outweighs gravel and pebbles, and the wadis fan out in wide, level mud fans and flats. On the regs made up of alluvial gravel and on the pebble fans of the wadis, the flora consists only of small shrubs and grasses. Acacia-dry shrub grows sparsely amongst them, particularly in the area south of Faynan, so that the landscape takes on a rather step-like appearance.

The first summary drawn from the material used for fuel in different periods shows, that generally woods from the direct surrounding of the smelting sites have been used. This is true in the Early Bronze Age, but even in the Late Bronze and Iron Ages the vegetation seem to have been similar to today's, as fuel had been used from the immediate neighborhood of the large smelting sites. It is noticeable that acacias do not yet appear in the charcoal of the Iron Age. This could be explained by the fact that the wadis still carried more water then (Baierle et al. 1989).

To sum up, the botanic research carried out by Frey and co-workers showed that the environmental impact of mining and metal production did not had the drastic impact on the environment as it is suggested e.g., for Cyprus and Rio Tinto in the southwest of the Iberian Peninsula (Craddock 1995) and where woodland management is required.

Chapter 4

The Raw Material Sources

4.1 Geological Overview of the Near East

The countries bordering the eastern part of the Mediterranean Sea have a rather complex geological history (Fig. 4.1). This area comprises a part of the African-Arabian Plate where crystalline Precambrian basement rocks are exposed in the south. Towards the north, sediments of increasing thickness have been deposited. Further to the north, the African-Arabian Plate borders the very extensive fault belt of the Taurides and Iranides in eastern Anatolia. Complex tectonic movements affected the geology of the Levant and are one of the decisive factors in the division of the Levant into different zones.

The fold belt, part of the so-called Tethyan Eurasian metallogenic belt ("TEMB" after Jankoviè 1997) contains numerous ore deposits. It forms an arch bounding the Arabic peninsula (Arabian Plate) in the north and northeast. The southern border of the belt runs along the Zagros Line from Cyprus initially in a northeasterly direction and turns in Turkey first to the east and then southeast. Folding episodes dated to the Upper Cretaceous are connected to obduction of ophiolitic rocks (Yilmaz 1993). The second folding dates to the Eocene-Miocene and was linked to the collision of the Arabian Plate with the Eurasian continent. Ophiolitic rocks occur in the Troodos mountain range on Cyprus, throughout the Taurides, along the Zagros-mountains in Iran and in Oman. Copper ores embedded in these rocks were rich metal resources in the past.

To the southeast the Mesopotamian depression is covered with a sequence of Mesozoic and Cenozoic sediments. Towards the south it turns into the broad zone of the mobile shelf of Tethys. Here marine, mainly neritic layers (Bender 1968) with predominantly carbonate rock have formed. The transition to the stable shelf of the Arabian-Nubian Shield runs roughly from central Sinai to the north-northeast along the Wadi Arabah and turns in Syria first east, then southeast towards the Persian Gulf. The hills framing the Red Sea at the Gulf of Aqaba/Eilat are exposures of the Precambrian basement rock of the Arabian-Nubian Shield.

The physical geography of the Levant Platform in the Mediterranean was geophysically studied, e.g., by Neev (1975); Ginzbourg et al. (1975). The platform consists of the Sinai Shelf, which is bordered by the Bardawil Escarpment. The Levant Platform is limited by a series of faults in the north, the so-called Pelusium Line.

The youngest and morphologically most impressive feature of the region is the rift system, which includes the Red Sea, the Gulf of Suez and the Rift Valley which extends from the Gulf of Aqaba via the Dead Sea over a distance of 1 200 km all the way north to the Zagros-Tauros subduction zone in southern Turkey (Bender

Fig. 4.1.
Geological structure of the Near East. The Rift Valley is a very prominent feature forming the course of the Jordan Valley and the Wadi Arabah. Precambrian rock of the Arabian-Nubian shield only protrudes in the south. To the north up to the Zagros Mountains it is covered with sediments and igneous rocks. Some copper ore deposits in southeast Anatolia, Cyprus and the region of the Precambrian basement are marked. *1:* Keban and other ore deposits of the upper Euphrates; *2:* Ergani Maden; *3:* Timna; *4:* Umm Bogma; *5:* Wadi Tar and other ore deposits of southeast Sinai (modified and complemented, after Horowitz 1979)

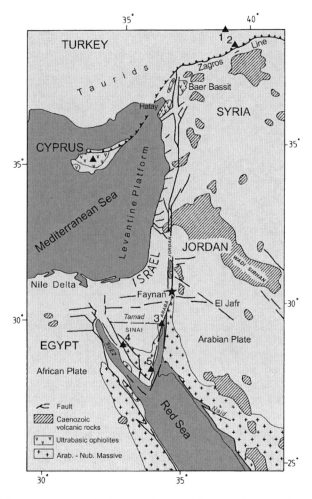

1974). It is part of a 6 000-km suture, which starts in the south in East Africa. The rift formation began in the Proterozoic and separated the African Plate from the Arabian Plate by about 200 km during the Mid Tertiary (Zak and Freund 1981). New oceanic crust was and still is being created along the central axis of the Red Sea. The Gulf of Suez extends over 600 km to the northwest and ends near the town of Suez at a system of minor east-west faults. It is basically an extensional feature in which the Precambrian basement and the Cambrian to Eocene sedimentary cover downfaulted and buried under some 2 500 m thick graben fill, as shown by extensive oil drilling. The most important of the east-west running faults is the Tamad or Themed Fault, which stretches laterally along the northern Sinai and runs up to the Jebel Umm Rousseis (Bartov 1974). The most conspicuous fault east of the Arabah Rift Valley is the Wadi Dana-Zakimat el-Hassa Fault, which has been a primary factor in the formation of the "niche" of Faynan. The Najd fault system was also created by the development of the above-described strike-slip fault Rift Valley-system in the late Proterozoic/Cambrian. It is a system of faults up to 300 km wide, which cross northern Saudi Arabia in a northwesterly direction up to the Hijaz (Sillitoe 1979).

Major strike slip faults along the Gulf of Aqaba and its northern continuation have shifted the Arabian Plate about 105 km to the north (Garfunkel 1981). The principal strike slip faults are located underneath the valley floor, which in the Wadi Arabah between Aqaba and the Dead Sea is only 5–15 km wide. They dissect recent gravel terraces and show that movements at an estimated rate of 0.5–1 cm per year still continue.

Crystalline rocks of the Precambrian basement are almost completely covered by sediments on the west bank of the Arabah. Outcrops are limited to the Timna valley. South of Elat, crystalline rocks extend in a narrow strip and extensive outcrops exist in the southern Sinai and in the Egyptian eastern desert (Fig. 4.1). The crystalline rock basement on the east bank of Wadi Arabah is exposed in a 50 km long, continuous strip, because the Transjordanian Block is uplifted in the southern Arabah to a higher altitude than its western counterpart. It forms the northwestern boundary of the Arabian Plate. Towards the south it stretches to the Hijaz in northwestern Saudi Arabia.

The petrography of the Precambrian Basement at Elat and Timna has been studied by Bentor (1985), Beyth (1987), that of the Sinai, e.g., by Shpitzer et al. (1989), and Shimron (1972). Precambrian igneous rocks east of the Wadi Arabah were studied, e.g., by Van den Boom and Rösch (1969) and Burgath et al. (1984). They can be separated into three units. The oldest unit consists of metamorphic rocks such as metasediments, gneisses and migmatites and others. The second unit is composed of an intrusive complex of alkaline and calc-alkaline granitic rocks (Garfunkel 1970). One of the largest northernmost outcrops is located at Faynan (Bender 1974). The Faynan Horst consists of andesites that have partly undergone a propylitic alteration. Various granitic rocks can be found in the wider surroundings of Faynan (Basta and Sunna 1972a), which are covered by Cambrian and Cretaceous sediments in the north, south and east (Fig. 4.2).

Basement rocks are intersected by latitic-basaltic and andesitic-rhyolitic dikes. Such composite dikes show rhyolitic cores and andesitic margins. Burgath et al. (1984) have studied these veins at the eastern edge of the Arabah and concluded that they are eruption channels for ignimbrites, tuffs and other calc-alkaline volcanic rocks. At the eastern edge of the Arabah, sandstones of the Lower Cambrian transgressively overlie the effusive rocks except where there are intrusions of rhyolites (Bender 1968). This dates the volcanic activities to the end of the Precambrian and early Cambrian. Radiometric measurements show K/Ar-ages of 471–542 Ma and $^{87}Sr/^{86}Sr$-ages of 510–560 Ma (Lenz et al. 1972). The dikes and volcanic rocks containing copper mineralizations can thus be seen as the primary source of ore deposits in the Arabah and the western part of Sinai. The volcanic activities can be associated with the formation of an island arch during the Precambrian subduction. This has been proposed by several authors for the Arabian-Nubian Shield (Al-Shanti and Mitchell 1976; El-Shazly 1979). The volcanic rocks of the Abu Khusheibah area can be compared lithologically with those of the Hutaymah/Hulayfa Group in Saudi Arabia, which have been mapped up to the Jordanian border.

Towards the north, the basement rocks are covered by a sequence of Cambrian to Quaternary sediments that are up to 1 900 m thick (Bender 1974). The layer in-

Fig. 4.2.
Geological map of Wadi Arabah and the central Negev. The copper ore deposits at the western and eastern edge of the Arabah are marked; so are the two largest ore deposits of Timna and Faynan (after data by Bender 1965; Jarrar 1984; Ilani et al. 1987)

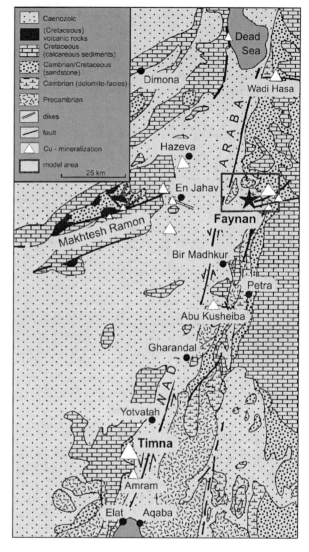

creases in thickness from south to north. Ball and Ball (1953) divided these sediments into three units using the Negev stratigraphy as a model. The lowermost unit consists of sandstones, traditionally known under the collective name "Nubian Sandstones" (Bender 1968). They include Cambro-Ordovician to Early Cretaceous clastics of fluviatile and shallow marine origin. Their lithostratography has been compiled by Segev et al. (1992). The sandstones of the Lower Cambrian are of specific importance, because the copper ore deposits of the Arabah and the Sinai are located within them (Bender 1965; Bartura and Würzburger 1974; Segev et al. 1992). During the Cenomanian a transgression took place. Limestone, dolomite, chalk and marl accumulated in a shallow marine environment. They form the second group of sediments. In the Negev but even more so in the area south of Amman, phosphate deposits were formed, which are economically viable and are being exploited (Abu-Ajamieh et al. 1988). The third group is composed of conglomerates, sandstones, and claystone, which have been formed during the Neogene. The sediments that had been deposited in the brackish water of the Lisan Lake are part of this group. The lake existed between Lake Tiberias and the Wadi Arabah (Bender 1968). Pleistocene Lisan marble from Lake Tiberias to Faynan show the original extent of the Dead Sea. Its present shape was formed later.

In the central Negev, but also in Timna and at other sites (Figs. 3.1 and 4.2) magmatic or volcanic activities occurred, which had a direct bearing on the formation of the graben. They date from the Triassic into the lower Cretaceous (Horowitz 1979; Segev et al. 1996). Large-scale volcanic activity is also reported from the Late Cenozoic. It produced numerous basaltic extrusions along the Jordan Valley and the Wadi Arabah. Cover basalts on the plateau belong to the same phase. These cover an area of some 42 000 km^2 stretching from Syria and the Hauran to the Azraq Depression in Jordan and the Wadi Sirhan in Saudi Arabia (Dubertret 1962). All these basaltic rocks erupted during the preglacial Pleistocene and date into the time span from 5.5 to 1.7 Ma (Siedner and Horowitz 1974; Heimann et al. 1996). The earliest human occupation is documented in the Levant by a large number of sites, which have been found mainly in the coastal areas of Israel and in the Jordan Valley, but also in the arid zones of the Wadi Arabah and eastern Jordan. The oldest site (Ubeidiya) dates back to the Lower Pleistocene period, i.e., 1.4 Ma (Goren-Inbar 1995), but most material is known from the Acheulien, which dates later than the African Olduvium. Bar-Yosef (1995) summarizes finds of human occupation in the Palaeolithic period. The first hominids, later the Homo sapiens, most likely came via the east African Rift Valley system into the region. Substantial permanent settlements have existed since 10 000 years ago.

4.2 Ancient Ore Deposits in the Near East

Ore deposits in the region (Figs. 1.2, 4.1, 4.2) clusters along favorable geological trends. They differ metallogenetically, and differences are also found in their shape and structure:

- Massive sulphide deposits with stockwork mineralizations occur in basic and ultrabasic rocks of ophiolite formations;
- Non-ferrous ore deposits concentrated in karst fillings of carbonatic rocks;
- Sedimentary copper deposits, possibly related to late Precambrian volcanism;
- Hydrothermal ores connected with rifting and/or magmatic-volcanic events of various ages.

Each of these groups shows a specific geochemical pattern and was formed at different ages, which is a very useful trait for discriminating between various ores and metal objects as it allows the use of lead isotope analysis. This will be treated in detail in the discussion of trade connections (Chap. 8). Here geology, geochemistry and mineralization of selected ore deposits will be discussed, which played a role in the early metallurgy of the eastern Mediterranean.

The northern concentration of the ore deposits is located in the orogenic belt of the Inner and Outer Taurid mountain range in southeastern Turkey (Öztunali 1989). Particularly in the region of the upper Euphrates, several non-ferrous metal deposits occur, e.g., those in the Keban-area (Figs. 1.2, 4.1). A large number of copper deposits can be found in the ophiolite rock sequence of the Outer Taurids, along the Zagros Line. The ore deposits of Ergani Maden, Siirt/Madenköy, Karadere, and Cüngüs-Midye are located here (Çatagay 1978). This series continues towards the southwest. In the ophiolitic rocks of Hatay and Baer Bassit, and of the Kizildağ (Dilek et al. 1991), at the Syrian-Turkish border, the entire copper mineralized ophiolitic sequence from the ultrabasic basement up to volcanic rocks is exposed. It forms an ensemble with the Troodos Ophiolite Complex in Cyprus (Delaloye et al. 1980).

The second concentration of copper deposits is in the south and closely connected with the formation of the Rift Valley. Copper- and copper/manganese-mineralizations occur in the Precambrian basement and in Lower Cambrian rhyolites. Predominantly larger ore deposits are located in Cambrian and Cretaceous sediments of the southern Sinai and in the Hijaz, but the largest are in the Wadi Arabah (Bender 1965; Bartura and Würzburger 1974; Burgath et al. 1984; Segev et al. 1992).

The Arabian Plate to the north of the Dead Sea is completely covered with sediments and does not contain any non-ferrous metal mineralizations. The rumors appearing sporadically in the literature that deposits of copper and tin in Lebanon as well as of copper and mercury in Syria have been found are based on spurious reports (Wainwright 1934; Burton 1872) and cannot be verified. Small iron ore deposits appear occasionally in Cretaceous limestone, e.g., in Mugharat el-Warda in the Aglun, roughly 50 km northwest of Amman in Jordan (Van den Boom and Lahloub 1962). There are signs of ancient mining there.

More than 30 copper ore deposits make Cyprus the most important copper district in the eastern Mediterranean. They have been exploited since at least the beginning of the Middle Bronze Age (Belgiorno 2000). The peak of copper production at Cyprus is dated between 1600 and 1200 BCE, and, hence, it lies between the decline of activities at Faynan at the end of the 3rd millennium BCE and a renaissance at the end of the 2nd millennium BCE. The estimates indicate that ca. 4 million t of slag has been produced on the island (Zwicker 1986). The deposits contain massive sulphides of volcano-sedimentary origin, which formed at a mid-ocean ridge and were moved, together with new oceanic crust, through plate tectonic activities (Sillitoe 1972). Therefore, the deposits are associated with (ultra-)basic rock and occur in all lithologic units of the Troodos ophiolite-complex. They date to the Cretaceous period. The prototype of the massive sulphide deposits (Constantinou 1972; Sillitoe 1972) belongs to the Pillow-Lava Unit. It is mineralogically and geochemically rather monotonous. The ore consists predominantly of pyrite with erratic occurrences of Cu-Fe sulphides and occasionally sphalerite. Cyprus provided the first sulphidic ores to ancient copper smelters and offers a model case to study the technology of smelting those ores in ancient times (Given and Knapp 2003). Ni, As, Co and Pb exist only in traces (Constantinou 1980). In contrast to Ergani Maden, in Cyprus bright red gossans exist, in which ancient mining was carried out intensively. Cementation zones developed locally with rich copper sulphides (covellite, chalcocite, bornite) or cuprite. Stos-Gale and Gale (1994) describe copper ores from the cementation zone of the Kambia mine as containing more then 30% Cu.

Some archaeometallurgical importance must be given to the so-called stockwork mineralization where chalcopyrite and Co-Ni-As-Fe ores appear in the rooting zones of the massive sulphide deposits, meaning in the gabbros and periodotites underneath the volcanic rocks (Constantinou 1980). Such ore mineralizations have been reported from Limassol (Panayiotou 1980; Foose et al. 1985). These have been mined at least since the Hellenistic-Roman period. The existence of manganese-containing and Cu-rich sediments played an important role in the smelting of ores. This so-called umber or devil's mud occurs in lenses together with radiolites and claystones formed above the pillow-lavas (see Sect. 7.3).

Ergani Maden was until the end of 1994 the largest copper mine in Turkey. Similar to Cyprus, massive sulphide deposits are embedded in a series of ophiolitic rocks of upper Cretaceous age (diabase, ultramafic rocks, gabbros, shales, radiolites; Helke 1964; Evans 1992). Two massive sulphide lenses overlie sulphidic impregnations in altered chloritic host rock. Ergani Maden has a geochemically similar pattern to the deposits in Cyprus. It is difficult to differentiate between them by means of their lead-isotope composition. The composition of the ore body is monotonous too, with pyrite, pyrrhotine, chalcopyrite, magnetite, hematite and subordinated sphalerite. Modern mining activities have destroyed any traces of ancient metal extraction. It has been suggested repeatedly that Ergani Maden was the most important prehistoric copper source in Anatolia. Reasons for this suggestion were the convenient location on a trading route between central Anatolia and northern Mesopotamia through the Tigris valley, the rich occurrence of native copper (Tylecote 1970; Wagner 1988) as well as oxidic Cu-ores and observations made on ancient mines in the beginning of the 20th century (Pilz 1917). Tylecote (1970) also reported old mines in the oxidation- and in the cementation zones, and the Heidelberg team described ancient mines and slag heaps in the surroundings of the modern open-pit mines (Seeliger et al. 1985). But as of yet, not one secure prehistoric date can be connected with Ergani.

Today there are no surface exposures of the ore body preserved and statements about the composition of the ancient ore body would be speculative. A cementation zone with Cu-Fe sulphides (bornite, covellite etc.) can be observed only in a few places. Tylecote's (1976) reference of As- and Sb-fahl ores in the cementation zone is slightly misleading. He must have meant cementation zones in general, because ore analyses by Seeliger et al. (1985) show that these elements including Pb and Ag are below 100 $\mu g\, g^{-1}$. In the opinion of Schmitt-Strecker et al. (1992) and (Pernicka (1995), it is questionable whether Ergani Maden ore was indeed the raw material for the Early Bronze Age Cu-As-Ni alloys, which are distributed all over southeast Anatolia and the Levant. It is possible that other sources of appropriate ores may still be found. Here, as in Cyprus, serpentinite rocks have been observed, which can be host rocks of Cu-Ni-As ores. Another potential area could also be in the Baer-Bassit-, in the Hatay- and in the Kizildağ-ophiolite-complexes (Fig. 4.1), where ophiolitic volcanic rocks are associated with ultrabasic rocks along the Zagros-line. Minor Cu-mineralizations have been observed there, too, according to Çatagay (pers. comm.).

A number of smaller Cu-deposits between the upper Euphrates and Lake Van (Seeliger et al. 1985; Wagner 1988) belong also to the series of ophiolitic rocks. However, most of them show no obvious signs of ancient mining and/or slag heaps. Nevertheless, some of these deposits might have been already exploited during the

Neolithic period (Schoop 1995), because they contain native copper (e.g., Kirmizitarla near Ergani Maden). All of these deposits are located in the southeast Anatolian Neolithic settlement area, where an early use of native copper has been proved (Muhly 1989; Maddin 1991).

Keban is an important Pb-Ag deposit in Turkey, situated ca. 50 km northwest of the town of Elazig on the upper Euphrates. It is possible to distinguish genetically and geochemically several types of ore formations in the Keban area (Öztunali 1989). They are associated with alkaline intrusions of Eocene magmatites. Some of the mineralizations are karstic ore deposits. As proven by finds of litharge and Pb-Ag slags from nearby Fatmalı Kalecik (Hess et al. 1998; unpubl. lead isotope data Bochum), lead and silver have been mined since the middle of the 4th millennium BCE from this polymetallic deposit, and possibly also (As-containing) copper and gold (Wagner et al. 1989; Palmieri et al. 1996).

Most ore deposits in Southern Jordan, in the Negev and in Sinai, differ fundamentally from those described above. They include mostly stratiform sedimentary mineralizations in sandstone and ore pockets in palaeokarsts. Secondary enrichment zones are not present. The ore is overwhelmingly oxidic. Sulphides are rare. The most important of these extensive copper (manganese) deposits occur in Cambrian and Cretaceous sandstone and dolomite. Between a number of smaller mineralizations, Faynan and Timna are the most substantial ore deposits of this kind (Fig. 4.2). They will be treated in detail in the following paragraphs.

Similar ore deposits occur in western Sinai (Fig. 4.1) in the vicinity of Umm Bogma and Serabit el-Khadim (El-Sharkawi et al. 1990; Abdel Motelib 1987, 1996). In addition, mineralized andesitic dikes and possibly hydrothermal veins in magmatic rocks of the Precambrian basement have been exploited in southern Sinai, e.g., at Wadi Riqueita. The erosion of such ores in the Precambrian basement may have partly been the source of the sedimentary ore deposits (El-Shazley et al. 1955; Beyth 1987; Segev et al. 1992). Most veins contained only small mineralizations with the exception of those of Umm Bogma, which were mined in prehistoric times (Hauptmann et al. 1997; Naim and Rothenberg, in prep.). Mineralogical and chemical compositions of ores are simple. At Umm Bogma there are Mn-ores and oxidic Cu-ores. Elsewhere, malachite, azurite, seldom cuprite and Cu sulphides occur. The available analyses so far indicate very low contents of trace elements.

The largest smelting site of the Sinai at Bir Nasib near Umm Bogna contains some 100 000 t of slag. This indicates a large-scale copper production at least since the Pharaonic period (Rothenberg 1987). At least as early as the beginning of the Old Empire in Egypt in the early 3rd millennium BCE the sedimentary copper deposits in Wadi Maghara and near Serabit el-Khadim have been mined (Beit-Arieh 1985; Weisgerber 1991). They are only a few kilometers away from Umm Bogma. Ancient inscriptions indicate as a main goal of the mining activities the extraction of turquoise, which was highly appreciated by the Egyptians.

It remains unclear whether a copper-arsenic mineralization in Wadi Tar in southeastern Sinai supported the early metal production in the region at all (Weissbrod 1987; Ilani and Rosenfeld 1994). Native copper and copper-arsenides such as koutekite (Cu_5As_2) and domeykite (Cu_3As) containing low concentrations of Zn, Mn, Pb, Ag and Sb have been found in a possibly ancient prospecting trench. Analyzed samples show highly dispersed ores in the brecciated host rock (consisting of meta-volcanic rocks cemented by calcite). However, richer massive ores may have existed. This is the only known deposit of its kind in the region. Native copper high in As is only known elsewhere from Talmessi in Iran. A comparison of lead isotope data from Wadi Tar ores with Chalcolithic, Early and Middle Bronze Age arsenical copper objects from the southern Levant seems to rule out the speculation that any of the objects were produced from Wadi Tar ore (Segal et al. 1999a).

A number of smaller Cu-mineralizations occur in the sediments of the central Negev and in Jordan along the Rift Valley (Figs. 3.1, 4.2). They are linked to tectonic faults and/or volcanic activities as well as hot brines (Ilani et al. 1987). Their ages range from Cretaceous to Pliocene. A very complex Pb-Ni-Cu-As sulphide mineralization has been described from Makhtesh Ramon (Fig. 4.2) and from Har Arif in the Negev (Itamar 1988). At these locations, no signs of ancient mining have been found. At Har Kahal on Mount Hermon, however, galena from a lead-zinc mineralization has been mined at least since the Roman period (Dar 1988/1989, 1993). Several finds of lead objects from nearby settlements and an ancient underground mine confirm this. Pieces of galena found in archaeological contexts at Numeira, Faynan and Beidha (Abu Ajamieh 1988; unpubl. data Bochum) may also have come from Har Kahal.

In northwestern Hijaz, ore deposits occur along with the Nadj fault system, which is linked geologically to the formation of the Rift Valley (Sillitoe 1979). Numerous ore deposits are located in the Ad-Dawadimi district in central Saudi Arabia. In the western part of Saudi Arabia, adjacent to the Gulf of Aqaba, ancient metal production seems to have concentrated on gold. This is at least the impression gained from a survey of 29 Roman-Islamic gold mines at Al-Wajh, Daba and Al-Jadida, which has been carried out by Kisnawi et al. (1983). Traces of early copper production have been found at three different sites: Shim at-Tasa (Shanks 1936), Imsayea and Az-Zuwaydiyah at Al-Disa (Kisnawi et al. 1983). They indicate ancient mining and smelting of small sedimentary ore deposits in sandstone.

4.3 The Copper Deposit of Faynan and its Relation to Timna

Geological Framework and Genesis

Copper ore deposits at the east side of the Wadi Arabah were initially described by Blake (1930), Shaw (1947) and Quenell (1951). More intensive studies were then carried out under the auspices of the German Geological Mission of the Bundesanstalt für Geowissenschaften und Rohstoffe and by the German Mining Consultant Company Otto Gold (Lillich 1963; Pauly 1964; Prokop and Schmidt-Eisenlohr 1966; Bender 1965, 1968, 1974). Later the Jordanian Natural Resource Authority (Abu-Ajamieh et al. 1988) continued the work. Copper mineralizations extend from the Faynan district some 70 km to the south (Figs. 3.1, 4.2). They occur in several geological units. At Wadi Huwar and Wadi Musa, west of Petra, minor secondary Cu-mineralizations occur in Precambrian dikes. Further to the southwest in the area of Wadi Abu Barqa, numerous minor Cu-Fe sulphides were found in andesitic rocks of the Upper Proterozoic. Locally, 0.77% Cu has been encountered. The region has been intensely prospected geochemically by Burgath et al. (1984), but no substantial mineralizations have been observed. Sedimentary copper ores appear locally in sandstones of the Lower and Middle Cambrian between Gharandal and Faynan and are particularly concentrated in the area of the wadis Abu Khusheibah and Abu Qurdiya (Pauly 1964; Prokop and Schmidt-Eisenlohr 1966). This region is the only one besides Faynan where ancient mining took place (KIND 1965; 8 mio metric t, www. nra.gov.jo, January 2004). In addition, 40 µg g^{-1} of gold was analyzed in felsic volcanic rocks in Wadi Abu Khusheibah, and visible gold was washed in heavy mineral concentrates from decomposed rhyolitic rocks. These gold mineralizations may be a continuation of the trend of famous gold mineralizations in the northwestern Hijaz (Burton 1878, Gow and Lozej 1986), where extensive gold mining was carried out in the past.

By far the richest mineralization of the region occurs in Cambrian dolostones and sandstones at Faynan, where secondary copper- and manganese-minerals are present. The mineralization is located within an area of roughly 20×25 km (Figs. 4.2, 5.1). It includes in the north the large ore outcrops at Khirbet el-Jariye, Khirbet en-Nahas, El-Furn, and Khirbet el-Ghuwebe (ca. 0.6 mio metric t ore, see www.nra.gov.jo) and then the slightly smaller outcrops at Umm ez-Zuhur near Al-Qurayqira. In the east are the plentiful ore outcrops at Ratiye and those in the wadis Khalid and Dana (ca. 19.8 mio metric t ore). In the south, the ancient mining area extends to the ore deposits on Jabal Mubarak and includes the small outcrop at Salawwan. These localities also mark the extent of the research area described in the following chapters.

Copper ore occurrences at Timna, small outcrops at Nahal Roded and Wadi Amram, on the western part of the Wadi Arabah Rift Valley are located about 105 km south of the Faynan area. Prior to the formation of the Rift Valley and associated lateral strike slip displacements, both copper districts had been joined to one coherent ore deposit (Segev et al. 1992). The lateral displacement along the strike slip fault can be measured rather accurately between the east-west trending Salawwan Fault at Faynan and its displaced western prolongation, the Themed Fault at Timna. Rifting and lateral movements began during the Oligocene and are still continuing at a rate of some millimeters per year.

A reconstruction of the palaeo-geographical position during the Cambrian period clearly shows the relationship between Faynan and Timna as the two most important ore deposits, as well as between mineralizations in the Eilat-basin (Nahal Shehoret) and in Wadi Abu Khusheiba/Abu Qurdiya (Fig. 4.3). Copper mineralizations have also been found in Timna in the crystalline basement, in the porphyrites and rhyolites at the transition to the Cambrian (Würzburger 1969). Here the most extensively mined ore formations are embedded in dolostones and sandstones of the Lower Cambrian. In contrast to Faynan, where copper mineralizations at the surface of dolomitic rocks have been the most important resource for the early miners, such mineralization shows only small outcrops at Timna, which was only mined in modern times. The ore formations in Nahal Shehoret south of Wadi Amram are comparable with those in the wadis Abu Khusheiba/Abu Qurdiya and Umm el-Amad south of Faynan. The deposit in Nahal Tuweiba southwest of Eilat, which has been reported by Rothenberg (1973), is not listed by Segev et al. (1992).

The Cambrian sediments, which are important for the copper mineralization at Faynan, can be stratigraphically divided in four units (Bender 1965, 1974; Van den Boom 1969; Bigot 1975; Heitkemper 1988; see Fig. 4.4):

4.3 · The Copper Deposit of Faynan and its Relation to Timna

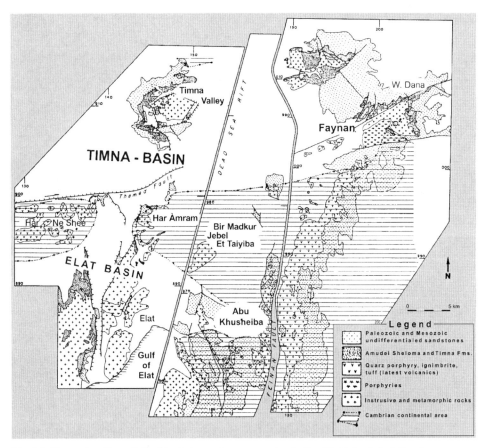

Fig. 4.3. Reconstruction of the Cambrian palaeogeography in the Wadi Arabah following a re-shifting of the Transjordanian and Palestinian Block by ca. 105 km. Timna and Faynan are shown, where the most important mineralizations are situated, as well as the Eilat-basin and Wadi Khusheibah, where Cu-mineralizations also occur (see Fig. 4.2) (from Segev 1986)

I. Arcosic Sandstone (cb1): A 60 m thick sequence of fluviatile, middle- to coarse-grained sandstones with coarse to clastic interbedded strata of rhyolitic and other rocks. This formation is not mineralized;

II. Dolomite-Limestone-Shale Unit (cb2) or "Burj Limestone" (Abu-Ajamieh et al. 1988): Throughout the mineralized area of Faynan, the thickness of the dolomites ranges between 20 and 40 m. The southern depositional limit is located about 6 km south of Faynan near Umm el-Amad, where sandy dolostones are grading to coastal sandstones. Further to the north, the shallow marine dolostone grades to open marine fossiliferous limestone, reaching a thickness of about 55 m near the Dead Sea. The brittle, weather-resistant dolomites and limestones are easily recognizable in outcrops all over the Faynan district due to their morphologically prominent cliffs, terraces and plateaus. The mineralized horizon at the top of this unit is therefore always easy to locate. The formation can be divided lithologically into three sections: sandstones (underlying bed), sandy dolostones (middle),

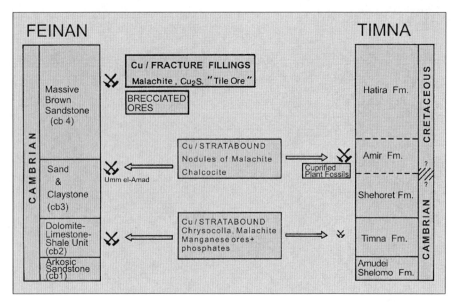

Fig. 4.4. Simplified lithostratigraphy and copper mineralizations in Faynan and Timna (the latter information from Bartura and Würzburger 1975; Segev and Sass 1989). Some geological and mineralogical characteristics, which play a role in differentiating between both deposits, are marked. It is difficult to date the transition from the Shehoret to the Amir Formation in the Kurnub Group. In Timna alone, one finds hints of early mining in the Timna Formation (Rothenberg and Shaw 1990). The hammer and wedge symbol close to the Dolomite-Limestone-Shale Unit (*cb2*, called *DLS* in the following) in Faynan indicates by comparison more then 100 mines (Hauptmann et al. 1992)

and silt and clay rock/shale (top layer). This marine sequence was formed in a lagoonal environment. The copper ore occurs in the upper part of the dolostones in pockets, as matrix mineralization and as vein fillings (Bender 1965). The most substantial copper and manganese mineralizations are in the overlying shale unit and in the upper part of the dolostone layer. This shale formation is 1–1.5 m thick. It is rather friable and has been the focus of mining activities during most of the periods studied (see Sect. 5.3). In the same unit, nodules and layers of phosphorite commonly occur and barite was occasionally noticed (Heitkemper 1988). The Dolomite-Limestone-Shale Unit is in the following descriptions abbreviated to "DLS," a term initially introduced at Timna and quite common in archaeological literature (Hauptmann et al. 1992).

III. Variegated Sand- and Claystones (cb3): the boundary between the DLS and this layer marks the change from marine to terrestrial sedimentation. The formation is some 50 m thick and characterized by rapid lithologic changes. In the basal part, hard conglomerate and coarse sandstones dominate. These include local copper and manganese ores. These minor mineralizations show rarely any signs of mining, probably due to their conspicuous hardness.

IV. Massive Brown Sandstone (cb4): there is a gradual transition to this formation. In the field, this is often indicated by a steeper morphology. The unit consists of fine- to medium-grained sandstones with intercalated conglomerates and claystones, which were deposited in a fluviatile environment. The thickness of this formation reaches 250 m at Ras Gebel Khalid in the Faynan area. Light brown to blackish-brown weathering crusts are typical, and erosion causes a honeycomb pattern. The sandstones show everywhere in the Faynan area a nearly vertical system of fractures, which are partly coated with oxidic and sometimes sulphidic copper ores in Qalb Ratiye and in Wadi Abiad. Efflorescence of salt can also be observed here. These copper ores were the focus of early mining activities. The joint planes controlled the technology of exploitation (see Sect. 5.3). This formation is abbreviated "MBS."

The lithostratigraphy of Cambrian sediments at Faynan and Timna are similar as reported by Segev and Sass (1989) and Segev et al. (1992) (Fig. 4.4). The Arcosic Sandstone (cb1) Unit corresponds to the Amudei Shelomo Formation. The DLS (cb2) is almost completely identical with the Timna Formation (Basta and Sunna 1972b). Following Segev et al. (1992) this formation is divided into the Hakhlil and the Sasgon Group, the latter having a dolomitic, sandy ("Zebra-Sandstone"), and clayish subfacies. In both areas, these are the formations with the richest occurrence of copper ore. But, with one exception, there is no evidence at Timna of ancient mining in this formation, because the largest part of this ore-bearing formation does not outcrop at the surface. Following Rothenberg and Shaw (1990), this ore had been mined in only a few small pits northeast of Har Timna at Givat Sasgon (site 250) and smelted on a nearby hill (site 149) during the Early Bronze Age IV. Outcrops of manganese ore have been found at Timna at Nahal Mangan and two other locations in Wadi Nehushtan (north of the quarries "F" and "G", Segev and Sass 1989). It is, however, impossible to ascertain whether outcrops of the Timna Formation south of Har Timna, shown by Segev and Sass (1989), were exploited during ancient times, because large-scale modern open cast mining has destroyed any archaeological evidence here. Ore-bearing silt and claystones on Har Timna have most likely been eroded entirely. Rothenberg and Glass (1992) found no archaeological sites there.

The series of the Variegated Sand- and Claystones (cb3) at Faynan corresponds to the Shehoret Formation on the western side of the Arabah. At Timna, this formation is only partly exposed. The sandstones of the Amir and the Avrona Unit, which are combined into the Kurnub Group, are not dated. They could be Mesozoic. They outcrop extensively at Timna and supplied copper ore for early metal production at Timna and Wadi Amram further south (Segev et al. 1992).

The origin of the copper ore deposits at Faynan and at Timna is complex and happened in several stages. Recent research suggests the following stages, which led to a gradual enrichment of copper and manganese (Segev and Sass 1989; Segev et al. 1992):

1. Primary Cu-Fe sulphide mineralizations in late Precambrian volcanic rocks are the source of the sedimentary ores (Burgath et al. 1984; Segev and Sass 1989). The age of the Precambrian/Cambrian volcanism has been dated to 471–542 Ma and 510–560 ±10 Ma (Lenz et al. 1972) or 523–535 Ma (Segev 1987):
2. Erosion of the volcanic rocks during the lower Cambrian led to synsedimentary (stratiform) enrichment of copper in sandy dolostones, which had formed in shallow marine and tidal flat environments (Basta and Sunna 1972; Bartura and Würzburger 1974; Bigot 1975; Khoury 1986). Keidar (1984) suggested a disso-

lution of copper from the Precambrian basement rocks in general. Migration and redeposition of the copper took place mainly as chlorides. Here, also synsedimentary Mn-ore-mineralizations were formed;
3. Epigenetic remobilization occurred during the formation of the Rift Valley and subsequent erosion and weathering since the Miocene (Khoury 1986; Segev 1986). Decarbonization starting from faults and fractures caused karst formation at the top of the dolostone unit. Subsequently, residual sandy-clayey components were deposited in this palaeorelief (Segev and Sass 1989). Enrichment of secondary copper and manganese ores in these sediments (the sand- and claystones of the DLS at Faynan and the "zebra-sandstone" at Timna) formed the ore deposits in their present form.

Stratabound ore formations and mineralized vein fillings of sulphides and oxidized copper ores in the hanging sandstone of the Amir/Avrona Formation have also been connected with an epigenetic remobilization of copper (Keidar 1984). Beyth and Segev (1983) and Segev et al. (1992) suggested that Lower Cretaceous volcanism at Timna (K/Ar-dated to 99 and 107 Ma) affected this mineralization. Enrichment of copper and manganese ores within the Timna Formation, which corresponds to the DLS at Faynan, occurred during Cenozoic (K/Ar-dated to 28 ±33, 12 ±8, 93 ±5, 47 ±27 Ma; Segev and Sass 1989). Similar to Faynan, the accumulation could also be connected with a hydrothermal alteration of the Cambrian sediments 13–15 Ma ago (Beyth et al. 1997).

The Cu-Mn-ores in the Wadi Arabah, which were targets of ancient mining activities, are all sedimentary, stratabound ore deposits (Evans 1992). "Stratabound," in juxtaposition to "stratiform," means that the ore is limited to narrowly defined stratigraphic intervals. The ore comprises predominantly Cu oxides and silicates.

The deposits show no vertical zoning and therefore no associated alteration of the mineralization, as it occurs in Anatolia's and Cyprus' sulphide ore deposits. Consequently not all hypotheses about single stages in metallurgical development (see Sect. 2.1), which stem from the study of these deposits, need to be taken *a priori* into consideration. It appears to be quite important for the development of early copper technology in the Wadi Arabah to stress this point, since only Timna, Faynan, western Sinai and a few other minor sedimentary deposits exist around the Eastern Mediterranean. The reporting of a piece of 'sandstone ore' in chalcolithic layers at Norsuntepe in southeastern Anatolia (Zwicker et al. 1985) has therefore repeatedly created confusion and led to speculations about possible trade connections between the region at the upper Euphrates and the southern Levant. The piece of ore itself has never been described, and its provenance has never been conclusively analyzed.

The Ores

The ore deposits of Faynan and Timna have a common geological history and were separated from each other by tectonic activities since the Oligocene/Miocene. The ore content of both deposits is hence comparable: the predominant ores are oxidic and silicate copper minerals. Principally, these minerals can be reduced to metal in a one-step smelting process and do not require a multistage process of roasting and smelting, as is the case with sulphide ores. In the sedimentary ores of the Arabah, only traces of sulphides occur, in contrast to ore deposits in Anatolia and Cyprus. Native copper is missing, if one disregards two minute grains, which have been observed under the microscope by Slatkine (1961) and Burgath et al. (1984). This lack of native copper in the Levant and its abundance in Anatolia and Iran may have been the reason that during the Neolithic period copper metallurgy developed in these regions rather than in the south (Schoop 1995). Petrographically, some differences do exist between the ore composition in Timna and Faynan. This is important as ore from both areas can only be separated with great difficulty geochemically and by lead isotope composition.

Unlike in other mining areas, the ores of the Arabah, especially at Faynan, can be precisely characterized because there is still the same ore available that was mined and smelted in ancient times. Elsewhere, this is not often the case. In the Wadi Arabah, both localities are in an unusually undisturbed and pristine state due to the fact that the remains and evidence of ancient mining have not been destroyed by modern mining activities. In contrast to most hydrothermal or volcanogenic deposits with secondary vertical zoning, it can be assumed that at Faynan and Timna extraordinary rich, fresh outcrops available today show the same quality and quantity of ore as that which was mined in the past. Prospection work by the Natural Resource Authority has created a number of fresh outcrops, which provided excellent conditions to take samples. It also facilitates access to ancient mines via tracks and prospection galleries.

The most important outcrops are mainly identical with the location of the ancient mines (Fig. 5.1), which adds up to much more than a hundred. The most important outcrops exploited in ancient times are bound to the DLS. They are located in the wadis Khalid and Dana, in the surroundings of Khirbet en-Nahas, in Wadi Ghuwebe and at Khirbet el-Jariye. The number of copper minerals is limited; they are mainly:

- Paratacamite, $Cu_2(OH)_3Cl$;
- "Chrysocolla", $CuSiO_3 \cdot 2\,H_2O$; this mineral exists after Heitkemper (1988) in at least five different amorphous to cryptocrystalline varieties (e.g., bisbeeite) and contains varying percentages of clay minerals (Fig. 4.5). Compatible ores in Timna have been examined by Würzburger (1970);

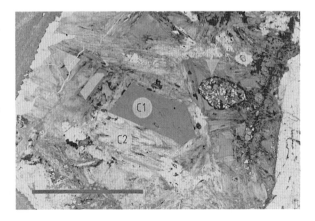

Fig. 4.5.
Wadi Khalid, sample JD-II/5 (Faynan DLS). Paragenesis of microcrystalline chrysocolla (*C1*) and fibrous Cu silicates (bisbeeite, *C2*). The silicates have partly been replaced by malachite, paratacamite and other Cu-minerals. Note the inclusions of quartz grains (*arrow*) and manganese oxides (*black*) (transmitted light; scale: 0.25 cm)

- Fine fibrous malachite, $Cu_2[(OH)_2/CO_3]$;
- Dioptase, $Cu_6[Si_6O_{18}] \cdot 6 H_2O$;
- Planchéite, $Cu_8[(OH)_2/Si_4O_{11}]_2$, in shiny blue crystals;
- Pseudo-malachite, $Cu_5[(OH)_2/PO_4]_2$ (Basta and Sunna 1972b).

The minerals are mostly 'masked' by abundant malachite (Heitkemper 1988). Only under the microscope traces of

- Bornite, Cu_5FeS_4;
- Chalcocite, 'Cu_2S';
- Pyrite, FeS_2.

have been observed.

These copper minerals are fibrous and give the appearance of a 'matted' texture. They form thin-walled coatings, veins and impregnations in the friable and soft silt- and mudstone. Locally they form fist-sized hard nodules (e.g., at the site of Wadi Fidan 4) and chunks up to the roofing arkosic sandstone and the underlying dolomite. The weathering of the silt- and mudstones causes the ores to accumulate in large layers and pockets on top the dolomite banks (e.g., at Madsus or Jariye) (Fig. 5.1). There, such ores were perhaps already collected in the Chalcolithic period or even during the Neolithic period. A typical characteristic is the replacement of quartz with copper silicates in the hanging layers of arkoses and conglomerates above the DLS (Fig. 4.6; Heitkemper 1988). This leads to a striking, macroscopically visible blue-green coloring of quartz grains. This has been observed to occur only at Faynan. It seems to be missing at Timna and elsewhere in the region. Such material was found, e.g., in the Chalcolithic-Early Bronze Age settlements in Wadi Ghazzeh west of Tell Abu Matar and must have been brought there from Faynan (Hauptmann 1989).

Copper ores in the silt- and mudstones of the DLS are intensively intergrown with oxidic manganese ores, which give them the qualities of a 'self-fluxing ore.' Towards the east lateral zoning can be seen, which leads to increasingly massive manganese ore at the expense of copper in the middle of Wadi Dana (Van den Boom 1969; Basta and Sunna 1972). This was not mined in ancient times. The

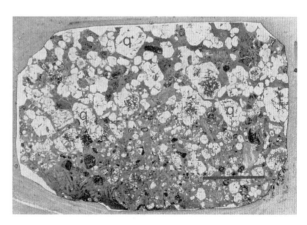

Fig. 4.6.
Wadi Khalid, sample JD-III/18-1 (Faynan, Variegated Sand- and Claystones, cb3). Matrix-mineralization of Cu silicates in conglomerate. Notice the replacement of quartz grains (*q*) by Cu silicates, which leads to a characteristic blue coloring of the quartz (transmitted light; scale: 0.7 cm)

manganese content of the ore averages 41–43%, while copper averages about 1.4% (Abu-Ajamieh et al. 1988).

The following manganese minerals occur:

- Pyrolusite, β-MnO_2;
- Hydroxi-manganomelane (psilomelane, $(Ba,H_2O)Mn_5O_{10}$);
- Cryptomelane or 'wad', $K_2Mn_8O_{16}$;
- Hollandite, $Ba_{1-2}Mn_8O_{16}$;
- Coronadite, $Pb_{1-2}Mn_8O_{16}$;
- Ramsdellite, γ-MnO_2;
- Manganite, MnOOH.

A geochemical separation of manganese and iron seems to have occurred only to a limited extent, although it can be typical for sedimentary deposits. Following Basta and Sunna (1972b) and Abu Ajamieh et al. (1988), the Mn-ore contains an average of 8–10% Fe_2O_3. A recognizable enrichment of Fe oxides and hydroxides (hematite, goethite) is noticeable at El-Furn (Fig. 5.1), an Early Islamic mining and smelting site. The question as to whether copper or iron was produced here will be discussed in Sect. 6.3.

Faults and fractures with a spacing in the order of one meter in the Massive Brown Sandstone (MBS, cb4) are locally (in the wadis Ratiye, Abiad and Khalid) coated with calcite and contain the following copper ores:

- Malachite, $Cu_2[(OH)_2/CO_3]$;
- Cuprite, Cu_2O;
- Chalcocite, 'Cu_2S', and covellite, CuS;
- Paratacamite, $Cu_2(OH)_3Cl$.

Cuprite occurs mostly intergrown with Fe-hydroxides, relics of chalcocite and quartz as 'tile ore'. It is sometimes brecciated by tectonic movements and cemented with malachite. Cu sulphides are replaced by a network of identified Cu minerals. Such a network and the brecciation of copper ores were not observed at Timna (Fig. 4.4). These petrographic characteristics made it possible for Hauptmann (1989) to determine Faynan as the provenance of copper ores found at the Chalcolithic settlement of Abu Matar.

There is no archaeological evidence that copper ores of the DLS and MBS had been beneficiated in any other way than by manual sorting. There was apparently no necessity for upgrading, because the analysis of 86 hand-picked ore samples showed a copper concentration ranging between 15% and 45% (Fig. 4.7, Heitkemper 1988; Hauptmann et al. 1992). All samples were obtained from either ancient mines or the above-mentioned modern prospection galleries at Faynan. This shows that in ancient times and today rather rich ores were available in sufficient quantities. At least for ores from the DLS their outstanding smelting qualities rendered further upgrading unnecessary. Ore samples that are representative of ancient mining output and furnace charges are equally available at Timna (Bartura et al. 1980).

Copper ores used for early metal extraction at Timna occur predominantly in the sandstones of the Amir/Avrona Formation, which has sizeable outcrops in the Timna

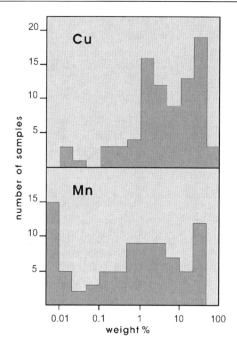

Fig. 4.7.
Histograms of Cu and Mn contents in 86 ore samples from the Faynan area. Samples of rich copper ore are available. Only ores with Cu > 10% and Mn > 20% have been chosen for further geochemical discussion. The histogram contains also – but not exclusively – the data from Table A.1 (Appendix) (after Hauptmann et al. 1992)

valley. Bartura et al. (1980) and Keidar (1984) differentiate between the following types of mineralizations in the Avrona Formation (Fig. 4.4):

- Replacement of floral remains;
- Concretions and nodules;
- Impregnations and cementations of sandstones;
- Vein fillings and layered mineralizations in sandstones.

While the latter two correspond largely to mineralizations at Faynan, the typical nodules and cuprified plant remains are lacking at Faynan. Nodules, with a diameter of up to 3 cm, have a core of copper sulphides with varying amounts of hematite and goethite interspersed with quartz grains. The crust consists of malachite. Cuprified plant remains are composed of malachite, cuprite, Fe oxides and -hydroxides, quartz grains and minor chalcocite. Data from Hauptmann et al. (1980), Leese et al. (1985/86) and Segev et al. (1992) show that copper of such concretions and nodules can reach 44% (Fig. 4.8). These rich copper ores indicate the type and grade of raw material that were available in the past. In contrast to the situation at Faynan, the Cu and Mn ores at Timna are mainly separated from one another. Low Fe contents (see, e.g., Leese et al. 1985/86) are astonishing. In earlier publications on Timna, the occurrence of mixed Cu-Fe ores even from archaeological contexts is explicitly mentioned (Slatkine 1961). Lupu (1970), Merkel (1990) and Segev et al. (1992) also reported mixed Cu-Fe ores. These ores, intergrown with quartz and calcite, play an important role in metallurgical processes, because they are 'self-fluxing', which means that they can be smelted without the addition of fluxing agents.

Fig. 4.8.
Histograms of Cu and Fe contents of 75 ore samples from the 'ancient' deposits at Timna (Amir/Avrona Formation). Samples of rich copper ores are widely available here as is the case at Faynan. In contrast to Faynan, the ores in Timna have a conspicuously high Fe-percentage, while Mn is under 0.05% (after data from Leese et al. 1985/86, Merkel 1990 and Segev 1992)

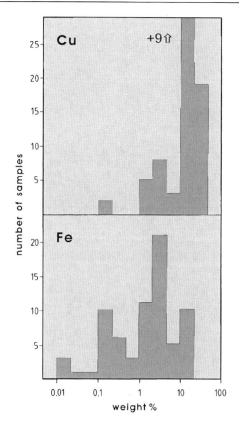

Geochemistry

Chemical characterization of ore deposits aim to establish analytical parameters, which can be used for provenance studies and for reconstructions of smelting processes. Analysis of major and minor elements in ore and gangue, as well as of trace elements in the ore itself, are preconditions for such studies.

Comprehensive geochemical analyses have previously not been carried out on copper ores from Faynan. Only analyses of Cu-, Mn-, Pb-, Ag-, and Au-concentrations and major elements of the host rock have been published (Bender 1968, 1974; Kind 1965; Basta and Sunna 1970/71; Khoury 1986; Abu-Ajamieh et al. 1988 and others). Extensive geochemical analyses of ores from Faynan have been carried out in this study (Hauptmann et al. 1985; Hauptmann 1989; Hauptmann et al. 1992). All data, which are used here for the discussion of the trace elements, are compiled again in Table A.1 in the Appendix. Fourteen ore samples have been analyzed for their content of host rock elements (meaning rock forming components), which do not end up in the copper, but are slagged (see also Fig. 2.4). They are listed in Table 4.1. Ores are presented in this table following the norm of rock analyses to emphasize their 'natural' character. It is common in archaeometallurgical literature to list elements not in the oxidic form, especially when metallurgical questions are discussed. We will adhere to this form in the following, too.

74 Chapter 4 · The Raw Material Sources

Table 4.1. Chemical composition of Cu and Cu/Mn ores from the Faynan area (ICP-analyses). Given are the host rock's components, which are usually slagged. They are calculated here together with copper as oxides in weight-%. Trace elements (Co etc.) are given as elements in µg g^{-1}. See Fig. 4.4 for the stratigraphic position. *n.a.*: not analyzed. The totals of the analyses are partly < 100 wt.-%, because organic and carbonatic C was analyzed together, and therefore the carbonate content was not calculated. Moreover, Mn and Fe exist in different valences. For analytical details, see Appendix

Component	Sample, JD-												
	3/1a	3/3	3/16	3/22b	13/3b	13/4	13/7	12/2	GR/2	WA/1	WA/4	41/5
SiO$_2$	62.4	6.26	27.3	75.7	45.7	72.0	22.6	73.8	46.5	59.2	61.7	50.7
TiO$_2$	0.08	0.05	0.16	0.13	0.43	0.12	0.16	0.24	0.20	0.56	0.70	0.25
Al$_2$O$_3$	2.64	1.54	4.19	4.92	10.6	3.93	4.75	2.02	4.96	3.25	6.76	6.71
Fe$_2$O$_3$	0.88	0.36	3.61	0.26	4.74	1.72	3.16	20.4	1.62	0.87	1.01	1.20
MnO	1.51	4.83	14.6	0.06	17.7	1.31	3.46	0.07	0.01	0.01	0.01	1.89
MgO	0.01	14.9	0.66	0.25	1.30	0.41	0.53	0.19	0.28	0.61	0.31	0.47
CaO	0.66	24.8	12.4	0.49	0.82	2.21	30.1	0.52	1.08	0.35	0.65	13.3
BaO	0.96	0.67	2.59	0.22	1.72	1.41	0.82	0.28	0.31	0.19	0.18	0.51
CuO	17.8	2.56	13.9	5.74	3.01	6.13	4.48	1.45	32.7	22.3	18.0	8.60
Na$_2$O	n.a.	1.50	1.86	0.88	1.43	0.96	1.34	1.00	0.92	1.01	0.81	1.02
K$_2$O	0.92	0.84	1.95	3.82	5.89	2.77	3.06	0.25	0.21	0.20	0.19	1.61
P$_2$O$_5$	1.18	1.24	8.68	0.21	0.14	1.40	22.5	0.27	< 0.02	0.21	0.21	9.90
S$_{ges}$	0.30	0.05	0.03	0.05	0.03	0.30	0.06	0.01	1.77	0.68	0.31	0.13
C$_{ges}$	0.04	14.3	0.16	0.02	0.01	0.04	0.18	0.12	2.59	1.34	1.67	0.13
H$_2$O$^-$	3.00	0.91	3.65	1.88	2.75	2.44	1.73	0.30	0.37	0.26	0.43	1.81
Σ	92.38	74.81	95.74	94.63	96.27	97.15	98.93	100.92	93.54	91.04	92.94	98.23
Co	n.a.	150	250	60	320	100	90	120	40	50	30	70
V	n.a.	120	270	30	150	60	120	210	80	90	160	90
Zn	300	135	1470	60	610	370	3 200	340	100	190	120	250
Ni	700	130	160	50	170	70	110	350	60	80	50	270
Pb	2 300	150	1.87%	40	6 700	4 300	6 000	310	790	540	560	1 150

Table 4.1. *Continued*

Location/Sample	Host rock
Wadi Khalid	
JD-3/1a	Cu- mineralized arkosic rock ("matrix mineralization"), footwall of the Variegated Sand- and Claystones (cb3)
JD-3/3	Sandy dolomite with copper ores, DLS (cb2)
JD-3/16	Manganese nodules and phosphorite with Cu-ores in shale, top layer DLS (cb2)
JD-3/22b	Cu-mineralized sandstone, bottom of the Variegated Sand- and Claystones (cb3)
Wadi Dana	
JD-13/3b	Manganese nodules with Cu-ores in shale, top layer DLS (cb2)
JD-13/4	Cu-mineralized sandstone, bottom of the Variegated Sand- and Claystones (cb3)
JD-13/7	Phosphorite with Cu/Mn-mineralization, top layer DLS (cb2)
Wadi Ratiye/Qalb Ratiye	
JD-12/2	Limonite with Cu ores in sandstone, Massive Brown Sandstone (cb4)
JD-GR/2	Vein filling with Cu ores in sandstone (cb4)
Wadi Abiad	
JD-WA/1	Vein filling with Cu ores in sandstone (cb4)
JD-WA/4	Vein filling with Cu ores in sandstone (cb4)
Umm ez-Zuhur	
JD-41/5	Phosphorite layer with Cu/Mn mineralization, top layer DLS (cb2)

Chip samples for analysis were taken from ore outcrops and from ancient mines. The analyses showed that one still can find high grade copper ores at Faynan (Fig. 4.7). That is seldom the case in other ancient ore deposits with the exception perhaps of Timna (Fig. 4.5). The composition of ores from the DLS (cb2) elucidates that they vary between pure Cu to mixed Cu/Mn up to nearly pure Mn ores. Cu ores from the MBS (cb4) contain only small amounts of Mn (<0.2%), but in parts up to 15% Fe. The sulphur percentage is low and reaches 3.5% in the vein fillings of the MBS.

The percentage of trace elements, which follow the copper (see Sect. 2.2), is low with the exception of lead (up to 6%) in the DLS. In the ores from the MBS it is exceptionally low (Fig. 4.9). It can therefore be expected that the copper produced from these ores will be pure. This means, however, that pure copper is not necessarily a characteristic for native copper, as e.g., Otto and Witter (1952) had assumed, but that it can also have been produced by *smelting* of ores (Tylecote et al. 1977; Maddin et al. 1980). It will be shown later that the existence of such pure ores in rather large amounts is of great consequence for the interpretation of Levantine archaeology (Chap. 8).

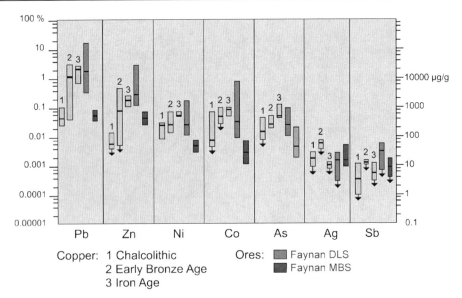

Fig. 4.9. Range of minor and trace elements of Cu ores from Faynan. Medians and interquartile ranges of ores from the Dolomite-Limestone-Shale Unit (DLS, cb2) are given and from the Massive Brown Sandstones (MBS, cb4). The generally higher percentage of elements in the ores of the DLS is clearly recognizable. The range of copper objects from different periods are given for reasons of comparison (from Hauptmann et al. 1992)

Ores from the DLS and the MBS formations, which have been mined in the past, differ from each other in their Mn- and Pb-contents (Fig. 4.10). Correlations between Mn, Pb, Zn and Co can be observed (correlation coefficients for the ores listed in Table A.1: $r_{Mn/Zn} = 0.90$; $r_{Mn/Pb} = 0.73$; $r_{Mn/Co} = 0.97$; $r_{Pb/Co} = 0.81$). Pb and Co are associated with Mn and less with the copper ore itself. Ag, however, is concentrated in the copper ore ($r_{Cu/Ag} = 0.75$). The elements with these correlations thus mirror the lateral change of facies in the ore deposit as observed in the mineralogical composition. Ores from the DLS are higher in Ni and As than those from the MBS, but as they show a large range of concentrations they have no pronounced correlation with Cu. It can be excluded that in the past at Faynan, ores with As- (and Sb-) percentages of above 1% were available. This has been repeatedly assumed to explain the raw material's origin of metal artifacts with a corresponding composition such as, e.g., the hoard from Nahal Mishmar, as coming from a local ore source (Hanbury-Tenison 1986; Shalev and Northover 1987; Ilan and Sebanne 1989). It can be stated with certainty – after numerous analyses of ores from a wide variety of outcrops from both deposits in the Arabah – that this assumption has no basis. But it is interesting to notice that at Faynan, As-contents in ore are lower than in the raw copper produced. It could be explained by the use, or better by the addition, of Mn ores with a higher As/Cu ratio.

Although the percentage of trace elements in the copper ores from Faynan is low, it shows variations. It might thus be more significant to use element ratios for determining provenance and for the comparison of different deposits. Seeliger et al. (1985) have suggested that Co/Ni and Sb/As are particularly interesting element

Fig. 4.10.
Lead and manganese concentrations in copper ores from the Faynan area, Wadi Abu Khusheibah/Abu Qurdiya and the ancient copper deposits from Timna (after data from Leese et al. 1985/86). The ores rich in Mn from the Dolomite-Limestone-Shale Unit (DLS) are higher in lead. Ores from the vein fillings of the Massive Brown Sandstone (MBS) contain little lead; they are compatible with Timna (from Hauptmann et al. 1992)

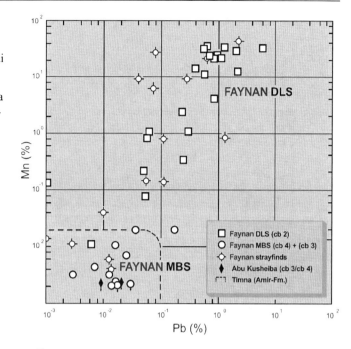

Fig. 4.11.
Sb/As- and Co/Ni-ratios in Cu-ores from Faynan and Timna (after data from Leese et al. 1985/86). The fields for Faynan DLS and Faynan MBS include ores with Cu > 10% and Mn > 20% (Table 2.2) (from Hauptmann et al. 1992)

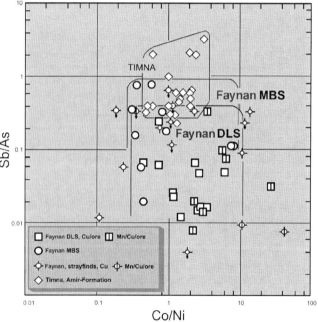

combinations. They suggested that the two respective elements would each be enriched and depleted during smelting in a similar way due to their similar geochemical behavior. These element ratios are shown in Fig. 4.11 for the ores from Faynan and Timna. It will be explained in Sect. 6.3 that in fact the four elements do not show the expected pattern of reaction, but that metal is relatively enriched in As and Ni. The ores from both formations in Faynan generally overlap, while the ores from Timna seem to diverge slightly and move towards a higher Sb/As-ratio. Measurements by Hauptmann et al. (1992) have shown that the Sb-concentrations of Timna ores published by Leese et al. (1985/86) are too high by one order of magnitude. The Timna and Faynan ore deposits cannot be differentiated from each other in this way as was expected. But at least it seems to provide a means of separating the Arabah ore province from other ore districts. In a larger geographic frame, however, a differentiation of ore deposits in this way does not seem possible any longer due to too much overlapping (Seeliger et al. 1985; Wagner 1988).

Host rocks of the copper ores can be classified geochemically into three groups independent of their stratigraphic position. They are mirrored in the slag. Host rocks with the highest quartz contents are sandstones of the MBS and quartzite and arkosic rock of the hanging wall of the DLS. The sandstone is partially carbonatized and contains Na_2O due to salt efflorescence (after Basta and Sunna (1971) up to 8.9%). This might have been advantageous for the smelting process in some cases. But the Fe_2O_3-containing ores in the sandstone, such as those in Qalb Ratiye, are more important in this matter. Such "self-fluxing" ores may produce Fe silicate slags during smelting without adding any further fluxes. Compared to Timna, they do not occur often at Faynan (Merkel 1985; 1990).

Copper ores from the middle of the DLS appear in the shape of pockets and vein fillings in the dolomite. The dolomite or dolostone, respectively, is high in Ca with a Ca/Mg-ratio of 1.4–1.6 (Table 4.1, see also data from Bender 1968 and Basta and Sunna 1971). After firing, this rock is a highly refractory raw material. Changing Mn contents and the transition towards sandy dolomite with up to 15% SiO_2 contribute quite considerably to the relatively easy smelting properties of this gangue.

The chemical composition of shales at the top of the dolomite/dolostones reveals higher Al_2O_3 contents compared with other host rocks and in addition varying Mn and Fe concentrations. This provides material with a low melting point whose melting properties are additionally improved by its K_2O contents.

Characteristic for the sedimentary origin of the ores is P_2O_5. which reaches in the MBS 0.03–1.14% (see also data from Khoury 1986). This is not above the average for sandstones, but the upper parts of the DLS have significantly high enrichments of phosphorite with >22% P_2O_5. These phosphorites and Mn-nodules are associated with uranium. Ayalon et al. (1985) found P/U-correlations in the Timna Formation, where up to 0.4% U were measured (Bar Matthews 1987). Otherwise, 3–25 µg g^{-1} U were analyzed in sandstones of the Amir-/Avrona formation (Segev et al. 1992). In Faynan, U concentrations were measured indirectly via Mn-rich silicate slag and showed P concentrations of 30–45 µg g^{-1} (Lorenz 1988). Studies on ancient copper metallurgy (in contrast to studies on ancient iron) have usually neglected the analysis of phosphorus. This element proved to be of importance not only for metallurgical techniques but also as a possible indicator of the provenance, because it can be enriched – depending on the temperature and CO/CO_2 ratio during the smelting – both in slag and in metal. Karageorghis and Kassianidou (1999), for instance, suggested that phosphorus in the form of bone ash would have been used as a deoxidizer for producing copper oxhide ingots in Late Bronze Age Kition

Table 4.2.
Medians of elements of ores from the "ancient" deposit in Timna (after data from Leese et al. 1985/86) and the ore formations in Faynan. Notice the similarity between the medians of Timna and Faynan DLS (except Pb, Ag and Sb)

Element	Timna	Fenan DSL	Fenan MBS
As	0.009	0.0016	0.0016
Ag	0.021	0.029	0.0045
Co	0.017	0.038	0.0035
Ni	0.025	0.029	0.0049
Zn	0.84	0.32	0.043
Pb	0.33	2.0	0.057
Sb	0.017	0.0035	0.001

(Cyprus). There is, however, no hard evidence yet for such a recipe. Neither copper oxhide ingots from the LBA shipwreck of Uluburun (Hauptmann et al. 2002) nor copper slags from Kition (Hauptmann et al., in prep.) contain phosphorus in concentrations that could support such a suggestion. Objects high in phosphorus from the Southern Levant may indicate Faynan as the source of copper, because other than in the Arabah and in Sinai there are no other significant sedimentary copper deposits in the Near East. Timna is a far less likely source, because the Timna Formation, which is the compatible unit to Faynan's DLS-containing phosphorite (Bar Matthews 1987), was mined only to a very limited degree during prehistoric times (see above).

Seen from a stratigraphical standpoint and judging from the Mn contents, (Fig. 4.10) the ore formation in the MBS in Faynan should geochemically correspond most closely to the "ancient" Timna ore deposit. Table 4.2 lists the medians of some trace elements from both regions and thus demonstrates that the trace element concentrations of the DLS ores correspond better. Ag, As, Co, Ni, and Pb at Timna are about five times higher than in the MBS ores of Faynan. It is difficult to distinguish ores from Faynan and Timna geochemically.

Lead Isotope Analysis

Copper ores from Faynan contain enough lead to measure the lead isotope abundance ratios. Fifteen samples from the DLS and the MBS have been analyzed in Mainz (Hauptmann et al. 1992). The results of the analyses are listed again, for reasons of a convenient summary, in Table A.2. Figure 4.12 shows a three-isotope diagram with $^{208}Pb/^{206}Pb$ vs. $^{207}Pb/^{206}Pb$ and $^{204}Pb/^{206}Pb$ vs. $^{207}Pb/^{206}Pb$. It should be mentioned that the analyzed samples had originally not been taken for studying the geochemical development of copper ore deposits in the Arabah. The initial goal had been the chemical and isotopic composition of ores mined at Faynan in order to use them later for provenance studies.

Figure 4.12 illustrates that the Cu-rich and the Mn-rich ores from the DLS plot in a narrow cluster. This means that the isotopic composition of the copper products would not have been affected by the conscious use of Mn ores as fluxing agents. Vein fillings in the Variegated Sand- and Claystones (cb3) above the DLS and those in the Massive Brown Sandstones (MBS, cb4) in the area of Qalb Ratiye (Fig. 5.1)

80 Chapter 4 · **The Raw Material Sources**

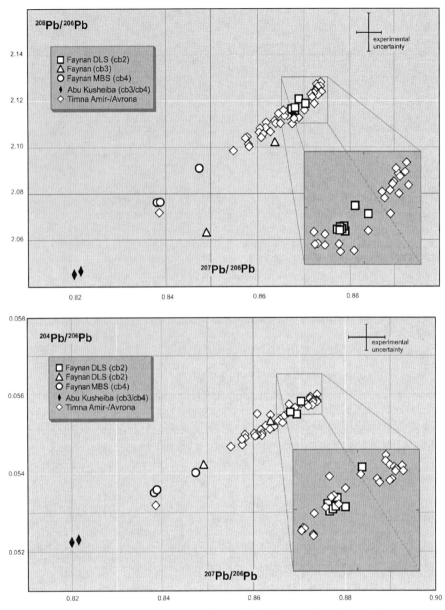

Fig. 4.12. Lead isotope abundance ratios of Cu- and Cu-Mn-ores from Faynan, Timna and the southwestern Arabah. Notice the wide range of compositions, which point towards a remobilization of the lead (after Hauptmann et al. 1992 and Gale et al. 1990; more data have been supplied by Dr. Z. Stos and Prof. Dr. N. Gale, Oxford)

have a distinctly different composition. They all appear in the lower left-hand corner of the isotope diagram and have a broad range of variation that is similar to isotopic compositions of other ore deposits in the Eastern Mediterranean.

A comparison of lead isotope ratios from Faynan with those from Timna is shown in Fig. 4.12. The latter are published by GALE et al. (1990), while another 36 analyses from Timna were analyzed in Oxford (Table A.3). The copper ores nearly all originate from the "ancient" deposits of the Amir/Avrona Formation (see Fig. 4.4 for stratigraphical setting). They reside mainly inside the range of the analyses of ore from Faynan only slightly jutting out with their ^{208}Pb/^{206}Pb, ^{207}Pb/^{206}Pb and ^{204}Pb/^{206}Pb-ratios. Differences do exist, and it might be possible to differentiate copper produced from both deposits, if field evidence and archaeological context are considered. No Cu ores from Timna have so far been observed that show such low values similar to those in the mineralizations of the MBS at Wadi Ratiye (Faynan) and in Wadi Abu Khusheibah (Figs. 3.1 and 4.2). Only *one* isotopic composition of an ore in Timna is known to correspond with those in the MBS at Faynan. It is an Fe ore sample from site 37, a remote small mining area in the south of the Timna Valley, which had been analyzed by Gale et al. (1990). Regrettably it is a scattered find and there is no chemical analysis given. But this isotopic data could have some bearing, as the mixed Cu-Fe ores with respectively varying concentrations of these elements are quite common in Timna (Fig. 4.8). A cluster which is represented by the ore compositions of the DLS in Faynan is missing. That is no surprise as it has already been mentioned (Fig. 4.4) that the Timna Formation, which is equivalent to the DLS from Faynan, has only one very small outcrop there.

Model ages of the lead in the copper mineralizations at Faynan are between 560 and 180 Ma and in Timna between 620 and 340 Ma (Fig. 4.13), and they have been calculated following the Two-Stage Model of Stacey and Kramers (1975). As explained in the section on the genesis of the ore deposits (see above), the actual age can only be approximated between ca. 620 and 470 Ma as the primary origin of the ores (in western as well as eastern Arabah); this approximation is closely connected to the volcanic activities at the end of the Precambrian.

Precambrian model ages of Timna correspond to the compositions of most ores, which have been found in the Cambrian rock of the DLS at Faynan. But with the exception of the Fe ore mentioned above, no ores have been measured from Timna, whose isotopic ratios would match the Cretaceous age of the Amir/Avrona Formation. This can only mean that the (Pre)-Cambrian lead has been incorporated nearly unchanged into the Cretaceous sandstone.

Epigenetic remobilization of copper and lead typical for sedimentary deposits is easily comprehensible at Faynan. Beyth and Segev (1983) and Segev et al. (1992) presume that there is a particularly close connection between volcanic activities during the Lower Cretaceous (99 to 107 Ma) and a remobilization of copper ores. It is possible to assign the model ages of two samples from the MBS at Qalb Ratiye (110 Ma) to such volcanic activities.

It seems likely that younger ages have no significance and represent mixed ages, which have been caused by the fact that pre-existent lead had been remobilized prior to the Miocene and was then redeposited together with radiogenic lead. Such a mixing has already been demonstrated by Hauptmann et al. (1992) using three samples from Faynan (*dotted line* in Fig. 4.13). In fact not all mineralizations are suitable for calculating model ages using the lead-lead method. This method requires a complete separation of lead from uranium and/or thorium during the deposit's formation so that the original composition of the lead will not be changed

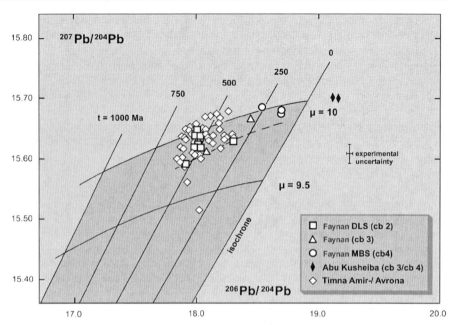

Fig. 4.13. Model ages of the Cu-ore from Faynan and Timna after the two-stage-model by Stacey and Kramers (1975). Different ages demonstrate a (repeated) remobilization of copper until the recent past. The *dotted line* connects three samples of varying ages, which indicate a mix of different leads

any more. As for ores containing uranium, it is possible that the original isotopic composition was changed by the addition of newly formed radiogenic lead. This would result in much younger ages as demonstrated by negative ages calculated for two samples from Wadi Abu Khusheibah. It has already been mentioned that ores from Faynan and from Timna contain uranium, which might have caused a drastic decrease in the Pb/U-ratios.

In summary, copper ores from Faynan and Timna belong to the most ancient so far measured in the Eastern Mediterranean as shown in Fig. 4.14 which includes data from other ore deposits (Stos-Gale 1993; Pernicka 1995). Data from Faynan and Timna cluster in the upper right-hand corner of the isotope diagram, while nearly all other ore deposit areas are younger and plot therefore in the lower left-hand quarter, such as the ore deposits embedded in ophiolite complexes in Cyprus and Ergani Maden. The lead isotope abundance ratios from Oman ores (not present here), which also belong to an ophiolite complex of comparable geological age and therefore are overlapping Cyprus, would also overlap in parts with the "Faynan/Timna-field" (Prange 2001). It should be kept in mind that the "Faynan/Timna-field" as defined here contains only the isotope ratios of (pre-)Cambrian age but not the ores' compositions from the MBS, from Wadi Abu Khusheibah and the Fe-ore samples from Timna.

In the western Mediterranean, only copper ores from Sardinia reach old ages that are similar to the ores from the Arabah (Brill and Barnes 1988; Gale 1989). But very old copper ores can also be found in the Sinai, in the Egyptian eastern desert

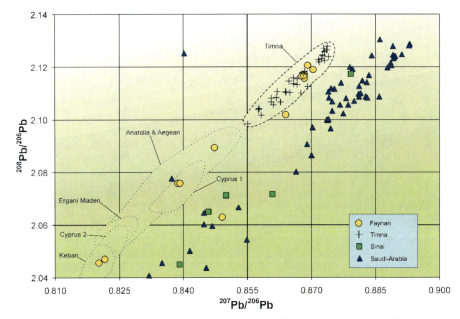

Fig. 4.14. Comparison of lead isotope abundance ratios of Arabah copper ores with ores from around the Eastern Mediterranean and the Arabian Shield. Evidence of ancient mining and smelting is known from Anatolia, the Aegean and from Cyprus (Stos-Gale 1993; Pernicka 1995) but is missing in detail from the Arabian Shield. The graph shows that the main copper mineralizations in the Arabah are among the oldest geological ages. Notice that the "Faynan/Timna field" contains only the data from the DLS and the Amir/Avrona Formation! Similarly old copper ores also occur in Saudi Arabia (Stacey et al. 1980; Bokhari and Kramers 1982; Brill and Barnes 1988) and on the Sinai Peninsula (Hauptmann et al. 1999; Segal et al. 2000)

and in northwest Arabia. Almost all these ore deposits are located in crystalline rocks of the Precambrian Basement. Hauptmann et al. (1999) and Segal et al. (2000) measured lead ages in copper ores from Sinai that were similarly early. Brill and Barnes (1988) detected compatible ages in polymetallic Cu-Pb-Zn ores from Umm Samiuki in the Eastern Desert of Egypt. Stacey et al. (1980) and Bokhari and Kramers (1982) published a number of lead isotope data on ores and rocks from the Arabian Shield in central Saudi Arabia and in northeastern Egypt (Fig. 4.14). In comparison to the "ore fields" mentioned above, ores from Sinai and from Saudi Arabia show a series of compositions from $^{208}Pb/^{206}Pb$ vs. $^{207}Pb/^{206}Pb$ ca. 2.01 to 2.15 and 0.837 to 897, with a gap between $^{208}Pb/^{206}Pb = 2.057$ to 2.105 that roughly corresponds to the gap of lead isotope compositions in the Wadi Arabah. They also show a tendency to higher $^{207}Pb/^{206}Pb$-values and are incompatible with those from the Wadi Arabah.

Chapter 5

Field Evidence and Dating of Early Mining and Smelting in the Faynan District

5.1 Landscape and Dating

In the framework of this research project an area of 27 × 35 km in the district of Faynan was surveyed in order to document sites, which have either a direct or indirect bearing on early mining and/or metal production (Fig. 5.1). Some sites were not accessible due to the occasionally very difficult conditions of the terrain; consequently, focal points had to be chosen and gaps had to be tolerated. The topographic maps that were available were not particularly useful; thus, aerial photos and satellite images were also used. Based upon data from satellite images (Ikonos Carterra Geo Ortho Kit), the coordinates of sites were determined with a positional accuracy of 15.0 meter (CE90) exclusive terrain effects.

The detailed plans in the site catalogue of Faynan and Khirbet en-Nahas (Fig. 5.4 and 5.33b) were mapped out from aerial photos and the plan from Wadi Fidan 4 (Fig. 5.40) is based on conventional ground surveying. The mining region of Wadi Abu Khusheibah/Abu Qurdiya, which had been described by Kind (1965) and from where some ore samples had been collected, was not included. The investigation of ancient mining was carried out in two steps. To begin with, all aboveground remains and signs of mining, such as shafts or openings ('Pingen') as well as tailings, were mapped and dated to an archaeological period with the help of surface finds (pottery, tools and coins). Most mines or pits, with a few exceptions, were found to be completely filled in with sediments. This required excavation and detailed documentation in order to reconstruct the technologies used to originally open a mine and to enlarge it following specific mineralization during the different periods. The excavations were carried out concentrating on a few selected areas in Qalb Ratiye and Wadi Khalid. Here the modern prospections carried out by the Natural Resource Authority had cut through a number of ancient mines. Keeping in mind that ancient mines normally provide very few finds, there have been fortuitously numerous finds of stone hammers and pottery. The latter were found in the waste dump of the mines. In some cases, even organic material such as timber or charcoal, useful for radiocarbon dating, were discovered. This made not only the dating of specific events possible, but also proved that the mines had been exploited over centuries, repeatedly being re-opened for further mining. The studies showed relatively quickly that the geological conditions decisively affected the system of mining technology, which also displayed the characteristics of the time period. This was the basis for a preliminary dating of altogether over 200 mines in the area of Faynan and for outlining the development of mining technology in the area, the main phases of which will be presented here. A comprehensive study of

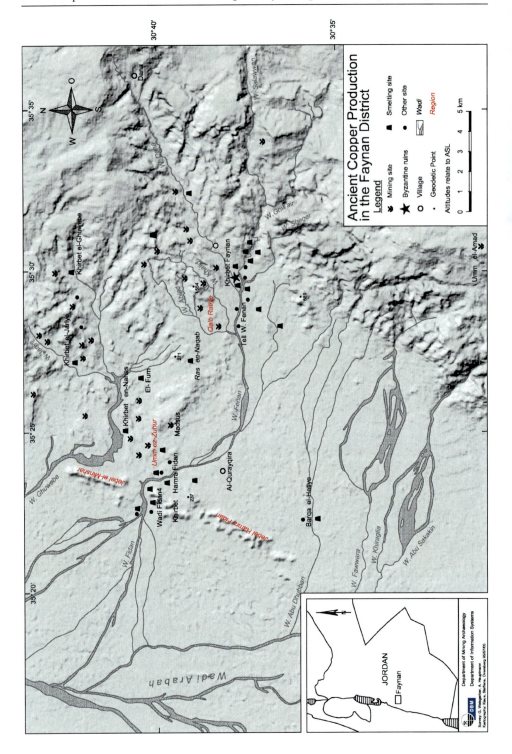

the archaeology of mining in Faynan, its dating as well as its social pattern and its political impact will presented elsewhere.

The copper district of Timna reveals a similarly dense accumulation of mining and smelting sites (Rothenberg 1988). It is important to point this out in order to show the intensity with which these deposits were mined in ancient periods.

The prospection, documentation and dating of slag heaps was relatively straightforward, because, unlike the situation in Europe, they were not covered with vegetation or soil sediments. There was thus no need to use geomagnetic prospection methods. It was particularly helpful that the slag heaps lay rather far apart from each other, which is a result of the dispersed locations of the outcropping ores and the settlement and organization patterns changing over different periods. The individual slag heaps were hence only little damaged through later activities. This made it easier to estimate quantities of slag produced and to define a typology; in particular, this helped the documentation of the Early Bronze Age smelting. It should be specifically emphasized that the entire copper district of Faynan shows a unique state of preservation. This is in contrast to the history of most other ancient mining districts in the Old World, which are as a rule heavily destroyed by modern industrial activities. How difficult investigations of ancient metal production can be is exemplified by studies carried out in Cyprus (Given and Knapp 2000) and Rio Tinto (Rothenberg and Blanco-Freijeiro 1981). Faynan, due to its limited reserves, escaped such a fate.

The dating that is presented is based on a chronology, which is itself founded on 51 radiocarbon-dates measured from timber and charcoal samples from waste dumps and slag heaps as well as from archaeological excavations. If possible several samples were taken and analyzed from each find spot in order to avoid problems with either contamination or mavericks, as can happen when particularly old timber is used for dating. The ^{14}C-dates are listed in Table 5.1, and their chronological position is displayed in Fig. 5.2. For comparison and as a basis for further discussion, calibrated ^{14}C-dates from the Timna district are given in Fig. 5.3.

In some cases, thermoluminescence has been used to date technical ceramics (furnace walls, tuyères, clay rods) and also some slag samples (Table 5.2). Although the age established by thermoluminescence is sometimes of low precision, the accuracy was sufficient to differentiate between the slag from the Early Bronze Age, Iron Age and Roman period (Lorenz 1988).

The studies of ancient metallurgical centers have made it repeatedly clear that archaeological surveys can only reveal metallurgical activities of those periods which are connected with the production of large amounts of slag. In Faynan, the Early Bronze Age III is just such a case, the period where the first 'industrial production' took place. In most other metallurgical centers in the Old World, the remains of early

◄ **Fig. 5.1.** Geographical overview of the Faynan-district at the foot of the Jordanian Plateau with the smelting sites and mining regions mentioned in the text as well as some other archaeological find spots. The regions around the town (Khirbet) of Faynan and around Khirbet en-Nahas are also presented as detailed maps in Fig. 5.4 and in Fig. 5.33. An overview map of the site Wadi Fidan 4 can be found in Fig. 5.40. Each miner's symbol represents a location where up to 50 mines have been discovered

Table 5.1. List of all dates of radiocarbon samples taken during the project, coming from different locations in the Faynan area, all of which are in some way connected to mining or smelting activities. The ^{14}C-dates were calibrated using the program CALIB 3 after Stuiver and Reimer (1993). There are some slight changes in comparison with earlier publications. They can be explained by the repeated measurements under slightly different statistical conditions, six to eight months after the ending of the actual analyses. All samples consisted of charcoal with the exception of HD-14926, which was timber. Not all samples come from a stratified context; some were collected from unstratified slag heaps. The calibrated dates are given with 1 σ-range

Sample	Location Find spot	^{14}C-age BP	Cal. ^{14}C-age BCE/CE	Reference
HD 12333	Wadi Fidan 11, N-slope settlement hill, −0.25 m	9157 ±112	8335 − 8040	Hauptmann 1991
HD 17219	Wadi Ghwair 1, 10/40 19, 36, −22.01	8812 ±61	7950 − 7705	Najjar et al., in prep.
HD 17220	Wadi Ghwair 1, 05/35 5, 10, −23.70	8627 ±46	7690 − 7540	Najjar et al., in prep.
HD 17221	Wadi Ghwair 1, 30/10 19, 15, −21.65	8528 ±89	7575 − 7485	Najjar et al., in prep.
HD 13777	Wadi Fidan 8 (A), wadi section 3 level 1	8220 ±117	7420 − 7035	Adams, in prep.
HD 13530	Wadi Fidan (A), trench 6 locus 11	8272 ±354	7575 − 6700	Adams, in prep.
HD 12334	Wadi Fidan 11, east slope ca. −0.2 m	7768 ±74	6610 − 6465	Hauptmann 1991
HD 12335	Tell Wadi Faynan, section edge of wadi, ca. −2.5 m	6370 ±42	5330 − 5265	Najjar et al. 1990
HD 10576	Tell Wadi Faynan, section edge of wadi, ca. −4 m	6408 ±114	5430 − 5250	Najjar et al. 1990
HD 12338	Tell Wadi Faynan square Fa II, −0.2 m	6105 ±68	5195 − 4930	Najjar et al. 1990
HD 13775	Tell Wadi Faynan square B2, locus 6, −0.8 m	6132 ±50	5195 − 4940	Najjar et al. 1990
HD 17471	Wadi Fidan 8 (A), 8/7, level 7 −26.04 m	6082 ±44	5055 − 4930	Adams, in prep.
HD 12337	Tell Wadi Faynan square A, locus 23, −1.4 m	5740 ±35	4675 − 4575	Najjar et al. 1990
HD 12336	Tell Wadi Faynan square A, locus 8, −1.05 m	5375 ±30	4330 − 4165	Najjar et al. 1990
HD 13776	Wadi Fidan 4 area A, locus 50	4684 ±50	3610 − 3365	Adams and Genz 1995
HD 16378	Wadi Fidan 4 area A, locus 22	4422 ±51	3255 − 2920	Adams and Genz 1995
HD 16379	Wadi Fidan 4 area A, locus 5	4576 ±44	3360 − 3165	Adams and Genz 1995
HD 16380	Wadi Fidan 4 area D, locus 4-14	4702 ±37	3610 − 3375	Adams and Genz 1995
HD 16327	Wadi Fidan 4 area D, locus 4-9	4718 ±25	3615 − 3380	Adams and Genz 1995
HD 13975	Barqa al-Hetiye house BH 1, 27/91, locus 13	4376 ±57	3080 − 2910	Fritz 1994
HD 13976	Barqa al-Hetiye house BH 1, 8/90, locus 3	4267 ±43	2910 − 2875	Fritz 1994
HD 10577	Faynan 9 ridge of slag heap, ca. −0.5 m	4140 ±109	2880 − 2500	Hauptmann 1989
HD 10573	Wadi Ghwair 4 ridge of slag heap, −0.5 m	4059 ±55	2840 − 2490	
HD 16533	Khirbet Hamra Ifdan trench 1, locus 114	4044 ±40	2585 − 2490	Levy et al. 2001
HD 16534	Khirbet Hamra Ifdan trench 2, locus 209	3914 ±45	2460 − 2320	Levy et al. 2001
HD 10574	Ras en-Naqab 1 flat slag heap, −0.3 m	3971 ±67	2565 − 2400	

Table 5.1. Continued

Sample	Location Find spot	14C-age BP	Cal. 14C-age BCE/CE	Reference
HD 10993	Faynan 9 furnaces nrs.7 and 8	3981 ±50	2560 – 2455	Hauptmann 1989
HD 10994	Faynan 9 furnace nr.25	3973 ±85	2575 – 2345	Hauptmann 1989
HD 10579	Faynan 16 slag heap, –0.3 m	3923 ±61	2465 – 2315	
HD 16529	Wadi Ghwair 3 slag heap, –0.5 m	3919 ±26	2460 – 2345	
HD 10584	Faynan 9 furnace nr. 24	3812 ±77	2395 – 2135	Hauptmann 1989
HD 14926	Wadi Khalid, mine 42 Backfilling 17 m away from opening	3197 ±39	1510 – 1415	
HD 10578	Wadi Dana, mine 13 waste dump before opening, –0.6 m	2949 ±63	1260 – 1025	
HD 16351	Khirbet el-Jariye, KJ2-4	2915 ±30	1150 – 1025	
HD 10990	Khirbet el-Jariye, slag heap wadi edge Wadj, –0.3 m	2886 ±56	1125 – 940	
HD 14308	Khirbet en-Nahas, KN-Eisen-5 sectioned slag heap, –1.15/–1.25 m	2876 ±38	1110 – 945	Steinhof 1994
HD 14057	Khirbet en-Nahas, KN-2 sectioned slag heap, –0.85/–0.9 m	2906 ±39	1150 – 1005	Engel 1993
HD 14302	Khirbet en-Nahas, KN-Eisen-2 sectioned slag heap, –0.1/–0.2 m	2880 ±28	1110 – 995	Steinhof 1994
HD 14336	Khirbet en-Nahas, KN-3 sectioned slag heap, –1.15/–1.35 m	2898 ±36	1120 – 1005	Engel 1993
HD 14113	Khirbet en-Nahas, KN-Eisen-6 sectioned slag heap, –1.65/–1.8 m	2864 ±46	1110 – 930	Steinhof 1994
HD 16530	Khirbet el-Jariye, KJ2-7 base of slag heap, –0.7/0.8 m	2839 ±22	1005 – 925	
HD 10581	Faynan 5 storage jar, locus 3	2726 ±102	985 – 800	
HD 14107	Khirbet en-Nahas, KN-1	2755 ±82	990 – 810	Engel 1993
HD 10575	Khirbet en-Nahas, section edge of wadi east, –0.6 m	2738 ±52	915 – 820	
HD 13977	Barqa al-Hetiye house BH 2, 208/110, locus 108	2743 ±23	905 – 835	Fritz 1994
HD 10991	Khirbet en-Nahas, slag heap NW, 1.5 m under hilltop, –0.3 m	2735 ±46	910 – 820	
HD 13978	Khirbet en-Nahas house 1	2704 ±52	900 – 805	
HD 10992	Faynan 5 slag heap, –0.3 m	2664 ±74	890 – 795	
HD 10582	Faynan 5 furnace, locus 2	2647 ±47	825 – 795	
HD 10580	Faynan 5 S furnace, locus4	2380 ±45	470 – 390	
HD 14307	Faynan 1, section L3 slag heap, –0.3/–0.6 m	2031 ±50	60 BCE–55 CE	Steinhof 1994
HD 14378	Faynan 1, section 3 slag heap, –0.25/–0.4 m	1991 ±72	50 BCE–115 CE	Steinhof 1994
HD 14380	Faynan 1, section 15 slag heap, –4.6/–4.8 m	828 ±34	140 – 245 CE	Steinhof 1994
HD 14066	Faynan 1, section 12 slag heap, –2.9/–3.3 m	1822 ±31	145 – 245 CE	Steinhof 1994
HD 14306	Faynan 1, section L9 slag heap, –1.05/–1.4 m	1801 ±34	220 – 320 CE	Steinhof 1994
HD 14097	Faynan 1, section R2 slag heap, –0.4/–0.8 m	1790 ±40	225 – 325 CE	Steinhof 1994

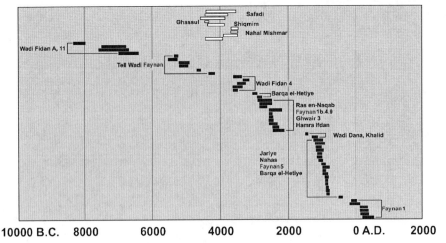

Fig. 5.2. List of the calibrated ¹⁴C-dates related to ore and metal extraction in the Faynan area. The values are given with 1σ-range. The dating sequence shows that the ore deposits in Faynan have been used from the Pre-Pottery Neolithic up to the Roman period (although at the present stage of information, with gaps in between). Later activities in the Mameluk period are not included here (see also Table 5.1)

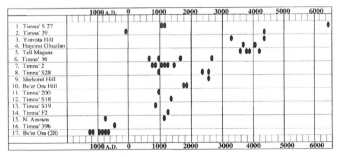

Fig. 5.3. Calibrated ¹⁴C-dates of mining and smelting sites in the Timna district and some other sites related to copper production in the Southern Arabah. The early date from Timna S27 is of uncertain archaeological context (Weisgerber 2006). Chalcolithic/EBA I copper metallurgy using Timna-ores is mainly based upon ¹⁴C-dates from Tell Magass, Tell Hujayrat al-Ghuzlan near Aqaba and Yotvata. Note the LBA-date from Timna F2 which underlines TL-dating of tuyère from this site (Hauptmann et al., forthcoming). Each symbol represents the mean value of one or more dates. Calibration based on OxCal 3.4 (after Avner 2002)

metallurgy have been so badly destroyed by later activities that it became very difficult to identify any period earlier than the Late Bronze Age. It seemed therefore reasonable to supplement the surveys with a number of clearly defined excavations and test trenches. These were carried out representatively in some settlements, where information about the use of ores and the development of metallurgy could realistically be expected. In addition, metallurgical installations representative either of the Early Bronze Age or the Iron Age were excavated. These excavations will only be presented in a brief summary in the following site catalogue, because detailed publications are in preparation.

Table 5.2. Natural dosage rate and TL-age of slag from Faynan (from Lorenz 1988). The U-, Th- and K-contents were measured with γ-spectroscopy. The TL-ages of the slag are given with 1σ-range. For reasons of comparison, the TL-ages of associated pottery finds are also given as well as the archaeological classification of the respective smelting sites. See also the ^{14}C-dates in Table 5.1 and Fig. 4.13. The samples marked with an asterisk show a heavily anomalous healing and had not been stored for a sufficient length of time between radiation treatment and measurements, so the calculated age can only be the earliest possible date for the sample. *FC:* furnace ceramic; *S:* pottery sherds; *TS:* clay rods

Sample HDTL	Find spot	U ($\mu g\,g^{-1}$)	Th ($\mu g\,g^{-1}$)	^{40}K (%)	Age (a)	TL-data for comparison	Archaeological phase
22b4	Wadi Fidan 4a	45	11	1.95	951 ±240	FC: 87 AD ±203 a S: 2772 BCE ±378 a 2901 BCE ±388 a 2727 BCE ±362 a	Iron Age II + Early Bronze Age II/III
22d2	Ras en-Naqab	33	7	1.67	1 740 ±571*		Early Bronze Age II/III
22d7	Ras en-Naqab	37	9	1.55	1 185 ±451*		Early Bronze Age II/III
22f4	Faynan 1	29	13	1.77	203 ±55*		Roman
22h5	Faynan 9	32	7	1.25	1 625 ±636*		Early Bronze Age II/III
22k3	Faynan 5	49	8	2.41	1 445 ±374	FC: 137 AD ±195 a S: 22 AD ±196 a 176 BCE ±183 a	Iron Age II
22m3	Wadi Faynan 4	36	9	1.26	2 172 ±588	TS: 2109 BCE ±477 a FC: 2332 BCE ±533 a	Early Bronze Age II/III

5.2 Site Catalogue and Archaeometallurgical Finds

The sites described in this chapter are presented in the geographical order shown below. The exact locations can be found in Fig. 5.1. If samples had been taken from a particular find spot, the site will have a JD-number or some other abbreviation.

I. Khirbet Faynan and its vicinity
II. Sites in Wadi Ghwair
III. The mining area of Qalb Ratiye
IV. The mining area of Wadi al-Abiad
V. The mining area of Wadi Khalid
VI. Sites in the lower Wadi Dana
VII. Ras en-Naqab
VIII. El-Furn
IX. Khirbet en-Nahas
X. The mining area of Umm ez-Zuhur and Madsus
XI. Khirbet el-Jariye
XII. Khirbet Ghuwebe
XIII. Sites in Wadi Fidan
XIV. Barqa el-Hetiye
XV. The mining area of Umm el-Amad

Fig. 5.4a. 3D perspective view of the Byzantine ruins of the town of Faynan (*middle*) generated by combining Ikonos satellite data and Aster DEM. *Red sticks* in the image show mining and smelting sites at Faynan and in its vicinity (see Fig. 5.4b)

5.2 · Site Catalogue and Archaeometallurgical Finds

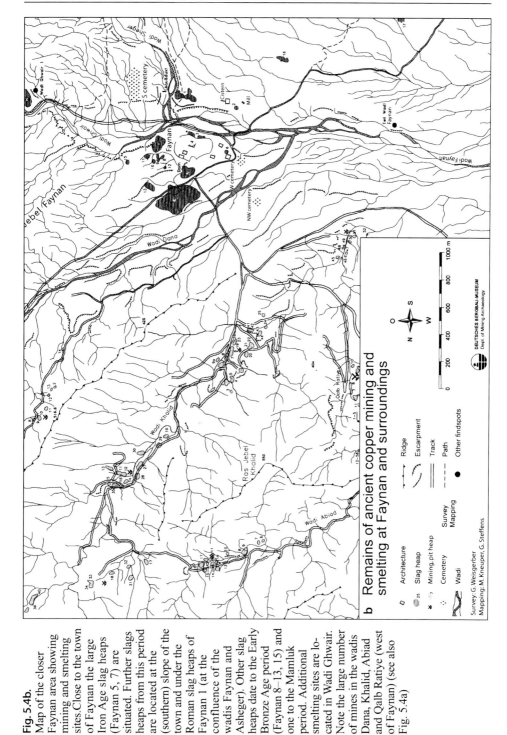

Fig. 5.4b.
Map of the closer Faynan area showing mining and smelting sites. Close to the town of Faynan the large Iron Age slag heaps (Faynan 5, 7) are situated. Further slags heaps from this period are located at the (southern) slope of the town and under the Roman slag heaps of Faynan 1 (at the confluence of the wadis Faynan and Asheger). Other slag heaps date to the Early Bronze Age period (Faynan 8–13, 15) and one to the Mamluk period. Additional smelting sites are located in Wadi Ghwair. Note the large number of mines in the wadis Dana, Khalid, Abiad and Qalb Ratiye (west of Faynan) (see also Fig. 5.4a)

I. Khirbet Faynan and Vicinity

This area stretches from the mouth of Wadi Sheger along Wadi Faynan towards the west up to Tell Wadi Faynan. The central point of the area is Khirbet Faynan, the Byzantine town ruins clearly visible from some distance (30°37.6' N/35°30' E), which has given its name to the region and been described by a number of early travelers (Musil 1907/8; Frank 1934; Glueck 1935; Kind 1965). The site has recently been studied (Barker et al. 1997) by the Centre for British Research in the Levant (then BIAAH). Only sites with largely undisturbed archaeological contexts will be presented here. The numerous slag scatters, on the other hand, will not be individually described. They spread over the fields and countryside to the west, badly disturbed by the agricultural activities and therefore impossible to date. Around the region of Khirbet, Faynan is the most substantial collection of smelting sites in the entire Arabah (Fig. 5.4). With the exception of the Chalcolithic period, all major eras of metal extraction can be found here.

Faynan 1 (JD-1)

- Geographical setting: 30°37'26.19" N/35°29'37.74" E

A large smelting site covering 300 × 150 m is situated opposite Khirbet Faynan on the southern bank of Wadi Faynan at the confluence of the Wadi Sheger and Wadi Ghwair (Fig. 3.3). The slag heaps have been severely cut by erosion; thus, sections up to 7 m height have been available for sampling. The entire heap contains architectural remains. Particularly noteworthy is the Roman water channel, which is the continuation of the aqueduct spanning the Wadi Sheger and continues on towards a cistern situated some 200 m west of the slag heap (Fig. 5.4). The slag heap has been dated with the help of pottery (which included not one piece of Nabatean pottery (Hauptmann et al. 1985)), coins (Kind et al. 2005), and ^{14}C-dates. This evidence points unanimously towards pre-Roman and Roman activities from the 2nd century BCE to the 3rd century CE with a focal point in the 2nd/3rd century CE (Table 5.1). The smelting site is linked with the mines at Umm el-Amad, several hundred meters higher up, (see below) via an over 7 km long, partially constructed road. The road starts to the southeast of the Roman slag heap of Faynan 1 at a Roman mansion with a few one-room houses and other structures.

The survey data from W. Lieder resulted in an estimate of a tonnage between 45 000 and 70 000 t of slag. The lower estimate is more realistic, because it became clear during the cleaning of several sections for the extraction of charcoal samples that the slag had been heaped upon a number of steps in the landscape. Several horseshoe-shaped stone structures with layers of mortar and pottery were discovered on the heaps themselves. The inner diameter is 0.6 m and the height of the rear wall is ca. 1 m (Fig. 5.5). They are not slagged. The question, if they are remains from furnaces, has to remain open. The bad condition of these remains allows no further elaboration about their exact shape, form and size. Only the use of tuyères is provable. Fragments of tuyères (Fig. 5.6) with an estimated internal width of 8 to 12 cm could also indicate the utilization of natural winds and not necessarily the application of artificial air supply through bellows (Weisgerber and Roden 1985).

Fig. 5.5. Faynan 1. Horseshoe-shaped stone setting on the Roman slag heaps, which is generally interpreted as the remains of a smelting furnace, although no further details can be given. The open, destroyed front of the furnace is oriented towards the east, to Wadi Ghwair. Height of the stone setting: 1 m

Fig. 5.6.
Faynan 1. Fragment of a tuyère from the Roman slag heap. It is made from hard fired pottery and has an interior width of 8–12 cm. It could possibly represent a 'wind-catching'-tuyère, which would not have been connected to a bellows but was meant to use the naturally available winds (Lz-Nr. 98; drawing: A. Weisgerber)

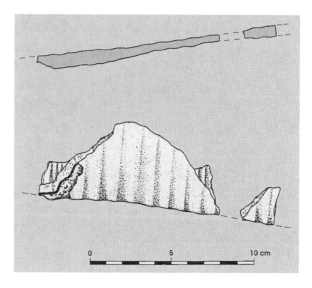

In this connection it might be of interest that the open side of the partially excavated furnace on Fig. 5.5 is directed towards the Wadi Ghwair and thus shows parallels to the furnaces at Faynan 9 (see below). A definite interpretation of this find is not possible without further excavations.

Roman slag is usually broken into fist-sized, up to head-sized, chunks. But sometimes blocks of 10 kg, even of over 100 kg have been found. The slag was obviously tapped as a viscous liquid into a deepened fore hearth in front of the furnace and solidified there to form compact, partly glassy slag-cakes. But porous 'conglomerates' full of bubbles (Fig. 5.7) and inclusions of charcoal and ore from the vein mineralizations of the 'Massive Brown Sandstone' (MBS) are more common. They hint at the ore material used during the Roman period and confirm the dating of the mines in Qalb Ratiye, in the wadis Abiad and Khalid as well as those in Umm el-Amad.

A chunk of ferritic iron of 15 kg containing over 15 wt.-% Cu has been found at the surface of the Roman slag heaps. It originates perhaps from the Mameluk metal production. Below the Roman slag heaps are meter-thick layers of crushed slag which dates from the Iron Age.

Faynan 1 is the only smelting site dating to the Roman period in the Faynan area. It was most likely the smelting center for the ores from all Roman mines in the entire region. Roman mining has also been documented in the wadis Abu Khusheibah/ Abu Qurdiya (Kind 1965) at a distance of some 80 km, but evidence of smelting activities directly on the site have not been recorded. The same is true for the mines in Wadi Amram south of Timna; they seem to have been suppliers of ore to Faynan 1 too. Beer Ora (Timna), the only smelting site which had been thought possibly to be Roman, can now be dated to the Early Islamic period (Rothenberg 1973; 1988b) due to a ^{14}C-date (1 390 ±50 BP = calib. 640 CE).

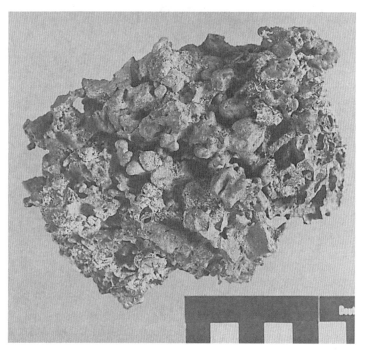

Fig. 5.7. Faynan 1. A large part of the Roman smelting site consists of only partially liquefied slag. It contains charcoal as well as ore inclusions, the latter coming from the ore mineralizations in the MBS, and are reminiscent of conglomeritic sediments

Faynan 2 (JD-1)

- Geographical setting: 30°37'28.98" N/35°29'24.23" E

It is a flat slag heap at the northwestern corner of the Roman-Byzantine cistern, ca. 200 m west of Faynan 1 (Fig. 5.4). The tonnage is only some 50 t. Cu-Mn-ores from the Dolomite-Limestone-Shale unit (DLS) can be found throughout the slag heap, which dates to the Mameluk period (13th century CE).

Faynan 3, 4 (JD-20)

- Geographical setting: 30°37'34.76" N/35°29'36.17" E

Two slag concentrations, clearly visible along the gravel slope, are located underneath the Byzantine buildings at the southern slope of Khirbet Faynan, the town ruin. They could be dated to the Iron Age II because of the pottery found there as well as the typology of the slag (see Faynan 5).

Faynan 5 (JD-1)

- Geographical setting: 30°37'37.39" N/35°29'47.61" E

This slag heap dating to the Iron Age (Fig. 3.5) is located directly adjacent to and east of the Byzantine town. It is separated from the latter by a wall and partly covered at the edge by a recent Islamic cemetery. Faynan 5 has been dated from its pottery to the Iron Age II (see also Knauf 1992 and HART 1992). Three ^{14}C-dates (Table 5.1) lie between 1000 and 800 BCE, a fourth dates to the time span from 520 to 400 BCE. Careful estimates put the tonnage at 30 000 t. This amount makes Faynan 5 the second largest smelting site (after Khirbet en-Nahas) of this time period in the entire Faynan region. It has to be assumed that the original extent was much larger and that this slag heap has to be seen in some connection with the slag heaps of Faynan 1, 3, 4, and 7. Due to the covering over of Khirbet Faynan, as well as its destruction by Roman activities and the Byzantine re-development of Khirbet Faynan, it is impossible to prove these connections without archaeological excavations.

Principally, two different kinds of slag type can be distinguished in the heaps of Faynan 5: a dense tap slag with the typical ropy structure, not dissimilar to the Pahoëhoë-lava on Hawaii, and remains of bubbled furnace slag with charcoal and ore inclusions. Immediately obvious on the surface of the heaps are tap slags formed in a semi-circular shape, which has flowed out from the furnace as low viscous, homogenous part-melts. It solidified in a flattish, rounded pit with a diameter of ca. 0.5 m (Fig. 5.8). This slag has a low copper content and sometimes shows remains of pottery, tempered with slag, sticking to its lower side. A section through such slag shows a pillow-like structure with cooling rims around each of the single glassy slag-bulges, which can be taken as a sign of the temporal intervals during the tapping. The weight of these tap slags formations can be up to 57 kg. They are mostly broken in the middle, actually at the transition to the more lumpy parts containing pieces of charcoal. This is also the transition point to the other kind of slag, the inhomogeneous

Fig. 5.8.
Faynan 5. **a** Fragment of an Iron Age tapped slag. **b** Plan and section of an Iron Age tapped slag. The slag, solidified in a round slag pit in front of the furnace, is comprised of several layers, which solidified next to as well as above each other. The slag is often broken in the middle where there is a transition to furnace slag containing pieces of charcoal. This slag is completely identical to the types found in Timna, site 30, stratum I (Bachmann and Rothenberg 1980), but the shape illustrated here is not the complete slag as had been assumed from the material in Timna

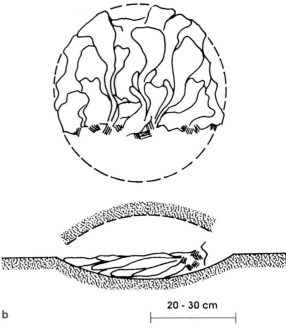

and badly liquefied 'restites' (see Piel et al. 1992), which in all probability had already solidified in the furnace (Fig. 5.9; Bamberger and Wincierz 1990). This furnace slag (and probably tap slag as well) has been crushed and churned in a comprehensive manner in order to obtain the remaining metal; the slag is therefore mainly existent as rubble and sand. These materials can be regularly seen on the Iron Age smelting sites in the Faynan region; one can thus assume that the mechanical beneficiation of metal-rich slag was a characteristic and recurring step in metal production (Bachmann and Hauptmann 1984). The metal prills, which would have been obtained in that way, would then have been melted together in the workshop quarters; evidence of this process has been found in Timna (Craddock and Freestone 1988).

Fig. 5.9.
Faynan 5. Iron Age furnace slag. The furnace slag body contains ore remains as well as charcoal or its negative imprints. This type of slag starts at the broken transition from the tap slag into the furnace

The tuyères with an inside width of 1.5–2.8 cm at the air hole were found in larger numbers, showing quite clearly that the furnaces were operated with an artificial air supply (using bellows). As added evidence, pottery-pipes have been found, which were most likely the connecting elements between these nozzles and bellows. Two differently sized tuyères were utilized as can be seen in Fig. 5.10. A comparison with corresponding artifacts in Timna (Rothenberg 1990, p. 38, 51) might lead to the conclusion that the two different types of tuyère date to two different time periods. In Timna the 'small' nozzles date to the 14th–12th century BCE, while the 'large' tuyères come from the 10th–8th century BCE. A precise dating is difficult for Faynan 5, as all the finds come from the surface.

Both types of tuyère are composed of a number of parts, for which different clays were utilized. The tuyère nozzles, always oriented towards the actual firing chamber, consist of whitish, highly refractive material. The main body however, at the outside of the furnace, is made of reddish clay tempered with crushed slag. The white ceramic contains 70–80 wt.-% SiO_2 and 10–20 wt.-% Al_2O_3 and is resistant up to over 1 100 °C (Klein and Hauptmann 1999). This ceramic is probably made from kaoline or china clay or from a siliceous rock. The reddish clay contains, with 7–9 wt.-%, far more Fe oxides, and the SiO_2 is therefore hardly above 65 wt.-% and is thus far less fire resistant (Schneider, pers. comm.). The same material was also used for the furnace walls; it is particularly typical of the Iron Age throughout the entire Arabah (Bachmann and Rothenberg 1980). Pendant drops of slag on the tuyères demonstrate that they had an inclination of 40–60° towards the furnace interior. The tuyères' nozzles are often quite thoroughly resorbed through reactions with the charge and heat effects, while the openings of the nozzles are blocked through slagging.

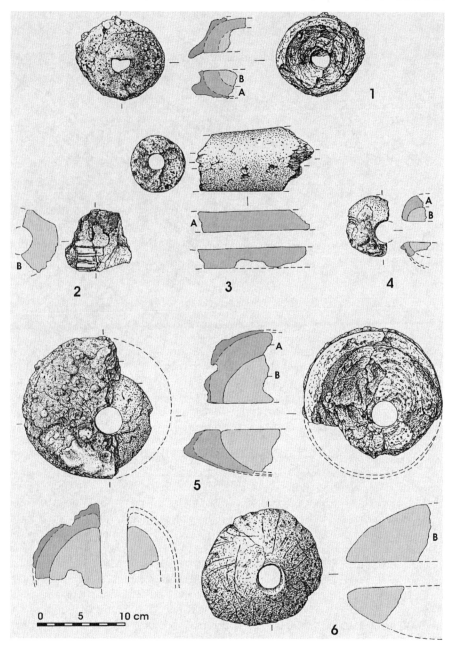

Fig. 5.10. Different types of nozzles (tuyères) and blow-tubes. *1, 4:* 'Small' tuyères, Faynan 5 (Late Bronze Age?); slagged heat shield made from refractory clay (A) with remains of packing material of slag-tempered clay (B). *2:* Nozzle with plant impressions, from 'northwestern cemetery' (not dated). *3:* Blow-tube. *5:* 'Large' tuyère, Faynan 5 (Iron Age), construction as for 1 and 4. *6:* 'Large' tuyère, Faynan 5 (Iron Age), back side of the heat shield made from slag-tempered clay (drawing: I. Steuer, Göttingen)

Several smelting installations were excavated in Faynan during the seasons of 1986 and 1988; in particular, the construction in locus 2 at the northern slope of the heap needs to be mentioned (Fig. 5.11 and 5.12). It dates to the early 8th century BCE (Table 5.1). The construction has a flat, circular depression in the middle, which is covered by a dome. A comparison with the Iron Age tap slags indicates that the depression is the slag-tapping pit of a furnace. The entire construction is made from slag-tempered clay and is only weakly fired. The inner diameter of the lentil-shaped pit is at most 1 m, the height in the middle up to 40 cm. The dome has in the middle an opening towards the top, where the impressions of vertically set slag or stone

Fig. 5.11. Faynan 5, locus 2. Iron Age furnace construction from the 8th century BCE. In the center is the domed slag-tapping pit visible with its opening to the top, which has been enlarged due to the erosion of the slag-tempered lining. The actual smelting furnace was probably located in the adjacent depression (*white arrows*). In front of the furnace, two storage containers plastered with mortar and an amphora have sunken into the floor. They were probably used for the holding of ore and fluxing agents as well as water. Photo taken ca. 6 m vertically above the uppermost level of the installation with a Reseau camera Rollei SLX for photogrammetrical documentation and recording

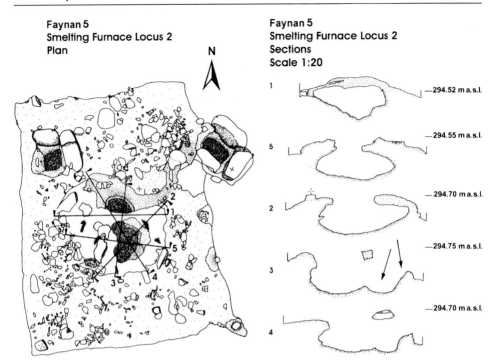

Fig. 5.12. Faynan 5, locus 2. Plan of and section through the Iron Age furnace installation shown in Fig. 5.11. Section nr. 3 shows particularly clearly the heavily sintered combustion-chamber (*arrows*), which is divided from the tapping pit by a threshold. The whole installation is identical to the 'pyrotechnical installation' of Timna 30, stratum I, locus 10 (Rothenberg 1988, 47) (drawing: C. Poniecki)

plates are very clearly recognizable. The opening can therefore not be interpreted as a secondary destruction. A second hole opens at the slope. The dome-covered tapping pit leads, via a small threshold, into a smaller hollow, which is almost completely vitrified and slagged due to intense heating (arrows in Fig. 5.11 and 5.12). This area is heavily impregnated with charcoal dust. The outer border is marked by vertically set slag plates and has a foundation of crushed slag. This hollow might possibly represent the last remains of the furnace combustion chamber, while the furnace upper walling is completely destroyed. Directly in front of the whole installation are two storage containers, made from partly sunken stone-settings and plastered with mortar. A flat bodied half amphorae had, most likely, the same function. But it was not possible to identify the actual use of these installations. The comparison with the corresponding findings of stratum I in Timna (Rothenberg 1980) suggests that ore and fluxing agents might have been kept there. However, the use of mortar points towards a small water basin.

This whole furnace installation shows remarkable similarities to that in Timna 30, stratum I, locus 10 (Rothenberg 1980, 202 pp.; Rothenberg 1990, 49 pp.), particularly concerning the tapping pit. Rothenberg interprets these findings quite correctly merely as a 'pyrotechnical installation' and is sufficiently careful not to fully reconstruct the actual smelting furnace.

Further excavations have been carried out in locus 1, which is 150 m south of locus 2 at the southern slope of the heap. It was possible to unearth the remains of a heavily eroded furnace installation, which showed the same division into two parts, the tapping pit and the actual firing-space, as had been found in locus 2.

Locus 3, only some 10 m away from locus 2, revealed a cylindrical storage pit with a diameter of 1.1 m and a depth of 1.2 m. A considerable amount of charcoal was on the floor. The area of locus 4 showed a construction compatible with the above-described dome-covered furnace installation.

It has to be emphasized that neither in the Iron Age smelting sites of Faynan nor at those in Timna have there ever been any finds or contexts which would have allowed a realistic reconstruction of the smelting furnaces. Bachmann and Rothenberg (1980) state this also quite explicitly when they present their suggestions for the reconstruction of Iron Age furnaces. It is therefore quite probable, also following comparisons with other smelting sites, that the smelting furnaces have systematically been destroyed after each smelting process (see also page 216).

Hauptmann et al. (1989) have already pointed out previously that the repertoire of finds from the Late Bronze Age/Iron Age smelting locations repeats itself not only in the sites in the Faynan region (see the description of Khirbet en-Nahas). Fragments of the furnace walls, clay tubes (tuyères), slag and the furnace installations are practically identical to corresponding material from Timna 30, stratum I (11th–8th century BCE, Bachmann and Rothenberg 1980; Rothenberg 1980, 1985, 1990). This provides evidence of clear parallels in the metal technology between Timna and Faynan, contrary to Rothenberg's opinion (Rothenberg, 1988, 17).

Faynan 6 (JD-1)

- Geographical setting: 30°37'39.40" N/35°28.41" E

At the southeastern corner of the so-called monastery (Frank 1934) is a flat, round heap with a few tons of slag. This mound proves smaller smelting activities from the Mameluk time period. That is indicated by the presence of five coins, which have been found in the direct surroundings (1300–1370 CE, Kind et al. 2004). In addition, the chemical composition of the slag is identical to those found in El-Furn (Ngeib Asiemer, see also Sect. 6.2): we have here Fe-Mn silicate-slags, which turn up in Faynan only during this time period.

Faynan 7 (JD-1)

- Geographical setting: 30°37'43.30" N/35°29'40.12" E

Directly northwest of Khirbet Faynan a heavily washed out Iron Age slag heap is situated between the reservoir for retaining water and soil and the track to the camp in Wadi Dana. The heap consists of roughly 5 000 t of slag. The metallurgical finds such as slag, ceramic fragments from furnaces, and tuyères are similar to those from Faynan 5 and Khirbet en-Nahas and therefore date to the Iron Age. Typically, the slag is mostly crushed to a finely grained sand or gravel.

Faynan 8-13, 15 (JD-23)

- Geographical setting:
  Faynan 8: 30°37'43.32" N/35°29'45.79" E
  Faynan 9: 30°37'40.14" N/35°29'40.44" E
  Faynan 10: 30°37'39.13" N/35°29'51.11" E
  Faynan 11: 30°37'36.47" N/35°29'45.96" E
  Faynan 12: 30°37'40.66" N/35°29'48.88" E
  Faynan 13: 30°37'40.94" N/35°29'45.74" E
  Faynan 15: 30°37'40.36" N/35°29'56.11" E

North of the Byzantine ruins, divided only by a small wadi on the opposite slope of the gravel terrace are seven partially merging slag heaps. They form flat heaps along the slope and are sparsely covered with some rubble. The slag found here is small (nut-sized to fist-sized) and clearly very different from the Iron Age slag. It is systematically crushed, as on all other heaps of the same period, which points to an almost complete reworking of the slag heaps. Finds of dimpled anvil stones with a diameter between 20 cm and more than one meter and numerous hammerstones indicate Iron Age activities (pers. communication Prof. Gerd Weisgerber 2006). This may suggest that reworking of the slag was carried out in the Iron Age. Test trenches showed a depth of 0.3–1 m (Fig. 5.13). We estimated the total tonnage at 1 000 t, but it can be taken for granted that the original amount of production was much larger, as it is obvious that large parts of the slag have been resmelted in the neighboring Iron Age copper production site due to its still high metal content (see also Ras en-Naqab).

Fig. 5.13. Faynan 9. Slag heap of the Early Bronze Age II/III. The heap is located, as are the six other heaps in the direct vicinity, on the Quaternary gravel terrace. On top of the heap are smelting furnaces (see Fig. 5.16). The test trench showed a section through the heap of up to 1 m depth

Faynan 9 and 15 have been dated by four ^{14}C-dates and pottery finds (Najjar et al. 1990) into the Early Bronze Age II and III, with a peak between 2600 and 2400 BCE.

The blackish slag has been pounded into small pieces, which have a reddish translucent sheen when they are very thin; this is due to the high content of copper oxide (cuprite). The slag is formed by single drop-shaped and cushion-shaped parts (Fig. 5.14), which indicates the cooling of highly viscous silicate melts that flowed vertically out of the furnace. This led to cone-shaped blocks, which are very different from the usual, flatly flowed, plate-like slags. In all, other than the slag, these heaps are composed, along with numerous pieces of ceramic furnace walls, of fragments of thousands of finger-thick clay rods ('ladyfingers'), seemingly only fired on one side, which have been made very carefully and fastidiously. They reach a length of ca. 10 cm and can sometimes be found stuck parallel to each other in the ceramic furnace wall or in huge slag lumps (Fig. 5.15). They are made from Ca-rich clay, strongly

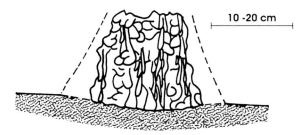

Fig. 5.14. Schematic section through Early Bronze Age smelting slag. The fragments collected from the different heaps allow the reconstruction of a nearly cone-shaped stump, which was made up from single, vertically elongated drops and pillow-shaped parts that had solidified in these shapes. The slag has probably flowed downwards out of a smelting furnace

Fig. 5.15.
Faynan 9. Roughly finger-thick clay rods, which were sometimes found in association with smelting slag or with slagged furnace walls, are the typical finds on Early Bronze Age smelting sites. Thousands of them have been found on slag heaps from this period. The sticks are in all probability the remains of reinforcement in the furnace wall

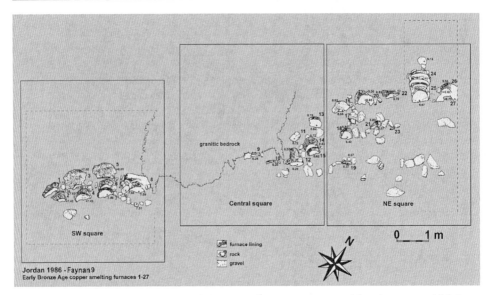

Fig. 5.16. Faynan 9. Location of the 27 Early Bronze Age smelting furnaces, which have been excavated at the head of the slag heap of Faynan 9. Their conspicuous position at a hilltop exposed to winds allows their interpretation as wind-operated furnaces (photogrammetry: Christoph Roden)

tempered with quartz. The clay has 65–70 wt.-% SiO_2 as well as 10–15 wt.-% CaO and therefore differs noticeably from the clay material used for the furnace walls themselves, which had 40–45 wt.-% SiO_2 and ca. 25 wt.-% CaO.

These clay rods are – next to the external typology of the slag types – the index find to date slag heaps into the Early Bronze Age period. They were apparently not part of a vertically fixed grill or bar at the furnace opening towards the wind, although that has been assumed in the past (Hauptmann and Roden 1988; Hauptmann 1989). The most recent experiments concerning the functioning of Early Bronze Age wind furnaces (Kölschbach 1999) illustrate that these rods must have been an integral component of the furnace wall and were constructed as a panel to stabilize it.

At the top of the slag heap from Faynan 9 the remains of 27 smelting furnaces, standing closely together, have been excavated (Fig. 5.16– 5.18). And in 1988 seven more furnaces have been uncovered at Faynan 15. The former group was built just below the crest of the heap and all of them face northeast towards the descending winds coming down from the Jordanian plateau. This location, like the sites of the other EBA furnaces in Faynan, was obviously carefully chosen. It was found to provide an optimal position for smelting processes over a long period of time: The furnaces were repeatedly built at the same spot over and over again, as can be seen by up to 20 different layers of furnace bottoms and rear walls.

The furnaces had most likely a dome-shaped or shaft superstructure. This is not only indicated by the results of a number of smelting experiments (see Sect. 7.3), but also by the amount of ceramic fragments from the furnaces, which have been found in the slag heaps. These fragments are partly very heavily vitrified due to intense heating. They show transitions from slagged ceramic and metallurgical slag in all possible stages.

Fig. 5.17. Faynan 9. Early Bronze Age copper smelting furnaces no. 24–27. The semi-circular section of the furnace base (no. 24, 25) is clearly visible as well as the rear wall of the oven, which has been renewed repeatedly and shows the individual ceramic layers. It proves that this and all the other furnaces existed over a long period at the same spot (from Hauptmann and Roden 1988)

The entirety of the different contexts and finds leads to the conclusion that Early Bronze Age smelting furnaces have been operated using the natural winds to increase the furnace temperature. It is characteristic that the lower parts of the furnaces, still surviving, show no signs of slagging as would be normal in the case of smelting furnaces. We assume therefore that the furnace bottoms were protected from a reaction with the liquid slag by a charcoal bed. A similar effect has been observed at the copper smelting sites of Batan Grande (Merkel and Shimada 1988; Epstein and Shimada 1983) and at the Late Bronze Age copper extraction in Trentino (Cierny et al. 1992).

The series of Early Bronze Age smelting sites from Faynan 8–13 and 15 continues eastwards into the ravine-like valley of Wadi Ghwair (see paragraph II), where four more slag heaps have been found, whose smelting furnaces were positioned on exposed hills or mountain edges. This choice of location for the smelting furnaces was again undoubtedly dictated by the prevailing wind conditions.

Fig. 5.18.
Faynan 9. Plan and section of the Early Bronze Age copper smelting furnace no. 24, 25. The repeatedly renewed rear wall led to the formation of a substantial accumulation of layers, which are very noticeable. From the top the furnace shows a semi-circular plan, which is bordered by a straight line towards the front (slopewards). The furnace front and the upper section of the oven are not preserved. The exact position of the clay rods could not be reconstructed (photogrammetry: Christoph Roden)

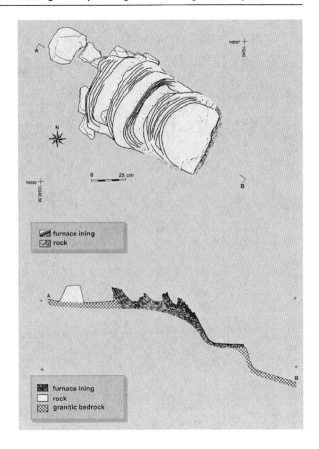

Faynan 14

- Geographical setting: 30°37'46.01" N/35°29'50.93" E

Close to the slag heap of Faynan 7 is another heavily alluviated and built over slagfield of ca. 100 × 15 m stretching between the Wadi Dana and the trail to the camp. The typology of a number of preserved slag cakes and some pottery finds allows a dating to the Iron Age II. The material is nearly completely crushed and reminiscent of the material in Khirbet en-Nahas (see there). The field is on both sides of a ringshaped wall, which comes down from a pebbled terrace in the vicinity (Fig. 5.4).

Faynan 16

- Geographical setting: 30°37'18.99" N/35°29'11.03" E

This heap with only a few tons of small-scale slag is situated roughly 1 km southwest of Faynan 1 at the outer reaches of a gravel terrace. Large amounts of clay rods already indicated the Early Bronze Age date of the heap, which was confirmed by a ^{14}C-date (2570–2330 BCE, Table 5.1). Pottery had not been found.

Faynan 17 (JD-28)

- Geographical setting: 30°37'15.85" N/35°29'31.34" E

About 1.5 km southwest of Faynan 16 and 1 km south of Tell Wadi Faynan in the direction of a chain of hills is an area of roughly 1 km², where numerous settlement remains and graves from different periods have been found. Some vertically set large rocks, clearly visible and eye-catching in the otherwise flat area, are characteristic of this spot. Some Early Bronze Age I pottery has been found here, in addition to widely scattered Nabatean pottery that must be presumably in connection with the fortress from the same time period, which is located on top of the hilly flanks. The earlier pottery is coarsely tempered and partly red slipped, and it can be compared with the material from Wadi Fidan 4, as well as some fragments of basalt vessels.

Clearly associated with the Early Bronze Age pottery, scattered nut-sized pieces of slag were found, which have a high copper content and sandstone inclusions that are not decayed.

Tell Wadi Faynan

- Geographical setting: 30°37'38" N/35°28'41" E

The ancient tell is situated ca. 2.5 km west of Khirbet Faynan at the southern bank of Wadi Faynan. It was discovered in the middle of the 1980s, when the water erosion cut into the archaeological deposits and thus showed a section through the flat settlement site, hardly recognizable from the surface. The tell sits in the middle of the Roman-Byzantine garden and field systems, which start at Khirbet Faynan and continue from there several kilometers to the west (see Fig. 3.8). The site stretches 120 m along the wadi (east-west-direction). The long section cut by the wadi showed stone foundations, ash-layers and large storage containers, and revealed all together 4 m of settlement deposits. The tell has been partly excavated during 1988 and 1990 by the Department of Antiquities, Amman, under the directorship of M. Najjar (Najjar et al. 1990).

The surface of the tell has a sparse scatter of Roman-Byzantine pottery, while here and there infrequent slag scatters can be found. This uppermost layer (stratum 1), created through the disturbance by recent agriculture, reaches only to a depth of 20 cm. Below it starts a phase dating to the Chalcolithic period (stratum 2). The third phase (stratum 3) dates to the Yarmukian and is thus contemporary with the Pottery Neolithic A. Three charcoal samples resulted in ^{14}C-dates from 5520–4910 BCE (Table 5.1) and therefore demonstrate a settlement in the second half of the 6th millennium BCE, for which for a long time a decrease in population had been assumed – at least for the larger settlements in Palestine (Weinstein 1984; Fritz 1985, but see Kerner 2001). Two other ^{14}C samples (4675–4165 BCE) date to the Chalcolithic period (Joffe and Dessel 1990). A small cultic/human figure, which had been found at the bottom of the section in some out of context soil, has to be dated to the Pottery-Neolithic period (Najjar et al. 1990). It closely resembles the so-called 'coffee-bean-eyed' statues of the Yarmukian.

The excavation revealed pieces of ore, slag and metal, exclusively found in stratum 1. Although fired pottery and several fireplaces were found in the Neolithic level, there was no proof for any metallurgical activity. The excavated ore had not undergone any heat treatment. This result substantiates the 7th millennium BCE slag finds from Çatal Hüyük (Neuninbger et al. 1964) and those from the 5th millennium BCE at Selevac (Glumac and Todd 1991) as still the earliest evidence of extractive metallurgy – at least for the moment. The assumption of copper metallurgy of the 6th millennium in Timna by Rothenberg and Merkel (1995) can certainly not be supported with any evidence in the finds from Faynan.

Wadi Faynan 100

- Geographic setting: 30°37'12" N/35°28'00" E

This site lies 500 m south of the main bed of the Wadi Faynan, and 1.5 km west of the Khirbet Faynan. It is located on a colluvial apron. The site was excavated by Karin Wright in 1997. According to Wright (1998), Wadi Faynan 100 is an extensive, shallow site in which the occupation layers were dated to the late fourth millennium BCE, or Early Bronze Age I. The extent of the site is much greater than any previously recorded late fourth millennium BCE site near a copper source.

In a clear EBA I context, numerous crucible fragments, ores, crushed slags and casting moulds (Fig. 5.19) were found.

Fig. 5.19.
Spoon-shaped casting mould from Wadi Faynan 100. The mould is made from coarse ceramic. The surface in the mould was probably lined with a fine layer of ash which is presently – after the heat impact during the casting of metal – vitrified

II. Find Spots in Wadi Ghwair

East of Khirbet Faynan runs a narrow gorge through the steep andesite horst block in which flows the Wadi Ghwair, which joins the Wadi Faynan at the ruins. The initial reason and starting point for the survey was to follow the Roman-Byzantine channel, which can be traced along the southern bank of the ravine for some 1.5 km. The following find spots have been found on the very steeply rising gravel terraces:

Wadi Ghwair 1 (JD-40)

- Geographical setting: 30°37'23" N/35°30'20" E

The first gravel terrace is located on the southern bank roughly 100 m above the narrowing of the actual wadi. The upper part of the terrace forms a slightly flattened plateau on which a prehistoric settlement had been built. It had been cut by water erosion caused by a creek on the southern side of the wadi. This section showed at this point stone walls up to 2 m in height. Along the slope, weighty mortars made from granite and different stone vessels made from 'white ware' (Kafafi 1986) as well as numerous flint tools such as borers, arrow-heads and microliths were found. The settlement has been excavated since 1993, first by the Department of Antiquities, Amman under the directorship of M. Najjar and now in cooperation with the University of Nevada under A. Simmons. It dates to the Pre-Pottery Neolithic B (8th millennium BCE, Table 5.1).

This excavation has produced important evidence for the understanding of the utilization of raw material as well as the development of pyrotechnology in the area of Faynan. It has been proven that copper ores were collected, but only for reasons of adornment as was common throughout the entire Levant at the time (Garfinkel 1987). The ores were not smelted, and no indications point to the use of native copper, as was the case in contemporary settlements in Anatolia such as, e.g., Ashiklihöyük and Çayönü Tepesi (Maddin et al. 1991; Esin 1996). But experience with pyrotechnology existed nonetheless, because it is likely that burnt lime-plaster was used in Wadi Ghwair 1 to plaster walls and floors. Afterwards, they were decorated with ochre. Pottery and small clay figurines were also fired in individual cases. The publication of the excavation is under preparation.

Wadi Ghwair 2

Directly next to find spot 1, an Early Bronze Age II/III smelting site is positioned on the slope leading to the east. The heap contains ca. 300 t of slag. The surface finds include the clay sticks so typical for this period and fragments of ceramic furnace walls. No indication of a settlement and no positions for any furnaces have been found.

Wadi Ghwair 3

100 m further from the last slag heap up the wadi is another smelting site situated at the southern bank of Wadi Ghwair on a raised gravel terrace. This slag heap also dates to the Early Bronze Age II/III and contains ca. 500 t of slag. Again no evidence of either a settlement or the position of the furnaces has been found.

Wadi Ghwair 4

On the northern bank, another 100 m up the wadi, the largest smelting site of this part of the wadi is located. It lies above a prominent, human-sized sandstone block, on which a Nabatean oil- or winepress was cut.

The heap holds ca. 800–1 000 t of slag, again including ceramic furnace walls and clay rods. The rear walls of ca. 10 smelting furnaces can be seen on the surface some 2–3 m above the highest point of the heap. The furnaces are of the same type as those excavated in Faynan 9 (see above). The type of slag, ovens and clay rods as well as the pottery indicate a dating to the Early Bronze Age II/III period.

III. The Mining District of Qalb Ratiye (JD-25, JD-GR)

- Geographical setting (pit 12): 30°38'42" N/35°28'41" E

This region is situated between the wadis Khalid and Ratiye and bounded in the north by a mountain range stretching along the Wadi Abiad. Towards the east, the area rises steeply up to the Ras Jebel Khalid. In this hilly landscape, Qalb Ratiye, copper ore coming from the vein fillings of the MBS (see Chap. 2) has been mined in 55 pits (Fig. 5.20). These mines are in a tectonically heavily affected region, which is permeated by three nearly orthogonally set fault systems. Contrary to other mining districts, here it was possible to quarry manganese-free copper ores (malachite with atacamite, cuprite as well as tile-ore with copper sulphides, chrysocolla). The

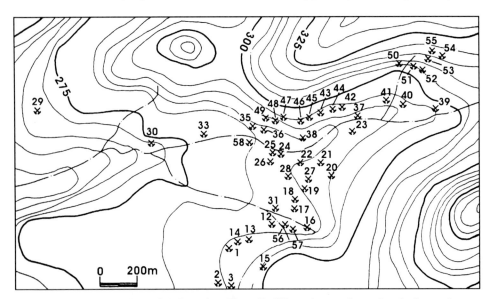

Fig. 5.20a. The mines of Qalb Ratiye. The vein fillings close to the surface in the sandstones northwest of Faynan were probably already mined during the Chalcolithic period. It was not before the Roman times that the ancient mines were reopened. This led to an intensive and systematic exploitation of the deposits, 55 mines being opened in a rather small area

Fig. 5.20b. The mines of Qalb Ratiye. 3D perspective view of the mines of Qalb Ratiye and Wadi Abiad generated by combining Ikonos satellite data and Aster DEM. Red sticks in the image show open mines and/or tailings (see also Fig. 5.4.a)

mines were sometimes opened very close to each other to follow the linearly arranged ore mineralizations, therefore forming rows of gravel pits.

The archaeological evidence allows us only to state the existence of Early Bronze Age and Roman mining activities in the sandstone, but the mines probably already began in the Chalcolithic period. There is no evidence of later Bronze Age and Iron Age mining at all. Coarsely tempered pottery and numerous stone hammers, which were manufactured in the Early Bronze Age settlement of Wadi Fidan 4 (see there), have been found close to the surface and in the waste dumps outside the mines (Haupt-mann et al. 1985). The peck marks from stone hammer mining are another supporting element for the dating of the mines to the 4th millennium BCE. Excavations in several mines showed that nearly all of them had been reworked in Roman times: pockmarks of hammers and wedges or picks illustrate the work and show that the initial openings were significantly enlarged (Fig. 5.21). The mineralizations have then been further explored by shafts and galleries, and the exploitation was carried out in a chamber-and-pillar system (Weisgerber 1996). The large amount of refilling with rock refuse suggests rather intensive clandestine mining. The region of Qalb Ratiye shows the highest density of mines that have been found in the entire area of Faynan.

In Qalb Ratiye, absolutely no traces of any smelting activities have been found. This leads to the conclusion that ore has neither in the 4th millennium nor in the Roman period been smelted in the direct vicinity of the deposit but was brought into the settlements or into the central copper smelter at Faynan 1. Smithing slag has been found, nonetheless, everywhere on the Roman waste dumps. These are the characteristic remains of mountain smithies, which were put up at the mine entrance in order to maintain the mining equipment that was easily worn out and manufacture more equipment when needed (Hauptmann et al. 1991). These kinds

Fig. 5.21.
Qalb Ratiye, mine 4. Reopening of an ancient mine. Remains are visible of, most likely, Chalcolithic mining behind the bedouin. In Roman times the pit was enlarged: steps were hewn into the rock for easier access, and chisel marks are clearly recognizable

of blacksmith activities were probably also quite common in other mining districts. They have been recently identified in medieval mining districts in Germany (Eckstein et al. 1994, Goldenberg 1993).

IV. The Mining District in Wadi Abiad (JD-26, JD-WA)

Wadi Abiad runs along a tectonic fault and is the east-west connection between the Wadi Khalid and Wadi Ratiye. Coming from the pass towards Wadi Khalid, where the mineralizations of the Dolomite-Limestone-Shale Unit (DLS) are exposed, the wadi cuts through the layers of the Colored-Sand-and-Claystones as well as the vein fillings of the Massive Brown Sandstones (MBS; Fig. 5.22). The mines in Wadi al-Abiad form another large mining district, which surrounds the mountain range of Ras Jebel Khalid, together with Qalb Ratiye and Wadi Khalid (see below).

The mines in the Wadi Abiad possess a key position in the understanding of prospection work and the exploitation of ores from single geological units in the Faynan district. They elucidate particularly clearly the differences between and development of mining techniques from the Iron Age to the Roman times (Heitkemper 1988): fourteen blackish waste dumps on the bottom of the valley show the shaft mining into the stratabound copper-manganese mineralizations of the DLS, which are present here below the surface at an unknown depth. It was possible to date all these waste dumps to the Iron Age II due to pottery finds.

Fig. 5.22. The mining district of Wadi Abiad. At the bottom of the valley, Iron Age mining can be observed, exploiting the ore mineralizations of the DLS (*dark arrows*), which were not exposed to the surface. The waste dumps to the left of the photo (*white arrows*) date back to the Roman period and are evidence of mining into the MBS (from Heitkemper 1988)

Roman mining activities, on the other hand, were concentrated along the northern and northeastern slopes of the Ras Jebel Khalid. Several waste dumps and heaps of reprocessed material prove large-scale exploitation from the vein fillings of the MBS. Galleries followed the mineralizations already mined in the Chalcolithic period, first from the surface and then in blind shafts. The ore winning was achieved by small room-and-pillar mining. In all, more then ten Roman mines have been documented.

V. The Mining District of Wadi Khalid (JD-3, JD-II)

A third large mining district is situated at the southeastern slope of the Ras Jebel Khalid range (550 m). It stretches from the upper course of the Wadi Khalid nearly to the mouth of the Wadi Ratiye. The ancient mining in this area concentrated exclusively on the copper-manganese-ore mineralizations in the DLS, which were exposed at several spots to the surface of the wadi. 56 mines have been counted here. Most of them have been refilled and can only be recognized by their waste dumps built up outside the (mostly filled) adits to the mine. The greatest density of ancient mines has been observed along the lower course of the wadi: The ore mineralizations were mined in 22 closely neighboring pits below the quaternary gravel terraces (Fig. 5.23). Six mines have been cut into sections due to modern prospection works carried out by the Jordanian Natural Resource Authority. They can be traced for up to 55 m into the rock. This offered the unique opportunity to study in detail the extent as well as the technique of ancient underground mining. The area in question has been taken as a model area and investigated with particular care. In addition, excavations were carried out in mine 42 in 1990 and 1993. Mine no. 7 was also excavated by Prof. Dr. G. Weisgerber and proved to have a lateral extent of ca. 30 m length and at least 5 m width. Parts of the mine are collapsed (Fig. 5.24).

The mining in Wadi Khalid can thus be described in detail: Following the seam-like ore mineralizations from the outcrops, which dip slightly towards the northwest, room and pillar constructions were dug. They had a height of 1–1.5 m and a lateral extent of ca. 30 × 55 m. The mines followed the ore deposit down to a depth of 15–20 m. It is quite surprising that the pits were carefully backfilled (Fig. 5.25). This practice resulted in comparatively small waste dumps being visible at the mines' adits, leading in the past to an underestimation of the real extent of the mining activities, which were thus incorrectly estimated by all earlier researchers. Altogether, the density of mining remains visible at the surface, and the extent to which the mines are accessible by excavation suggests an almost quantitative exploitation of the DLS mineralization under the gravel terrace. There are no parallels of similar EBA mining activities known in the entire Near and Middle East.

Several of the mines could be dated through pottery finds on the waste dumps as well as from the backfilling in the mines themselves. The dates show that the deposit exploitation started generally in the Early Bronze Age. This makes sense of the large number of contemporary smelting sites, which have been discovered in Wadi Khalid. But pottery from the Middle Bronze Age I has also been found

5.2 · Site Catalogue and Archaeometallurgical Finds

Fig. 5.23. The mining district of Wadi Khalid. In particular, the ore mineralizations in the Dolomite-Limestone-Shale Unit along the lower course of the Wadi Khalid were intensively mined. Each single mine, which is indicated by a pink stick, can only be recognized by its waste dumps in the field (*dark spots*). The distribution of the mines is characteristic: the Early Bronze Age examples sit in a row along the outcrop of the ore formation in the wadi. The heaps at the foot of the cliff (left) date to the Iron Age (for further explanations, see text) (see also Fig. 5.4a)

118 Chapter 5 · Field Evidence and Dating of Early Mining and Smelting in the Faynan District

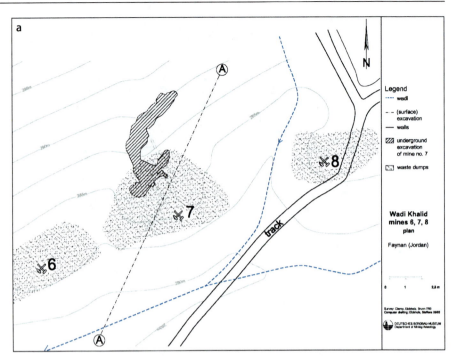

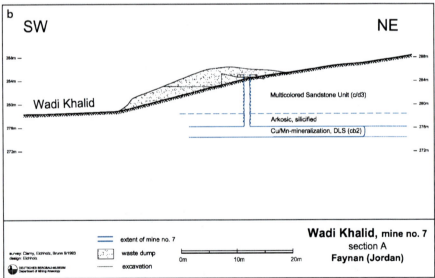

Fig. 5.24a–c. Wadi Khalid, mine 7. Plan of the mine. The ore seam in the DLS was reached by a shaft and than exploited for a length of ca. 30 m. Part of the mine is collapsed and backfilled. We assume that the width of ca. 5 m could have originally been much larger. Note the downfall near the shaft which was reinforced in Roman times. The mine was excavated by Prof. Dr. G. Weisgerber

5.2 · Site Catalogue and Archaeometallurgical Finds 119

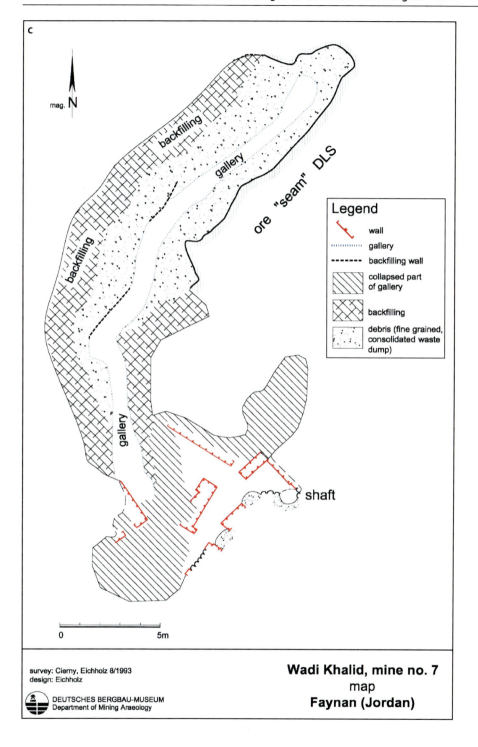

Fig. 5.25. Wadi Khalid, mine 42. This Early Bronze Age mine has been cut by a modern prospection tunnel of the Natural Resource Authority. It is difficult to localize the original mine as it has been almost systematically backfilled. The original height of the now backfilled mining chamber can be recognized with the help of the scale. The bottom and roof of the mine are marked by the *hatched line*. The lateral extension of the pit, constructed as a room and pillar mine, is roughly 30 × 50 m

(Hauptmann et al. 1985), particularly on the heaps of mines 11 and 18. A radiocarbon date (1510–1415 BCE), coming from wood spills out of the backfill from mine 42 (Table 5.1), confirms mining activities until the Late Bronze Age. The wood spills were used to light the mines. The botanical analysis (Engel 1992) showed that juniper wood (Juniperus phoenicea) had been purposely chosen for this function. The Bronze Age tailings can be quite clearly distinguished from those of the Iron Age. While the former are nearly without exception positioned close to the outcrops of the DLS, the Iron Age heaps are normally located at the foot or even at the slope of the steep mountain ranges, thus illustrating an exploitation of the ore mineralizations using much deeper mining shafts.

These mines demonstrate, in this particular part of the deposits, the long duration and continued re-initiation of mining activities, which slowly penetrated deeper and deeper into the mountains. In the Iron Age, they were thus forced to exploit those parts of the DLS, where previously the ore had not been completely removed, via deep shafts in the steeply rising valley sides. This phenomenon can not only be observed in Wadi Khalid, but is also noticeable in the entire Faynan area, e.g., in Khirbet en-Nahas, south of Khirbet el-Jariye and above the Wadi Ghuwebe. In nearly all observed cases, double shafts had been constructed, one being sunk for climbing and the other for raising the ore. On the upper course of the Wadi Khalid, a venti-

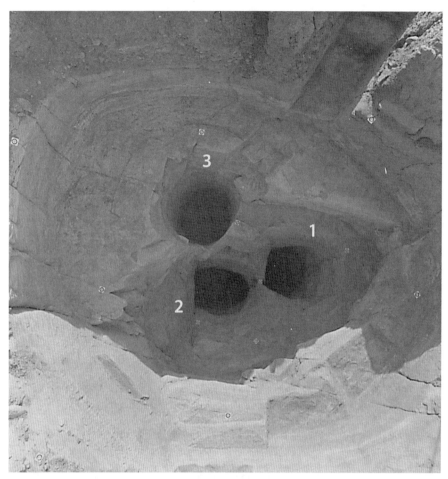

Fig. 5.26. Wadi Khalid, mine 1. The construction of the 'triple shaft' is most likely a unique technical monument in the history of mining at Faynan. Two of the shafts (1, 2) have been sunk in the Iron Age, while shaft 3 as well as the upper platform were built in Roman times and thus form a good example of the prospection activities and the re-opening of ancient mines in the Roman period. Photo taken ca. 6 m vertically above the upper edge of the installation with surveying points for photogrammetrical measurement (from Hauptmann et al. 1985)

lation shaft with a depth of 50–60 m was found (Weisgerber 1989). Iron Age mining shafts have also been observed on the western side of the Wadi Arabah in Timna (Conrad et al. 1980).

Mining in Wadi Khalid was taken up again in Roman times. This is proven by a Roman prospection shaft, which was sunk parallel to the two Iron Age shafts of mine 1 (Fig. 5.26, 'triple shaft', see Hauptmann et al. 1985; Hauptmann et al. 1991) and has been dated by pottery finds as well as coins (Vespasianic minting, 69–79 CE, KIND et al. 2005). Roman masonry around the shaft has also been observed inside the Bronze Age mine 7.

VI. Sites in the Lower Wadi Dana (JD-13, JD-III)

The 'Northwest cemetery' is situated on the lower gravel terrace to the left-hand side of the track leading from Khirbet Faynan into Wadi Khalid directly after the crossing of Wadi Dana (Fig. 5.4). Slag scattered over a considerable area is observable here but is concentrated in six different locations. The distance to the Iron Age slag heap of Faynan 7 is roughly 150–200 m. Thin plate slag found here has not been detected on any other find spot in Faynan. Similarly conspicuous are the tuyères which are also found here (Fig. 5.27). With their interior width of ca. 1–1.5 cm, they are much smaller than their Iron Age counterparts. The impressions of reed on the tuyères' exterior are also very noticeable. They are identical to the type of 'small tuyères' from Timna, site 30, stratum 2 (Rothenberg 1990, 41, 74), dated back to the Late Bronze Age. The tuyères seem to be contemporaneous there with the mining activities, demonstrated for mine no. 42 in the LBA (see above).

Fourteen mines can be recognized from their waste dumps along the southeastern slope of the mountain ridge between Wadi Khalid and Wadi Dana at a height of ca. 400 m above sea level. Here again the ore mineralization in the DLS was exploited. Pottery finds allow a dating of the mines back to the Early Bronze and Iron Age. It was not possible to study this area in detail.

An Early Bronze Age smelting site with several dozen pieces of slag is located roughly 3 km north of the camp at the pass towards Wadi Khalid. An interesting find from this site consists of a grille fragment made from small clay-sticks (Fig. 5.28).

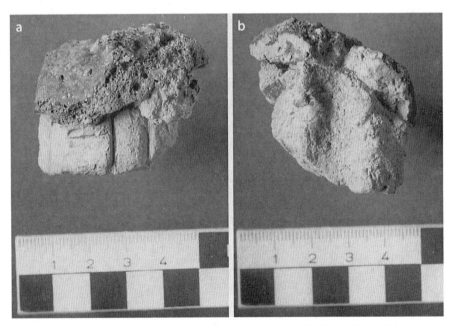

Fig. 5.27. Wadi Dana, slag heap in the 'Northwest cemetery'. Tuyères with reed impressions on the outside (*A*) as well as the interior view (*B*). The comparison with compatible finds from Timna (Rothenberg 1990) points to a dating back to the Late Bronze Age. This would be in line with the mining activities in mine no. 42

Fig. 5.28.
Pass between Wadi Dana and Wadi Khalid. Fragment of a grille made from clay rods. The piece is slagged on one side and was vertically oriented. It must have been part of a reinforcement in the smelting furnace wall, although the wall itself is weathered away

It is slagged only on one side and therefore obviously part of an Early Bronze Age wind-powered smelting furnace. This grille fragment is, as well as the numerous finds of clay rods, part of an installation in the furnace wall, although most of the furnace wall itself has been weathered away.

VII. Ras en-Naqab (JD-5)

- Geographical setting: 30°39'7" N/35°27'48" E

The smelting site is located 5 km northwest of Faynan in an exposed position on a cliff towards Wadi el-Ghuwebe, directly next to the former track from Faynan to Khirbet en-Nahas. The slag heap could be dated to 2570–2400 BCE by the ^{14}C-date of a charcoal sample (Table 5.1). The TL-dating provided a minimum age of 1 740 years (Table 5.2). There was no pottery found. The shape and chemical composition of the slag as well as the typical find of clay rods is identical to Faynan 8–15 and confirms the placing of this smelting site in the Early Bronze Age II/III.

Already visible from far away is a large slag cone, piled up by a bulldozer, of some 100 t. It can be dated back to the Iron Age II through the pottery finds and the exterior typology of the slag (see Faynan 5). This material was most likely produced via reprocessing and remelting of the Early Bronze Age slag in order to extract the remaining metal inclusions (see Sect. 7.3). Rectangular shaped houses can be observed only a few meters south of the heap.

Fig. 5.29. Ras en-Naqab. Early Bronze Age slag heap at the edge of the slope towards a tributary of Wadi Ghuwebe. The stone settings (*arrows*) show the positions of the smelting furnaces, which are directed towards the wind (see Fig. 5.16). The slag crushed into small pieces is characteristic of all heaps of this time period

The Early Bronze Age slag heap is directly next to and on some rocky terraces of the steep acclivity. In all areas of the slag heap, the slag cover is only up to 10 cm deep. Amongst the uppermost heap are eleven stone settings of dolomite. They are the remains of smelting furnaces, which were partially lined with clay (Fig. 5.29). Clear signs of firing can be seen here. Large parts of the slag heap have been tossed down the slope, and having been further washed down, they are now over 100 m further away. The entire amount of slag has been estimated at roughly 100 t. Southeast of the smelting site systematic reprocessing of the slag, in order to gain the remaining metal inclusions, has taken place on a number of small plateaus. These activities led to hundreds of cup marks in the sandstone (Fig. 5.30). The slag waste, which would have been produced during this reprocessing, was probably remelted in later times (see above).

The slag is typically crushed into nut-sized pieces and, similarly to the slag from Faynan 8–15, it has solidified into a blackish, quite compact glassy material that has a reddish sheen in its thinnest splinters. Head-sized lumps have been found in the rubble along the slope; they consist of vertically tapped droplets of the liquid slag (see Fig. 5.14), which mostly solidified in the crystalline state. They are rich in inclusions of copper prills that can only be retrieved with difficulty, as the slag is extremely tough. Other find material includes fragments of furnaces, which consist of several layers of sandstone slabs with only a small quantity of ceramic material in between them (Fig. 5.31). These sandstone fragments appear to be brightly shining green, sometimes reddish, bluish or black glazes on the surface. They were caused

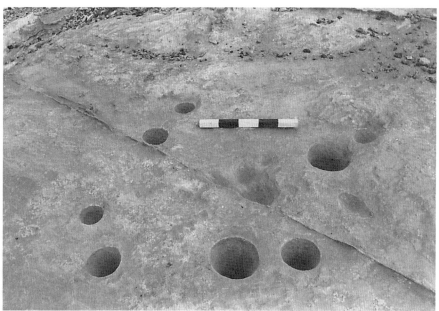

▲
Fig. 5.30.
Ras en-Naqab. Slag has been reprocessed in order to gain the small metal prills; this was carried out on a plateau directly next to the Early Bronze Age slag heaps. It led to hundreds of cup-shaped depressions in the rock, which are now filled with slag rubble. It is not possible to date these activities, but an Iron Age slag heap close by might be a hint of the dating

Fig. 5.31. ▶
Ras en-Naqab. The building material for the smelting furnaces here are sandstone slabs, which were, possibly intentionally, only loosely layered and lined with clay. They are slagged heavily on the furnace interior (**b**)

by the reaction of copper, iron and manganese, being part of the charged ore, with Na- and Ka-containing sandstone (Hauptmann and Klein 1999). They indicate that a large part of the charge was made up of very finely grained material. Glazes like this are an important indication of a possible origin of fayence technology from copper metallurgy as it is known from Egypt since the 4th millennium BCE (Tite et al. 1993; Schlick-Nolte 1999). Clay rods have occasionally been found. The ore samples found are in the majority of the mineralizations in the DLS and point therefore to deliveries from the mines in the wadis Abiad, Khalid or Dana.

VIII. El-Furn (Ngeib Asiemer, JD-6)

- Geographical setting: 30°40'31" N/35°26'48" E

The site of El-Furn contains the remains of a small hamlet for miners and smelters, which had grown around the smelting site from which the name is taken. The site is roughly 1 km southeast of Khirbet en-Nahas. The pottery found there has already been described (Hauptmann et al. 1985) and dates this site, along with six coins, to the Ayyubid and Mameluk period (13th century CE that is between 1206–1236 CE, Kind et al. 2005). The remains of ca. fifteen buildings and other architectural fragments have been found in El-Furn and have already been partially described by Glueck (1935). Forty meters south of the largest house, which is located next to the slag heap, remains of a mosque were discovered. In a valley to the north, a cemetery is located. The settlement was systematically surveyed in the year 2002 as part of the Wadi Ghuwebe field work by T. E. Levy. These settlement remains with their mining and smelting context are reminiscent of similar patterned settlement systems described by Weisgerber (1978) and Costa (1978) from Early Islamic smelting sites in Oman.

The eye-catching smelting site with ca. 1 000 t of slag is located at the very western edge of the settlement, directly in front of the gates of a 17 m long, rectangular house (Fig. 5.32). This is built from naturally rectangular broken dolomite stone with a double-faced wall and has two rooms, each of which is in turn also divided into two compartments. An addition of 5 × 5 m was built to the north. Outside the two gates the slag is piled up in two heaps, which contain primarily slag chunks up to 6–8 kg but also tuyères and furnace fragments. Directly towards the wall of the house the slag heaps are covered with three waste dumps, which prove the mining of ores from the DLS. In the rear part of the building, several kilograms of nut-sized lumps of copper-iron-alloys have been discovered, which might indicate further stages in the metal processing. The high content of iron in the copper is nevertheless confusing, and it is problematic to conclude that copper production is a clear interpretation of this finding. A number of tailing dumps, which might be taken as an indication of mining, are present in a small valley west of the building towards Khirbet en-Nahas (see there). To the south of this house are waste dumps pointing to a beneficiation of ore.

The building is obviously linked to copper (or iron?) production. It can also not be excluded that the building itself housed a shaft and a windlass. Arguments for such a hypothesis can be found in the waste dumps directly outside the two gates and

Fig. 5.32. El-Furn. The buildings from the Ayyubid-Mameluk 'industrial complex' are some of the most notable architectural remains in Faynan. The slag heaps directly in front of the houses represent a small echo of earlier metal production, which has been evident in the entire Arabah since Early Islamic times

the ore-bearing mineralizations, which can be reached at a depth of only 5–7 m below the building. But little is known about the mining technologies and traditions in the Near East during the Ayyubid and Mameluk periods. But it should be pointed out that 'shaft-houses' are a frequent find from Germany's Late Medieval period and are described impressively in words and pictures by Agricola (1556).

The constructions in El-Furn can be counted among the most remarkable in the entire Faynan district. Other activities from the same time period have been found at Faynan 1 and below the Dana village. There a few 1.5 m high shaft furnaces have been discovered, which are the same kind used in Oman during the Early Islamic period (Weisgerber 1981, 1988). An argument for iron production could be substantiated by the block of iron found in Faynan 1 (see there).

IX. Khirbet en-Nahas (JD-2)

- Geographical setting: 30°40'50" N/35°26'10" E

The smelting site of Khirbet en-Nahas ('ruins of copper') is the second largest smelting center in the southern Levant, next to Faynan. It was already described several times in the first half of the last century by several travelers (Musil 1907; Frank 1934; Glueck 1935). In a depression, surrounded by hills and mountains in the east, west and south, directly south of Wadi Ghuwebe are dozens of slag heaps with a tonnage of some 50 000–60 000 t (Fig. 5.33). They have accumulated in huge heaps around the entire smelting area and particularly impressively in the southeastern corner of the site.

Fig. 5.33a. View of Khirbet en-Nahas. The 'ruins of copper', as the name can be translated, are ca. 6 km north of Faynan. The quantity of Iron Age slag has been estimated at ca. 50 000–60 000 t. This indicates copper production on an industrial scale. The smelting site of Khirbet en-Nahas along with that of Faynan, is one of the largest in the Levant. The square fortification complex is clearly visible at the left-hand corner. The houses are surrounded by slag heaps. Along the slopes are the heaps of reprocessed slag (*arrows*). Wadi Ghuwebe shown in Fig. 33b is here outside the photo to the left

The copper mines supplying the smelting sites were in the immediate surroundings of Khirbet en-Nahas in a small valley southwest of the site and in the southeastern valley towards El-Furn. A number of enormous waste dumps have also been discovered here.

At the smelting site itself is an Iron Age settlement with over 30 houses as well as a square fortification with two gate towers. The walls are still standing up to 1 m height. The site was studied intensively by Bachmann and Hauptmann (1984), Hauptmann et al. (1985), and was surveyed subsequently by Knauf and Lenzen (1987) and MacDonald (1992). Pottery from the Iron Age I has been found, but the site dates mainly to the Iron Age II, or in other words, the Edomite period. Nine charcoal samples were taken from the sections cut through some of the slag heaps for physical dating (Table 5.1). Charcoal samples from the lower part of some of the slag heaps produced a date equivalent to the Iron Age I period. The earliest dates cluster in the 9th century BCE. Fritz (1996) dated a house from his excavations in 1990 at the smelting site to the 9th century BCE but pointed out that a large part of the pottery dates to the period of Assyrian supremacy in the 7th and 6th century BCE.

T. Levy conducted excavations at Khirbet en-Nahas in 2002 and 2006. He counted over 100 building complexes (Levy et al. 2004) and reported on a series of eight stratified high precision radiocarbon datings by Accelerator Mass Spectrometry, which range from the 12th to the 9th century BCE. The gateway of the fortress, directed to the wide plains of the Wadi Arabah, was constructed at the beginning of the 10th century BCE. The presence of Midianite pottery, a polychrome decorated pottery usually dated to the 14th to the middle of the 12th century BCE documents a possible Late Bronze Age occupation of the site. This is supported by finds of Egyptian scarabs which are dated from the New Kingdom. Midianite pottery obviously continues to appear in 10th century BCE contexts at Khirbet en-Nahas (Levy et al. 2004). According to petrographic and chemical analyses, part of the Midianite pottery was

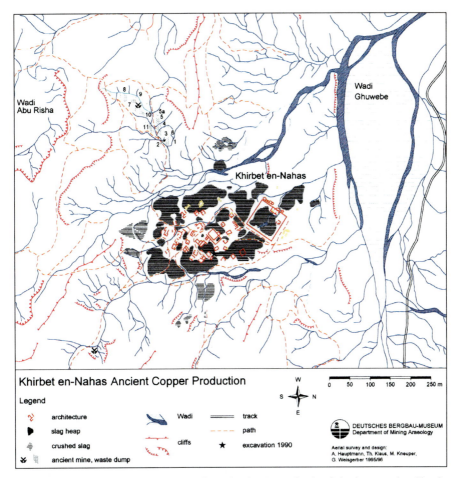

Fig. 5.33b. Khirbet en-Nahas. Cut out of map in Fig. 5.1 on basis of air photographs. Clearly visible are numerous houses and the fortification surrounded by slag heaps. Ancient mines which at least partly provided the ore for the smelters were discovered west and southeast of Khirbet en-Nahas (see Fig. 5.33a)

manufactured at Khirbet en-Nahas itself. Levy et al. (2004) did not find any evidence of metal production for the 7th and 6th century BCE.

In Khirbet en-Nahas, the same kind of recognizable tap slag is present as is found on the Iron Age heaps of Faynan 5. As the ceramic furnace fragments and the tuyères are also comparable to those from Faynan 5, they will not be described in detail here. A large part of the heaps, however, consists of broken and finely crushed slag. This material forms vast heaps (several 1 000 t) along the slopes around the smelting site (Fig. 5.34a,b). The solidly baked granular slag is piled up to 1.5 m height; the individual slag grain size can be up to 0.5 cm. Flow patterns similar to the 'flute casts' known in sediment-petrology indicate a possible impact of water during the formation of these heaps. These finds imply that the mechanical reprocessing of slag in order to extract the metal prills was a standard procedure used in copper extraction even in the Iron Age.

Fig. 5.34a. Khirbet en-Nahas. Several thousand tons of crushed slag has been found along the slopes around the smelting site. Slag that still contained copper was mechanically reprocessed here and the metal prills were washed out by water

Fig. 5.34b.
Khirbet en-Nahas. Large amounts of hammer stones and anvil stones were found for crushing slags (photo: Gerd Weisgerber, Deutsches Bergbau-Museum, Bochum)

X. The mining district of Umm ez-Zuhur and Madsus (JD-41)

- Geographical setting: 30°40.3' N/35°24.4' E and 30°40'18.38" N/35°24'37.90" E

The district of Umm ez-Zuhur is located 1.5 km east of Khirbet Hamra Ifdan (Wadi Fidan 3) and therefore is the closest ore outcrop to the smelting sites in Wadi Fidan. It is in an area south of Wadi Ghuwebe, which is characterized by a cuesta landscape. It is covered with extensive quaternary gravel terraces out of which single monadnocks rise steeply towards the northeast. The whole area consists of a system with terraces on three different levels. In Umm ez-Zuhur, reddish and grey-violet sandstones of the Lower Cambrian are exposed, and the mineralizations of the DLS as well as the rock strata such as the Cretaceous limestones are stratigraphically situated above it. They are all heavily decomposed and salt bearing. The Quaternary gravel terraces are cemented by a calcareous matrix and break off above the decomposed underlying bedrock in enormously large boulders.

The mining district can be easily recognized by 24 tailings from the DLS. The dark rock debris contrasts with the light gravel cover around them. Some of the ancient galleries have been cut by erosion and are thus exposed in the wadis, very similar to those observed in Timna (Hauptmann and Horowitz 1980). But the original technique of exploring the ore mineralizations was shaft mining from the surface. The waste dumps were dated back to the Iron Age II period on the basis of archaeological evidence. Thus they conform completely with the development of mining technology as it has been observed in other mining districts.

A few hundred meters further east, the morphology of the landscape becomes more diverse. The monadnocks of the DLS, being quite resistant against weathering, are the predominant feature in the landscape. Here, one can find widespread outcrops of ore. Without any effort, nut-sized pieces of pure malachite and copper silicates can still be collected today, which demonstrates clearly that the remains of mining activities found in the vicinity can be interpreted as a kind of placer mining close to the surface. In the small valley Madsus, the name used by the local Bedouins, 20–30 refilled shafts ('Tellerpingen') have been observed as well as the remains of what is probably an EBA I house (Prof. Tom Levy, pers. communication November 2006). It is difficult to date the mining activity at Madsus. The existence of grooved hammerstones compatible to those from Wadi Fidan 4 (see there) point towards a Chalcolithic or Early Bronze Age I date for the mining activities. This seems to be a credible interpretation in the light of the general development of technology as it has been observed in Faynan. Albeit Craddock's (1995) interpretation of a stone construction close to the aforementioned house as a Chalcolithic smelting furnace was probably a bit hasty.

Not far away from Umm ez-Zuhur, a small smelting site from the Early Bronze Age III is situated. It will be dealt with under the Wadi Fidan heading (no. 13).

XI. Khirbet el-Jariye (JD-11)

- Geographical setting: 30°42'21.62" N/35°27'06.92" E

This smelting site is located in the Wadi Jariye and is most likely identical to the site mentioned by Kind (1965) as being situated in a Wadi Sel es-Sasar. It is easiest to reach the site from Wadi Arabah along the Wadi el-Mahash. In Khirbet el-Jariye, 15 000–20 000 t of slag is distributed over a number of slag heaps. Layers a few meters deep with some 100 t of crushed slag can be found particularly at the southern end of the site. In the center of the heaps, the ground plans of ca. two dozen houses are recognizable. The site has been dated archaeologically, and the available ^{14}C-dates also put it in the Iron Age I and II. The small tuyères, which can be found on the slag heaps (see those described under Faynan 5, Fig. 5.10), would nevertheless allow the beginnings of the smelting activities to have been already taking place in the Late Bronze Age.

East of the smelting site, the Wadi Jariye is bordered by a number of steeply rising sediment packets of Cambrian to Tertiary date. At the edge of this steep rise, ca. 50 mines can be identified by their widely visible waste dumps; most likely all of them were meant to exploit the mineralizations in the DLS.

From the smelting site, a track leads to the south into a moderately rising wadi. At the highest point, after ca. 500 m, one has an extraordinary view over the Wadi Ghuwebe. All along the rising track a number of copper tailings and remains of ancient mines (directed to the DLS) are visible at the bottom of and along the western flank of the wadi. The wadi was surveyed for the first time in 2002 together with Prof. T.E. Levy. Seven adits of modern prospection galleries from the Natural Resources Authority, Amman, can be easily distinguished. Similar to the situation in Wadi Khalid, these are constructed at the top of the Bronze Age tailings and thus cut into the ancient mines. Altogether. twelve Early Bronze Age, Iron Age and Roman-Byzantine mines were dated by pottery. Stone tools were collected on top of tailings nos. 544 and 548. Typologically, they are identical to those from Wadi Fidan 4 and date the mining activities into the middle of the 4th millennium BCE or even earlier.

Westwards, the landscape opens up towards the Arabah and takes on the character of a cuesta landscape, where the formations of the DLS, being resistant against weathering, form a steep descent on one side, and plateaus, gently dipping towards southeast and south are on the other side. The uppermost layer, an ore bearing claystone, has eroded away, thus leaving an enrichment zone of up to 10 cm size with very hard ore chunks of malachite or copper silicate at the surface. These plateaus, covering an area of some km^2, are strewn with Neolithic and Chalcolithic flint artifacts. A great deal of weathered ore can also be found amongst the gravel in various wadis. This ore would certainly have originally been collected everywhere.

XII. Wadi el-Ghuwebe (JD-27)

- Geographical setting: 30°41'34.90" N/35°28'18.89" E

This wadi is one of the few west-east-connections from the Wadi Arabah up to the Jordanian plateau. However, in comparison to the Wadi Dana, which is only
walkable with difficulty, the wadi seems to have formed a strategically important connection between Khirbet en-Nahas and the core area of the Edomite kingdom, which is proven by two Iron Age fortresses in the middle course of the wadi. From these forts and the fortifications in Khirbet en-Nahas, it was possible to control the entire wadi and all the mines situated in it. The position of the mines can be observed, when coming from en-Nahas, first on the banks of the wadi and later far above the northern bank by a row of waste dumps and also by a number of typical double shafts (Fig. 5.35). The positions of the fortresses between Khirbet el-Jariye and Ghuwebe are marked in Fig. 5.1 with two points.

A few finds of tap slag and crushed slag of possible Iron Age origin have been found in the oasis of Khirbet el-Ghuwebe.

Fig. 5.35. Double shaft constructions are typical features of Iron Age mining. Note the three holes cut in the stone for using a wooden winch with a central beam as constructed for hauling with a rope. Wadi Ghuwebe (after Weisgerber 2006)

XIII. Sites in Wadi Fidan

Close to the settlement of Ain el-Fidan, the wide valley bottom of the Wadi Fidan narrows down to ca. 30–50 m. It meanders in a ravine-like cut over 4 km through the Precambrian igneous rocks of the Jebel Hamrat Fidan, first in a northwesterly direction, later towards the west to the Wadi Arabah. The morphology of the land south and north of the wadi is characterized by huge Quaternary banks of conglomerates, which irregularly overlie the crystalline basement with a thickness sometimes up to 10 m. These banks are very resistant against erosion and form either recurring 'islands' in the wadi or break away in enormous blocks above the crumbly crystalline rock. Settlement remains can often be found on these terraces of conglomerates.

Wadi Fidan is of strategic importance. Not only can a significant water source be found here, the wadi cut is also actually the most important route to the Wadi Arabah and thus to the centers of the different cultures in the west. The potential settlement area of Faynan is otherwise barred from the Arabah because of the north-south running Jebel Hamrat Fidan. Wadi Fidan had already been surveyed by Frank (1934) and one year later by Glueck (1935). But both described only two archaeological sites. The first more detailed surveys were published by Raikes (1980) and MacDonald (1992). They identified numerous sites from the Neolithic to the Early Bronze Age I and thus showed the relatively early settling of the wadi provided the impulse for the renewed fieldwork of the University of Sheffield and the University of California, San Diego. Levy and Adams (1999) counted no less than 133 sites in the area of Wadi Fidan. Their team focused on settlement sites, etc., rather than focusing on metallurgical and mining sites.

Wadi Fidan 1 (JD-7)

Coming 700 m from Al-Qurayqirah towards the spring at Ain el-Fidan, a flat slag heap with crushed slag is situated directly left of the track. The locale had already been described by Kind (1965, 'Platz Nr. 14'). Some stone tools used for crushing the slag have been found in the slag heap. The age of the heap is unclear.

Wadi Fidan 2 and 10

This is a Nabatean tower at the corner of the northern terrace above the wadi. Raikes (1980) 'site G' describes the same site. Not far away from the tower a slag heap is located (Wadi Fidan 10). Hauptmann et al. (1985) found a Middle Bronze Age II pottery sherd here.

Wadi Fidan 3 (Khirbet Hamra Ifdan, JD-9)

- Geographic setting: 30°39'41" N/35°23'33" E

This settlement and smelting site, is located on a naturally defended island-like "inselberg" (Fig. 5.36), and described as 'site F' by Raikes (1980). It consists of a remnant of a solidified Pleistocene gravel terrace that rises up ~25 m above the wadi. It is ca. 1 km northwest of the oasis of Ain el-Fidan at the western edge of the wadi and catches one's eye from far away with its distinctive slag heaps. Several house ground plans can be recognized on the plateau and have been systematically documented for the first time by Adams (1992).

The majority of the slag dates back to the Iron Age II, easily recognizable by its typical plate-like shape and its flow structure. Some tons of Early Bronze Age slag – in small pieces – can be found on the northern part of the terrace. In all probability,

Fig. 5.36. Khirbet Hamra Ifdan. View of the smelting site from the south. On the plateau slag heaps are recognizable, which date from the Early Bronze Age II/III and the Iron Age

Fig. 5.37.
Khirbet Hamra Ifdan, stratum III (Early Bronze Age III). Cache of 16 T-shaped bar ingots of copper. This type of ingot was a common shape in which metal was traded during the Early Bronze Age III/IV in the Southern Levant (photo: O.Teßler, Staatliche Museen Preußischer Kulturbesitz Berlin)

Fig. 5.38.
Khirbet Hamra Ifdan, stratum III (Early Bronze Age III). Ceramic casting moulds for making different sized ingots. These moulds are typical forms of T-shaped bar ingots as shown in Fig. 5.37 (photo: O.Teßler, Staatliche Museen Preußischer Kulturbesitz Berlin)

the ore came from the mining district of Umm ez-Zuhur (see above), which is located roughly 2 km northeast of Hamra Ifdan.

The first excavations were carried out by Russell Adams (Adams 1992). Subsequently, Thomas Levy (Levy et al. 2002) conducted systematic excavations. Three main occupational phases have been identified. Radiocarbon dates, coupled with ceramic and stratigraphic data provided a solid chronological framework for evaluating changes in metal production during the last phases of the EBA. Stratum I represents later occupations from the Islamic, Byzantine and Iron Age periods. Stratum II dates to the EB IV occupation and stratum III to the Early Bronze III period when the site was occupied most extensively. There are indications of an Early Bronze II occupation in stratum IV; however, the excavation sample size is too small to make meaningful observations. These excavations have revealed the largest Early Bronze Age III metal workshop in the Middle East. The excavations yielded almost 3800 finds related to ancient copper smelting and processing mainly in stratum III (ca. 2700–2200 BCE). This unique assemblage of archaeometallurgical remains includes smelting and melting crucible fragments, prills and lumps of copper, hundreds of kilograms of (crushed) slag, ores, copper tools (e.g., axes, chisels, pins), a cache of copper ingots (Fig. 5.37), a few smelting furnace remains, and casting moulds for tools (Fig. 5.37). Of special interest is an extensive collection of ceramic casting moulds for 12 to 15 cm long, T- or crescent-shaped bar ingots (Fig. 5.38).

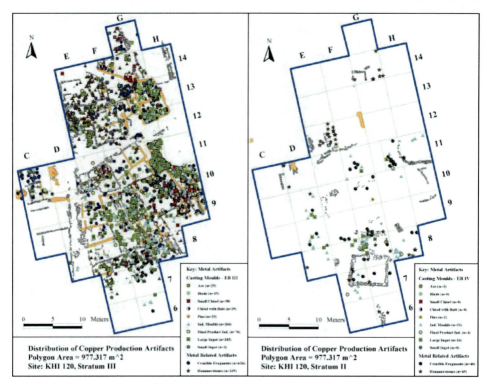

Fig. 5.39. Khirbet Hamra Ifdan, stratum III (Early Bronze Age III). The enormous number and distribution of archaeometallurgical finds at Khirbet Hamra Ifdan clearly demonstrates that the site was probably the largest factory for copper smelting and processing of raw copper in the Near and Middle East

Similar moulds were found (that were not as well preserved) at Barqa el-Hetiye. The making of these ingots will be discussed in more detail in Chap. 7 and their trade to the northwest in Chap. 8.

As can be seen in Fig. 5.39, the metal processing activities were concentrated in some 80 rooms, courtyards and other spaces within the excavated area.

Wadi Fidan 4 (JD-8)

- Geographic setting: 30°40'15" N/35°28'53" E

One of the most marked features in the landscape at the southern edge of Wadi Fidan, 1.7 km down the wadi from Hamra Ifdan, is the table mountain or butte already described as 'site E' by Raikes (1980). It is the penultimate rise of ground before the wadi reaches the valley of the Wadi Arabah. This hill has steeply descending flanks similar to those of Khirbet Hamra Ifdan and sits therefore quite isolated in the wadi bed, connected to the mountains only by a saddle on its southern side. The plateau is situated some 10–15 m above the wadi bottom and has a length of 200 m and a width of 50 m (Fig. 5.40).

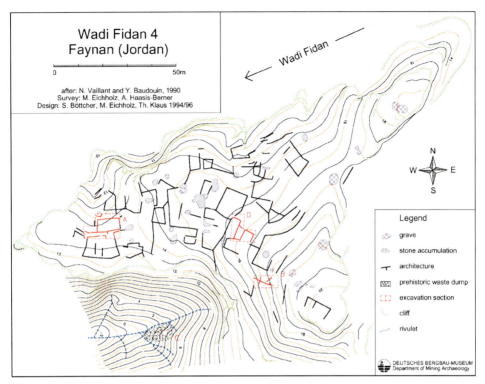

Fig. 5.40. Wadi Fidan 4, plan. The settlement is located on the plateau of a small butte in the lower course of the Wadi Fidan. It dates back to the middle of the 4th millennium BCE (Early Bronze Age I). The excavation areas of 1993 are marked

Raikes (1980) already described the numerous house structures on the plateau as belonging to a Chalcolithic settlement. His chronological assignment was based on the flint tools characteristic of this period as well as the very coarse pottery, which he compared to material from Tell Maqass near Aqaba (see also Kerner 2002, Khalil 1988, 1992, 1995). The site has been further studied in the framework of the project presented here (Hauptmann et al. 1985). An enormous amount of (semi-)finished stone hammers were collected from the surface, which proved the substantial production of mining tools for 'Chalcolithic' mining in the region. The type of these stone hammers varied from drilled, notched and subordinated grooved hammerstones and picks (Fig. 5.41). These tools were also identified in quantities on many tailings in front of copper mines in Qalb Ratiye, Wadi Abiad, Wadi al-Jariye, and Madsus. The grooved hammerstone does not seem to be the most typical working tool at Faynan. It is more commonly distributed in other mining areas (Weisgerber 2006). Other finds included stone vessels typical for this period as well as large quantities of hammer and anvil stones.

After the site had been surveyed in the year 1990 by members of the Institut Français d'Archéologie du Proche Orient (IFAPO), Amman, it was partly excavated in 1993 by the Deutsche Bergbau Museum (Adams and Genz 1995). The excavation

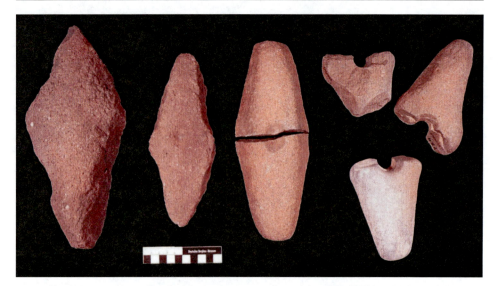

Fig. 5.41. A variety of typical hammerstones collected from Wadi Fidan 4. Most of them are semifinished or broken. Such stone tools were found in quantities at mine tailings at Qalb Ratiye, Wadi Abiad, Madsus, and in the Wadi al-Jariye (photo: G. Weisgerber)

showed quickly that the site of Wadi Fidan 4 was a single phase settlement, consisting of a very thin occupation deposit and thus being different from the multi-layered tells, e.g., in the Jordan valley. In three areas, five houses as well as house walls have been excavated, of which the largest unit measured 9 × 4.6 m (Fig. 5.40). Scattered splinters of copper ores, metal prills, some handfuls of centimeter-sized slag pieces, and more than twenty fragments of crucibles have been found, the majority in area D, which demonstrates the existence of small-scale, metallurgical craft activities inside the settlement. The slag was systematically reworked and crushed into small pieces. The slag quantity is minute compared to the slag tonnage in other sites. Waste had obviously been dumped in a small wadi between the plateau and the mountains leading to the south (area C in Fig. 5.40). Here some small waste heaps consisting of ash, ceramics, bones, fragments of stone hammers, slag, etc. have been found. A small Iron Age slag heap is situated at the eastern foot of the plateau and suggests perhaps reworking of EBA I slags.

Copper ore was also used for the production of beads. In the courtyard of one house a series of eight fireplaces bordered by clay have been excavated, which are most likely hearths or *tannurs*. They must have been used for food preparation, but it is unclear whether they would also have been used for metallurgical purposes. They show no similarities to the Chalcolithic smelting furnaces in Timna, site 39, postulated by Rothenberg (1978), nor do they resemble the installations excavated in Tell Abu Matar, which have been interpreted as smelting furnaces by Gilead et al. (1991) and Gilead and Rosen (1992).

The twenty fragments of crucibles (Fig. 5.42) can be divided into two different groups. One group involves thick-walled vessels made from a lightly colored, Ca-rich clay, which are slagged/vitrified on the interior and show the remains of a handle.

Fig. 5.42.
Wadi Fidan 4. Fragments of Chalcolithic crucibles from a house courtyard in area D, square 4. Thin- and thick-walled vessels, each with a different quantity of slag adhering to them, can be seen. Compatible finds from Tell Maqass and Tell Hujayrat al-Ghuzlan, (unpublished data Amman, Bochum) and Meser (Dothan 1959) imply that the crucibles in Wadi Fidan 4 had a handle (drawing: R. Adams)

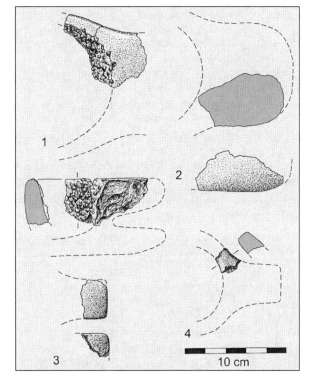

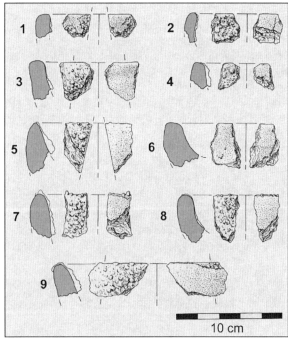

The other group includes thin-walled vessels, hard fired, which are reddish on the inside and show a partial, thin cover of slag crust. Typically enough no indication was found of the material with which the contents of these crucibles were heated. Nozzles of blowpipes, as are known from, e.g., the Neolithic period of the Alpine region (Roden 1988) have not been found nor have any remains of reeds used for such a purpose (see Sect. 7.1).

The ^{14}C-dating of charcoal from areas A and B produced a date between 3500 and 3100 BCE (Table 5.1) and thus puts the site in the transitional period towards the Early Bronze Age I (Adams and Genz 1995). In opposition to the opinion of Rothenberg and Merkel (1995) as well as of Gilead (1990), no sign of a settlement or or of metallurgical activities from the Qatufian phase (6th/5th millennium) have been found at this site.

Wadi Fidan 5

Chalcolithic pottery and nut-sized pieces of slag similar to those from Fidan 4 were collected on a mudflat with acacia bushes at the southern side of the wadi east of the settlement of Wadi Fidan 4. The site was heavily disturbed by bulldozers and thus no further investigations were carried out there.

Wadi Fidan 6, 7

These sites consist of a tower and some other archaeological remains without any recognizable connection to metal extraction in general. The find spots are visible from the Early Bronze Age smelting site at Wadi Fidan 13.

Wadi Fidan 8 (JD-38)

This settlement, forming a tell of roughly 1 ha size, had been called 'site A' by Raikes (1980). It rises from the wadi bottom at the northwestern side of the alluvial fan at the mouth of the Wadi Fidan into the Wadi Arabah. Neolithic and Chalcolithic flint tools, Early Bronze Age pottery and some scattered copper prills have been found on the surface. Adams (1991) carried out two test trenches, which revealed houses from the Pre-Pottery Neolithic B period with well preserved floors made from partly colored lime plaster. Two ^{14}C-samples from the excavation showed dates between 7400–6700 BCE (Table 5.1).

This shows, together with the partly contemporary settlement of Wadi Ghwair, evidence of the usage of pyrotechnological processes before the actual knowledge of extractive metallurgy.

Wadi Fidan 9

This cemetery, called 'site D' by Raikes (1980) is located opposite the Chalcolithic settlement of Wadi Fidan 4. Levy et al. (1999) excavated here and dated the graves back to the Iron Age of the 10th/9th century BCE; the site is now labeled Wadi Fidan 40.

Wadi Fidan 10

See Wadi Fidan 13

Wadi Fidan 11 (JD-32)

The site is situated west of the Early Bronze Age I settlement of Wadi Fidan 4. It consists of a small, steeply rising slope, which is completely covered with ruined wall lines. Two test trenches on the summit and at the western flank resulted in evidence of a further settlement from the Pre-Pottery Neolithic (Adams in prep.). Copper ores have been collected here, as in Wadi Ghwair, possibly for cosmetic use, and lime plaster was produced here, too.

Stone borers and arrowheads made from flint have been found in erosion gullies at the northern slope of the settlement. One ^{14}C-sample (charcoal) with an age of 8300–8000 BCE (Table 5.1) supports the Pre-Pottery Neolithic dating.

Wadi Fidan 12

Between Wadi Fidan 4 and 6 a wide alluvial fan, coming from the south, enters the wadi, and a sparse rush cover is apparent at this point. Raikes (1980) had already mentioned this locality.

One can pick up isolated walnut-sized slag from the granite detritus; these form loose slag scatters further up the wadi (max. 1 t). The slag lies between heavily disturbed architectural remains. The typical assemblage of slag droplets, ceramic furnace walls and clay rods indicates, along with the pottery, a dating to the Early Bronze Age II/III and possibly later (Levy et al. 2001: EBA IV).

Wadi Fidan 13

This smelting site is situated on the slope of a limestone ridge towards a gravel terrace, some 1.5 km, southwest of the mines at Umm ez-Zuhur. The site can be dated to the Early Bronze Age period due to the typology of crushed slag and the discovery of scattered clay rods. The slag heaps (ca. 100 t slag) are clearly visible from the track from Al-Qurayqira to Khirbet Hamra Ifdan in an easterly direction.

XIV. Barqa el-Hetiye (JD-31)

- Geographic setting: 30°35'53" N/35°22'54" E

Barqa el-Hetiye is a settlement site where not only ore was smelted but also copper further processed into rod-shaped ingots and other items, just as in Hamra Ifdan. The site is roughly 5 km south of Al-Qurayqira. This location, equally distant from the outcrops of ore deposits and the other settlements in the area, was in all probability chosen because of its favorable conditions for transport towards the Arabah and due to the advantageous conditions for the running of smelting furnaces.

The chain of mountains, consisting of Tertiary limestones of the Umm Rijam formation, rises directly north of the Wadi Abu Dubbana out of a plain, which is covered with shifting sand dunes and grown over with salt trees or white saxaul (Haloxylon persicum). The hills are striped with layers of chert, which have seemingly not been exploited. On top of one of the biggest hills, a black slag heap with 500–1 000 t of slag is situated in an exposed position facing towards the north. The top of the slag heap is at 130 m height. The remains of a number of smelting furnaces set into the ground in a row can be seen at the upper end of the heap, just below the knoll of the hill. They are of the same kind as observed and excavated in Faynan 9 (Fig. 5.16). A second slag heap with far less tonnage is situated ca. 250 m away at the southwestern flank of the adjacent chain of hills. Clay sticks are a very frequent find in the slag heaps, and ceramic fragments from the lightly colored, Ca-rich clay, being parts of slagged/vitrified furnace walls, also commonly appear. On the mountain ridge, the slag had been reworked with stone hammers and anvil stones into slack in order to extract the remaining metal inclusions, and in the plain below the hill slag was systematically crushed. The find assemblage in the heap is as typical for smelting sites of the Early Bronze Age as is its position exposed to the winds. Barqa el-Hetiye is thus one of the largest smelting sites from this time period in the region of Faynan.

Some 400 m north of the mountain slope, the remains of four houses sit on a low hillside, two of which have been excavated in the years 1990 and 1993 (Fritz 1994a,b). One house can be dated to the Early Bronze Age II by pottery and the ^{14}C-dating (Table 5.1). The pottery found here can be compared with the material from Arad, strata II and III and demonstrates, just as the metal objects themselves, a close connection between Faynan and Arad (Hauptmann et al. 1999, see Sect. 8.3). The other house dates to the Iron Age I. It is very interesting that Midianite pottery has been found here (Adams 2003, Fritz 1994b). This is a pottery made in northwestern Arabia between the 13th and the 11th century BCE (Knauf 1988). The material appears in Timna as well as in Khirbet el-Msas in the Negev in connection with smelting or processing of copper (Rothenberg 1973; Fritz 1994a).

In the vicinity of both houses at Barqa el-Hetiye, settlement remains spreading over 1.5 km^2 can be observed, where extensive copper production had been carried out (Fig. 5.43). They are heavily eroded and partly covered with shifting sand dunes. This provides the context for definite signs of extended processing of copper. Everywhere crushed slag, fragments of crucibles, smelting furnaces, casting moulds, metal prills, crushing stones and pieces of ore have been found. The metal produced on the hilltops and the metal extracted from the crushed slag were cast here into larger lumps. Fragments of casting moulds, identical to those from Khirbet Hamra Ifdan (Fig. 5.38), prove the production of crescent-shaped ingots. They are from the Early Bronze Age. Together ca. 10 kg of metal pieces were collected, which are partly reminiscent of fragments of plano-convex ingots, similar to one found at a smelting site in En Yahav (Fig. 4.2) further west (McLeod 1962; Rothenberg 1990; Yekutieli et al. 2005).

The ore found inside the Early Bronze Age house makes it plausible that the ore transport during this time was managed via the settlement. The exploited mines could have been at Umm ez-Zuhur, located 7 km to the north.

5.2 · Site Catalogue and Archaeometallurgical Finds

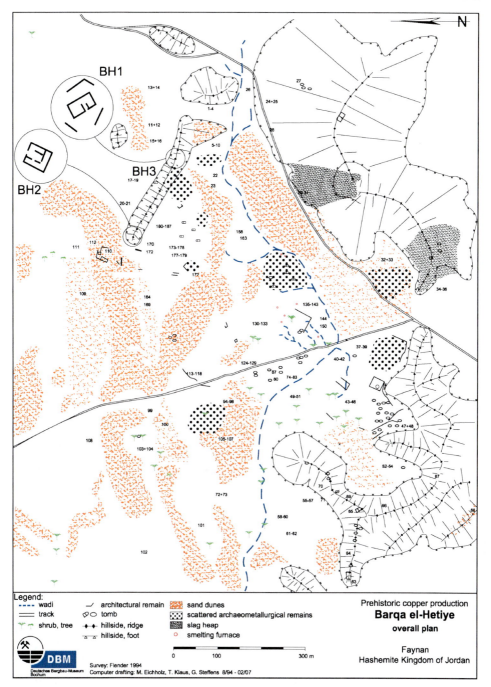

Fig. 5.43. Barqa el-Hetiye. Survey map of the Early Bronze Age II/III smelting site on top of the hill and surroundings. Indicated are areas with slag and copper scatter, and two houses from the EBA II (*BH1*) and from the Iron Age (*BH2*).

XV. The Mining District of Umm el-Amad (JD-10)

- Geographic setting: 30°34'18.6" N/35°29'43.88" E

In the mines of Umm el-Amada, six kilometers south of Khirbet Faynan, a mostly stratabound copper mineralization existing also in veins in the MBS was exploited. Four Roman mines are located in the very rugged steep incline towards the Jordanian Plateau at a height of 700–800 m above sea level. The most prominent of these mines, which gave the whole area its name, was the only mine visited by Glueck (1935). And it was first documented more precisely by Kind (1965), who described it as a room and pillar mine (Fig. 5.44) with a lateral extent of 30 × 30 m. Measurements made by the Deutsche Bergbau Museum were able to correct these measurements: the mine is from the working face to the entrance completely preserved and measures 120 × 55 m. It is thus the only known Roman mine in the entire empire that is still fully preserved. The shape of the entrance (Fig. 5.45) indicates that the mine, as with so many others in Qalb Ratiye, had already been opened in the Chalcolithic. The complete lack of any Bronze or Iron Age activities also corresponds very well with the developments observed in the other areas of the Faynan district.

Three further mines in the direct vicinity of Umm el-Amad might possibly be of a comparable size, but as they are very heavily collapsed, a conclusive study of these mines is impossible. The pottery and coins (Kind et al. 2005) date them to the Roman period. The mines were full of very finely grained sand from decomposed or more likely from crushed sandstone. We have ruled out that these fillings would have been washed in from outside.

Fig. 5.44. Umm el-Amad. The horizontally running, stratabound ore mineralization was exploited in Roman times through room and pillar mining leading to an extent of 120 × 55 m. It is therefore one of the largest preserved mines in the entire Roman world. The scratch marks from hammers and wedges are very impressive

Fig. 5.45. Umm el-Amad. View towards the adit of the mine already described by Glueck (1935) and Kind (1965). Its shape indicates that the openings might date originally back to the Chalcolithic period, as can be demonstrated in Qalb Ratiye

5.3 The Development of the Mining District of Faynan: Field Evidence

The fieldwork in the Faynan district led to the discovery of over 50 sites and areas directly related to mining and metal production. Altogether, over 200 mines from very different periods have been found. The ^{14}C-dates of charcoal coming from the slag heaps, waste dumps and actual settlement layers, as well as the general archaeological analysis, demonstrate that the deposits have been exploited for some 9 000 years (Fig. 5.2), although there are some chronological gaps in their usage. High points of activity are also indicated by the chronological pattern, which will be discussed here in detail. It has to be kept in mind that these results are based mainly on surface studies and that test trenches and excavations have been carried out only to a very limited extent. This means that the results can only be a chronological framework and do not amount to a finely tuned chronological study.

Firstly and most fundamentally, one has to differentiate between the general use of ore on the one hand and the production of metal via pyrotechnological processes on the other. The latter started in Faynan much later than the former. Since the Palaeolithic period at Faynan, the green secondary copper ores, which are strikingly exposed at numerous spots on the surface, were certainly known as well as the local water supply. Finds of stone axes (unpublished data DBM) indicate that it was settled by hunter-gatherers in this time period as was the entire southern Levant (Byrd 1994). During the Pre-Pottery Neolithic period, the green copper minerals were used for pigments and beads (see Chap. 8). The conclusion is that the ore deposit was never

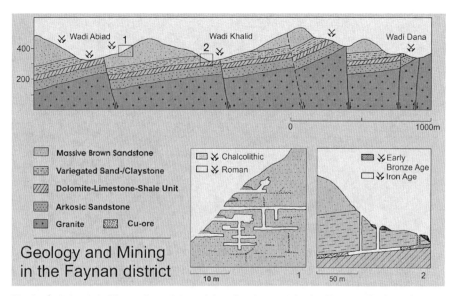

Fig. 5.46. Schematic illustration of the mining development in the Faynan district. The geological section Wadi Abiad – Wadi Dana shows both ore bearing formations. They have been exploited during different time periods, and thus different techniques were used in each respective period (modified after Heitkemper 1988 and Weisgerber and Hauptmann 1988)

intentionally prospected, i.e., it was never searched for as a deposit. This is probably also true for Timna, where the geological and archaeological frame is comparable to Faynan (see also Avner et al. 1994).

The different ages of the mines demonstrate that the mineralizations in the DLS and the MBS underwent their most intense exploitations at different time periods (Fig. 5.46). In the 5th/4th millennium, ore was collected out of the erosion rubble of the DLS, but it was also mined from the vein fillings of the MBS. In the Early Bronze Age and the Iron Age, mining meant, nearly exclusively, exploitation of the DLS, while in Roman times a large number of the very ancient mines into the MBS were re-opened and enlarged.

The area of Faynan shows a similar density of sites as at Timna on the western side of the Arabah. But if the quantity of slag is taken into account, it becomes clear that Faynan's economic role was much greater than that of Timna. In Faynan an estimated quantity of 150 000–200 000 t of slag was produced by copper smelting (Table 5.3), while in Timna the total amount of slag hardly surpasses a few thousand tons (Bachmann and Rothenberg 1980). This balance of importance is not changed when a numerical comparison of mines is observed, because the count of ca. 9 000 shafts in Timna (Conrad et al. 1980) is mainly due to geological reasons. It can therefore not be taken as evidence of more extensive mining activity. The large number of mines is a result of the extraction techniques in flat-layered mineralization in soft sandstone, running in wide areas directly underneath the surface (Bartura et al. 1980). Shallow pits and shafts have been sunk, thus leaving circular, densely clustered and filled shafts, so-called 'plates' on the gravel terraces in the Timna-valley. In Faynan, however, the spatially limited outcrops of the ore mineralizations in the DLS and in

Table 5.3. Estimated quantities of slag and copper produced in the Faynan district from the 4th millennium BCE to the Mameluk period. These are rough estimates, as only a few slag heaps could be precisely measured. The approximate calculations for copper are based on an idealized relationship metal: slag = 1:10 to 1:15, which roughly follows the experimental results of Tylecote et al. (1977) and Tylecote and Boydell (1978). It has to remain an open question as to whether the Cu-content of the ores always has to be so high

Period	No. of smelting-sites	Amount of slag	Copper
Early Bronze Age I	4	2 kg	x kg
Early Bronze Age II ff.	13	5 000 t	300 – 500 t
Iron Age	4	100 000 – 130 000 t	6 500 – 13 000 t
Roman period	1	40 000 – 70 000 t	2 500 – 7 000 t
Early Islamic period	2	1 500 t	100 – 150 t

the vein mineralizations of the sandstone rapidly led to extended underground mining activities. Particularly the seam-like mineralizations in the DLS, either flatly running or gently inclining, are of a shallow thickness but large extent, which made the room and pillar mining the most logical extraction technique. In this technique the hanging wall has to be supported by pillars, which have been left unexcavated during the mining process (such as in Umm el-Amad), or which have been artificially built (such as in mine 42 in Wadi Khalid). Mined tailings were backfilled into the mines, which left relatively little visible waste material on the surface, thus giving a misleading impression of the real extent of the miners' work.

The earliest evidence of use of the spectacularly colored ores comes from the Pre-Pottery Neolithic B (PPNB). The ore was transported from the outcrops of the DLS at Khirbet el-Jariye, in Wadi Khalid and at Umm ez-Zuhur, to the nearby settlements in Wadi Fidan and in Wadi Ghwair and converted there in the settlements into adornments of different kinds. Faynan was obviously part of the long-distance trade pattern at this time, because it can be shown in a number of contemporary sites in the Levant that the ore found on these sites came from Faynan (see Sect. 8.1). But there is no evidence of mining activities at this time in Faynan (Hauptmann and Weisgerber 1996). The earliest indication of copper mining in the Arabah is an unpublished ^{14}C-date from a gallery at Timna (mine S27, No. Bonn 2360: 7 680 ±120 BP, 6650–6410 cal. BCE; see Avner 2002 and Fig. 5.3). In the light of some PPN-artifacts from Nahal Issaron which were manufactured from copper ores from Timna, this does not seem to be impossible. In addition, PPNB flint mines were discovered in northeastern Negev (Taute 1994) and near Ain Ghazal (Qintero 1996).

Copper, as a metal, seems still not to have been known in the Pottery Neolithic in Faynan, or in the entire Southern Levant. At least no evidence has been found in the excavations in Tell Wadi Faynan (Najjar et al. 1990). Only the use and working of copper ore can be proven, but again without any evidence of mining. The knowledge of pyrotechnical processes was limited in the PN as in the PPNB to the production of lime plaster. This ability to produce quicklime or lime plaster by firing limestone at 800–900 °C and slaking it in order to get a first workable then hardening material is considered one of the hallmarks of the PPNB (Mellaart 1975; Gourdin and Kingery 1975). Lime plaster has been found – as in Wadi Ghwair 1 or in Tell

Fidan A – in nearly all ancient settlement sites of this time period in the Eastern Mediterranean, where it was often used to make the floor- or wall-plaster. It was also used for sculptures and vessels such as white ware (Kingery et al. 1988). But we have no indication at all of the smelting of ores.

In the neighboring Timna (site F2), the earliest extraction of copper from its ores was said to begin in the 6th/5th millennium BCE, the Qatufian period (see above), along with the use of fluxing agents to improve slag properties (Rothenberg and Merkel 1995; Merkel and Rothenberg 1999). The authors base their (indirect) dating of slag on comparisons of their material with pottery from Wadi Fidan 4, for which they then use an incorrect chronological assumption (Gilead 1990). It has been shown by Adams and Genz (1995) and Genz and Adams (in prep.) that this assumption has no basis (see also Kerner 2001). Additionally, a thermoluminescence analysis has been carried out by Hauptmann and Wagner (2007) of one of the (slagged) tuyères from Timna F2, which confirms the Late Bronze Age date of the site (Burleigh and Matthews 1982; Stuiver et al. 1998) already established by radiocarbon dates.

In Faynan, no precursor state of metallurgy existed, such as that which has been found in Anatolia in the Pre-Pottery Neolithic, and that is simply because the deposits along the Wadi Arabah contain no native copper. This precursory state, following the stages of development in metallurgy by Strahm (1994), is characterized by the occasional use of native copper to produce, via cold and warm treatment, small artifacts such as beads or hooks. These items were only used as exotic ornaments next to other colorfully conspicuous materials such as carneol, galena, or oxidic copper ores as is shown by the evidence from Asikli Höyük (Yalçin and Pernicka 1998), Cayönü Tepesi (Özdogan 1990; Maddin et al. 1998) and Nevali Cori (Hauptmann et al. 1993) in eastern and southern Anatolia.

The earliest evidence of extractive metallurgy in Faynan itself dates – with the excavation in Wadi Fidan 4 (WF 4) – to the middle or the second half of the 4th millennium. This, however, is not the earliest evidence of copper production in the Levant. Numerous archaeometallurgical finds from Nahal Mishmar, Tell Abu Matar, Bir Safadi, Shiqmim and other sites in Israel (Tadmor et al. 1995; Perrot 1955; Shalev and Northover 1987; Shalev 1991; Golden 1999) demonstrate that copper ores from Faynan were traded into the Beersheba basin and there smelted into metal (Hauptmann 1989; Shugar 2000). The rich finds from the one-phase settlements of Tell Maqass (Khalil 1988, 1992, 1995; Khalil and Riederer 1998) and Tell Hujayrat al-Ghuzlan at the Gulf of Aqaba (Khalil and Eichmann 2001, Brückner et al. 2002) also show a vividly abundant copper production during the corresponding period. In both settlements copper ores, slag, metal objects, and crucibles have been found. The ore for the copper production used in these settlements came from Timna (Hauptmann et al. 2004). Rectangular moulds found in Tell Hujayrat al-Ghuzlan will have far reaching consequences for the understanding of copper trade during that time period (see Chap. 8). The significant result of this evidence is the fact that the technological step towards extractive metallurgy in the Southern Levant has not been taken in or close to the ore deposits but far away in the complexly organized settlements.

Adams and Genz (1995) had considered WF4 still contemporary with the Beersheba-Ghassul-culture, dating it thus to the Chalcolithic period, but Genz (1997) corrected this in favor of an Early Bronze Age Ia dating. This suggestion is backed up with a number of recently carried out re-evaluations and discussions of the already existing ^{14}C-dates by Avner et al. (1994), Gilead (1994) and Joffe and Dessel (1995). These and other studies (Kerner 2001, Lovell 2001) demonstrate that the development of Chalcolithic cultures in the Southern Levant might have already started at the beginning of the 5th millennium. The transition towards the Early Bronze Age I is presently considered to be around 3500 BCE. This is the period when arsenical copper begins to be distributed in the Old World. The finds and contexts of the mining and smelting in WF4, however, remain in the tradition of the Chalcolithic period. This is demonstrated clearly, e.g., by the type of stone hammers in Wadi Fidan 4, which have been manufactured inside the settlement. This particular type of stone hammer has been found everywhere in and outside the mines exploiting the MBS in the Faynan district and also in Madsus and in Wadi al-Jariye. They are completely nonexistent in the mines of the Early Bronze Age II/III, which quite clearly illustrates, together with the mining technology and the metallurgy, a technological progression. But the earliest era of copper production in Faynan itself, however, has to be dated to the very beginning of the Early Bronze Age.

The excavation of the settlement showed that nevertheless at this time ore was still transported from the deposits to the settlement in order to be further processed there, at some distance from the actual deposit. In the settlement, the smelting of ore was then organized on a level of 'household or workshop metallurgy' (see also Costin 1991). The ore source can probably be found in the nearby mines at Umm ez-Zuhur and Qalb Ratiye, where specifically malachite was exploited from mineralizations in shallow pits close to the surface. 'Tile-ore', Cu silicates and chalcocite were apparently acceptable up to a certain degree. It is rather striking that a relatively high amount of Cu silicates have been found in WF4. This seems at first to form a contradiction with the previous statement (Hauptmann et al. 1985) that the mines providing the material for WF4 were situated mainly in the MBS. This is certainly the case, but obviously other ore outcrops were also exploited especially in order to get material for making green and bluish beads. Helping to clarify this is the important point that the MBS ores nearly exclusively form the material which was found in the settlements of the Beersheba valley. This point will be discussed in detail in Sect. 8.2. The ores found there are indirect dating evidence of the mining into the MBS at the transition from the 5th to the 4th millennium, because the Beersheba settlements are older than WF4.

Metal was only produced in a rather limited quantity in WF4, as was the case in all other, Early Bronze Age I sites that are known so far in the Faynan area (Wadi Fidan 5, A, Faynan 17, Wadi Faynan 100). This shows a certain disproportion to the extent of mining activities from earlier periods than Early Bronze Age II in the district of Faynan. But the situation becomes immediately understandable when the obviously quite extensive ore exports into the Beersheba valley are taken into account as well as the quantity of metal artifacts, which can generally be found in the Levant during this time period.

Intriguing and quite unusual is the small amount of slag produced in WF4. This reminds one of the slightly dogmatically formulated technical term 'slagless metallurgy' by Craddock and Meeeks (1987) and Craddock (1990). This concept was used as an approach to explain the archaeological situation in the British Isles during the Early Bronze Age. There, extensive mining activities could be proven, while no evidence of contemporary smelting in the form of slag was available. The most convincing corollary was that high-grade ores exploited from the mines were smelted (in crucibles) without the production of any sizeable slag. This can also be proven for Faynan when compared with the quantities of slag from later periods. Strictly speaking, the term 'slagless metallurgy' is probably not correct, but it describes the archaeological evidence quite aptly. At Wadi Fidan 4, we can not exclude that a small Iron Age slag heap, located just below the plane of the village may have resulted from reworking ancient slag.

The social pattern of mining and metal production with the transport of ores away from the sources to the settlements and the smelting and further processing happening inside them, as exemplified in Faynan (WF4) and Beersheba, seems to be characteristic of the 4th and probably even more so for the 5th millennium. And that seems to be the case, not only in the Southern Levant, but also in other areas of the Near East. Parallels become particularly clear on the upper Euphrates (Zwicker 1980; Hauptmann et al. 1993; Palmieri et al. 1993; Müller-Karpe 1994; Pernicka et al. 2002). This is also the reason why it has proved so difficult to show the existence of metal production during the Chalcolithic, respectively Early Bronze Age I, from the ore deposits in Timna (Rothenberg 1978; Muhly 1984; Rothenberg 1990): The utilization of Timna ores for metal production in this time period could only be verified with metal finds in the settlements of Tell Maqass (Khalil 1988) and Tell Hujayrat al-Ghuzlan (Hauptmann et al. 2005).

There is certainly no reason for speaking of an 'industry' during this period, as Levy and Shalev (1989) and Ilan and Sebanne (1989) have framed it. And this holds true also for the period of EBA I/II, which is represented by the settlement of Wadi Faynan 100, west of Khirbet Faynan (Wright 1996; pers. comm.). The Chalcolithic/ Early Bronze Age I periods led to the earliest specialization of craftsmanship (Kerner 2001), but the social preconditions for an industrial production simply did not exist at this point in time. It is also unnecessary to discuss the problem of fuel supply for metallurgy, as even normal life required a large amount of wood for a variety of different purposes.

The later Early Bronze Age is a period of time when rather dramatic cultural and material development and changes happened in the Near East. This can be seen in the development of metallurgical technology. In addition to copper, other metals such as gold, silver and lead were widely distributed at this time (Muhly 1980). The most important achievement during this period, however, is the use of tin bronzes, which led quickly to the first intentional alloying techniques (see Sect. 8.4). In the copper producing district of Faynan itself, there is no evidence of the use of tin bronze at that time. But at Faynan, probably the most extensive and best preserved evidence of Early Bronze Age copper mining and production in the Old World was discovered. This is testified by numerous mines especially in the Wadi Khalid, by smelting sites distributed all over the mining district of Faynan, and by settlements such as the one at Barqa el-Hetiye (EBA II) and the only recently excavated site of Khirbet Hamra Ifdan (EBA III/IV).

The area of Faynan was thus settled continuously from the middle of the 4th millennium BCE until the transition from the 3rd/2nd millennium BCE, during which time ore was mined and smelted. In the middle of the 3rd millennium, roughly during the Early Bronze Age II/III, there was an initial peak in mining- and smelting activities. This continues into the Early Bronze Age IV. The discovery of twelve smelting sites with a few thousand tons of slag, not covered with later activities, was in the area of Faynan during the surveys carried out up until now. The smelted slag has been carefully reworked to extract any remains of copper as can be seen in Ras en-Naqab, at Faynan and especially at Khirbet Hamra Ifdan. It is a firmly established step in the production of copper.

In Faynan, radical technological changes occurred during the third millennium BCE. The step from a domestic mode of production in the Early Bronze Age I to the mass production of copper in the 3rd millennium is generally representative of the entire history of mining and metallurgy: it reflects the confrontation of mankind with continuously decreasing resources which, in turn, necessarily leads to the invention of new technologies. At Faynan, high-grade copper ores were no longer available in the quantities required for mass production; therefore, ores with a lower copper content had to be smelted. With caution, we might propose the use of ores with not more than 15–20 wt.-% of copper. Subsurface mining developed and was practiced to a large extent. The so-called chamber-and-pillar mines with lateral extents of at least 30 × 15 m were preferably put into the Cu-Mn-mineralizations of the DLS in the Wadi Khalid and Wadi Dana but probably also at Umm ez-Zuhur. The easy smelting of malachite no longer played a role, as it was rather the mineralizations consisting of copper silicates and copper-chlorides with manganese oxides that were mined. The smelting was no longer a small-scale activity carried out inside the settlements, but – based on the development stages defined by Strahm (1994) – the first actual 'industrial' production started. The slag heaps are evidence of the local production of metal, while the trading of unprocessed ores comes to a stop. Fritz (1994a) assumed that these rather extensive activities required the existence of larger settlements. Weisgerber (1996) inferred from the widely scattered location of the smelting sites and the existence of several Early Bronze Age settlements in the area that the metal production was not organized centrally but was carried out by several independent bodies.

This question has to remain open, because the location of the mines as well as that of the smelting sites was primarily determined by geography and geology. The distribution pattern of the Early Bronze Age mines depended on the actual position of the ore deposit itself, while the location of the smelting sites depended on the availability of natural winds for the operation of the smelting furnaces. They were deliberately positioned at the edges of slopes and escarpments in the landscape. The same phenomenon can be observed in Timna, where furnace remains and slag have been found – to a much smaller extent – in similarly exposed situations (Bachmann and Rothenberg 1980), as well as in other parts of the Arabah. This point will be discussed in detail in Sect. 7.2. But the positioning of Barqa el-Hetiye and particularly of Khirbet Hamra Ifdan has to be especially mentioned: both settlements were apparently centers for the further processing of copper from the Faynan district. Copper, cast into ingots or finished objects, was exported from these checkpoints. Finds in the metal importing settlements, such as, e.g., Arad, reveal that the metal in those sites was only further processed there by remelting and recasting (Amiran 1978). There is a high probability, in fact, that settlements in Palestine chronologically comparable to those in Faynan imported their copper from Faynan (see Sect. 8.2). The flourishing of Arad (stratum II) is contemporary with Barqa el-Hetiye, and the settlements of Har Yeruham, Ain Ziq, Beer Resisim and Lachish (see Fig. 8.9) can be correlated with Khirbet Hamra Ifdan. It is evident that the trade with Faynan copper was again oriented towards the west and organized through the Wadi Fidan, the 'Gateway to the West' (Levy et al. 2002). The Beersheba basin was a "Metallurgical Province." From the Early Bronze Age (ca. 3600–2000 BCE), the Faynan district was a center of copper metal production that inexplicably ended around 2000 BCE. In Timna, mining and smelting activities are proved for the developed EBA only to a limited extent by three ^{14}C-dates (Bonn 2362, Bonn 2363, HAM 215, see Fig. 5.3). In contrast to Faynan, the cultural hinterland is missing, and the largest settlements nearby, Tell Hujayrat and Tell Magass, were already given up in the EBA I.

In the Middle Bronze Age the extraction of copper in Faynan diminished significantly and can only be proven by a few isolated finds. All of these have been found in the mines of Wadi Khalid and Wadi Dana, which have been cut by modern prospection galleries. Mining is verified here for the Middle Bronze Age I and II and later again for the Late Bronze Age. There are no clues though, as to the smelting. Probably any possible existing slag heap has been covered over or destroyed by the enormous heaps of the later Iron Age.

The situation in Timna is compatible, as here the metal production decreased in the Middle Bronze Age, too. This means that the entire copper production area in the southern Levant was interrupted for several hundred years. It only started again with the transition from the Late Bronze Age to the Iron Age I, somewhere in the 13th century BCE (Rothenberg 1988a). Copper production was then reactivated by Egyptians of the New Kingdom obviously in partnership with Midianites (i.e., immigrant specialists in metal production from the Aegean, Rothenberg 1998). Midianite pottery, a polychrome decorated pottery usually dated from the 14th to the middle of the 12th century BCE, which shows close relations to pottery from Qurayah in the Hedjaz area in northwest Arabia, was found along with Egyptian items at Timna and at Faynan. It was excavated at Barqa el-Hetiye (Fig. 5.47, Fritz 1994a) and at Khirbet en-Nahas (Levy et al. 2004). At least at Faynan, part of the Midianite pottery did not originate in the Hijaz area, but was manufactured locally, close to the smelting sites (unpublished results Bochum).

Fig. 5.47.
Barqa el-Hetiye, house 2. Midianite pottery showing a human figure. Part of the sherd was already published by Fritz (1994)

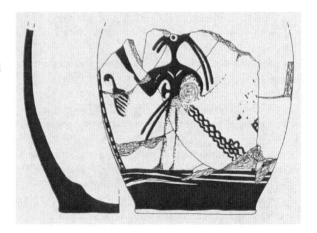

5.3 · The Development of the Mining District of Faynan: Field Evidence

One probable reason for the breakdown of copper production in the Arabah during the Middle Bronze Age might be found in the increasing copper export from Cyprus. A lively exchange of goods between the island and the Eastern Mediterranean, particularly the Levant, can be established for the time period between 1800 and 1200 BCE (Muhly 1991). In the Late Bronze Age, about 1600–1200 BCE, in a time of growth, prosperity and further social complexity (Knapp 1986; Muhly 1986), production and export of Cypriot copper reached its peak. Copper from Cyprus was the "world bulk metal" and was traded in large quantities to Egypt and many other localities in the Eastern Mediterranean and beyond. A relatively large metal workshop, dated to the beginning of the Middle Bronze Age where copper ores were smelted and further metal processing was carried out, has been excavated at Mavrorachi near Pyrgos (Belgiorno 2000). It shows striking similarities with Khirbet Hamra Ifdan. Written sources even prove that copper from Cyprus was – next to copper from Dilmun (most likely Bahrain) – traded all the way to Mesopotamia in the 18th century BCE. The restarting of copper production in the Arabah in the 12th century BCE has in all probability to be perceived in relationship with the breakdown of the Late Bronze Age long-distance trade in the Eastern Mediterranean and the closely connected dwindling copper exports from Cyprus (Muhly 1991). This seems to have made the use of local resources economical and necessary again.

In Faynan, the copper production took on real industrial proportions with the late 2nd / beginning of the 1st millennium BCE. Most likely caused by the employment of iron tools, an enormous technical progress can be observed in mining (Weisgerber 1996), which is predominantly visible in the newly built shafts. The opening of double shafts up to 70 m in depth enabled the miners to reach new parts in the DLS deposits (Fig. 5.35), which had simply not been accessible during the Bronze Age. For the first time in the history of mining, hoisting ropes and the ventilation of mines can be demonstrated.

Differing from the spatial distribution of Early Bronze Age slag heaps where more or less similar quantities of metal were produced, the pattern of the largest smelting sites is during this time concentrated in a few places: Faynan 5, Khirbet en-Nahas ('ruins of the copper') and Khirbet el-Jariye were the most important centers of Iron Age smelting. In addition, finds of Iron Age slag, pottery, hammerstones etc. in numerous Early Bronze Age slag heaps indicate that copper bearing remains of older smelting sites were systematically processed and remelted during this period, too. According to Levy et al. (2003), Iron Age settlements and cemeteries were concentrated north of a geographic borderline along the Wadi Fidan to Ras en-Naqb and formed an "Iron Age Landscape in the Edomite Lowlands." Such a chronological division demonstrates that a geologically defined ore district must not necessarily be fully identical to a mining district as defined by Stöllner (2003).

The overall tonnage of Iron Age slag heaps has been estimated at more than 100 000 t (Table 5.3). Ore from the mines in the wadis Khalid, Dana, Ghuwebe, Jaryie as well as from Umm es-Zuhur was brought to Faynan or Khirbet en-Nahas where it was smelted to metal. Similar to the mining techniques, the copper smelting technology of this time, using new furnace constructions with special refractories and multiple composed tuyères, new kinds of slag tapping arrangements and the production of tap slags with a weight up to 60 kg, was a result of new technical achievements. A firmly established step in Iron Age metal production was a careful and systematic reutilization of metal which not only lead to a recycling of older materials, but also caused a large-scaled slag crushing at the sites mentioned above to extract metal prills. Here, thousands of tons of crushed slag accumulated.

There are technological parallels only in the New Kingdom smelters at Bir Nasib, Sinai (Rothenberg 1987). The question remains where and when these achievements developed. A technological transfer from Cyprus via Egypt seems possible. For instance, Knapp et al. (1998, 1999) excavated a LBA smelter at Politiko Phorades, Cyprus, where they found ideally composed cakes of tap slag from matte smelting with a weight of 15–20 kg. The furnaces had a diameter of more than 40 cm and were operated by bellows. Kassianidou (2003) found over 500 fragments and 30 almost complete tuyères. The slags from Politiko Phorades was not as intensively crushed as the slag we observed at Faynan 5 or at Khirbet en-Nahas.

The mines and the smelting sites seem to have been under the central (state-) control of the Edomite kingdom (Levy et al. 2004). Unlike Timna, where metal production was organized by Ramesside Egyptian expeditions (Rothenberg 1973, 1988), no Egyptian pottery was found at Faynan. A scarab with a "walking sphinx" and hieroglyphs from the New Kingdom (12th century BCE) and another scarab from the 12th / 10th century BCE are supporting evidence of a dating of LBA / IA activities at Faynan, but not for any involvement of Egyptians in mining and metallurgy. The formation of the Edomite state, in turn, is redated now by radiocarbon dating using Accelerated Mass Spectrometry to the 10th century BCE and was closely connected with the metal production in the lowlands. The significance of the region and military importance are very clearly demonstrated by the fortification wall in Khirbet en-Nahas and the two fortresses above Wadi Ghuwebe.

The complete time span of this production peak extends in Faynan from the Late Bronze Age to the Iron Age IIIc. This includes therefore the period before the beginning of the Edomite state, when Faynan was still controlled from the west and south (Knauf 1992) up to the end of the Edomite kingdom. The high point of copper production, however was reached in the Iron Age IIbB and IIC (900–587 BCE; Hauptmann et al. 1985; Knauf and Lenzen 1987; Fritz 1996) based on archaeological finds and ^{14}C-dates. In Timna the Iron Age copper production comes to an end in the 8th century BCE (Rothenberg 1980), meaning that metal had only been produced for a relatively short time period in both deposits.

The mass production of copper in Faynan in the first millennium BCE starts at a time when iron, respectively steel, began to be accepted as a new material ('people's alloy') having great economic success. Muhly (1980) suggested that the breakthrough of this new technology may have been caused by a supply crisis of copper or tin as a consequence of a general breakdown in international trade relations. Waldbaum (1989) argues on the other hand that there may have been an increasing shortage of fuel resources. Both arguments are hardly convincing, because the giant slag heaps produced by copper smelting, which can be observed all over Europe and all the way to eastern Anatolia and Iran (Steinberg and Koucky 1974; Rothenberg and Blanco-Freijeiro 1981; Weisgerber 1986; Voss 1988; Cierny and Weisgerber 1992), demonstrate quite visibly that after the appearance of iron the demand for copper was larger then ever before. The production of large quantities of tin-bronze is proven by the excavation of numerous bronze workshops in Egypt (Pusch 1990) and Greece (Cooke and Nielsen 1978; Rostoker et al. 1983; Schneider and Zimmer 1984) as well as by artifact analysis (Craddock 1976).

5.3 · The Development of the Mining District of Faynan: Field Evidence

From the 2nd century BCE to the 5th century CE, copper production was taken up again and continued, mainly under Roman control. The presence of the Nabateans is proved by numerous finds and other archaeological evidence, e.g., several towers in the west and southwest of Khirbet Faynan. Nabatean pottery and 40 coins have been found in the spacious fields there (Kind et al. 2004). Weisgerber (1996) argues that this evidence might express the actual presence of Nabateans and their involvement in agriculture, but it is no proof of their participation in mining and smelting as it was assumed by Wenning (1987).

Roman mining was a second harvesting mining. Roman prospection work in the Bronze and Iron Age mines in, e.g., Wadi Khalid would have shown quite plainly that these would not have been profitable and not worth the effort to reopen them, even by means of new Roman technology. Mining concentrated thus on exploitation of vein mineralizations in the sandstone. In Qalb Ratiye and Wadi Abiad, a large number of the Chalcolithic/EBA I mines were reopened (Fig. 5.17). In the period between the 2nd and the 4th century CE some 120 mines were operated. The Romans' most impressive achievement in the region in mining technology though was the mine at Umm el-Amad. We know from written sources (as listed in Geerlings 1985) that criminals and mainly Christians, prosecuted for their religion, formed the workforce in the mines.

The ores from all mines in the Faynan district were brought to a central smelting site at Faynan 1 and smelted there to metal. A road, over 7 km long was built for this purpose from Umm el-Amad to Faynan, partly paved with broken stone and militarily secured. The Roman smelting site of Faynan 1 was evidently not just the center of copper production in the area of Faynan itself but of the whole of the Arabah, because Roman mining activities have also been documented in the wadis Abu Khusheibah and Abu Qurdiya, some 40 km to the south. Kind (1965) reported several chamber-and-pillar-systems and extensive tailings of white sand that point to mines and a beneficiation of ore. Roman mining was also observed in the south of Timna in Wadi Amram (Rothenberg 1988a; Willies 1991). Surprisingly, no Roman smelting slag has been found at any of these locations. At Wadi Amram, small-scale smelting activities are indicated by one ^{14}C-measurement that gave an age of 730–700 ±36 CE (BM 1163, Willies 1991; see Fig. 5.3). There are also no other signs of any Roman activities on the western side of the Arabah, since the original dating of the smelting site of Beer Ora at Timna to the Roman period (Rothenberg 1973) has since been corrected. The smelting site is now assigned to the Early Islamic period (Rothenberg 1988b). The inscriptions in Nahal Tuweiba south of Elat do not deal with possible mining activities of the Legio Tertia Cyrenaica, which was stationed in the province Arabia at the beginning of the 2nd century CE, but simply describe stone quarrying (pers. comm. Prof. Weisgerber 1997).

The strategies described above in the mining and smelting activities correspond with the pattern of a centralized organization as is usually characteristic of the Roman administrative system. The ^{14}C-dates of the charcoal samples collected from the Roman slag heap have a tendency to be slightly older then the majority of the Roman coins, of which 90% (600 coins) date to the period between 312 and 420 CE (Kind et al. 2005). This, however, is no inconsistency as the timber used for smelting might well have had growth ages of 100 years and more. We might also assume that a military unit arrived at Faynan, as suggested by the steep increase of coins after

312 CE. As there is no reference in contemporaneous texts about a garrison at Faynan, we suggest that a temporary task force may have been in charge of the copper production for only a few decades. At the time of these activities in Faynan, the Roman empire had numerous other copper mines operating (Healy 1978; Davies 1979); therefore, Faynan was in all probability of importance only in the Levant or at the most in the Eastern Mediterranean.

The role of Faynan as one of the large copper production centers in the Southern Levant is generally finished by the end of the 5th century CE. But written sources and the remains of churches and monasteries in Faynan itself (Geerlings 1985) illustrate the importance of the site during the Byzantine period, when it was even a bishop's see. Faynan, a copper district exploited since the 4th millennium BCE, changed from metal production to agriculture (Weisgerber 2006).

At Beer Ora, and at Timna itself, mining and smelting activities continued during the Early Islamic period. Only in the 12th/13th century CE, in the Ayyubid and Mameluk periods, can another phase of second-harvesting be proven (Hauptmann et al. 1985). This goes along with a flowering of mining and metal production in the entire Islamic world at this time (Allan 1979). Numerous mines and smelting sites from this period have been found in Oman (Costa 1978; Weisgerber 1978), in Iran (Weisgerber 1990), in Afghanistan (Weisgerber 2003) and in Turkey (e.g., Wagner et al. 1986, 1989). Mining activities from the Mameluk period can also be shown in other parts of the Southern Levant such as with iron smelting sites at the iron ore mine Mugharat el-Wardah and some other smelting sites in the Ajlun Mountains, Jordan (Gordon et al. 1986; Al-Amri, in prep.).

For the time being, it remains an open question whether gold was ever exploited in Wadi Arabah. Written sources and questionable field evidence were the basis for discussion of 8th century CE gold winning on a small scale or at least prospection activities in Wadi Tawahin northwest of Eilat (Frank 1934; Amar 1997). Based upon the Onomasticon of Eusebius, there should have been gold mines at one time near the copper mines of Faynan (Meshel 2006). The descriptions do not exclude the ancient mines found in the mining district of Wadi Abu Khusheibah/Wadi Abu Qurdiya. In fact, a few fine gold particles have been washed in this area recently by geologists of the Natural Resource Authority Amman.

Chapter 6
Study of Archaeometallurgical Slag and Metal

6.1 Slags of the 4th Millennium BCE

The 'Most Ancient' Slags

Ancient slag from non-ferrous metallurgy has mostly been divided into two groups based on technological grounds: smelting slags and crucible slags (melting/refining/casting slags). Samples analyzed from the former have been mainly from sites dating into the (Late) Bronze Age, Iron Age or even later periods and were collected from large slag heaps, which indicated a mass production of metal. Crucible slags, on the other hand, were classified as waste products from the processing of non-ferrous metals e.g., by remelting or alloying. They acquired their name from the small crucibles in which these processes were carried out. Crucible slag has been analyzed, e.g., from Late Bronze Age tin-bronze casting workshops in Kition (Cyprus) (Zwicker et al. 1985), Nichoria (Cooke and Nielsen 1978), Isthmia (Rostoker et al. 1983), and Olympia and Athens of the classical period (Zwicker 1984).

Cooke and Nielsen (1978) and later Craddock and Freestone (1988) worked out criteria for the definition of crucible slag based on suggestions for the differentiation between smelting and crucible slag (Tylecote 1976). The remains of the pottery crucibles still adhering to the slag are one unmistakable attribute. The slag is often attached to the rim of the crucible and shows transitions to the pottery (Hauptmann et al. 1993, Klein and Hauptmann 1999). Chemically, crucible slags are alkali-aluminum silicates, which develop by reactions between metal and oxide melts with crucible material, Fe-contents in the copper, charcoal ash and silica-rich fluxes. Their $SiO_2 + Al_2O_3/FeO + MnO$ ratio is high. Cu and/or Cu_2O concentrations are higher than those of the smelting slags. Crucible slags are heterogeneous, largely glassy and porous, while smelting slags are well crystallized and show only a few large gas bubbles. Fayalite appears typically in smelting slag but not necessarily in crucible slag, while the latter contains higher-valent oxides, e.g., delafossite.

This definition becomes less clear-cut when one looks at the slag from the beginnings of extractive metallurgy. Several studies demonstrate that at least up to the beginning of the Early Bronze Age I, pottery crucibles were used for most of the metallurgical processes (Maggetti 1990; Palmieri et al. 1993; Hauptmann et al. 1993; Müller-Karpe 1994; Rehren 1997b, Craddock 2002). For a long time, this fact hindered the precise understanding of the initial stages of metallurgy, as can be shown by the example of the slag from Çatal Hüyük (Anatolia), an early Neolithic settlement from the middle of the 7th millennium BCE. The minute slag granules were published by Neuninger et al. (1964). The unusually early date and the rather limited sample size made it impossible for the authors to sufficiently explain the formation of the samples. Tylecote (1976) thought it improbable that the Çatal Hüyük slag was a product of smelting, because it contained no Fe silicate. Pernicka (1995) even stated that it would have been the product of an accidental domestic fire, which might have caused the creation of the earliest known droplet of molten copper. The proof presented by Sperl (1997) for this molten copper refers only to a ca. 1–2 mm large copper prill. Now it seems to be more than likely that the slags from Çatal Hüyük are the first evidence of ore smelting in crucibles, because the material shows a remarkable similarity to the slag from settlements of the 4th millennium BCE like Abu Matar, Wadi Fidan 4 (Hauptmann 1989; Hauptmann et al. 1993), Tell Maqass, and Tell Hujayrat al-Ghuzlan near Aqaba (unpublished results Bochum/Berlin). Experimental studies have shown that it is not difficult to produce metal in crucibles (Donnan 1973; Tylecote 1974; Gale et al. 1990). It is therefore not very surprising that Glumac and Todd (1991) interpret slag from Selevac, in former Yugoslavia, dating from the Vinèa period (mid 5th millennium BCE), as crucible slag produced by smelting ore. Bachmann (1978) studied what was assumed to be Chalcolithic slag from Timna, which was supposedly already produced in smelting furnaces. He interpreted the slag in general terms as coming from a technological 'trial and error' stage. But he also thought it possible that slag found in the Sinai and dating from the 3rd to 1st millennium might have come from smelting ore in crucibles (Bachmann 1980). Shalev and Northover (1987) assumed moreover that the slags excavated in the Chalcolithic settlement of Shiqmim (Israel) were also produced by smelting ore in crucibles. The same opinion is put forward by Yalçin et al. (1992), who have analyzed the slag from Norsuntepe (eastern Anatolia) from the end of the Chalcolithic period (4th millennium).

All these slag samples have weight and size in common. They hardly weigh more than a scant ten grams and reach nut-size at the most. They have a heterogeneous structure and generally a high cuprite and higher-valent content of Cu-Fe oxides. These characteristics would mark them as crucible slags. But their high concentration in Fe oxides, which is often at 20–40 wt.-%, does not fit the original definition. This, as well as the often accompanying finds of ore, points, on the other hand, towards smelting processes. These facts indicate that the differentiation based on the definitions usually used for crucible and smelting slag does not apply to the oldest finds.

The studies presented below have at their heart the goal of detecting those metallurgical processes, which have produced the slag finds from Wadi Fidan 4 (WF4), as an exemplar for all sites of the second half of the 4th millennium at least in the Faynan area.

The Material from the Faynan District

The slags were found in area D in direct association with high-grade secondary Cu ores and fragments of partly slagged crucibles in the settlement of WF4 (see Sect. 5.2). This slag is, so far, the earliest found in the Faynan area, although its dating does not mark the actual beginning of metallurgy in the Southern Levant. Already visible to the naked eye, it shows stages of transition from decomposed ore to slag that had been affected by high temperatures with adhering ceramic fragments. This identifies the slag *a priori* as products of ore smelting. The size of the different pieces (being only a few centimeters), the archaeological context (where they have been found together with the crucibles), as well as the negative evidence of missing smelting furnaces make it very likely that this slag was generated in small crucibles. It can therefore with all due consideration be called crucible slag.

Thin surface-polished sections and mounted samples were prepared from 33 slag samples in order to analyze them by light microscopy using polarized (transmitted and reflected) light, for metallographic studies and microanalyses under the scanning electron microscope as well as electron microprobe. The impetus was on analyzing slag with adhering crucible lining and showing the fabric and texture of completely preserved slag via thin polished sections.

From ten samples, chemical bulk analyses were carried out using atomic absorption spectrometry; the results can be found in Table A.4. The heterogeneous nature of the slags called for further study and the bulk analyses were therefore complemented by 50 semi-quantitative micro-plane scans (each ca. 4 mm^2) on thin sections, performed by an energy-dispersive system attached to a scanning electron microscope.

Chemistry and Melting Behavior

The totals of the bulk analyses are partly far below 100 wt.-%. The reason is that Cu to a degree exists in its oxidic state but was calculated as Cu$^{\pm 0}$. The Fe concentrations were calculated as FeO although several different Fe oxides, partly even Fe-hydroxides, are present. Finally it must be noted that the slag could contain sulphur in the form of Cu sulphide, which has not been analyzed.

The earliest slag in Faynan is relatively low in MnO, which is not above 6 wt.-% and mostly even below 1%. This is one characteristic that divides them from slag of later periods, whose MnO concentration is at 20–40 wt.-%. This might be explained by the fact that mining in the 4th millennium was predominantly directed at the mineralizations in the sandstone, which were poor in Mn content, but also that copper ores from the DLS were carefully separated from the dolomitic host rock. But the most important characteristic of the slag is their extremely high Cu content, which is mainly present as cuprite, which was already recognizable macroscopically by the dark red hue of the slag. If the total copper content is calculated as oxide, then the samples contain 13–60 wt.-% Cu$_2$O, which shows this particular oxide to be one of the main components. These slags are far removed from the original definition of smelting slag, as it is valid for later periods and recent times. The high metal content is inconsistent with the essential quality of (modern) slag: here it is *not* the necessary means to separate the gangue from the metal in the liquid state. It proves rather that the finds from WF4 are slagged ore (with percentages of Fe-(hydr-)oxides and quartz), which was already understood from the macroscopic studies. It will be further verified in the following chapter with analyses of the texture.

Such high Cu oxide contents don't seem to be unusual in slag from the beginnings of extractive metallurgy in the Near East. This can be seen in Fig. 6.1, where the Cu$_2$O-content of such slag is plotted with characteristic slag-forming oxides in common silicate melts (SiO$_2$, Al$_2$O$_3$, Fe oxide, CaO). The Cu-content of slags from

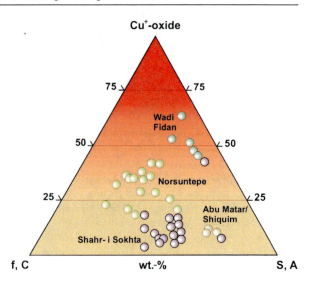

Fig. 6.1.
Chemical composition of the slag from Wadi Fidan 4 and some other sites of the 4th millennium BCE shown in the system Cu_2O–f–S,A. Notice the development of extremely Cu-rich slag to iron-rich slag with low Cu losses, which date into the Early Bronze Age I and later periods. *EBA:* Early Bronze Age; *f:* total iron oxide content (as FeO) +MnO+MgO; *S:* SiO_2; *A:* Al_2O_3 (after Hauptmann et al. 2003)

the transition towards the 3rd millennium and later periods decreases drastically. Early Bronze Age slags therefore would plot along the lower side of the triangle. This is a clear indication of the increasing technological improvement in the later stages of metallurgy (see also Fig. 6.16). Figure 6.1 shows Early Bronze Age I slags from Nevali Çori and Çayönü Tepesi, both in eastern Anatolia, (Hauptmann et al. 1993; unpubl. data DBM), as well as slag from Murgul (Lutz et al. 1994), Timna 39 (Bachmann 1978) and Shahr-I Sokhta (Iran) (Hauptmann et al. 2003). The low copper content in the slag from the Late Chalcolithic Murgul can only be explained by an unusually developed smelting technology there, while the interpretation for the samples from Timna 39 is still unclear, because the dating of the slag into the Chalcolithic is not secure.

The bulk analyses of the slag from WF4 (Table A.4 and A.5) show alkali-concentrations of max. 2 wt.-%, while P_2O_5 and BaO are in the range of a few percent and Pb in the range of 1/10th percent. The erratic concentrations of the main elements SiO_2 (25–40 wt.-%), CaO (1–45 wt.-%), MgO (<1–7 wt.-%) and Fe oxide (<1–30 wt.-%, calculated as FeO) are noticeable. They vary even more in detailed plane scans. When combined with the textural analyses, these variations can be compared directly with the compositions of the Faynan copper ores and can therefore be easily explained and systematized. Leaving aside the Cu_2O-content, two groups can be observed:

1. Silicate slags with varying Fe oxide contents
2. CaO-MgO-rich silicate slags

The first group with the main components SiO_2, Fe oxide and CaO shows the beginning of most of the iron-rich silicate slag of early copper smelting. We therefore attempted to estimate their melting behavior and the firing temperatures respectively by a projection of analyses reduced to these components. For this study, the ternary system CaO–FeO–SiO_2 that is usually applied to copper slag was utilized, whose configuration has been determined by Eugster and Wones (1962). Small

6.1 · Slags of the 4th Millennium BCE

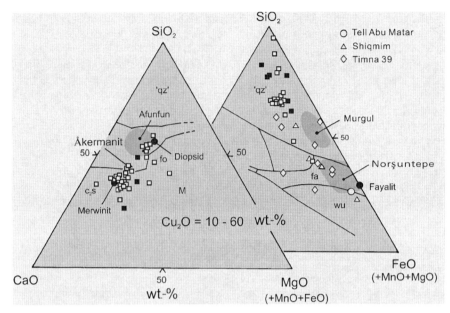

Fig. 6.2. The chemical composition of slag from Wadi Fidan 4 is presented here in the systems CaO–SiO$_2$–MgO(+MnO+FeO) after Osborn and Muan (1960) and CaO–SiO$_2$–FeO(+MnO +MgO) after Eugster and Wones (1962). The low melting region around the liquidus of fayalite (*fa*) is indicated in the diagram on the *right*, which is limited by the steeply rising isotherms of the SiO$_2$-liquidus ('*qz*'). For reasons of comparison, Chalcolithic/Early Bronze Age slags different from those of WF4 are shown. In the diagram to the *left*, the analyses also cluster in the indicated low-melting area of the system. The composition of the early copper smelting slags from Afunfun and Ikawaten (Niger) are illustrated in the dark shaded area. They have also been melted from sandy dolomites (Killick et al. 1988). Bulk analyses (*black squares*) as well as semi-quantitative analyses of plane sections on the micro-scale (*hollow squares*) of the slag are plotted here. FeO always means the total of iron oxides. *wu:* wuestite; *fo:* forsterite; *M:* periclas; *C$_2$S:* modifications of Ca$_2$SiO$_4$

percentages of MgO and MnO have been added to the Fe oxides, based on crystallo-chemical considerations. Figure 6.2 shows in the diagram the bulk and semi-quantitative analyses of sections on the micro-scale of these iron-rich compositions. One can recognize a development from quartz-rich compositions with strongly varying FeO/SiO$_2$ ratios moving only haltingly towards the fayalite composition. Melts with such a composition would require, under equilibrium conditions in a pure system, liquidus temperatures of 1 600 °C or more. This is an unrealistic assumption, as it is known that in early metallurgy the formation of slag melts was possible only up to an upper temperature level of about 1 100–1 300 °C. The reason for the high SiO$_2$ contents are rather (as will be clarified further by petrographic analyses) found in un-decomposed quartz inclusions, which are embedded in the actual liquid slag matrix. These are relics of sandstone from the MBS. Compared with the sandstone (e.g., sample JD-12/2, Table 4.1), the slag has a higher CaO-content that can in all likelihood be explained by calcite in vein fillings. Some other slags, e.g., those from Norsuntepe, Shiqmim and Tell Abu Matar, seem to come very close to the ideal composition.

The high Cu_2O, Na_2O, P_2O_5 and Pb concentrations probably caused lower liquidus temperatures. In the system $Cu-O-SiO_2$, several binary and ternary eutectic points exist at ca. 1 100 °C, e.g., the eutectic Cu_2O-SiO_2 with 8% Cu_2O is at 1 060 °C (Berezhnoi et al. 1952). This temperature is at a level, which seems far more likely, as was the case with the Fe-rich slag. This will be underlined later when the mineralogical phases are studied. In general, statements made about the melting behavior of slags, based upon the application of phase diagrams, are not very helpful without further mineralogical studies. They usually lead to unrealistically high temperatures. In addition, the liquidus of fayalite and 'quartz', as indicated in Fig. 6.2 are only valid for reducing conditions up to $p_{O2}= 10^{-13}-10^{-9}$ atm. The phase diagram strongly depends on the composition of the gas atmosphere. If p_{O2} is higher, which may be suggested here due to the formation of cuprite, then the fayalite field will disappear, and SiO_2 is precipitated along with either magnetite or hematite, leading to a drastic rise of the liquidus temperature (Muan and Osborn 1965). In such a case, Cu^+ und Cu^{2+} would have to be included in the calculation of the oxidic phases (see below). More realistic is the discussion of analyses of plane sections on the microscale. One of these analyses (Lz-276/2c in Table A.5, Appendix) plots for example at the borderline between the liquidus of wollastonite and fayalite. Here, the liquidus temperature in the pure system is 1 150 °C. The analyzed area shows a texture that was created through the solidification of a melt and that contains no inclusions. Mineralogically it consists of glass and clinopyroxene.

The second group of slag consists of the main components SiO_2, CaO, and MgO. The CaO/MgO ratio is 2.0–6.3 ($\varnothing$ = 3.96), the ratio (CaO + MgO)/ SiO_2 = 0.66 – 2.69 ($\varnothing$ = 1.49). The bulk chemistry of these slags is similar to that of modern shaft furnace slags. Here, the quaternary system $CaO-MgO-SiO_2-Al_2O_3$ may be used to discuss it further. This diagram is of fundamental importance in steel making. It has been studied in great detail by Osborn et al. (1954). For our issues, however, the ternary subsystem $CaO-MgO-SiO_2$ is sufficient, because the Al_2O_3 content of the slags from WF4 is only at ca. 3%. When calculating the reduced analyses, FeO- and MnO-concentrations were added to MgO. The geometry of the different liquidus fields in the low melting area, as also indicated in Fig. 6.2, is only slightly repositioned by it. This ternary system has the advantage that it is not dependent on the composition of the gas atmosphere, because the metal oxides involved here are stable under the existing redox conditions. The liquidus temperatures rise sharply in the precipitation fields of SiO_2, periclase, forsterite, and the Ca_2SiO_4 modifications around the indicated eutectic area. The ideal compositions of the slags should therefore lie in the indicated sector.

The projection of the bulk analyses of the second slag type, as well as the analyses of plane scans, shows that a number of them fulfill these conditions. But a number of the plots are outside the eutectic area; the corresponding samples (e.g., JD-8/24, see Fig. 6.4) show a lower degree of liquefaction. It needs to be mentioned that in the system $CaO-MgO-SiO_2$, the formation of slag depends essentially on the SiO_2 content of the charge. Dolomite with a SiO_2 content of only 10 to 15%, e.g., is a highly fire resistant material, which has a very high melting temperature (>2 000 °C) after calcination and is therefore employed in modern metallurgy as a refractory. Only with a content of 40–60% SiO_2, eutectic compositions are reached with melting temperatures around 1 400 °C.

The liquidus temperatures in the low-melting area of the system $CaO-MgO-SiO_2$ are 150 to 200 °C higher than in the above-mentioned system $CaO-FeO-SiO_2$. This does not allow us to presuppose that the Ca-Mg silicate slag from WF4 was produced by higher temperatures. The limitations mentioned above, which are caused by the contents of further minor constituents, are also valid here.

The chemistry of these slags can be easily explained by the use of ores from the Dolomite-Limestone-Shale Unit (DLS). A typical source material, of all the ore bearing rocks studied so far (Table 4.1), could be, e.g., the sandy dolomite JD-3/3. This rock has a comparable CaO/MgO ratio of 1.66 and additionally contains Pb, Na_2O and P_2O_5 in a range of 0.1 to 1%. The SiO_2 content of 6% is too low when compared with that of the slag, but marly dolomites can contain up to 15% SiO_2 and then also several percent Al_2O_3. Furthermore, the smelting of high-percentage Cu silicates can lead to a rise in the SiO_2 content in the slag.

Only one parallel of the chemistry of this slag has so far been found among (pre-)historic examples in general, and that is from Afunfun and Ikawaten in the Agadez region in Niger (Tylecote 1982; Keesmann 1985; Killick et al. 1988). The smelting activities there date into the 3rd until the late 1st millennium BCE. Tylecote doubted that the material would be 'real' smelting slag, because it does not contain the expected amount of iron oxide of other archaeometallurgical slag. He interpreted the slag as a product of reactions between dolomite, soil environment and an uptake of CaO from charcoal ash (to a magnitude of over 20 wt.-%!), which would have originated during the melting of native copper. Killick et al. (1988) suggest on the other hand (a far simpler and more realistic proposition) that it is slag produced by the smelting of oxidic and silicate copper ores containing some native copper. The composition of the slag could then be explained by the sandy-dolomite host rock of the ore mineralization as is the case in Faynan.

The fact that the slag from WF4 can be described as direct products of highly heated but not completely liquefied ore bearing rock permits an interpretation of the slag chemistry without a (deliberate) addition of fluxing agents, a resorption of crucible lining or an uptake of charcoal ash. Tylecote et al. (1974) have explained the chemistry of very similar slag from the Chalcolithic settlement of Tell Abu Matar as a result of the uptake of dolomite rock fragments from the crucibles, at that time surely without any knowledge of the regional geology in Wadi Arabah and the prehistoric trade connections. In the meantime, it has become clear (Hauptmann 1989, Shugar 2000) that ores from Faynan have been smelted in Tell Abu Matar. The CaO and MgO contents in the slag probably originated from the dolomite in the DLS. Consequently, Tylecote's interpretation needs to be reconsidered. That there is not always an intensive exchange of material between slag and crucible lining can be shown by the slags from WF4 with their low contents of Na_2O und K_2O (always < 1 wt.-%), which Tylecote (1987) understands as typical elements of crucible slag. Such an exchange of material would be surprising, considering the low degree of liquefaction.

These observations make it quite clear that the usefulness of equilibrium phase diagrams for the interpretation of very early slag is limited, as they are too inhomogenously composed. Useful statements about melting behavior and firing temperatures can only be suggested when plane sections, which have truly solidified from the melt, are analyzed on the micro-scale. It is advisable to supplement chemical analyses with studies of the texture and the mineralogical phase content.

The Petrography of the Slag

The heterogeneous texture and fabric of the slag pieces from Wadi Fidan 4 (WF4), already observed macroscopically, becomes even clearer in the thin section. Complete liquefaction was not the rule; it happened only in small parts. The slag has formed in a semi-liquid state from several pieces of ore that are sintered together or gangue material (Fig. 6.3). The use of finely grained, powdered ore might be possible but has not been recognized. Any inclusions of charcoal are also missing, such as that which can be found regularly in the (completely liquefied) slag of later periods. The chemically distinguishable groups show a varying phase content (Table 6.1) plus a varying texture as was expected.

The Fe-rich silicate slags form small cakes with a solid framework of quartz grains, which are embedded in varying amounts of the flowable slag matrix (Fig. 6.3). There are still structural constituents of ores from the MBS left in these pieces. Sandstone remains dissolve through a reaction with the surrounding silicate melt successively into a mixture of tridymite and cristobalite. The matrix contains, in contrast to these light-colored relics of the charge, a high percentage of opaque components: Cu and Cu-Fe oxides such as cuprite, magnetite and delafossite, as well as glass. Fe-rich diopside was observed as the only iron silicate.

Relatively light-colored slags, assigned as Ca-Mg silicate slags, can easily be identified within the sample collection. They still have, to a degree, a heterogeneous texture and show relics of the only partially decomposed charge (Fig. 6.4); in part they are much better crystallized and consist of coarse crystals of diopside, merwinite and åkermanite in a matrix of glass with finely dispersed cuprite. The copper oxide also forms spectacular ruby red inclusions that are the size of around 1 cm (Fig. 6.5). These slags stem from the smelting of DLS ores.

The heterogeneous composition of slag with partially very high volume contents of (refractive) rock fragments seems to be a fairly widespread characteristic of many slags from the beginnings of extractive metallurgy (Lutz et al. 1994; Hauptmann et al. 1993; Keesmann et al. 1997; Hess 1998). The reason for this mainly lies in the use of small reaction vessels for the smelting, as has already been shown with the example of WF4 (Hauptmann 1989). The charged ore was exposed only briefly to the maximum firing temperatures, which was hardly sufficient to make it reach a complete liquefaction. The slags of WF4 show all the transitional stages between thermal decomposition of the original material up to the formation of proper melts. This also means that we have to expect, as has already been observed by Hauptmann (1989), phase formations by reactions in the solid state. The interaction of solid constituents with liquefied parts principally shows parallels to the sintering process, e.g., in pottery or the mineralogy of refractories. It is also comparable with the effect of slag melts on alkaline refractive material or silica bricks. The formations of phases under such conditions cannot reach a thermodynamic equilibrium, which would require high temperatures and a sufficiently long reaction time. The formations are mainly controlled by kinetic processes, as they are typical of the dynamics in firing processes (Brownell 1976). The formation of melts with a tendency towards phase equilibrium can be observed, but it occurs often only in very small areas of the slag, setting out from the grain boundary of different minerals.

Table 6.1. Overview of slag phases occurring in the slag of Faynan, in a simplified, idealized formula script

Slag phase	Formula
Cu-phases in all slags	
Cuprite	Cu_2O
Chalcocite	$Cu_{2-x}S$
Covellite	CuS
Copper	Cu
Mn-free slags of the 4th millennium BCE	
Åkermanite	$Ca_2MgSi_2O_7$
Alite	Ca_3SiO_5
Calcite (?)	$CaCO_3$
Celsian	$Ba[Al_2Si_2O_8]$
Delafossite	$Cu^+Fe^{3+}O_2$
Diopside	$CaMgSi_2O_6$
Fe-rich diopside ('salite')	$Ca(Fe,Mg)Si_2O_6$
Glass	Fe-Al-Ca silicate
Magnetite	Fe_3O_4
Merwinite	$Ca_3MgSi_2O_8$
Periclas	MgO
Quartz/tridymite/cristobalite	SiO_2
Schreibersite	Fe_3P
Mn-bearing slags of later periods	
Baryte	$BaSO_4$
Bixbyite	$(Mn,Fe)_2O_3$
Lead-silicate (glass)	$Pb-SiO_2$
Braunite	$Mn^{2+}Mn_6^{3+}[O_8/SiO_4]$
Celsian	$Ba[Al_2Si_2O_8]$
Crednerite	$CuMn_2O_4$
Fe-phosphate	$Fe_2P_2O_8(?)$
Glass	Mn-Ca-Al silikate
Hausmannite	Mn_3O_4
Hilgenstockite (?)	$Ca_4P_2O_9$
Jakobsite	$(Mn,Fe)_3O_4$
Johannsenite	$(Mn,Fe,Ca)[SiO_3]$
Knebelite	$(Mn,Fe)_2SiO_4$
Partridgeite	Mn_2O_3
Pyrolusite	MnO_2
Pyroxenoids (bustamite, pyroxmangite, rhodonite)	$(Mn,Fe,Ca)[SiO_3]$
Schreibersite	Fe_3P
Tephroite	Mn_2SiO_4

It is not correct, however, to deduce from these facts a 'solid-state'-melting, meaning a reduction of ore in the solid state (Tylecote and Merkel 1995). This is not necessary, not least because the reduction of ore to (liquid) metal takes place much faster than the formation of slag melts. Therefore, the aim of the Chalcolithic copper

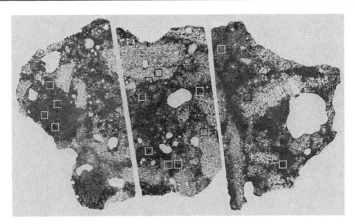

Fig. 6.3. Wadi Fidan 4, sample JD-8/13a. Section through Fe-rich silicate slag of the Early Bronze Age I period (entire slag body, composed of three thin sections). The light colored, undecomposed sandstone relics are clearly visible, embedded in a darker melt. The framework of quartz grains dissolves successively while forming tridymite and cristobalite. Those sections, whose chemical compositions have been measured with EDS, are marked (see Table A.5) (width of piece: 5.5 cm)

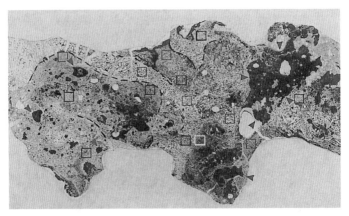

Fig. 6.4. Wadi Fidan 4, sample JD-8/24. Section through an Early Bronze Age I Ca-Mg silicate slag with a low degree of liquefaction. One can recognize the partly decomposed (sandy) dolomites (*marked areas*); in between are crystallized areas and dispersed inclusions of cuprite (*arrows*). Those sections, whose chemical compositions have been measured with EDS, are marked (see Table A.5) (width of piece: 3.7 cm)

production could have only been to 'liquate' the metal itself from a mixture of ore and host rock through high temperatures and under fairly reducing conditions. Slag was thus not a deliberately produced material as it was the case in the smelting processes of later periods (Hauptmann et al. 1996).

Slags with light, mostly high-melting rock or mineral inclusions have generally been named – not very fortuitously – 'free silica slags,' and their formation has been interpreted very varyingly. In consideration of the above-described observations, these slags will be treated here in more detail.

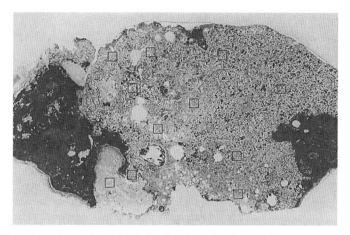

Fig. 6.5. Wadi Fidan 4, sample JD-8/29. Section through a Ca-Mg silicate slag from the Early Bronze Age I as an example of completely liquefied slag. The black areas are individual drops of cuprite, which is also finely dispersed in the crystallized slag. Those sections, whose chemical compositions have been measured with EDS, are marked (see Table A.5) (width of piece: 3.8 cm)

'Free Silica Slags'

Tylecote and Tylecote (1981), while describing the early copper and lead/silver technology in southern Spain, created this designation. Keesmann (1991, 1993) maintains that this would be a petrographically misleading term. Even disregarding the fact that one often finds other inclusions than quartz, the term 'free silica slags' would invoke a crystallization of cristobalite and/or tridymite as SiO_2-surplus phases from a (homogeneous) melt, independent of the quartz inclusions. However, his studies showed that the respective slags are mostly fayalitic or pyroxenitic and not necessarily SiO_2-oversaturated.

The formation of 'free-silica' slag has been explained in a variety of ways. In the case of the Spanish slag mentioned, it has been suggested that quartz pieces were deliberately added materials to liquid slag. Tylecote (1987) presumed that these would have been added to cool down the slag in order to facilitate the removal of the slag cake from the molten metal bath underneath – following the principle of 'Scheibenreißen'. At Monte Romero 'free-silica slag' was seen in connection with the recovery of lead from its oxidic waste (Keesmann 1993), and rock pieces are interpreted as a kind of sieve, which would facilitate the 'absaigern' (liquation) of the lead (Kassianidou et al. 1995). Lutz (1990) proposes a similar model for Chalcolithic copper slag from Murgul (Turkey). All these suggestions of deliberate technical optimization methods are, however, based on descriptions of historical metal technology from the 19th century (Percy 1870).

It has not so far been considered that these inclusions might be remains of a charge coming from a quartz-rich host rock, as is the case in WF4. This could be true, e.g., for the Spanish slag, because 'free silica slag' has been often described there in connection with quartz-bearing ores. And it is more plausible to interpret inclusions of xenomorph inclusions of quartz/cristobalite described from a very inhomogeneous Late Neolithic slag from Brixlegg/Austria as a "resister" instead of primary precipitates from a melt (Bartelheim et al. 2002). This is also true for the copper slag from Oman (Hauptmann 1985). Piel et al. (1992) and Metten (2003) describe fayalitic slag from Trentino, representative for numerous sites with Late Bronze Age copper smelting in the Alps, as having a large volume percentage of quartz and rock inclusions with a size of a few centimeters, which also come undoubtedly from the host rock of the ore mineralization. Low melting Cu-Fe sulphides were separated from it as matte, presumably with portions of metallic copper and a eutectic (fayalitic) slag melt. The purpose of the whole process was to liquate the metal-bearing phase, at relatively low process temperatures, from the refractory gangue. The quality of the slag was peripheral so that SiO_2 or SiO_2-rich gangue material stayed undecomposed in the slag. Recent research showed that in the Alps these techniques can be dated all the way back to Neolithic copper production (Metten 2003). This appears to be plausible, because it must be assumed that the techniques for metal extraction during the Alpine Neolithic were rather simple – as was the case in WF4.

The different possibilities for the formation of 'free silica slag' in lead and copper metallurgy cannot always be discerned unambiguously by petrographic texture analysis. When quartz is present in fragments, it is not possible to determine if they come from the original charge or have been added afterwards. They always show phenomena of resorption due to non-equilibrium with the silicate melt, mostly in the shape of rounded edges and cavities comparable to early precipitated quartz in rhyolites, being surrounded by a rim of glass with high SiO_2 contents or pyroxene.

It proved to be helpful to study reactions between different oxides with quartz, such as, e.g., those in the binary phase diagrams SiO_2-PbO and SiO_2-Fe oxide (Fig. 6.6). The first one shows that already small amounts of PbO (and of alkali oxides, Al_2O_3 and BaO) are sufficient to dissolve SiO_2 and to lower the liquidus/solidus temperatures and with it the viscosity to a level of ca. 700–750 °C. During smelting, this leads to a premature formation of slag with low viscosity before the necessary reactions are completed to precipitate metal from ore (Tafel 1929). In such a case, an addition of

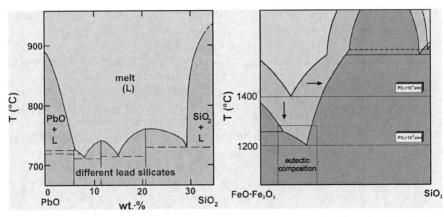

Fig. 6.6. The system SiO_2–PbO shows that already small amounts of PbO lead to liquefaction at relatively low temperatures, meaning that SiO_2 will be dissolved. In the system SiO_2–Fe oxide, high amounts of FeO are necessary; on the other hand, relatively small amounts of SiO_2 can cause an iron-rich silicate melt to freeze. This is particularly true under more oxidizing conditions (the diagram is modified after Muan and Osborn 1965)

quartz might be helpful, and as a matter of fact, quartz inclusions sometimes occur to a large extent in old lead slag (Hauptmann et al. 1988; Kassianidou et al. 1995).

Iron-rich silicate melts of copper metallurgy are different. FeO and other divalent oxides (CaO, MgO, MnO), which occur in such slag, have a far more limited ability for forming a melt with SiO_2 due to their higher liquidus/solidus temperature. To put it a different way, when SiO_2 is added to a (fayalitic) slag melt, the danger arises that the liquidus/solidus temperatures rise very drastically, leading to a solidification of the melt and cause the furnace to freeze. This is particularly true for working under more oxidizing conditions. The adding of quartz would be very difficult to understand in such a case and could hardly be explained with the reasons valid in lead metallurgy.

The Development of Mineralogical Phases

Texture analysis has shown that the slags from WF4 have been only partly liquefied. Accordingly, phases appear that have been formed through solid-state reactions and others, which have been formed by a crystallization of the melt. This explains why thermodynamically incompatible phases can exist together in one sample.

Slag, which was liquefied from sandstone-bearing ores of the MBS, shows solid-state reactions primarily with undecomposed quartz. *Tridymite* and *cristobalite* were identified here, both being high-temperature modifications of quartz. The reconstructive transformation of quartz to tridymite and cristobalite results in this case most likely from a reaction of the finely grained quartz with Fe-Ca-Al-rich partial melts. This conforms to the observations made by Flörke (1959) and Kienow and Seeger (1965), who showed that the formation of tridymite is speeded up by non-lattice ions, particularly alkalis. Indeed, cristobalite and tridymite analyzed in the slag from WF4 with a microprobe reach a SiO_2 concentration of often only 98.6 wt.-% with additional percentages of Na_2O and K_2O.

Tridymite crystals are formed perpendicular to the surface of the quartz grains and form dense felt-like structures, while the original roundish shape of the quartz granules survives at the beginning of the process (Fig. 6.7). The grains are fissured

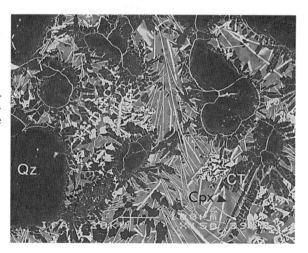

Fig. 6.7.
Wadi Fidan 4, sample JD-8/20. Reaction between liquefied slag (*grey matrix*) and quartz (*Qz*), the latter being a relict from the mineralized sandstone. Quartz grains have light-colored fissures; they are transformed progressively from the outer edge to the core into a mixture of cristobalite and tridymite (*CT*). At the contact point with the melt, sections of tabular tridymite. In the glassy slag melt columnar delafossite, magnetite and clinopyroxene (*Cpx*) (SEM-picture, backscattered mode)

by cracks starting at the outer rim, through which the melt penetrates further into the grain and speeds up the transformation. This causes a separation of the grains into different sections, which seems to be a typical process that has already been observed by Flörke (1959) during the transformation of tempered quartz into Na-rich silicate melt. The quartz grains are dissolved successively. The matrix between the tridymite crystals solidifies as a glass, or the tridymite crystallizes partly intergrown with magnetite or Fe-rich diopside.

The 'Ballengefüge' typical of cristobalite occurs as a corona around the quartz grains as well as in the tabular tridymite crystals. Here is, after Flörke (1959), a structural mixture of both phases occurring. Kienow and Seeger (1965) observed a zonation of cristobalite as well as tridymite, depending on temperature, in silica bricks used in modern shaft furnaces. Cristobalite forms favorably at the hottest zone of the firing vessel at a temperature range of 1 200 to 1 400 °C (Baumgart 1984), while tridymite crystallizes in the area of low t-gradients and after a longer exposure to heat. It would, however, be a mistake to try and deduce from these observations the exact temperature achieved during the formation of slag melts, as the inversion temperature of the two modifications of SiO_2 cannot be estimated due to the complex composition of the melt.

The formation of tridymite/cristobalite during the decomposition of quartz is obviously more likely to occur in slag, which develops from finely grained parent material, because these phases have not been observed in the slags with (lumpy) quartz inclusions described so far (see above). Tridymite has been identified only in finds from Timna (Bachmann 1978), which is not too surprising as sandstone-bearing ore has been smelted there.

The most distinctive characteristic of Chalcolithic/Early Bronze Age I slags is a phase content, which was formed by the crystallization of (partial) melts under a moderate to weak reducing gas atmosphere. This is confirmed by the recurring appearance of *cuprite* in all slags. It occurs in all transitions from the beginning solid-state reduction of secondary copper minerals up to the crystallization from a melt via a finely dispersed liquid-liquid separation. This intergranular exsolution causes the macroscopically visible red tinge of the slag that was already mentioned. Even copper prills, which contain cuprite in eutectic intergrowth, become oxidized again – through intermittent oxidation during the (s)melting process (Fig. 6.8). The texture of the cuprite clearly shows that it precipitated from the liquid state, because had it been a product of corrosion it would have taken shape as a stack of single layers, which would lie concentrically around the metal core.

In the group of iron-rich slag, (Cu-)Fe oxides (delafossite, cuprite, magnetite) + clinopyroxene + SiO_2 + glass can be observed as typical phase associations in liquefied sections of the slag. Their crystallization is directly dependent on the oxygen's partial pressure. Under increased oxidizing conditions, (Cu-)Fe oxides + clinopyroxene crystallize at the expense of silicates and oxides with Fe^{2+}, represented by the two reactions:

$$3\,Fe_2SiO_4 + O_2 \rightarrow 2\,Fe_3O_4 + 3\,SiO_2 \tag{I}$$

$$2\,Cu_2O + Fe_3O_4 \rightarrow 3\,Cu^+Fe^{3+}O_2 + Cu \tag{II}$$

Fig. 6.8.
Wadi Fidan 4, sample JD-8/24. Copper prills (*center*), which were oxidized into cuprite (*grey rim*) during the smelting process (see text). The eutectic intergrowth of cuprite and copper is dimly visible along the boundaries of the copper grains. The copper has a white inclusion of lead (SEM-picture, backscattered mode)

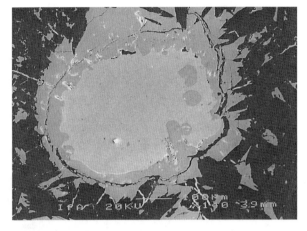

Fig. 6.9.
Wadi Fidan 4, sample JD-8/20. Skeletal magnetite (*white*) in eutectic growth with a mixture of cristobalite and tridymite (*black*). In between is a glassy slag matrix (*grey*) and columnar delafossite (SEM-picture, backscattered mode)

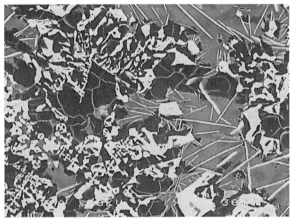

Magnetite develops through crystallization from the melt into idiomorphic individuals, but exists also in (eutectic) intergrowth with cristobalite (Fig. 6.9), thus representing the right-hand part of the reaction I. Under more reducing conditions it would be fayalite, which developed from the same bulk composition (left-hand part of formula I). Irregular agglomerates of magnetite are of importance for the reconstruction of the smelting process. They represent partially decomposed ore or gangue material, as do the quartz inclusions, and have formed via reduction from limonite. They almost regularly contain inclusions of copper. Such inclusions have sometimes been a bit rashly interpreted as deliberately added fluxes (e.g., Merkel and Tylecote 1999). This is very difficult to assess, particularly in those cases when adequately composed raw material exists or when slag has not been completely liquefied.

As the solubility of Cu in magnetite is – after studies by Luraschi and Elliott (1976) – very low, it dissolves in small droplets. The solubility of Cu^{2+} in magnetite, on the other hand, rises exceptionally and leads under slightly reducing conditions at 900–1 000 °C to the formation of a $CuFe_2O_4$–Fe_3O_4–solid solution (Gadalla et al. 1966; Schaefer et al. 1970). During cooling down to a temperature < 980 °C, nearly

Fig. 6.10.
Wadi Fidan 4, sample JD-8/13a. Skeletons of magnetite with rims and exsolution lamellae of delafossite (*arrows*) along//(111), which also appears as plate-like crystals in the glassy matrix. In between quartz (*black*), glass and single clinopyroxene crystals (SEM-picture, backscattered mode)

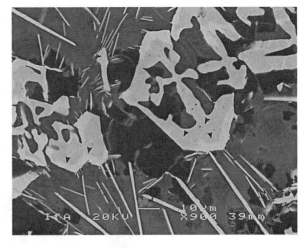

all compositions of this solid solution are subjected to an eutectoid decay forming delafossite + magnetite (+ hematite) (Yamaguchi and Shirashi 1971). In Fig. 6.10 magnetite and exsolution lamellae of *delafossite*//(111) can be recognized. In addition, the Cu-Fe oxide forms rims around the magnetite and continues to crystallize there. Delafossite and magnetite are often the major components of the slag. On the other hand, eutectic intergrowth of delafossite with cuprite develops in extensive areas, which Luraschi and Elliott (1976) interpreted as being a sign of temperatures of just below 1 100 °C, the same range of temperatures that is involved in the formation of the spinel-solid solution.

After the phase equilibria in the system Cu–Fe–O at 1 000 °C, the phase content in the slags represents two stable parageneses:

magnetite + delafossite + Cu   (I)

and

Cu + cuprite + delafossite   (II)

All oxides are associated with a clinopyroxene, which could at the beginning (due to its pleochroism varying from orange-brown to green) not be differentiated from hedenbergite, which often occurs in old slag. Electron microprobe analyses proved that it was actually a Ca-Fe-Mg clinopyroxene, an *Fe-rich diopside* (the outdated name is 'salite', with up to more than 11% MgO, 10–13% FeO, 22–23% CaO, see Table A.6). It is roughly in the middle of the solid solution series diopside–hedenbergite (Fig. 6.11). Although the clinopyroxene contains 2.7–6 wt.-% CuO, the name will be continued to be used here, because such Cu-rich variations do not exist in nature and could therefore not be included in the nomenclature of Morimoto et al. (1988). A small content of Al_2O_3 up to just above 2 wt.-% might have raised the thermic stability of pyroxene and consequently facilitated a crystallization at a temperature range between 1 000 and 1 200 °C (Bowen et al. 1933; Schwarz 1980), just as it does with hedenbergite.

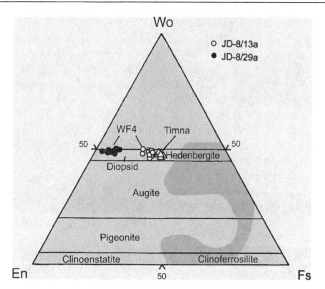

Fig. 6.11. Composition of clinopyroxenes in different slags from Wadi Fidan 4 (WF4), shown in the system enstatite (MgSiO$_3$)– ferrosilite (FeSiO$_3$)–wollastonite (CaSiO$_3$) (after Morimoto et al. 1988) (see Tables A.6 and A.7). For reasons of comparison clinopyroxene that is in all probability from a Chalcolithic or Early Bronze Age slag from Timna (Krawczyk 1986). The area of pyroxene analyzed so far from old, iron-rich silicate slag is also shown (*dark shaded area*) (after data from Hauptmann 1985; Hezarkhani and Keesmann 1996; Metten 2003)

The intergranular '*glassy matrix*' consists of hyalophane parts and shows all transitions of micro- to cryptocrystalline phases, which can be distinguished neither with an electron microprobe nor by X-ray diffractometry. Quantitative measurements with a scanning electron microscope illustrate that the glassy matrix consists mainly of Ca-Al-Fe silicates.

The presence of Cu silicates, as had been proposed, e.g., by Oelsen (1952), has not been reaffirmed. This backs up Gadalla et al. (1962/1963), who had found no indication of the presence of copper silicate in slags. Their studies of the system CuO–Cu$_2$O–SiO$_2$ showed rather that an exsolution of cuprite in the liquid state happens if Cu oxide is not substituted in spinel or crystallized as delafossite. High Cu oxide contents lead to a considerable decrease of the liquidus (see above).

Cu oxide can only be substituted in iron-rich diopside up to 6 wt.-% (Table A.6); it is much less in the other silicate phases, mostly below 1%. Copper is calculated in its divalent form in the calculation of the chemical formula. The existence of copper in the lower %-range in clinopyroxenes and fayalite has been attested in the slag from the Chalcolithic settlement of Norsuntepe (Turkey) by Yalçin et al. (1992). However, the nature of the Fe-Cu and Cu-Ca-Fe silicates described by Glumac and Todd (1991) in the slag of the Vinèa culture is not completely clear. These silicates should contain more than 20 wt.-% CuO judging from the chemical formula ('CuO · FeO · SiO$_2$'; 'CuO · CaO · FeO · SiO$_2$'). The description of the material suggests that very finely dispersed cuprite in a glass matrix has been measured. Cu contents in silicate phases have obviously not been observed in the later slag.

The above-described phase content of the slag from WF4 seems to be common for the early stages of extractive metallurgy. The crystallization of (Cu-)Fe oxides is particularly typical. It indicates high-temperature processes under relatively strong oxidizing conditions, indicating melting processes below only a thin layer of charcoal. In modern metallurgy these kinds of processes occur primarily during the converting of matte (converter-process) (Trojer 1951; Jacob et al. 1977). During prehistoric times, they can generally be found in the Chalcolithic and the beginning of the Early Bronze Age and on an international level from the Near East (Hauptmann et al. 1993; Hauptmann 1985) to the Iberian Peninsula (Keesmann 1994; Craddock 1995). The oxides are often the main components in such slags (Yalçin et al. 1992); in Tell Khuera (Syria) even 'slags' with roughly 90 vol.-% of cuprite were found (unpubl. data DBM). It is not constructive to try and draw on the existence of such (Cu-)Fe oxides as a criterion for a definition of refining slag (Shalev and Northover 1987), because, so far, any proof of the deliberate implementation of such a process is missing. The occurrence of Fe-rich diopside should be remarked upon. While one of the characteristic phases of old copper and iron slag is clinopyroxene of hedenbergitic composition (Hauptmann 1985; Keesmann 1989), Fe-rich diopside ('salite') has been proven only in a few cases. Krawczyk (1986) discovered these phases in a slag from Timna, site 189-A3 and 250B as well as in Yotvata, site 44, north of Timna. In a slag from Timna, site 250B, and only there, are Cu-containing exsolution lamellae in magnetite, supposedly delafossite, compatible with those from WF4. Interestingly enough Tylecote (1988) dates Timna 189 and 250 into the 4th millennium BCE. This would make them roughly contemporary with Wadi Fidan 4. Krawczyk (1986) interprets the slag as a product of reaction with the furnace lining, because she found quartz granules within the slag. This interpretation is not necessarily correct because Timna has, just like Faynan, Cu ores in finely grained sandstone, so the quartz grains could be relictic inclusions from this rock.

In the group of Ca-Mg silicate slag, in those parts which were crystallized from a melt, the main components melilite, diopside and merwinite were observed optically and by X-ray diffraction (Fig. 6.12). Interstitial drop-shaped and dendritic dissolutions of cuprite and copper can be seen, next to celsian and a 'glassy

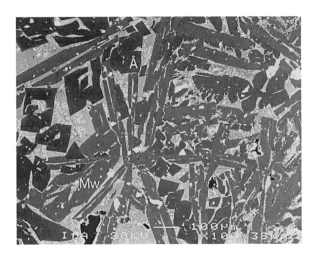

Fig. 6.12.
Wadi Fidan 4, sample JD-8/29. Tabular crystals of merwi-nite (*Mw*) and rhombohedric åkerma-nite (*Å*). In the matrix, dissolution of dendritic cuprite in glass (SEM-picture, secondary electron mode)

matrix' of unidentifiable phases. This phase content is unusual, because Ca-Mg silicates are more characteristic of the medieval slag coming from iron metallurgy (Kresten 1984; Yalçin and Hauptmann 1995) as well as of modern shaft-furnace and non-ferrous metal slag (Obenauer 1954), which are produced by smelting dolomite and quartz.

This is an unusual phase content as the Ca-Mg silicates are more characteristic of Medieval slags from iron production (Kresten 1984; Yalçin and Hauptmann 1995) and modern furnace shaft and ferrous metal slags, which are also produced from a mixture of dolomite and quartz.

In all slags from WF4, *melilite* is the predominant phase. It belongs mineralogically to a complex solid solution with the general formula $X_2YZ_2O_7$ or $(Ca,Na,K)_2(Mg,Fe^{2+},Fe^{3+},Al)(Si,Al)_2O_7$ (Nurse and Midgley 1953). In iron-rich silicate slag produced by the old copper and iron metallurgy, melilite normally occurs only as an accessory constituent. But melilite is the main component in the Ca-Mg slags from Afunun, Niger already mentioned (Keesmann et al. 1985). Microlites of Fe-gehlenite–Fe-åkermanite, formed by devitrification, have been found in medieval iron smelting slag from the Schwäbische Alb (Yalçin and Hauptmann 1995).

Melilite crystallized in slags from WF4 in long prisms or in planar slabs and occasionally shows the typical 'Pflockstruktur'. An opaque phase perpendicular to the surface can be seen. Microprobe analyses showed it to be pure *åkermanite* with the formula $Ca_2MgSi_2O_7$ (Table A.7). It contains up to 2 wt.-% Al_2O_3, while the alkalis are <1%; little iron contents (<1.6%) exist as Fe^{3+}. SiO_2 can reach 44.7 wt.-%. Åkermanite can co-exist with either diopside or merwinite, depending on the CaO/SiO_2 ratio.

Merwinite occurs as a primary phase in long, needle-like to tabular crystals. Polysynthetic lamellar twinning (parallel to the crystallographic 110- and 100-planes) is typical. The lamellae form a herringbone pattern and thus allow an unproblematic identification of this phase if associated with diopside and åkermanite. Merwinite has only a very limited liquidus field in the system $CaO–MgO–SiO_2$ (Parker and Nurse 1943; Osborn 1943), but it belongs to those phases that crystallize spontaneously. Gutt (1967) observed that merwinite also crystallizes from melts with a melilithic composition as a primary phase. In WF4 very pure merwinite is present; the FeO and MnO contents are below 1 wt.-% (Table A.8), and occasionally some Cu_2O has been substituted. Merwinite has so far not been described in archaeometallurgical slag.

Clinopyroxene is the final Mg-rich member of *diopside*. Figure 6.11 shows that it is rather poor in FeO (mostly < 2 wt.-%, see Table A.9), and Si is substituted by Al at below 5%. The occurrence of Fe-poor diopside is unusual in old slags. Clinopyroxenes here are generally rich in iron as are, e.g., hedenbergite and Fe-rich diopside. Keesmann and Hilgart (1992) assume that Mg-rich clinopyroxene could be formed by a resorption of Mg-rich furnace lining in the melt. However, the diopside here originated from SiO_2-containing dolomite, as it is in the case in modern blast furnace slag and in alkaline refractories.

The phase associations observed in the thin section including åkermanite–merwinite and merwinite–diopside along with the glassy matrix correspond with a number of projection points in the system $CaO–MgO–SiO_2$. This becomes obvious in Fig. 6.13, where the phase relations are discussed in the system already mentioned. Most of the bulk and micro-plane analyses cluster in the sub triangles merwinite–åkermanite–larnite

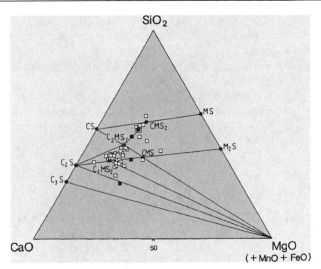

Fig. 6.13. Chemical bulk and selected micro-plane scans (*hollow symbols*) of Ca-Mg silicate slags in the system CaO–MgO–SiO$_2$ (after Osborn and Muan 1960). The clusters in the sub triangles merwinite–larnite–åkermanite (C$_3$MS$_2$–C$_2$MS$_2$–C$_2$S), merwinite–åkermanite–monticellite (C$_3$MS$_2$–C$_2$MS$_2$–CMS) and åkermanite–diopside–wollastonite (C$_2$MS$_2$–CMS$_2$–CS) can be partly proven microscopically. The technical abbreviations mean: *S:* SiO$_2$, *C:* CaO, *M:* MgO; the formula for diopside (CaMgSi$_2$O$_6$) is therefore, e.g., CMS$_2$

(C$_3$MS$_2$–C$_2$MS$_2$–C$_2$S), merwinite–åkermanite–monticellite (C$_3$MS$_2$–C$_2$MS$_2$–CMS) and åkermanite–diopside–wollastonite (C$_2$MS$_2$–CMS$_2$–CS). This leads to the conclusion that particular areas of the slags must have crystallized nearly under equilibrium conditions. Other projection points do not fulfill this condition; here we observe most likely incomplete reactions, which can be explained by undecomposed or only partially composed charge material.

Certain spots optically show non-distinguishable mixtures of different phases, which consist of material that was possibly only very briefly involved in the smelting and formed through a solid-state reaction. The research by Taylor and Williams (1935) and Brownell (1976) showed that a number of silicates can be formed through sintering, once a temperature of 800–900 °C is reached, because at this temperature range a highly reactive mixture exists, containing CaO, MgO, SiO$_2$-particles and possibly alumosilicates.

One of the identified elements was, for example, *calcite*, though it is not always recognizable whether it crystallized as a secondary mineral by deposition in the soil or if it is thermally undecomposed calcite. Apart from the primary lime content in the original material, calcite can also be formed by thermal decay of dolomite. The formation already starts quite quickly at 550 °C, and a mixture of MgO + CaCO$_3$ is formed, these days called semi-fired dolomite. Calcite itself decomposes under oxygen influx at 600 °C and up and is not detectable anymore above 800 °C. In a reducing fire it only starts to decompose at ca. 800 °C (Noll 1991). It may happen that calcite might well be preserved in a brief smelting process.

The existence of *periclase* can likewise be explained by the decomposition of dolomite; it can be seen in many samples in the shape of miniscule spherical to

Fig. 6.14.
Wadi Fidan 4, sample JD-8/24. Slightly polygonal crystals of periclase (*P*) in a matrix of merwinite (*Mw*). Remains of cuprite (*white*) (SEM-picture, back scattered mode)

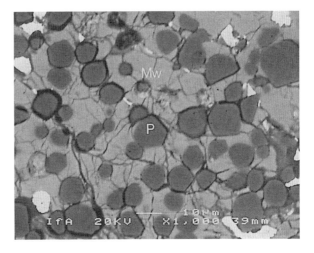

polygonal crystals (Fig. 6.14). This is even more likely as periclase dissolves in Ca-Mg silicate melts only with great difficulties (Parker and Nurse 1943).

In the samples JD-8/24 and -29, *alite* (Ca_3SiO_5) has been found by X-ray diffraction; a typical phase in Portland cement clinker. The extremely high melting point (>1700 °C) ensures that it cannot be a primary precipitation. It is far more likely a crystallization at ca. 1200 °C, because alite has, for example, been observed during technical sintering in SiO_2-containing dolomite, which is carried out on an industrial scale at these temperatures. Alite is only stable at a temperature above 1175 °C (Trojer 1963). Below this temperature, it decomposes under conditions of equilibrium into α-Ca_2SiO_4 + CaO. Eckstein et al. (1994) described alite in Ba-rich medieval slag from the Erzgebirge in Germany; they assumed the crystallization occurred through the decomposition of a complex calcium fluorosilicate.

In the phase content of the slags from the Early Bronze Age I, fayalite was not observed. It is particularly important to mention this again, because this phase appears frequently in later smelting slag and has even be named a 'conditio sine qua non' for the identification of smelting slag (Rothenberg 1991). This statement is most likely based on Tylecote's (1976) opinion, who did not believe the slag from Çatal Hüyük (6th millennium BCE), comprising cuprite, delafossite and tenorite, to be a product of the smelting of ores, because it did not contain any fayalite. The crystallization of fayalite or Fe-rich olivines in crucible slag is perfectly possible but can only be expected when the smelting occurs in a reducing gas atmosphere ($p_{O_2} \sim 10^{-10}$ atm) and with the original ore containing Fe + Si. The formation of fayalite can therefore hardly be expected from the kind of sandy-dolomitic host rock, as has been utilized in WF4. Fayalite in Chalcolithic/EBA I slag does not seem to occur regularly. The phase is proven in Norsuntepe (Yalçin et al. 1992, unpubl. data DBM), at Tell Abu Matar (Shugar 2000) and Tepe Sialk (Schreiner 2002). The occurrence of fayalite in allegedly Chalcolithic slag from Timna 39 has to be questioned on two grounds; firstly the dating of the site itself is rather doubtful and secondly fayalite has been found several times with delafossite (Bachmann 1978). This might be conceivable in view of the possibly inhomogeneous texture of slag, but it would still be surprising and has not been detected anywhere else. It is questionable if the fayalite described from Shiqmim has actually been (optically) correctly determined (Shalev and Northover 1987). The actual phase might be laihunite or ferrifayalite in intergrowth with oxides. Both phases, which can be described approximately with the formula $Fe^{2+}Fe_2^{3+}(SiO_4)_2$ (Fu et al. 1982; Kitamura et al. 1984; Kan and Coey 1985), are varieties of fayalite with 10–40 wt.-% Fe_2O_3. The conditions of their formation are still rather unclear. Following Kondoh et al. (1985), laihunite develops from fayalite with oxygen access at ca. 700 °C. Moesta et al. (1989) identified ferrifayalite for the first time by Mössbauer-spectroscopy in Bronze Age slag from Mitterberg; Metten (2003) could prove ferrifayalite/laihunite in slags from Trentino. She suggests that the formation of these phases was enabled by an excess of oxygen at the end of the smelting process at low temperatures. Such a phenomenon has not been observed in WF4.

Sulphide Inclusions

Hauptmann (1989) had already described occasionally occurring inclusions of copper sulphides in the iron-rich slag from WF4. They consist of *chalcocite* with varying compositions ($Cu_{2-x}S$), mostly of lamellar chalcocite, low in iron, sometimes also of *covellite* (CuS). The sulphides can be observed in undecomposed ore as well as in the shape of small embedded prills in the slag, where they surround metallic copper in a crescent or ring shape. The sulphide inclusions consist of characteristic dendrites of ($Cu_{2-x}S$) in a $Cu_{2-x}S$–Cu_2O eutectic. They indicate the existence of oxygen in the liquid matte that has separated during cooling as Cu^+ oxide. This observation corresponds with those of Oelsen (1952) and Schmiedl et al. (1971). They found that an area of limited miscibility exists in the system Cu–S–O between 1 200 and 1 400 °C, where 20 mol-% Cu_2O are soluble in the sulphide melt and where copper therefore does not necessarily precipitate. The oxygen content in the sulphide from WF4 can thus be interpreted as the beginning of desulphurization of the matte by oxygen excess during melting. This would also be compatible with the crystallization of $Cu^+/Fe^{2+,3+}$ oxides in the slag. This would form the initial stage of the roast reaction following the equations

$$Cu_2S + 3/2\,O_2 \rightarrow Cu_2O + SO_2 \qquad (I)$$

$$Cu_2S + 2\,Cu_2O \rightarrow 6\,Cu + SO_2 \qquad (II)$$

Due to the nature of the mineralizations in Faynan, the use of mixed oxidic and sulphidic ores is plausible, and this would also result in comparable phenomena.

Sulphide inclusions or matte with varying Cu/Fe ratios have been found in several slags from the 4th/3rd millennium BCE (Lutz 1990; Hauptmann et al. 1993; Keesmann 1997; Ryndina et al. 1999), often even as the only phase containing copper. This seems to be a surprising observation, because the smelting of oxidic and carbonatic ores is put at the beginning of pyrotechnological development (see Sect. 2.1). Only at the second stage, contemporary roughly with the Late Bronze Age, does smelting of sulphide ores start, for which the traditional models of the matte-production would have required several processual steps to be taken. This possibility, however, cannot be demonstrated with the material at hand. But there are other approaches to explaining sulphide inclusions along with other Cu phases such as delafossite, cuprite and metallic copper in the earliest slag, which also show that the denoted development model might be too much of a simplification.

1. In hydrothermal sulphidic as well as in sedimentary ore deposits close to the surface, intergrowth of Cu or Cu-Fe sulphides with malachite regularly occurs (Beyschlag et al. 1914/1916; Locke 1926; Smirnov 1954). For this reason, the use of ores with a corresponding composition can be expected even in the earliest stages of extractive metallurgy. The formation of matte can thus be explained, although it was the intention of the smelting process to reduce the oxidic part of the ore. This solution, the easiest from a mineralogical standpoint, could be an explanation of phase associations such as chalcopyrite/bornite + copper in the slag from Murgul and Nevali Çori (Anatolien) (Lutz 1990; Hauptmann et al. 1993), because due to the equilibrium conditions in the system Cu–Fe–S (Schlegel and Schüller 1952), copper cannot be precipitated during a smelting process from a matte with a composition close to chalcopyrite-bornite.
2. On the other hand, the thin charcoal cover of metallurgical operations in crucibles might lead to an oxygen surplus so that the sulphide components in the ore (after the roast reaction) get roasted. This is at least hinted at by smelting experiments carried out by Zwicker et al. (1985). The oxides formed here also lead, in pure Cu matte, to Cu_2O-contents, before the copper finally separates. In Cu-Fe mattes, magnetite separates from a high-temperature solid solution.
3. Experimental work of Rostoker et al. (1989) illustrated that reactions between Fe sulphides and Cu oxides lead to the formation of metallic copper during a 'co-smelting' of mixed ores in crucibles. Next to the reaction

$$3\,Cu_2O + FeS \rightarrow FeO + SO_2 + 6\,Cu$$

a number of other reactions occur, which depend on the composition of the ore as well as on thermodynamic conditions. The result of the smelting process is in all cases matte + copper. If ores contain a surplus of sulphides, then after smelting, matte will be formed in excess so that hardly any metal can be found in the slag.

The comparisons between Early Bronze Age slag from Maysar (Oman), Shahr-i Sokhta (Iran) and Çayönü Tepesi (Anatolia) show that two-phase copper reguli with larger parts of very pure Cu sulphides have often been produced – in contrast to what has been observed in material from Faynan (Hauptmann 1985a,b; Helmig 1986; Hauptmann et al. 2003). The reasons for this are manifold, as has been demonstrated, but they all indicate comparable technologies under relatively high oxidizing firing conditions and point to the use of mixed ores. But one ought to test in every single case whether the designation of an independent technological process (e.g., 'roast reaction process') can be verified at the beginning of the Bronze Age.

6.2 Manganese-Rich Silicate Slag from the Early Bronze Age to the Mameluk Period

Manganese-Rich Slag

With the full impact of the budding Early Bronze Age at the turn of the 3rd millennium BCE (EBA II), a rather marked step in the development of the technology of copper production can be observed in the Near East. This is obviously noticable in the sharp increase of metal finds over the entire region.

It is more difficult to pin down this step in production centers themselves. There are only a few examples where large slag heaps act as witnesses to the first more extensive metal production. Such examples come from the island of Kythnos in the Aegean Sea as well as Chrysokamino on Crete (Gale et al. 1985; Muhly pers. comm.). In Timna there also seems to be a small slag heap from the Early Bronze Age (Rothenberg and Shaw 1990). In Faynan, however, this early time period is characterized by a surprisingly large number of slag heaps or smelting sites, respectively. While the ore was smelted during the EBA I in the settlements, at that point and in all subsequent periods, raw copper was produced in close proximity to the actual ore deposits. The positioning of smelting sites was thus based upon technological considerations (see Sect. 5.2).

The first studies by Bachmann and Hauptmann (1984), Hauptmann and Roden (1988) and Hauptmann (1989) showed that the earliest slag from the 4th millennium BCE at Faynan has a completely different composition from that produced in all later periods. During all different times, except the Mameluk period, the latter composition mainly consists of manganese silicates. The reason that can be found is the exploitation of Cu-Mn mineralizations in the upper part of the DLS. Mn-rich silicate

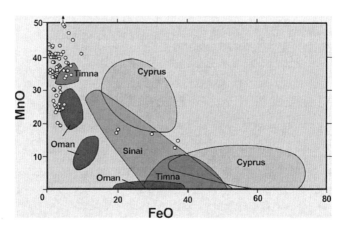

Fig. 6.15. FeO- vs. MnO concentrations (wt.-%) in Mn-rich slag from different (pre-)historic smelting sites. When more than one field from one site is shown, then compositions richer in Mn are more recent. The FeO and MnO contents of the slag from Faynan are also marked (*hollow circles*) (after data from Bachmann 1978; Bachmann and Rothenberg 1980; Bachmann 1982; Hauptmann 1985; Krawczyk 1986; Lupu 1970; Lupu and Rothenberg 1970; Steinberg and Koucky 1974)

6.2 · Manganese-Rich Silicate Slag from the Early Bronze Age to the Mameluk Period

slag is a rarity in archaeometallurgy as most (pre-)historic smelting sites of the Old World produced only slag consisting of Fe-rich silicates with varying portions of CaO, MgO, BaO und Al_2O_3. The substitution of Fe by Mn oxides occurs in the early non-ferrous as well as in iron production only at the lower percentage level.

Slags with MnO as one of the main components are known only from very few sites. One of these is Timna (Lupu 1970; Lupu and Rothenberg 1970; Bachmann and Rothenberg 1980), where they appear after a long phase of production of Fe silicate slags, but not before the Iron Age (Timna 30, stratum I) (Fig. 6.15). In the Sinai, Mn-rich slag also seems to occur in the Early Bronze Age and can be found up to the Islamic period (Bachmann 1980; Bachmann et al., in press). The same kind of slag was also found in Cyprus, where they can be dated only with great difficulty. In all probability, the Mn-rich compositions come from the Late Bronze Age, from the Phoenician, Hellenistic, Roman and up to the Byzantine period (Steinberg and Koucky 1974; Bachmann 1982; Kassianidou 2003). In Oman, Mn-rich slag appears only in the 12th century CE and is produced up to the 19th century (Hauptmann 1985). A comparison of the FeO and MnO contents in the slags of these sites illustrates a tendency, which shows a growing use of Mn ores in the later stages of metallurgical development. The chronology of these examples plays an important role in the understanding of the development of metallurgy, because the occurrence of Mn-rich slag has often been used as an argument for a deliberate utilization of fluxing agents.

In the following paragraphs the characteristics of the Mn-rich slag from Faynan will be investigated. Later the question will be discussed what impact Mn-containing ores had on early smelting technology (Sect. 7.2) and if or when Mn-ore was recognized as a fluxing agent (Sect. 7.3).

The Material from the Faynan District

In opposition to the material from the 4th millennium BCE, the slag from all later periods has not been obtained through archaeological excavations. Rather, they have been collected from the surface of the slag heaps or taken from sections, which have developed in the slag heaps through erosion.

The slag has been mineralogically and chemically analyzed. A number of thin sections and mounted samples have been prepared for this purpose (see Sect. 6.1). Unlike the material from the Early Bronze Age I, the slag blocks of these periods are much larger and only selected parts were chosen for the analysis. They included representative sections of predominantly homogenous slag that solidified from melts, but also relics from the charge that were not completely melted were analyzed to obtain information about different reactions of smelting processes.

Seventyfour bulk analyses by X-ray fluorescence, as well as atomic absorption spectrometry were carried out, which are presented in Table A.4. For each analysis, ca. 200–300 g of powdered material was used. It became clear during the studies that the analyzed samples include a piece of furnace lining (JD-6/2b) and smithing slag (JD-12/3a), which stand out due to their high content of Fe oxides. As the respective technology is not discussed here, both samples will not be discussed any further. Analytical details and measurements are explained in the Appendix.

Chemistry and Melting Behavior

The chemical bulk analyses (see Table A.4) show that the MnO content in the slag of all time periods from the Early Bronze Age II on rises rapidly to 20 up to over 45 wt.-% (Fig. 6.15); only in the Mameluk period do the concentrations decrease again a little. The high MnO content is the decisive difference in the compositions of the 4th millennium BCE. The slag is now composed of up to roughly 80% of the oxides SiO_2, MnO, CaO, Al_2O_3, and partly FeO. Minor elements are MgO, P_2O_5 and BaO. The total of the chemical analyses is often considerably below 100%. This is because parts of the metals such as Cu and Pb are present in the oxidic state and sometimes FeO can be present in the trivalent and MnO in its tri- or tetravalent form.

The analyses also illustrate that through successive periods the Cu content in the slag decreases. The extremely high percentage of Cu in slag from the Early Bronze Age I has already been pointed out. Figure 6.16 illustrates the decrease of Cu by two decimal places during the following periods, which verifies continuing technological progresses in the metal production. This development is here compared with the slag from Timna, whose values were presented by Rothenberg (1990). There are sound correspondences between the Late Bronze Age and Iron Age slag. They con-

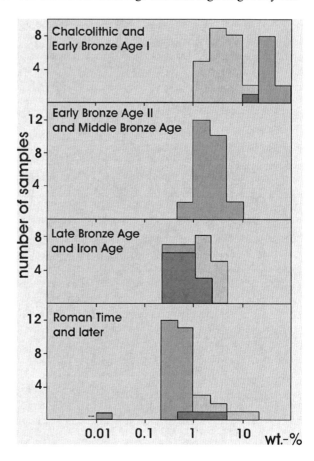

Fig. 6.16.
Copper contents in slag from Faynan from different periods. The chronological development shows an improvement in metal extraction by two decimal places. The measurements from Timna are shown for comparison (following data from Rothenberg 1990). *Middle grey:* Faynan; *light grey:* Timna; *dark grey:* overlapping of both

6.2 · Manganese-Rich Silicate Slag from the Early Bronze Age to the Mameluk Period

firm again that in Faynan and Timna metal has been produced with the same technological processes during these time periods. The Cu contents of between 0.2–1.5% in Faynan also coincide with those published by Craddock (1995), who generally suggests Cu contents of ca. 0.1–1% in slag produced in the Late Bronze Age or later. His research showed that only since the Industrial Revolution was it possible to decrease the metal contents in slag any further. It is noticeable, but not very surprising, bearing in mind the doubtful dating of Timna 39, that the so-called Chalcolithic slags from there have metal contents far more compatible with those of the Early Bronze Age II/III in Faynan. Further fieldwork in Timna might lead to firmer correlations meaning better dating for Timna 39. The unusually high Cu values from later periods in Timna might perhaps be caused by sampling differences and are statistically irrelevant.

The bulk compositions reduced to main components of Mn-rich slag can be illustrated in the quaternary system SiO_2–CaO–MnO–FeO (Fig. 6.17). The variations in the analyses are best expressed when Fe oxide is included, while Al_2O_3, which is in all samples just between 3 and 6 wt.-%, is neglected. It becomes clear that all samples plot into an area just above the ortho-silicate level, with a MeO/SiO_2 ratio between 1:1 and 2:1. In this area, clusters can be recognized with varying MnO/FeO, MnO/CaO and MnO/SiO_2 concentrations. They actually allow classification into four groups, which correlate with the chronology of metal production.

The Mameluk slag is the only one that shows Fe oxide concentrations in the level of several tens of percent. This fact can be explained by the exploitation of ore in the DLS formation at El-Furn, where Mn was substituted by Fe via lateral zoning. It cannot yet be explained why ores from this particular area were chosen to be smelted.

If these Fe-rich slags are disregarded and all other analyses are reduced to the ternary system SiO_2–CaO–MnO (as shown in Fig. 6.18), more conclusions can be

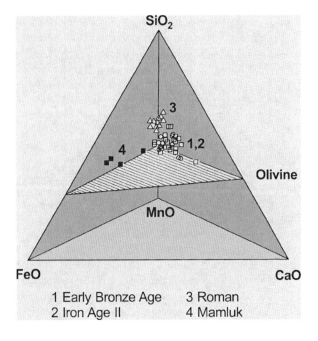

Fig. 6.17.
Chemical bulk analyses of slag from Faynan in the system SiO_2–FeO–MnO–CaO. One can recognize clusters, which correlate with certain time periods during the history of metal production

1 Early Bronze Age
2 Iron Age II
3 Roman
4 Mamluk

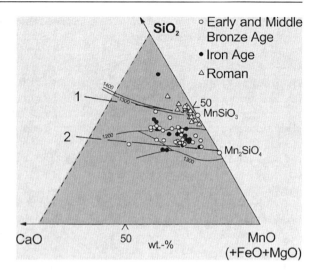

Fig. 6.18.
Mn-rich slag, dating from the 3rd millennium BCE to the beginning of the 1st millennium CE, displayed in the system SiO$_2$–CaO–MnO (+FeO+MgO) after Glasser (1962). The Early Bronze Age slags clearly show a larger variation in their composition than the later Iron Age ones. In the Roman period, there is a change towards a higher SiO$_2$ content. *1:* Line of metasilicates; *2:* line of orthosilicates

drawn. Based upon crystallochemical considerations it seems to be useful to add the relatively low quantities of FeO and MgO to MnO. It becomes clear that slag from the Early Bronze Age has a broad range of compositions, particularly concerning their Ca/Mn ratio. But the picture slightly changes for the Iron Age: the analyses show less variations and a tendency to Mn-richer compositions close to tephroite. In the Roman period, a cluster of compositions is visibly shifted to higher SiO$_2$ contents.

These results are of great value for the understanding of the metallurgical development in Faynan. They reveal that the smelters of the 3rd millennium BCE utilized all copper ores available in the Dolomite-Limestone-Shale Unit (DLS). It cannot be quantified, if and how much of the CaO content of the slags might be due to an uptake from the lime-rich (not Mg-rich!) clay of the furnace wall, caused by overfiring. Taking the numerous fragments of vitrified furnace walls that have been found on the smelting sites into account, this might indeed have been an important factor. This is pointed out by Hezarkhani and Keesmann (1996) in their analysis of nonferrous metal slag from Iran. Kronz (1997) uses medieval iron slag from the Dietzhölztal in the Lahn-Dill-area to suggest that resorption of furnace wall or lining could control an automatic tuning of liquid slag melts to produce cotectic compositions during smelting.

In the Iron Age, some evidence points to a more careful choice of the raw material. The slag composition has shifted to being more Mn-rich, which allows the assumption that Mn-rich Cu ores were deliberately selected in order to increase their fluxing properties.

A careful controlling of the smelting process by an equally careful selection of raw material can also be assumed for the Roman period. The Roman slag has an astonishingly high level of SiO$_2$, which has yet to be explained. The plot of the analyses in the ternary system SiO$_2$–CaO–MnO (Fig. 6.18) shows that the samples must have liquefied at temperatures not below 1 400 and 1 600 °C. This was decidedly above firing temperatures necessary for the production of Bronze and Iron Age slag.

Otherwise, all analyses results plot into a rather wide range, in which the liquidus fields towards MnO are below 1 200 °C. This clearly shows that MnO had a strong

6.2 · Manganese-Rich Silicate Slag from the Early Bronze Age to the Mameluk Period

effect on decreasing the melting point and was thus an excellent fluxing agent. It cannot be expected however, that Mn silicate slag is liquefied at lower temperatures than Fe silicate slag, as has been presumed by Bachmann and Rothenberg (1980). This becomes clear by comparing the respective phase diagrams, which show, e.g., that the temperatures of the liquidus fields of tephroite, rhodonite or wollastonite/bustamite in the system SiO_2–CaO–MnO are more than 100 °C higher than the compatible iron-rich compositions in the system SiO_2–CaO–FeO (e.g., tephroite 1 345 °C compared with fayalite 1 205 °C).

The general problems that exist in the estimation of melting temperatures for slag from equilibrium phase diagrams have already been mentioned (Sects. 2.2, 6.1). In the case of the samples analyzed here, temperatures taken from phase diagrams for the liquefaction of the slags might generally be a bit too high, because FeO, K_2O, Na_2O, BaO, Pb, Cu and the chloride-contents in the ore will have undoubtedly led to a further decrease of the melting temperature. But these temperature estimates are in general far more realistic than in the discussion of the Wadi Fidan 4 samples, because these samples have clearly less undecomposed inclusions.

In order to prove this point, some samples have been investigated on a heating stage microscope to determine the beginning of softening, the temperatures of the hemispheric and the flowing points (T_E, T_{HK}, T_F) according to DIN 51730 (Table 6.2,

Table 6.2. Beginning of softening, of hemispheric and flowing point temperatures (T_E, T_{HK}, T_F) of some slag samples from Faynan, measured on a heating stage microscope according to DIN 51730

Sample	T_E in °C	T_{HK} in °C	T_F in °C
Early Bronze Age II/III			
JD-5/1d	1 131	1 193	1 216
JD-5/1g	1 123	1 149	1 170
JD-5a/2b	1 110	1 138	1 179
JD-5a/2d	1 105	1 124	1 158
JD-9/1	1 139	1 141	1 146
Iron Age II			
JD-2/2c	1 096	1 111	1 160
JD-2/5a	1 106	1 123	1 158
JD-2/6	1 107	1 123	1 150
JD-11/2	1 102	1 117	1 163
JD-13/1b	1 160	1 185	1 207
Roman period			
JD-1/11	1 172	1 189	1 217
JD1/21b	1 136	1 149	1 216
JD-1/24b	1 133	1 153	1 197
JD-1/38b	1 153	1 169	1 214
JD-12/3b	1 138	1 152	1 164

Fig. 6.19.
Intervals of softening, hemispherical and flowing point temperatures (*upper*, *middle* and *lower horizontal line*) from some slag samples (from Faynan), measured with a heating microscope (see Table 6.2)

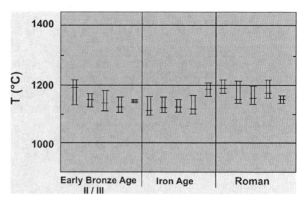

Fig. 6.19). Although these measurements have been carried out in an oxidizing atmosphere, their results can be compared with the melting temperatures of the ancient slag, even though these were produced in a reducing atmosphere. This is due to the wide range of stability of manganese silicate melts from strongly reducing to oxidizing atmospheres. The T_F-values (mostly between 1 150 °C and 1 170 °C) obtained for slag from the Early Bronze and Iron Age agree very well with the conclusions reached from the phase diagram. The Iron Age samples show, with the exception of one sample from a small smelting site in Wadi Dana (JD-13/1b), the lowest melting temperatures. That agrees again very well with the analytically proven optimization of metallurgical processes at that time period. The Roman slag reached the highest smelting temperatures of around 1 200 °C.

The Petrography of the Slag

Slag fabric and texture allow us to conclude that it has developed largely through the solidification of fully liquefied melts, quite the opposite of older finds. The slag can be found as small-scale fragments up to flat or cone-shaped bodies, which might reach a weight of up to several tens of kilograms. Their typology has been described in Sect. 5.3.

Optical studies and X-ray diffractometry show a grouping of the slag following varying phase associations ('critical parageneses' after Krawczyk and Keesmann 1988):

1. Tephroite slag (mainly Iron Age);
2. Knebelite slag (exclusively medieval);
3. Tephroite-pyroxenoid slag (mainly Early Bronze Age II/III and Roman);
4. Pyroxenoid slag (mainly Roman).

Figures 6.17 and 6.18 illustrate that the plots of bulk analyses of the slags in the two systems concur with the observations of the optical analyses: slag with tephroite (Mn_2SiO_4) or knebelite (($Mn,Fe)_2SiO_4$) as main components group around these points along the lines MnO–SiO_2 or (MnO, FeO)–SiO_2 or along the level of olivine. As do the analyses of slag-containing pyroxenoid along the line with metasilicate compositions $MnSiO_3$-$CaSiO_3$.

A common feature of all slag is that they mostly have solidified in the glassy or microcrystalline state. This is of interest as it occurred far less frequently in iron-rich silicate slag of ancient non-ferrous or iron smelting. Glassy slags mainly occur in Faynan at Early Bronze Age and Roman slag heaps. Evident even to the naked eye are differently colored, streaky areas with flow structures. The slag is in general grey-green colored although occasionally strikingly orange to red colored pieces can be observed. Iron Age slag normally shows a markedly higher degree of crystallization.

Hyaline textures with a prevailing volume of glass occur preferentially at the surface of single layers or droplets, which have cooled down very quickly. Towards the middle of the slag bodies where cooling is slower, tephroite- and pyroxenoid form inclusions in the glass (Fig. 6.20). Glassy areas often show flow structures with a layered-streaky pigmentation caused by copper or cuprite (Fig. 6.21). Microlithic devitrification with fibrous spherolites of pyroxenoid or tephroite can also dominate (Fig. 6.22). Sometimes they are so closely spaced that at the end the glass disappears completely. Pyroxenoids intergrown like tufts of grass or skeletons of tephroite can form successively. In Iron Age slag, spinifex-textures are very typical.

The thin sections show very clearly how the slag is built up with single rope-like structures or drops, which are sharply separated from each other by cooling rims.

Fig. 6.20.
Ras en-Naqab, sample JD-5A/2b. Vertical flown slag drops, glassy solidified at the rim (*1*) and changing towards the middle into an intersertal texture of tephroide (*2*). The more rapidly cooled rims are also marked by the crystallization of Mn oxides. Early Bronze Age II/III (width of the specimen: 2.7 cm; transmitted light; Hauptmann and Roden 1988)

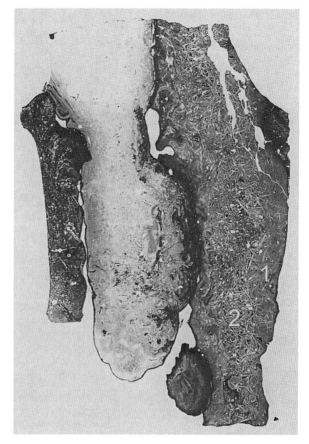

Fig. 6.21.
Wadi Ghwair 3, sample JD-21/1a. Yellowish-brown glass with laminar flow structure and pigmentation through cuprite and copper prills in single streaks. Early Bronze Age II/III (scale: 0.2 mm; transmitted light)

Fig. 6.22.
Faynan 1, sample JD-1/32. Pyroxenoid-slag with laminar flowing and sphaerolithic devitrification. Roman (width of the sample: 3.5 cm; transmitted light)

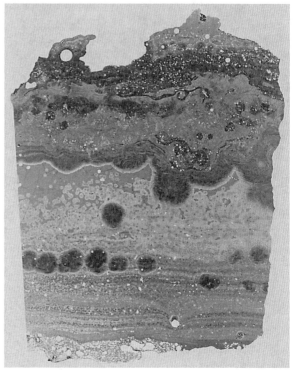

Here, hausmannite crystallizes in the shape of dendrites (Fig. 6.23). In the section, a pillow-like structure is sometimes observable, where solidification skins around single slag 'ropes' or slag drops can be interpreted as evidence of temporal intervals during the solidification of the melt. This is characteristic of free flowing slags, which have formed during the tapping from the smelting furnace and have solidified very rapidly. Comparable cooling rims consisting of Fe oxides are well known from ancient iron-rich slag (Keesmann and Heege 1990; Yalçin and Hauptmann 1995).

Fig. 6.23.
Ras en-Naqab, sample JD-5A/2b. Cooling rims of single slag drops with clearly visible dendrites of hausmannite. Early Bronze Age II/III (transmitted light; from Hauptmann and Roden 1988)

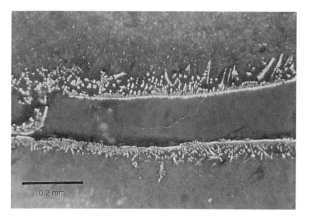

It has already been mentioned (Sect. 2.2, 5.2) that such slags are partial melts of "ideal," i.e., eutectic composition. Depending on the degree of optimization during smelting, their volume may predominate. This slag is always, although in varying quantities, accompanied by so-called furnace slag that remained in the furnace after the smelting process was finished. The furnace slag cooled much more slowly than tap slag. It regularly contains inclusions or imprints of charcoal and inclusions of charge components, which have been only partially decomposed during smelting. Such slag shows similar characteristics to the ones from Wadi Fidan 4. Skeleton growth of individual phases normally observed in tap slag is less often noticeable here. Instead, a porphyric texture with (hyp-)idiomorphic crystals of tephroite or knebelite and pyroxenoids form, which indicates a slower cooling process.

As an example, one type will be introduced here, which is often but not exclusively found amongst the Roman slag of Faynan 1. It is a material, which is baked together similarly to a breccia (erroneously often called "furnace conglomerate"). The single components have been only shortly exposed to higher temperatures and thus have just sintered together. Inclusions of millimeter- to centimeter-sized Cu-bearing sandstones as well as Mn ores and their water-free decomposition products are common.

On the micro-scale, one can observe a sequence of Mn^{4+} hydratoxides to pyrolusite (MnO_2), partridgeite (Mn_2O_3) and hausmannite (Mn_3O_4), that is from higher-valent to lower-valent Mn oxides. Also observable are reactions with silicates, which led to the formation of braunite, pyroxenoids and sometimes also tephroite (see also Fig. 6.25). Mn oxides can be identified within the smallest pores; they fully opacify the material. Pb silicates, barite and a Ca-P phase ($Ca_4P_2O_9$? = hilgenstockite, known from modern iron smelting slag) also occur.

During its formation this slag was exposed, at least at times, to a relatively strong oxidizing atmosphere in the furnace, which led to an oxidation of Mn silicates within largish bubble cavities and along fissures. Here crusts of Mn oxides, sometimes of crednerite, have formed through recrystallization, normally much thicker than the cooling rims observed in tap slags. The association with Mn compounds containing $(OH)^{2-}$ or H_2O is probably caused by the soil environment. The formation conditions of such crusts and inclusions are not always absolutely definable, and it is probable that reduction and retrogressive oxidation of Mn ores overlie each other

Fig. 6.24.
Faynan 1, sample JD-1/15. 'Slag conglomerate' with inclusions of recycled slag (*arrows*). The surrounding matrix is opacified by finely distributed Mn oxides. Roman (width of the sample: 2.8 cm; transmitted light)

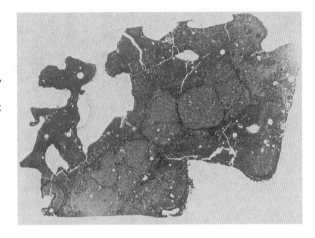

successively. It seems clear that reduction of mixed Cu-Mn ore happened, where crednerite occurs because the enrichment of copper, which is connected with the formation of this phase, can hardly be explained by diffusion of the metal from the slag body.

Inclusions of thermally affected and partly decomposed slag fragments can frequently be found. They differ from the surrounding matrix due to a different phase content (Fig. 6.24): tephroite is the main component, while everywhere else pyroxenoids prevail. The latter ones are the main components in Roman slag, while tephroite occurs as the predominant phase in Iron Age slag. The original glass content has nearly completely recrystallized to an extremely fine-grained, granoblastic texture. This underlines the archaeological interpretation and can be used as further indication that reworked Iron Age slag from Faynan 1 has been used in Roman times as retour slag.

The Development of Mineralogical Phases

Generally, slag and its phase content have formed under reducing conditions, although in some cases it is not possible to definitely determine if the formation of a phase has been affected solely by reduction or also by re-oxidation. During smelting, the Mn-containing copper ores from Faynan were transformed by reduction to Mn^{3+}- and Mn^{2+} phases with lower oxygen valences. These reactions would have happened, on the basis of minor and trace elements, roughly at the following temperatures, after the data compiled by Roy (1968):

$$MnOOH \rightarrow 380\ °C \rightarrow MnO_2 \rightarrow 500\ °C \rightarrow Mn_2O_3 \rightarrow 800–900\ °C \rightarrow Mn_3O_4$$
manganite　　　　　　pyrolusite　　　　　　partridgeite　　　　　　　　hausmannite

When the Fe content in the parent material is sufficiently high, bixbyite $((Mn,Fe)_2O_3)$ forms instead of partridgeite; jacobsite (Mn_2FeO_4) instead of hausmannite or vredenburgite if the Mn_3O_4 content < 54%. The above-mentioned transformation temperatures are furthermore dependent on the oxygen partial pressure in the gas atmo-

Fig. 6.25.
Reduction process and phase content in the system SiO$_2$–MnO$_2$–MnO (modified after Abs-Wurmbach et al. 1983)

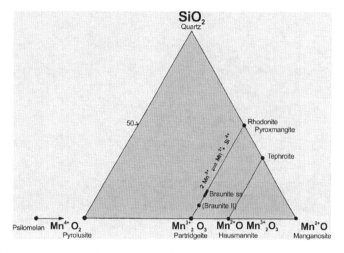

sphere of the furnace. Oxides with lower valences react with SiO$_2$ and form ortho- as well as metasilicates with Mn^{2+} and Mn^{3+}. The idealized process of the reduction and the composition of the Mn phases can be seen in Fig. 6.25.

The following paragraphs will deal with the most important slag-forming silicates as well as some more common oxides; their composition and conditions of formation will be studied. Cu-containing compounds will only be mentioned very briefly, because they do not affect the slag formation in any way, unlike in the samples from Wadi Fidan 4 (Sect. 6.1). After the smelting, Cu is existent in the metallic state, sometimes also in the sulphide state. The composition of the metal will be discussed in Sect. 6.3.

Tephroite is the main component of Iron Age slag, sometimes also of Bronze Age slag. It also occurs in Roman slag, although the bulk composition indicates that pyroxenoid slag is typical of this period. This can be explained by a pronounced tendency of olivine to crystallize even from SiO$_2$-rich liquids at fast cooling rates (Milton et al. 1976). But more likely inhomogeneities are caused by inclusions of retour slag, in which tephroite is identifiable. The frequent occurrence of tephroite in Faynan corresponds with the general distribution pattern of this phase, which has been commonly identified as the main component in Mn-rich slag (Bachmann and Rothenberg 1980; Bachmann 1982; Steinberg and Koucky 1974; Hauptmann 1985).

Electron microprobe analyses demonstrate that the olivines show a relatively monotonous composition in the Faynan slag (Fig. 6.26; Tables A.10–A.12). In all samples except the Mameluk period samples, tephroite exists with a rather low content of larnite, fayalite, and forsterite. Compositions with up to 20 mol-% forsterite appear in Early Bronze Age slag that has a corresponding bulk composition. No zoning from Mg- to Fe-rich compositions, as in the case of Fe-rich olivines (Keesmann 1989), has been observed. Larnite contents in tephroite are always low (5–15 mol-%), even when CaO in the bulk composition is relatively high. In sample JD-21/1a, for example, tephroite was identified with a larnite-content of ca. 7 mol-% in a glassy matrix of 40 wt.-%.

The comparison of the Mn-rich olivines from Faynan with such from Oman shows that there is probably another miscibility gap between the well-known solid solution of forsterite-tephroite and the assumed solid solutions of monticellite-glaucochroite

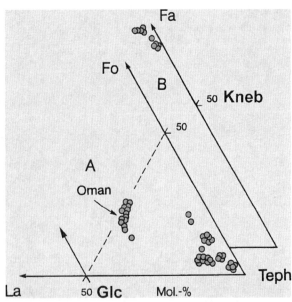

Fig. 6.26. *A:* EMP-analyses of tephroite from various slag samples in the Faynan area, plotted in the system forsterite–tephroite–larnite. The comparison with data of Ca/Mn-olivines from slag in Oman (Hauptmann 1985) makes it clear that a miscibility gap exists between the solid solution forsterite–tephroite and glaucochroite–monticellite.
B: EMP-analyses of Fe-knebelite from the sample JD-6/3 (El-Furn), plotted in the system fayalite–tephroite–larnite. *Fa:* fayalite (Fe_2SiO_4); *Fo:* forsterite (Mg_2SiO_4); *La:* larnite (Ca_2SiO_4); *Teph:* tephroite (Mn_2SiO_4); *Mc:* monticellite (($CaMg)_2SiO_4$); *Glc:* glaucochroite (($CaMn)_2SiO_4$)

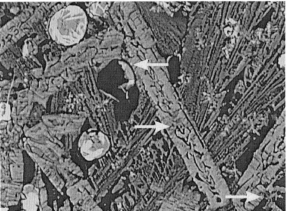

Fig. 6.27.
Khirbet en-Nahas, sample JD-2/28. Ingrowth of tephroite with Mn oxides in a slag that had solidified inside a furnace (*arrow*). It can be most likely explained by the oxidation of tephroite, caused by an oxygen input into the embers of the charcoal bed (SEM-picture, secondary electron mode)

(Brown 1982). A similar miscibility gap exists in Fe-rich compositions and was observed in iron-rich slag by Hauptmann et al. (1984). It is caused by the exsolution of Ca- and Fe-rich compositions in the subsolidus. Due to the very rare occurrence of Mn-rich magmatic rock in nature, very little is known about the relevant phase associations of the isomorphous series monticellite ($CaMgSiO_4$)–glaucochroite ($CaMnSiO_4$) and forsterite (Mg_2SiO_4)–tephroite.

Occasionally, tephroite is intergrown with one (?) opaque phase. The latter enters the silicate perpendicular to the surface, in a wedge shape (Fig. 6.27). The phase is brownish-pink to grey and shows slight effects of anisotropy. EDS-measurements have revealed that it is probably a mixture of Mn oxides and a Mn silicate. It was not possible to identify them due to their very small grain size. But it is likely that the phase(s) result from an oxidation of tephroite, which is reminiscent of the occurrence of ferrifayalite or laihunite with fayalite (see Sect. 6.1). The oxidation in our case might have also happened via a sudden oxygen input, because Muan (1959b) demonstrated that tephroite is oxidized into Mn_3O_4 + braunite or to Mn_3O_4 + rhodonite just below its liquidus at decreasing temperatures (<1 204 °C) and oxygen access ($p_{O2}=10^{-7}$ atm). This is of some significance for a possible reconstruction of the process technology, because the oxidation of tephroite can only happen under conditions of a slow cooling, as is hardly imaginable when the slag solidifies outside the furnace. This phenomenon is actually only observable in those slags that have solidified inside a furnace.

A second observation was that tephroite was stable in the slag smelts even when the oxygen partial pressure was relatively high. In Early Bronze Age slag the Mn silicate exists next to finely dispersed drops and dendrites of cuprite (Fig. 6.28). This differentiates tephroite noticeably from its Fe-rich pendant fayalite. Fayalite decomposes, according to its stability ranges in the system Fe–Si–O (Eugster and Wones 1962), e.g., at temperatures of 1 100 °C above $p_{O2} \sim 10^{-10}$ atm, to quartz and magnetite, therefore under fairly strong reducing conditions. Tephroite, however, decomposes not before $p_{O2} \sim 10^{-2}$ atm to rhodonite and hausmannite (Abs-Wurmbach et al. 1983); that is at a level when copper would be re-oxidized again. The positive influence of the Mn silicate's ranges of stability on the smelting will be discussed in more detail in Sect. 7.2.

In Mameluk slag, Fe-rich olivine (Fig. 6.26, Table A.12) instead of tephroite occurs due to the high Fe oxide content in the bulk composition. The olivine shows weak zoning with slightly richer Mn compositions in the core and Fe-and Ca-rich compositions in the rim ($Fa_{70-77}Teph_{19-28}La_{1,3-5,9}$). This olivine has to be seen as part of the solid solution fayalite–tephroite and ought to be called *Fe-knebelite*, following the nomenclature given by Deer et al. (1982).

According to the MeO/SiO_2 ratio of the bulk analyses, crystallization of metasilicates was discovered in Roman and sometimes older slag. In the material here, unlike in iron-rich compositions where generally pyroxenes crystallize, pyroxenoids form. The difference between the two lies in the chain structure of the SiO_4 tetrahedra. In pyroxenes, identical chain periods are repeated after two SiO_4 tetrahedra (2-periodic single chains). The pyroxenoids consist of 3-, 5- and 7-periodic chain struc-

Fig. 6.28.
Ras en-Naqab, sample JD-5/1e. Columnar-like skeletons of tephroite with interstitial precipitations of light-colored cuprite dendrites. The sample contains >5% Cu, which exists as oxide and thus shows a crystallization in relatively weak reducing conditions (SEM-picture, secondary electron mode)

Fig. 6.29.
Faynan 1, sample JD-1/11. Pyroxenoid crystals of a tuft-like habitus in Roman slag (scale: 400 μ; transmitted light)

tures. The Mn-richer pyroxenoids are rhodonite, bustamite, and pyroxmangite with the general formula $(Mn,Fe,Ca)[SiO_3]$. Johannsenite, with a compatible composition, is actually a pyroxene. Pyroxmangite and johannsenite are not of interest for the questions at hand because pyroxmangite is a low-temperature-/high pressure phase and johannsenite is transformed from ca. 800 °C onwards into bustamite (Abrecht and Peters 1975).

The pyroxenoids form long, fine needles, which grow in broad radial tufts in the analyzed slag (Fig. 6.29). These tufts show a spinifex texture resulting from fast cooling. Electron microprobe analyses of pyroxenoids from Roman slag Table A.13 and A.14) illustrate that the crystals have a large range of variation in their MnO/CaO ratio; they contain 5–43 mol-% CaO, although the Mg and Fe oxide contents stay consistently low (<10 mol-%). In opposition to the pyroxenes from the slag samples in Wadi Fidan 4, the pyroxenoid here contains CuO only approximately some tenth of one percent. Using the classification of natural pyroxenoids after their CaO, MnO, (FeO + MgO)contents from Abrecht and Peters (1980), the pyroxenoids under study here are rhodonite, while those with a higher Ca-content are bustamite (Fig. 6.30).

Mn-rich pyroxenoids have been observed repeatedly in ancient slag. Bachmann (1978) identified bustamite using X-ray diffraction in a slag from Timna 189a. This is archaeologically quite interesting, as Rothenberg (1988) dates this site into the Chalcolithic period. The occurrence of Mn-rich slag could be considered evidence that during this time period not only the Mn-poor ore from the sandstone of the Amir Formation was smelted (Bachmann 1978), but also the rather rarely occurring Cu-Mn ore of the Timna-Formation. Regrettably no secure dating evidence has been published for this site so far. Lupu (1970) determined johannsenite in a slag from the same site. This is mineralogically a surprise, because this phase is only stable up to ca. 800 °C. The high-temperature modification of this phase is bustamite (Abrecht and Peters 1975).

Hauptmann (1985) identified bustamite in ancient slag from Oman through microanalysis. Krawczyk (1986) and Krawczyk and Keesmann (1988) observed bustamite in a slag from Yotvata close to Timna, site 205. The identification was also based on X-ray diffraction. The microanalyses, however, demonstrated that this pyroxenoid had to be considered rhodonite (Fig. 6.20), which also conforms better to the classification by Abrecht and Peters (1980). Sáez et al. (2003) identified ferrobustamite and pyroxmangite in slag from the 3rd millennium BCE.

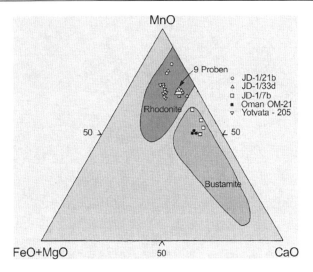

Fig. 6.30.
EMP-analyses of pyroxenoids from various slag samples of the Faynan area, plotted in the system MnO–(FeO+MgO)–CaO in mol-%. The analyses, compiled by Abrecht and Peters (1980), are shown here as separated fields. Following this classification, rhodonite and bustamite exist in Faynan

Pyroxenoids are also often appearing phases in modern shaft furnace slag and have been described in a number of publications at the beginning of the last century. The latest state of research shows that some of the old designations ought to be changed (Deer et al. 1982). Vogtite, diagnosed in acid slag from steel making, has been so far called Fe rhodonite (Hlawatsch 1907; Hallimond 1919), but should be named ferrobustamite. The phase called Fe rhodonite by Hallimond (1919) should on the other hand better be counted as pyroxferroite.

The iron-rich slag from El-Furn does not contain pyroxene or pyroxenoid. This is due to the low CaO-content in this slag. A precondition for the crystallization of an iron-rich pyroxene is the existence of CaO, because ferrosilite ($FeSiO_3$) is not stable. Only clinopyroxene with a composition close to hedenbergite could be formed here. The analyses of knebelite in sample JD-6/2 have shown that CaO, which was available in the melt, was substituted in the olivine or enriched in the glass-like residual melt (see above). This confirms the observation by Krawczyk and Keesmann (1988) that the formation of fayalite or pyroxene slag is not only dependent on the SiO_2 but also on the Ca content of the melt.

It has been pointed out in the paragraph on texture analysis that slag solidifies in large parts to a dull *glass*. Chemistry and cooling rate are mainly responsible for this. Figure 6.31 shows, for example, the relatively high SiO_2 concentrations of the glass in comparison to the bulk composition of the slag. The Al_2O_3 content in the glass is also roughly 2% higher (Table A.15), while no conspicuous alterations of the alkali contents are noticeable. Such melts become very viscous during the extremely fast cooling rate, so that the crystallization is suppressed. Another fact, which obviously facilitated the formation of glass in the slag, is the high stability of MnO in comparison to 'FeO' against fluctuations of temperature and oxygen partial pressure. This suppressed the nucleation of oxides in the Mn silicate melts, while in iron-rich slag a primary precipitation of wuestite and magnetite is enhanced and thus a crystallization of the melt during solidification. This whole phenomenon is obviously confirmed by the studies of Bachmann and Rothenberg (1980), who have found in ancient slags from Timna, glass in Mn-rich compositions but not in Fe-rich samples.

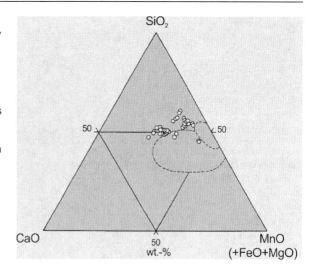

Fig. 6.31.
EDS glass analyses of Early Bronze Age and Roman slag from the Faynan area, plotted in the system SiO_2–CaO–MnO (+FeO+MgO). The bulk compositions of other slag from these time periods are shown for reasons of comparison (*hatched fields*). The glass is visibly richer in SiO_2 (see also Table A.15)

The different coloring of the glass is only partly caused by variations of the chemical composition, but can be mainly explained by different valences of Mn and Fe ions and by Cu_2O contents. This effect is particularly evident in manganese and lead as a consequence to the utilization of Mn ores for glass production. Mn^{4+} oxides, which frequently occur in nature, are already reduced at 500 °C to tri-valent oxides plus oxygen. They have therefore been used in the past as oxidation material, e.g., to decolorize glass melts ('glasmacherseife'). During this process, MnO_2 dissolves in the glass to Mn^{2+} silicates and oxidizes sulphides, coal and Fe^{2+} compounds, e.g., according to the following equation:

$$MnO_2 + 2\,FeO \rightarrow MnO + Fe_2O_3$$

because MnO is not stable next to FeO. This leads to a 'bleaching' of all colors originally caused by these phases. Particularly, Fe-containing glass shows distinct changes of color: Fe^{2+}-containing glass, that is typically blue-green colored (just like the slag glass often found in ancient smithies) becomes yellowish brownish with an increasing Fe^{3+} content; if at the same time the Mn oxide content is high it becomes increasingly yellow-brown. This color is conspicuously typical of the Roman period, sometimes also for slag from the Early Bronze Age.

The red and orange colored opaque slag glass can, in Faynan, most probably be explained as being primarily caused by a coloring through the Mn^{3+} ion, because Mn oxide is the main component in all these samples. Studies by Weyl (1959) demonstrated that Mn^{2+} ions react with oxygen to become Mn^{3+} and bring about a red hue in the glass. Finally, it comes to a crystallization of Mn_3O_4, as can be observed along cooling rims. But it also has to be considered that the high Cu content in the Faynan slag, which is between 1–4% and sometimes even above 5% (Fig. 6.21, Appendix, Table A.4), might have had a coloring effect on the slag both by metallic copper and Cu^+ oxide (cuprite). Metallic copper can occur in the colloidal state, as Hezarkhani and Keesmann (1996) have observed in Iranian glassy slags. But finely dispersed droplets and dendrites of cuprite are much more common (Fig. 6.28). They

6.2 · Manganese-Rich Silicate Slag from the Early Bronze Age to the Mameluk Period

are very similar to those which have been analyzed in copper-ruby glasses from the late 2nd millennium BCE in the Near East by Freestone (1987) and Brill and Cahill (1988). It is quite likely that the technological roots of this glass production come from copper metallurgy, particularly because red slag with precipitations of cuprite is well known from other Early Bronze Age smelting sites e.g., Shahr-i Sokhta/Iran (Hauptmann et al. 2003). The metallurgical aspect the red tint of the slag glass indicates that the melt must have already had a relatively high oxygen content during its formation in the furnace.

The existence of *braunite* must also be mentioned. It regularly and exclusively appears in Roman 'slag conglomerates' along the transition between Mn-ore inclusions and Mn^{2+} silicates. The single fine crystals are subhedral and nearly opaque. They are greyish-white with a hint of brown; the optical anisotropy is weakly developed and shows dark to light slate-blue hues. A secure differentiation from the visually very similar Mn oxides is difficult but proves quite successful with the help of microanalytical methods. The composition of a braunite crystal in sample JD-1/33d is as follows: 9.44% SiO_2; 0.62% TiO_2; 4.61% Fe_2O_3; 70.3% Mn_2O_3; 11.1% MnO; 0.23% CaO; 1.66% CuO.

Noll's (1991) studies of phase formation in black-colored ancient pottery showed that braunite forms only under slightly reducing conditions between 800 °C and 900 °C via thermal decomposition of bixbyite according to

$$3\ Mn_2O_3 \rightarrow 2\ MnO \cdot Mn_2O_3 + \tfrac{1}{2} O_2$$

and the subsequent reaction with SiO_2. The structural and chemical similarity (braunite and bixbyite form a solid solution) is most likely beneficial to this chemical reaction. In consideration of the above-described reactions, it is not surprising that braunite does not occur in slag that has formed through the solidification from a liquid, but only in material that was exposed to high temperatures but relatively strong oxidizing conditions for only a short time.

Mn oxides are relatively rare phases in the slag from Faynan, while the crystallization of Fe oxides is a common occurrence in ancient iron-rich silicate slag, in the non-ferrous as well as in iron metallurgy. The limited occurrence of these Mn oxides is undoubtedly due to their limited stability at high oxygen partial pressure. They are rapidly reduced to lower-valent Mn oxides. This is particularly true for Mn^{3+} oxides. Higher-valent Mn oxides are present in the slag only as metastable phases, which is not surprising when one considers its comparatively narrow range of stability in the p_{O2}/T-diagram. Figure 6.32 illustrates that in the p_{O2}/T-diagram the equilibrium curve Mn_2O_3/Mn_3O_4 cuts the lines of constant oxygen pressure in a very acute angle. This means that Mn_3O_4 forms preferably below 1 000 °C and that already very small fluctuations of oxygen in the furnace atmosphere can change the temperature of formation significantly. Above this temperature, Mn_3O_4 is rapidly reduced to Mn^{2+} and Mn silicates are formed.

Partridgeite (α-Mn_2O_3) and (through semi-quantitative microanalyses) also *bixbyite* (α-$(Mn,Fe)_2O_3$) were discovered only occasionally in the slag and if so only in context with ore remains.

Slightly more common than braunite is *Mn-rich spinel* in the slag from Faynan. This Mn oxide consists of a tetragonally deformed lattice of the spinel type. It also

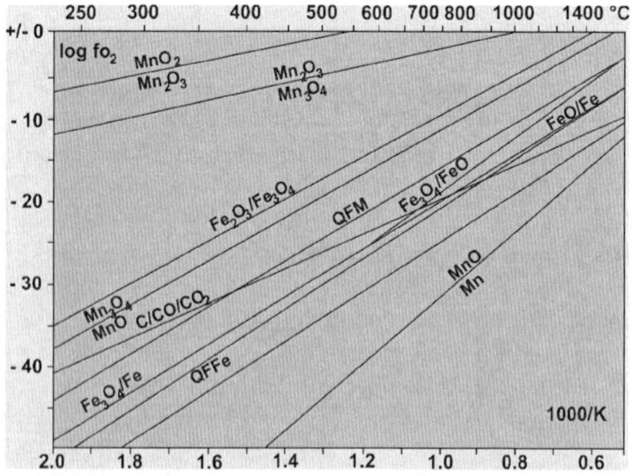

Fig. 6.32.
Stability ranges of Mn and Fe oxides as a function of temperature and oxygen partial pressure. The diagram shows that higher valent Mn oxides develop only under relatively strong oxygen influx conditions. Slag forming MnO is stable in a much larger range than FeO (after Huebner 1969)

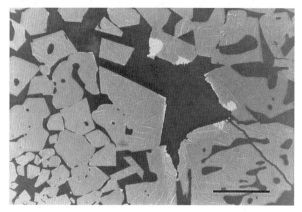

Fig. 6.33.
Faynan 1, sample JD-1/33a. Idiomorphic crystals consisting of crednerite (filling) and Cu-bixbyite (lamellae, rim, and inclusions) at the rim of a gas bubble. The crystals were most likely formed through the reduction of Cu-Mn ores (scale 0.1 mm; reflected light, oil immersion)

crystallizes in context with ore remains and along the cooling rims due to re-oxidation of Mn^{2+} phases and continues from there into the core of the slag body (Fig. 6.23). Mn-rich 'spinel' has only been identified in Early Bronze Age and partially liquefied slag from Roman times.

'Spinel' along cooling rims forms nearly opaque crystals; it shows rather seldomly the deep red internal reflections and the moiré-like shine typical of hausmannite; effects of anisotropy are brown to grey-yellow. Lamellar twinning, typical of hausmannite, has not been observed. Microprobe analyses showed the phase to contain up to 6 wt.-% Al_2O_3; furthermore, Mn oxides can be substituted by up to 27 wt.-% Fe_2O_3. The solid solutions did not show any exsolutions but remained in the homogeneous state. This is remarkable because comparable high-temperature solid solutions are not stable in nature but decompose during cooling to a mixture of hausmannite (Mn_3O_4) + jacobsite $((Mn,Fe)_3O_4)$ (= vredenburgite) (Ramdohr 1975).

Another variety of spinel, which occurs exclusively in 'slag-conglomerates' from Roman periods, is characterized by conspicuous exsolution lamellae. This spinel forms inclusions in bixbyite or idiomorphic crystals, which are embedded in glass. The crystals consist of a dens lamellae structure with filling material in between (Fig. 6.33). Inclusions and homogeneously composed margins are common. They show the same lively anisotropy from grey to blue like the lamellae, while the matrix is orange-brown to blue-grey. Microprobe analyses (Table A.16) revealed in both cases Mn-Cu-Fe oxides with Cu contents of up to 13 wt.-% in the lamellae and over 20 wt.-% in the filling.

The exact analyses of single components are rather problematic due to their small size; for that reason, only approximate formula calculations are possible. These show the lamellae probably consist of Cu-rich bixbyite with the composition $(Mn_{1.7}Cu_{0.2}Fe_{0.1})O_3$. The filling material shows a chemistry of approx. $Cu_{0.75}Fe_{0.25}Mn_{2.0}O_4$ and thus comes close to a Fe-bearing crednerite ($CuMn_2O_4$), whose composition can be very variable (Doelter and Leitmeier 1926; McAndrew 1957).

6.3 The Composition of Copper

Trace Element Content

The spatial distribution of datable smelting sites as well as the uncovering of smelting remains in the excavated settlements offered the possibility of recovering small metal prills from slag of well-defined time periods. Thus the opportunity arose to analyze the composition of 'unadulterated' primary raw copper from varying time periods. Larger metal pieces were also excavated, which must have been re-melted at least once and have thus undergone a second metallurgical process. Over 10 kg of bar and ingot fragments have been collected in the vicinity of the houses in Barqa el-Hetiye, and the excavations at Khirbet Hamra Ifdan (Adams 1992; Levy et al. 2002) also brought numerous metal objects such as axes, chisels and ingots.

With these finds, excellent material was available for an archaeometallurgical characterization of the Faynan copper. It offered a good geochemical and isotopic average of the material and thus complemented the analytical investigation of the ores. The composition of the copper could thus be analyzed in chronological order. Those effects, which influenced the trace element pattern, could thus be isolated, being mainly the variations in the ore composition and technological factors.

The fact that metallurgical processes might affect the behavior of trace elements in different ways and thus change the original element pattern of the ore has long been known. Intensive studies about this topic have been carried out, e.g., in Timna (McKerell and Tylecote 1972, Tylecote and Boydell 1978, Merkel 1990). The change of the trace element pattern from ore to metal has led to numerous debates about the classification of copper objects and their identification with particular ore deposits. The latest knowledge about early metal extraction, for instance, was one of the reasons for a reassessment of the most all-inclusive project of analyses so far carried out: 'Studien zu den Anfängen der Metallurgie' ("SAM"-project) by Junghans, Sangmeister and Schröder (Junghans et al. 1960, 1968, 1974). Pernicka (1987, 1995) has published a comprehensive systematization about this topic. He discusses smelting experiments concerning the simulation of prehistoric copper extraction on the basis of thermodynamic data from modern metallurgy and transferred these onto archaeological material. Craddock and Meeks (1987) also studied these questions and indicated the possibility of comprehending the chemical-physical parameters of archaeometallurgical processes with the help of the metal composition. They concentrated specifically on iron in copper. But no comprehensive study about the behavior of the trace elements within archaeological material such as ores, raw copper and re-melted copper had so far been carried out.

The impact of technological innovations on the trace element content in copper produced in Faynan has hitherto also only been studied superficially (Hauptmann et al. 1992). The results of the archaeological work in Faynan demonstrated quite quickly that these innovations could have had a larger impetus, because the size, shape and structure of the 'reaction vessels' used for the smelting have greatly changed over time (see Sect. 5.2 and 5.3). These alterations must have also changed the thermo-dynamic parameters such as temperature (-distribution) and redox conditions, which themselves control the distribution of elements between the metal and the slag.

The result of the geological and archaeological fieldwork also offered the chance for a differentiated, chronologically relevant way of looking at geochemical non-homogeneities, because the ore formations of the Dolomite-Limestone-Shale Unit (DLS) and of the sandstones (MBS) have been mined with varying intensity in the different time periods (Fig. 5.46).

The Material from the Faynan District

Metal prills and lumps, sometimes several millimeters or even centimeters in size, have been extracted from the slag through simple manual crushing. They were used for the analysis of trace and minor elements in the copper. A characterization of such copper has already been published by Hauptmann et al. (1992). The 48 analyses from this publication are included here in Table A.27.

Further analyses of Faynan copper are compiled in Table A.18. Twelve of those come from metal samples, which have been taken by drilling from ingot fragments from the vicinity of the Early Bronze Age settlement of Barqa el-Hetiye. Table A.19 shows the analyses of samples coming from copper inclusions, which have been sampled together with the surrounding slag at the various smelting sites. Nearly all analyses have been carried out using atomic absorption spectrometry, some with instrumental neutron activation analysis. Details about the analytical and measuring methods are explained in the Appendix.

Chemical Composition

The concentrations of some trace elements in Faynan copper from different time periods are shown in Fig. 4.9. These statistical data have been calculated from 48 metal samples and are compared with 33 (of altogether 70) ore analyses, which had been published in Hauptmann et al. (1992).

The elemental content in this figure as well as in Table A18 and A.19 does not sum up to 100 wt.-%. This is due to cuprite inclusions caused by corrosion in the metal, which have not been analyzed, just as the sulphide and phosphor contents have not been analyzed, because they are founded on corresponding concentrations in the ores.

The elements Sb as well as Bi, Se and Ir are at or below the detection limit, in the range of 10 µg g^{-1} or even lower. Ag and Sn are mostly below 100 µg g^{-1}. The contents of As, Co, Ni, and Zn are one unit of measurement larger. This partly very low level of trace elements shows that the metal produced in Faynan is quite compatible to modern raw copper. This is particularly true for the earliest copper-producing periods. It also became clear that the differentiation between smelted and native copper might be difficult. 'Pure' copper is not necessarily an indicator of native copper as had been suggested by, e.g., Otto and Witter (1952), Tylecote et al. (1977) and Maddin et al. (1980). It can also have been smelted from pure ores, as they occur in Faynan particularly in the ore formations of the MBS.

Without exception, the very low contents of As, Sb, Ni, and Ag differentiate the ancient copper from the Arabah very palpably from the metal that has been smelted from As-containing or complex sulphide ores. It is also possible, without any difficulty, to distinguish the metal from any intentionally or unintentionally produced As or Sn bronzes. Non-regional metal can thus be easily recognized in the Levant. It is, e.g., possible – without any doubts – to distinguish between Faynan copper on one side and the copper finds with a Cu-As-Sb-alloy composition like the Nahal Mishmar hoard (Tadmor et al. 1995) and other Chalcolithic sites on the other side of the Wadi Arabah. Any assumption that such alloys might have been of local origin in the Arabah, e.g., through the mining of ores, which would have had a different composition than the ones known to us today (Hanbury-Tenison 1986, Ilan and Sebanne 1989), has to be rejected after taking into account the analyses at hand. Faynan copper is also clearly different from the occasionally occurring Cu-As-Ni alloys, which exist in some Chalcolithic and Early Bronze Age sites (Kfar Monash: Hestrin and Tadmor 1963; Nahal Mishmar hoard: Tadmor et al. 1995; Tell esh-Shuna: Rehren et al. 1998; other Chalcolithic mace heads: unpublished results Bochum/Jerusalem).

Lead and iron are the only elements reaching some weight percent in the analyses. The frequency of higher Pb concentrations – 31 samples of altogether 87 analyzed metals include more than 2% Pb, the median is 1.15% – is not only caused by tiny agglomerations of the metal due to a complete insolubility of lead and copper, but also points to the smelting of relatively Pb-rich ores from the ore horizon since the Early Bronze Age II/III. The Faynan copper is thus one of the lead-richest metals in the Eastern Mediterranean. This fact is by itself not a sufficiently dependable indication of a certain provenance, because lead as well as iron concentrations can be easily decreased by any further processing of raw copper. But it has some significance of the origin of the metal, as similarly lead-rich copper ores only occur elsewhere in eastern Anatolia (unpublished data: Consiglio Nazionale delle Ricerche, Rome). Copper ores and artifacts from neighboring Timna contain 0.3% Pb (median) – less than those from Faynan (Leese et al. 1985/86). The same is true for ores from the Sinai, which include lead only in the range of 10–100 µg g^{-1}; only the ore deposit at Wadi Tar contains lead up to just above 1% (Hauptmann et al. 1997). The analyses of more than 460 copper ore samples from Cyprus (Stos-Gale and Gale 1994) resulted in Pb contents of only 1–50 µg g^{-1}. Therefore, copper coming from such ore must be very low in lead, as demonstrated by the analyses of Early Bronze Age copper objects from Lapithos, which have Pb contents < 0.05%. Early Bronze Age copper artifacts from Hassek Höyük on the Upper Euphrates also have only 0.001–0.01% Pb (Schmitt-Strecker et al. 1992), and the copper ores from Ergani

Maden would have allowed, at the most, metal production with 100 µg g⁻¹ Pb (Seeliger et al. 1985). Copper objects from the Aegean hold noticeably more Pb, e.g., the finds from Poliochni with 2–4% (Pernicka et al. 1990). Artifacts in Thermi have Pb concentrations up to 3% (Begemann et al. 1992).

The Fe concentrations show a rather unusual tendency in the Faynan copper. They increase from a range of 0.01% up to the percentage level over time, until they finally lead, in the Mameluk period, to the formation of two-phase alloys, which can even consist predominantly of iron. This development will be discussed in the following paragraphs.

The Homogeneity of Copper

Figure 6.34 displays some trace elements in copper in chronological order, which clearly illustrates different patterns. The concentrations of Pb, Zn, Fe, Co, Ni, and As in the raw copper are generally lower in the Early Bronze Age I and II/III and show larger ranges of variations. The Iron Age copper has higher trace element contents but they vary within more restricted fields.

One reason for this chronological development is the choice of raw material. As the fieldwork evidence has already shown, it is firstly the use of the 'purer' MBS ores and then the increasing utilization of the DLS ores, which causes the development of the trace elements' contents. This tendency is further verified by the results from the lead isotope analysis (see below). And even slag analyses have demonstrated that a deliberate choice of raw materials was carried out during the Iron Age. The clusters formed by these analyses (see Fig. 6.17 and 6.18) mirror that development. Technological influences must also have played a role. Their part is particularly clear with those elements which react sensitively to fluctuations in the oxygen content: Fe, Zn and Pb. The enrichment of these elements in Iron Age copper indi-

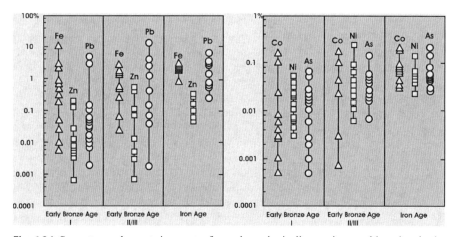

Fig. 6.34. Some trace elements in copper from chronologically varying smelting sites in the Faynan area. Notice the bunching of elements in the Iron Age copper, which indicates a more reducing atmosphere in the furnace and a better controlled smelting process (partly from Hauptmann et al. 1992)

cates a stronger reducing atmosphere in the smelting furnaces than in earlier periods. This was also noticed by Craddock (1986, 1995): he observed a drastic increase of the Fe contents in metal artifacts (from 0.0X to 0.X wt.-%) from the Early Bronze Age to the 1st millennium BCE on an "international" level, which he attributed to increasingly improved smelting techniques.

In studying the copper from Khirbet Hamra Ifdan (KHI), it becomes clear that another factor also had a strong influence on the composition of copper objects. In Sect. 7.2 the complex way in which the raw copper smelted from DLS ores was further processed will be described in detail. The crucial point is a repeated smelting and re-smelting of small amounts of melted raw copper, which had come to KHI from different smelting sites in the Faynan area. The goal was to produce large amounts of metal, which was sufficient and in good enough condition to cast crescent-shaped bar ingots or axes etc. The copper was homogenized through these processes. In Fig. 6.35 the percentages of some trace elements of DLS ores are compared with those of raw copper from Faynan as well as with those of items from KHI. Particularly, the elements Pb, Zn and Ni show a clear tendency to cluster. It becomes obvious that discrepancies in the composition of the raw material and finished products can more easily occur when individual cases are compared instead of statistically far more viable series of data.

As a first summary, one can state that better controlled process techniques led to an increase in metal production (see Fig. 6.16) but that the copper obtained in this way had a higher level of impurities. This 'line of development' can possibly – with all due caution – even be used as a help in dating. How this could work can be demonstrated with the example of the fragments of ingots, which have been col-

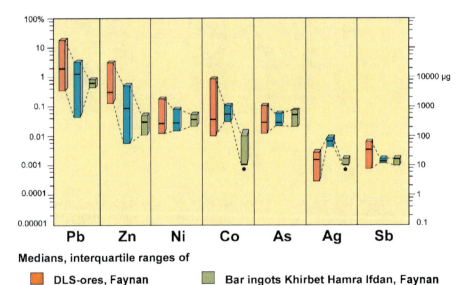

Fig. 6.35. Medians and interquartile ranges of some trace elements of copper ore (DLS) and of EBA II/IV raw copper in the Faynan area compared with crescent-shaped bar ingots from Khirbet Hamra Ifdan. Note the bunching especially of Pb, Zn, Ni, and Ag

Fig. 6.36.
Some trace elements in ingot fragments from the surroundings of Barqa el-Hetiye. Although the finds date mostly into the Early Bronze Age II/III, the close concentration of elements is more easily comparable with Iron Age copper (see Fig. 6.34)

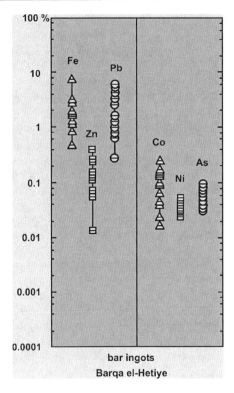

lected in Barqa el-Hetiye in the vicinity of the Early Bronze Age houses. Their pattern of trace elements (Fig. 6.36) shows a much better correspondence with the Iron Age metal than with the Bronze Age material. It is very unlikely that the trace element content of this raw copper would have been changed that severely through simple re-melting. Such drastic alterations of elements occur only through deliberately blowing air into the liquid copper, thus through refining (Merkel 1982; Pernicka 1987). The results of the analyses in the case of Barqa el-Hetiye could hence give a reason for studying the archaeology of the site again more carefully.

Partitioning of Trace Elements between Metal and Slag

The influence of technological changes through different shapes of furnaces and modified firing techniques can be, in the case of Faynan, clearly shown through the partitioning of trace elements between metal and slag (Table A.20). This partitioning coefficient is calculated by

$$D_{Cu/S} = (\% \text{ M in copper}) / (\% \text{ M in slag})$$

The formula is based on the principle that impurities will be enriched either in the metal or the slag phase depending on the oxidation potential. If $D > 1$, then the relevant element will concentrate in the metal; if $D < 1$ it will be in the slag. Figure 6.37

Fig. 6.37.
Partitioning coefficient $D_{Cu/S}$ of some trace elements between copper and slag of samples from chronologically different smelting sites in the Faynan area. Notice the width of variation for $D_{Cu/S}$ of Fe, Pb, Zn, Co, and Ni in the Early Bronze Age samples. The high partitioning coefficients of As and Ni are particularly interesting, as they indicate the strong tendency of the elements to concentrate in the metal

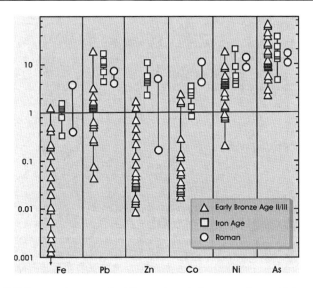

displays the partitioning of different elements in 27 metal and slag samples, which come from smelting sites of the Early Bronze Age to the Roman period in the Faynan area. This demonstrates that the order of enrichment of the different elements in the metal phases is roughly As > Ni > Pb > Co > Zn > Fe.

The $D_{Cu/S}$-values of Zn, Fe, Co, and Pb show the most marked chronological differences. The first three have partitioning coefficients of $D < 1$ in the EBA II/III; they become oxidized and slagged, while lead already shows a strong tendency to accumulate in the metal. In later periods, we observe $D_{Cu/S} > 1$ for these elements in most cases, which means that these elements accumulate in the metal due to stronger reducing atmospheres. This can be clearly confirmed by the occurrence of lead and iron in the copper of this time period.

The partitioning coefficient of Ag stays, through all time periods, constantly in an area between one and ten; the noble element obviously accumulates within the metal. The elements As and Ni seem to be stable under varying oxygen pressures and are quite independent from particular firing conditions during smelting. As shows the highest partitioning coefficient; it accumulates in the copper more strongly than all other studied elements. This is quite surprising, as the element should easily sublimate and oxidize (sublimation point As: 613 °C, As_2O_3: 193 °C) during heating due to its volatility. But conversely, a cementation takes place during the smelting of As-bearing ores, during which – quite similar to the production of brass – gaseous As is absorbed into the copper in its liquid or solid state (Pollard et al. 1990, Craddock 1995). This causes an enrichment of As in the copper, leading to As contents in the Cu appearing to be higher than in the ore, as has been previously often observed (Merkel 1985; Leese et al. 1985/86). The enrichment of As is important for the understanding of the early As copper that has led to some controversy (see Sect. 2.2). This question will also be dealt with in Sect. 7.1 when crucible smelting is examined.

The relative enrichment of Ni in comparison with Co in copper is of interest. The difference is particularly noticeable in the Early Bronze and Iron Age samples, while the partitioning coefficient of both elements is similar in Roman times. This is an

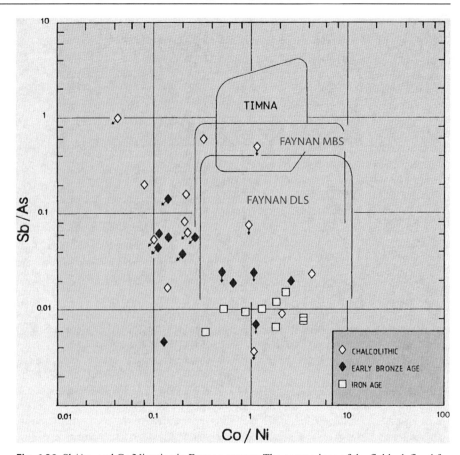

Fig. 6.38. Sb/As- and Co/Ni ratios in Faynan copper. The comparison of the fields defined for the ores (see Fig. 4.11) shows a shift towards lower Sb/As ratios and particularly for the older samples, toward low Co/Ni ratios. This coincides with the data in Fig. 6.37, where a stronger accumulation of Ni compared to Co in the metal can be recognized

addition to the initial studies by Seeliger et al. (1985), who had assumed that based on the similar geochemical behavior of both elements, the Co/Ni ratio would not change greatly during the smelting of ore to metal. Schmitt-Strecker et al. (1992) maintained, on the other hand, that the Co/Ni ratio decreases on the way from ore to metal. That this latter observation is correct can be demonstrated by the comparison of Faynan copper with ore (Fig. 6.38). Co is easily oxidized and thus transferred more readily into the slag than Ni, if the smelting does not happen under strong reducing conditions (see also Fig. 2.4). The comparison of partitioning coefficients of Early Bronze Age and Iron Age samples illustrates, furthermore, that the smelting in the 3rd millennium BCE was carried out under stronger oxidizing conditions than in the 1st millennium BCE.

The order established here corresponds principally to the results of Yazawa's (1980) experimental studies, which he carried out on a laboratory scale under equilibrium conditions and a defined oxygen partial pressure. He used the partitioning of the trace elements to compile a classification of modern copper smelting processes, based on oxygen partial pressure. With increasing oxygen pressure he distinguishes between smelting processes such as matte smelting, direct reduction of copper ores in the Noranda-process, converting and refining. His results can generally be compared, with the exception of lead, with the sequence of the oxidizability of the elements, as shown in the Richardson-Ellingham-diagram (see Fig. 2.4).

The partitioning coefficients determined in the material from Faynan cannot, however, be compared directly with the results from Yazawa, since a sulphide phase existed in all his analyses with metal and slag. It is thus not possible to give any quantitative estimates about which redox conditions were present during the smelting processes in the different time periods.

Iron in Copper

Iron has quite often been observed in copper and in greatly varying quantities and has also been described several times (Cooke and Aschenbrenner 1975; Sperl 1979; Charles 1980; Craddock and Meeks 1987). The study of numerous analyses of ancient copper artifacts has demonstrated that the iron concentration in copper increases in many different cultures with the onset of successive time periods when at the same time metallurgy also undergoes technological innovations. In Europe, such a metallurgical step forward can be pinned down to the spread of the slagging process in the middle of the 3rd millennium BCE (Craddock and Meeks 1987). In this process, iron got into the charge through the intergrowth of ore with gangue or the use of fluxing agents. Iron contents in copper are always a sign of higher temperatures and stronger reducing conditions in the (each time larger) furnaces.

Table A.17 illustrates that the Fe concentrations in Faynan copper can rise from the level of 0.001 up to 76%. Metallographic investigations of some copper from Faynan showed that the iron contents are due to inclusions of (Cu-rich) α-Fe as well as to iron phosphides.

The earliest inclusions of α-Fe have been observed in Iron Age slag, where the Fe content in the copper is around 1–2.6%. This is quite common for ancient copper from this time period. No α-Fe had been identified in Early Bronze Age copper. Here, Fe phosphides occur.

In the Mameluk period one can even find two-phase Cu-Fe alloys (Fig. 6.39), e.g., in the shape of nut-sized metal prills from El-Furn or the 15-kg-heavy chunk of iron (probably a bloom) found at Faynan 1, which had already been mentioned (Sect. 5.2). The former can contain far more than 50% Fe. Microanalyses showed that both metals contain parts of the other when dissolved: iron regularly contains 5–7% Cu, additionally $P < 1\%$; Cu contains 1.5–3% Fe that can also be observed as minuscule inclusions. This corresponds roughly with the phase equilibria in the system Cu-Fe (Hansen and Anderko 1958), which says that Fe is, for a large part, dissolved in Cu at temperatures of ca. 1450 °C. Caused by a nearly total miscibility gap, iron precipitates during cooling, initially as γ-Fe, which is transformed at 835 °C to α-Fe. Iron contents in copper are thus already microscopically visible in the lower percentage area. The precipitation of Fe from Cu and vice versa is based on a lowered solubility during the cooling process.

Fig. 6.39.
Sample JD-6/4, El-Furn. Two-phase Cu-Fe alloy with accumulations of lead (Pb) along grain boundaries and manganese sulphide (MnS). Copper (Cu) contains miniscule inclusions of α-iron; this (Fe) has in turn, inclusions of Cu (SEM picture, backscattered electron mode)

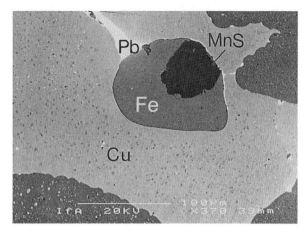

The Cu-Fe alloys from El-Furn are obviously waste material, which was no longer needed. But it cannot be unambiguously stated which kind of metal the intended product was. Experience from smelting experiments suggests that quite possibly 'pure' copper has been produced in El-Furn. For this interpretation, the repeated observations by Tylecote and Boydell (1978), Bamberger et al. (1986), Merkel (1990) and Bamberger and Wincierz (1990) that during the smelting of copper, over-reduction or a too strong of an air supply can lead to the formation of Cu-Fe alloys along with the pure copper can be applied. The alloys often get stuck in small particles in the furnace slag. This will be commented upon in Sect. 7.3 in more detail. The iron can agglomerate into larger units and often forms, during the smelting of non-ferrous metals, blooms at the furnace bottoms consisting of iron. This is well known and often mentioned in lead and copper metallurgy (Gale et al. 1990). We therefore could interpret the chunk of iron from Faynan 1 as resulting from a similar process. Sperl (1979) is also convinced that such high Fe contents in copper are caused by technical 'accidents' during the introduction of new smelting techniques or new metallurgical installations.

During the medieval period Fe-rich copper, such as in El-Furn, was widely distributed in the Near East, e.g., in the southeast of the Arabian Peninsula (Hauptmann 1985), at smelting sites in central Iran (Hezarkhani and Keesmann 1996) or at Beer Ora in the southern Arabah (Craddock and Freestone 1988). The alloys seem even to have worldwide parallels in early metallurgy (a compilation can be found in Craddock and Meeks 1987; see also Hauptmann 1989b).

It is not too difficult to separate both metals in alloys such as the ones in El-Furn. Tylecote and Boydell (1978) managed to isolate iron with only 0.5% Cu from a Cu-Fe alloy via simple re-melting in open crucibles, and to rework it. It cannot be said whether that had also been practiced in El-Furn: if so, it is unlikely that the material would have been treated as waste to such a large extent. Finds of iron artifacts that could have been produced from such Cu-Fe alloys, as they are known from Timna (Gale et al. 1990), have not been found in Faynan.

Inclusions of *Fe phosphides*, which occur during all periods in Faynan copper (Hauptmann et al. 1992, 1996), are characteristic of this region. They can be explained firstly by P-bearing ores, which may form up to 30% in the DLS. The min-

Fig. 6.40.
Sample JD-2/19, Khirbet en-Nahas. Fe$_3$P-dendrites (P), prills of α-Fe (Fe) and of lead (*white*) in a copper inclusion of an Iron Age slag (SEM picture, backscattered electron mode)

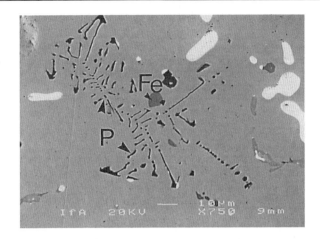

Fig. 6.41.
Sample JD-2/23, Khirbet en-Nahas. Tabular Fe$_3$P-crystals (P) intergrown with Fe phosphate in a copper inclusion of an Iron Age slag (scale: 100 µm; SEM picture, secondary electron mode)

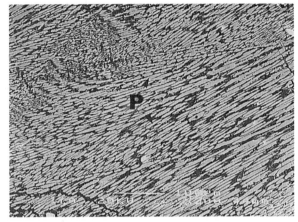

eralizations in the MBS also contain some P$_2$O$_5$. Another source could have been charcoal, which in the Arabah includes 2–6%, sometimes even up to 10% P$_2$O$_5$ in the ash (Merkel 1990).

Fe phosphides, which are formed due to a higher affinity of Fe to P than of Cu to P, occur often in copper as discrete individual crystals in idiomorphic prisms or needles as well as in dendrites (Fig. 6.40), or they form spheroidal agglomerates. Microprobe analyses show them to contain 79–85 wt.-% Fe, 13.7–15.4 wt.-% P as well as traces of Cu, Co, Ni, Pb, and Mn. This matches the composition of Fe$_3$P. In nature, this compound occurs as schreibersite or rhabdite in meteorites. Figure 6.41 displays Fe$_3$P intergrown with a phase not exactly identified, which is obviously an Fe phoshate (Fe$_3$P$_2$O$_8$?).

Phase relations between Cu, Fe, P, and O are illustrated in Fig. 6.42. Here an inclusion of α-iron is visible, which contains 4–5% Cu as well as several percent P. The solid solution is surrounded, in a crescent-like fashion, by Fe$_3$P + Fe phosphate. The phases form through segregation from an initially homogenous melt, as the system Cu-Fe-P has an extensive miscibility gap (Vogel and Berak 1950; Raghavan 1988). The coexistence of a Fe phosphate with α-Fe is possible (Trömel and Schwerdtfeger 1963),

Fig. 6.42.
Sample JD-1/55, Faynan 1. Exsolution in the quaternary system Cu–Fe–P–O. Copper (Cu), iron (Fe) with rope-shaped Fe$_3$P-exsolutions and a mixture of Fe$_3$P + Fe phosphate (P) precipitate from a homogenous solid solution. Miniscule lead droplets (*white*) can also be seen. Metal inclusion in a Roman slag (SEM picture, backscattered electron mode)

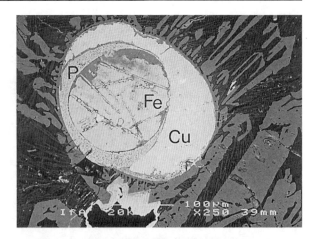

Fig. 6.43.
Sample JD-1/55, Faynan 1. Tabular Fe$_3$P-crystals with Fe$_3$P/Fe phosphate eutectic. The texture is very similar to pig iron inclusions in copper from Timna (Meeks and Craddock 1987) or from Sa Sedda in Sardinia (Tylecote 1983) and had still been interpreted in that way by Hauptmann et al. (1992) (SEM picture, backscattered electron mode)

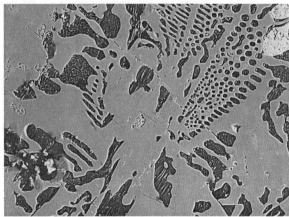

because phosphor is more easily oxidized than iron below 1 100–1 200 °C. Figure 6.43 shows another eutectic intergrowth of Fe$_3$P + Fe phosphate in Fe$_3$P, which displays an astonishing similarity to ledeburite (cementite + perlite). The same texture has been described for inclusions of cast iron in ancient copper from Sa Sedda in Sardinia (Tylecote 1983) and from Timna (Craddock and Meeks 1987). In our case, C-bearing alloys can be excluded on the basis of microprobe analyses. This corresponds with the stability relations in the system Fe-C-P (Schürmann 1958), following which the carburization of P-bearing iron phases is suppressed.

Fe phosphides in ancient copper seem to be rare. They were described for the first time a few years ago in (undated) copper pieces from En Yahav on the western side of the Arabah and from Deir Alla in Jordan (Roman 1990). An origin from Faynan had already been considered possible. The existence of Fe phosphides in ancient copper, which had been produced in the surroundings of sedimentary ore deposits such as Faynan or Timna, is not surprising because the P-contents are so high in these sediments that they are commercially mined in the so-called phosphorous belt between Amman and Maan in enormous quantities (Abu-Ajamieh et al. 1988). The reason Fe phosphides are probably not known from ancient copper in other sites is because phosphor hardly occurs in copper ore deposits of magmatic origin. Phosphorus as an indicator in provenance studies has not as yet been investigated in detail. The element becomes theoretically, during the further processing of copper, easily oxidized and slagged, but the Fe phosphides from En Yahav, described by Roman (1990), have nonetheless been found in ingots, which had been at least once remelted. This proves that phosphor can definitely be found in finished products.

Fe phosphides have been found in iron more often than ancient copper, where it was gained from P-rich iron ore of sedimentary origin (Yalçin and Hauptmann 1995; Ganzelewski 2000). The medieval smelting of P-bearing lead ores in the Black Forest (Germany) also led to the formation of Fe phosphides (Goldenberg 1996).

Lead Istotopy

The discussion about the metal composition showed that Faynan copper contains a sufficiently high quantity of lead for the analysis of lead isotope abundance ratios. Metal inclusions in slag and slag rich in metal from chronologically varying smelting sites have been analyzed for their isotopic composition and published by Hauptmann et al. (1992). These results are compiled again in Table A.21 (Appendix) and displayed in Fig. 6.44 together with the results from Faynan ores (Table A.2). The lead contents of the metal are impurities derived from the ores, because all analyzed metal samples were raw copper, which excludes the possibility of a deliberate addition of lead from another source. This would be a serious problem, e.g., whenever metal objects from the Middle Bronze Age (2nd millennium) are analyzed, because here the intentional addition of lead has to be expected (Philip 1991).

In the following, the results of the isotope analyses of metal from Faynan will be compared with ores from this mining district and with data from Timna. It will be demonstrated that it is possible in certain cases, contrary to the assumption by Budd et al. (1995), to trace individual small ore 'districts' that have supplied the smelters during different time periods. Similar conclusions have also been reached for the Bronze Age copper production at Cyprus. Stos-Gale et al. (1997) were able to locate different production centers for oxhide ingots with the help of isotopic analyses of carefully chosen ore samples.

All copper prills from the slag in Faynan as well as three of the slag samples themselves plot into the same narrow field as the ores from the Dolomite-Limestone-Shale Unit (DLS) in Wadi Khalid and Wadi Dana. They form a cluster which will here be called group 'A', in conformity with a suggestion by Gale et al. (1990) for ore and objects from Timna. It is again proven, as had already been observed for the ore, that the lead isotope abundance ratios of Cu, Mn and Cu-Mn ores are not different and that the copper produced has not been affected in its isotopic composition through Mn-rich fluxing agents. These copper prills are from varying smelting sites, which themselves had been supplied from different mines, which obviously means that the Cu-Mn ores from the DLS in the entire Faynan area have a (nearly) identical isotopic composition. It is telling, compared with this fact, that particularly the Early Bronze Age I slags from Wadi Fidan 4 have an evidently lower isotope ratio, which comes close to those from the MBS ores and probably from their own group 'B' in all three isotope ratios.

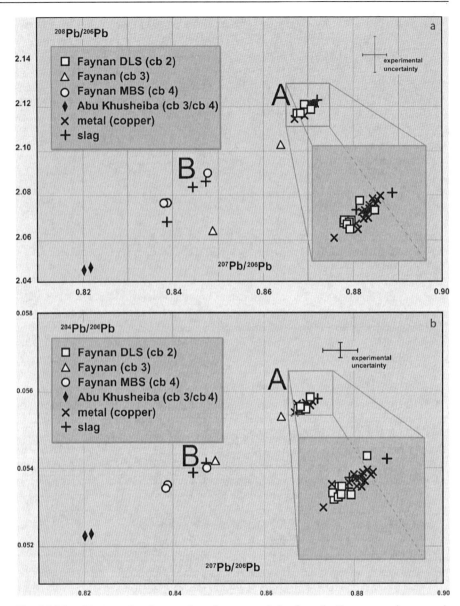

Fig. 6.44. Lead isotope abundance ratios of copper and slag from the Faynan area in comparison to the ore. Most samples cannot be differentiated from the ore samples of the Dolomite-Limestone-Shale Unit (DLS) = group 'A'. Notice that the three slags (EBA I, Wadi Fidan 4) are in the lower left-hand corner of the diagram and resemble the ore from the sandstone (MBS) = group 'B'

This signifies firstly that during the EBA I not only ores from the DLS but also those from the MBS have been utilized. This is confirmed by archaeological results, which demonstrated that many mines were opened in the vein fillings of the sand-

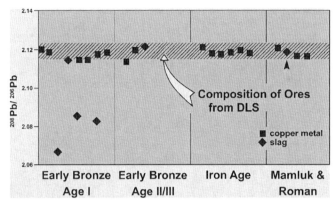

Fig. 6.45. ^{208}Pb/^{206}Pb ratios of Faynan copper and some slag samples, plotted in chronological order. The values are compared with those of ores from the Dolomite-Limestone-Shale Unit (DLS). It becomes clear that ore and metal have the same lead isotope ratio, meaning that this ore was the main raw material source for smelting during these periods. Only the slag from the Early Bronze Age I differ and indicate the utilization of ore from the sandstones. A black arrow marks the sample JD-1/9 from the Roman smelting site Faynan 1 (modified after Hauptmann et al. 1992)

stone in the Qalb Ratiye area. But it shows secondly that it was, through all time periods, obviously easier to smelt metal from the ores in the DLS than from those in the sandstone. These observations have been substantiated by mineralogical investigations of slag. They showed that both kinds of ore were smelted and that ore from the DLS had a tendency to form slag more readily, thus easing the separation of metal from slag. The data that exist so far indicate that the space between the groups 'A' and 'B' is empty. This division is emphasized by data from Timna, as will be explained below. The configuration of the Timna/Faynan field as suggested by Stos-Gale (1993) and Pernicka (1995) should therefore not be maintained in that form.

Figure 6.45 shows that the ores from the DLS were the most important raw material source during all time periods. The isotope ratios of ^{208}Pb/^{207}Pb in the ore are compared with those in the metal and slag samples of the different periods. It becomes clear that (nearly) all slag and metal samples from all time periods are identical to the ore samples from the DLS. For the Early Bronze Age II/III, the Iron Age and the Mameluk period it thus confirms the results of the trace element analyses and the archaeological findings that the mining during these periods concentrated solely on this ore formation.

The isotope ratios of the Roman copper sample JD-1/9 from Faynan 1 are not easily understandable as they correspond with the ore samples from the DLS. Although it is problematic to come to far-reaching conclusions on the basis of one case, the sample causes problems of interpretation because the analysis of a metal inclusion provides a more representative composition of the parent material than a single ore sample. The archaeological evidence showed most discovered Roman mines to be opened in the ore formations of the MBS in the area of Qalb Ratiye, which would lead to the conclusion that the ore smelted in Faynan 1 came from there. The slag should thus have an isotopic composition similar to the material from Wadi Fidan 4, which corresponds with ores from the MBS and differs from those of

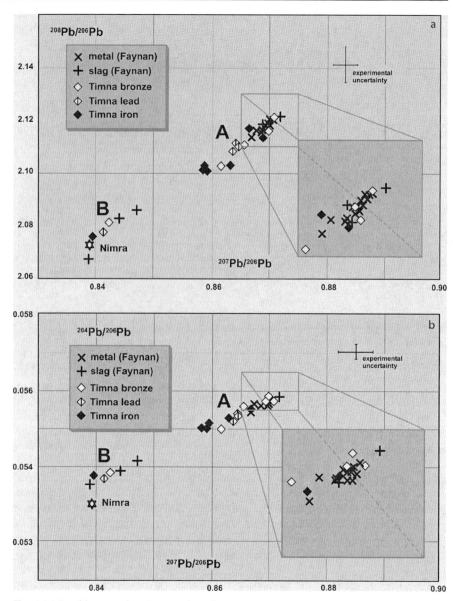

Fig. 6.46. Lead isotope abundance ratios of Faynan copper and some slag samples compared with metal objects from Timna (date after Gale et al. 1990). The samples of both localities fall into two groups. Also shown is a slag sample from Wadi Nimra (Timna) after measurements by Brill and Barnes (1988)

the DLS. As this is not the case, it makes the mines in Qalb Ratiye a very unlikely candidate for the raw material for sample JD-1/9. Other ores from other Roman mines in the Arabah are also not consistent with its composition, neither those in Abu Khusheibah/Abu Qurdiya (sample JD-24/1; Table A.2) nor those in the Wadi

Amram south of the Timna valley (sample 23082S/D33a; Table A.3). This only leaves the possibility that the smelting site at Faynan 1 was mostly supplied with ores from the mines at Umm el-Amad, as had already been suggested by the archaeological evidence (see Sect. 5.2). The isotopic characterization of the corresponding ore formation is still missing. But Fig. 4.2 clearly shows that the stratabound ore formation in the hanging wall of the Colored Sand- and Claystones (cb3) might well be compatible to the Amir-/Avrona-formation in Timna. In such a case, the high isotope ratios would be explicable.

In Fig. 6.46, metal and slag samples from Faynan are compared with metal objects from Timna, which were published by Gale et al. (1990). The above-mentioned division into two groups becomes even more pronounced here. Most samples of both localities plot as expected into group 'A'. Group 'B' contains three slag samples from Faynan and one artifact each, of bronze, lead and iron. It has already been stated that in Timna no Cu ore with a corresponding isotopic ratio has been measured, but a Fe ore has been that falls together with these three objects into group 'B'. It is possible that the supposed lack of such an ore composition is more a sampling problem than really proof of the nonexistence of such ores in Timna, because this particular isotopic pattern of the lead is known in the MBS ores in Faynan (see Fig. 4.12). This is underlined by the analysis of a slag from Wadi Nimra in the southern part of the Timna valley. Although Brill and Barnes (1988) state a considerable experimental uncertainty in its composition, the slag can be sorted undoubtedly into group 'B'. That would thus mean that ore with this particular composition has also been smelted in Timna. Looking back it is hence not necessary to assume an import of these objects from Arabia or Egypt as was suggested by Gale et al. (1990).

Chapter 7

Copper Smelting Technology

7.1 The Earliest Stage: Crucible Smelting

The Archaeological Evidence

According to the notions of Scott (1912) and Rickard (1932, the initial idea about copper smelting technology was that smelting the first copper happened accidentally in a campfire. Rickard based this on observation from Katanga, where he had found copper prills adhering to chunks of ore in a campfire made by the local population. Theoretically it is possible to reduce copper in a campfire. This observation alone, however, does not mean that this principle has been the universal origin of metallurgy, as it has already been proven for the Neolithic period of the Near East that pyrotechnological processes had been carried out in appropriate reaction vessels (Voigt 1985). It is only in exceptional cases that the thermodynamic conditions for the reduction of copper can be maintained in a 'normal' bonfire. Strong winds can help in some cases and for a short time, temperatures of 1 000 °C can be reached, but it is most unlikely that a strongly reducing atmosphere could be maintained over a long period, as would be necessary for the reduction of copper.

The earliest archaeological finds of 'metallurgical reaction vessels' are crucibles. Those from Tepe Ghabristan in northwestern Iran date into the 5th millennium BCE (Majidzadeh 1979). Two hearths with a quantity of oxidic copper ores were also found there. The crucibles are heavily slagged on the inside. The contemporary Tal-i Iblis (Iran) yielded more than 300 crucible fragments (Dougherty and Caldwell 1966), one of which was found together with large amounts of charcoal (Smith 1967). The crucibles from Merhgarh III in Baluchistan (Pakistan) date into the first half of the 4th millennium (Jarrige 1984). The Chalcolithic settlements on the Upper Euphrates (Fig. 1.2) also rendered crucibles, e.g., from Değirmentepe, Tülintepe, Tepecik, Arslantepe, and Norsuntepe (compilation in Müller-Karpe 1994). Most of them are in context with 'furnace installations'. The number of crucible finds rises swiftly during the following Bronze Age. All of them were found in settlements; not a single discovery came from the immediate surroundings of ore deposits or ancient mines. This can only mean that the earliest stages of extractive metallurgy were carried out in workshops inside the settlements (Müller-Karpe 1994), which could have been located at further distances and not necessarily in close vicinity to the mines.

These early crucibles show certain typological characteristics, at least all over the Near East: they are normally shallow bowls with a diameter of 10–15 cm, several centimeters high and have a volume of 200–400 cm^3. They are all made of ceramic. Regrettably it is still not clear in most cases, if these crucibles were only used for metal melting or casting, or whether they were also employed for smelting. The main reason for this uncertainty lies in the continuous attempts to transfer earlier ideas about ancient smelting methods (Tylecote 1976) onto the most ancient processes. Following these trains of thought, ore was smelted at the beginning of extractive metallurgy in furnaces with a volume of several tens of liters. Only the subsequent steps of metal processing, e.g., refining and casting would have been carried out in small crucibles. On the basis of such a model, it is not surprising that the furnace installations in Değirmentepe, with outlets, airing vents and an inner diameter of far above one meter, have been interpreted as smelting furnaces (Esin 1986). However, this is very unlikely based on technical considerations, even though some slag has been found. It might not be impossible that crucibles were heated inside of these "furnaces," but they could just as well have served as simple fire hearths, which have been used for many purposes in the houses. The other archaeological evidence of smelting furnaces from the 4th millennium is also not convincing; that is true for Norsuntepe (Hauptmann 1982) as well as for Timna (Rothenberg 1978). This does not altogether mean that no smelting furnaces at all existed in that time period. Bachmann assumed that the early stages of metal production in Sinai saw smelting of ore in small hearths (personal comm., October 1997). The archaeological record seems to be uncertain.

But the present state of knowledge tends to maintain that the first steps of metal extraction were performed in crucibles (Craddock 1995, 2000). The study of metallurgical finds has also demonstrated that the earliest crucibles had been used to smelt ore in them (Yalçin et al. 1992; Hauptmann et al. 1993; Craddock 1995). This is true not only for the Near East but also for the Iberian peninsula (Delibes et al. 1991; Hook et al. 1991; Keesmann et al. 1994; Rovira 1999), for central Europe (Maggetti et al. 1991) and probably for the British Isles, too (Pollard et al. 1990). Even in South America, where the beginning of pyrometallurgy has to be dated much later, it happened through smelting in crucibles (Donnan 1973; Shimada and Merkel 1991). The same can also be assumed for Wadi Fidan 4, where smelting crucibles have also been found (see Sect. 5.3, Fig. 5.36). There is no evidence at all in WF4 that smelting furnaces ever existed there, but slags and finds of copper ores clearly indicate that these were smelted on site.

Crucibles are also known from a number of other Chalcolithic or Early Bronze Age sites in the Levant (Figs. 8.6 and 8.10), particularly from the Beersheba valley, where so far the best evidence of Chalcolithic metallurgy exists. Examples come from Shiqmim (Shalev and Northover 1987), Arad (Amiran 1978), and the slightly later Meser (Dothan 1959) and Lachish IV (Turner 1958); and in the south, Tell Maqass as well as Hujayrat al-Ghuzlan (Khalil 1988; Brückner et al. 2002). The crucibles from the last three sites have the closest similarities to those from WF4. That these are not just crucibles for melting and casting but smelting crucibles was made apparent by the finds of copper ore and small slag fragments in Shiqmim and Maqass. In Tell Abu Matar it was even possible to distinguish between melting crucibles for casting with a diameter of 8–10 cm (Hauptmann et al. 1993) and slagged ceramic fragments, which can be reconstructed into reaction vessels with a diameter of 30–40 cm (Perrot 1955; Tylecote 1974; Golden 1998; Shugar 2000). At Tell Abu Matar Golden (1998) and Shugar (2000) could distinguish between different sorts of slagged technical ceramic, which they assigned to (smelting) furnaces on the one hand and smaller (melting) crucibles on the other hand. The reconstruction of a tuyère that would have been used for blowing air into a furnace as proposed by Shugar (2000), however, is hardly convincing.

If one considers practical and technical factors, the utilization of crucibles becomes quite logical. It is much easier to keep the valuable raw material, ore or metal in a small container under close control during the smelting process than in an (even very small) furnace. Particularly, the management of the firing process is much easier to handle using this method, not only for the control of the temperature but also the redoxconditions (Rehren 1997a). These advantages were appreciated even in later periods. The use of crucibles is substantiated in later times not only for ore smelting and the melting of metal, but also for metal processing (remelting of metal, alloying techniques for the production of bronze and steel) and also processes where a precipitation of metal occurs via condensation during the vapor phase (mercury, zinc) or by cementation (brass). In the medieval period, crucibles played an important role in assaying for the evaluation of precious metal (Rehren 1997a).

Previous Experimental Studies and Thermodynamics of Crucible Smelting

One of the first experiments to smelt copper ore in a crucible was carried out by Coghlan (1939/1940). He thought that the necessary parameters for the reduction of copper would be easiest to meet in a two-chamber or reverberatory furnace with a separated heating chamber, a type of furnace known from ancient pottery production. Coghlan built such a furnace with a simple stone construction, which he filled consecutively with ca. ½ m^3 charcoal. The ore charge was separated from the charcoal bed by a closed crucible. He published no precise information about his experiments and briefly described the resulting metal regulus as compact, finely grained and malleable. It thus becomes unclear as to whether, and if so how much, copper was reduced (in the liquid state). Coghlan's experiments inspired Tylecote (in Perrot 1955) to reconstruct a two-chamber furnace of the kind which he thought was used in Chalcolithic Tell Abu Matar for copper smelting in crucibles. The furnace has never been tested. Coghlan's idea was taken up again by Moesta (1983), who reduced malachite to metal at a temperature of 700–800 °C in a closed crucible, surrounded by charcoal during a reaction period of half an hour. During this solid state reaction sponge metal was formed, which could subsequently be worked through hammering.

However, the shape of the oldest crucibles suggests that this technical principle was most likely not employed. Hence, for the time being, it has not been possible to create the desired connection to pottery technology. The signs of heating and the slagging pattern on the inner side of the crucibles suggest rather that they were heated deliberately from the top. Their content was exposed to temperatures that are higher than the melting point of copper. The exterior generally shows only minor traces of heating.

Donnan (1973) carried out smelting experiments based on the evidence from the Moche culture in South America (4th–8th century CE). He filled small crucibles with finely grained oxidic copper ore and powdered charcoal. The crucible was then put into a glowing charcoal bed, which had a diameter of over one meter, and covered it completely with charcoal. With the help of five to six blowpipes, a temperature of 1 300 °C was reached after ca. 15 minutes. After one hour of constant blowing the ore was reduced and a small regulus with a little slag had formed. It was necessary to break the crucible in order to get to this regulus. It is very interesting that Donnan got successful results only with cuprite ore and that 64 wt.-% of the charge was reduced to metal. In using malachite, a small copper lump formed, which was more porous and of an inhomogeneous composition.

Zwicker et al. (1985) also used relatively large quantities of charcoal in order to smelt ca. 1 kg of malachite to copper in a crucible. They achieved an output of nearly 90% with a charcoal covering of ca. 50 cm and the use of three to six blowpipes. There was no slag. Rostoker et al. (1989) managed to produce copper even from mixed oxidic-sulphidic ores, but similar to Zwicker et al., only under laboratory conditions.

The experiments by Merkel and Shimada (1988) carried out in Batan Grade (Peru) are of importance for a better understanding of the assumed smelting furnaces or crucibles in Tell Abu Matar. They used no crucibles for the smelting of oxidic as well as sulphide copper ores with As contents, but charged the ore directly in a niche-shaped furnace. The use of sufficiently large quantities of charcoal prevented the formation of slag on the furnace lining. The authors had no problems in reaching 1 200 °C with the help of blowpipes, but only for a spatially very limited area and for a short period of time. The ore was thus only partially slagged. The copper and arsenic contents of the ore had been reduced and formed small metal prills, which could be manually pried from the slag and then further processed.

The finds of blowpipes are common. They were operated by human breath. It is more than likely that they had clay tuyères at their tips, as they have been found in several sites from the Neolithic and Early Bronze Age in Europe and the Near East (Esin 1986; Roden 1988; Müller-Karpe 1994; Martinek 1996). In more recent times, the clearest and most substantial complex was unearthed in the Pre-Columbian Batan Grande in Peru (Epstein and Shimada 1983; Shimada and Merkel 1991). The inner diameter of the tuyères varies between 2 and 6 mm. The same technique is illustrated in an Egyptian wall painting from Saqqara (Egypt) dating into the 6th Dynasty (2350–2350 BCE). Six people are visible, using blowpipes with tuyères to melt metal or to smelt ore in a crucible (Fig. 7.1). It is not of great importance that these tuyères have not always been found in context with the crucibles, as is the case in WF4. Blowpipes made from reed work very well without the extra tuyères, as long as their

Fig. 7.1. Egyptian depiction of a smelting or melting operation in a crucible using blowpipes with clay(?) tuyères. Saqqara, tomb of Mereruka, 6th Dynasty, 2450–2350 BCE (after Duell 1938)

7.1 · The Earliest Stage: Crucible Smelting

points are slightly narrowed and kept wet during the process. The material is easily decomposed while buried in the soil for thousands of years.

Firing using blow pipes creates neither constant conditions inside a crucible nor uniform physical conditions inside the charcoal bed (Rehder 1994), similar to smelting furnaces run with tuyères and bellows. In order to bring the air necessary for the combustion of the carbon into the middle of the charcoal bed, the speed of the air intake needs to be increased, which can be achieved through the narrowing of the blowpipe's tip. This brings about an injection of atmospheric oxygen into the glowing charcoal, which leads to an initial spot ignition according to

$$C + O_2 \rightarrow CO_2 \quad \Delta H\,(298.15\ K) = -94.05\ kcal\ mol^{-1} \tag{I}$$

The energy input of this exothermic reaction is required to start, for example, the conversion of malachite to cuprite. The high gas velocity thus created can lead – particularly when small quantities of charcoal are used – to a shortening of the time of the reaction

$$CO_2 + C \rightarrow 2CO \quad \Delta H\,(298.15\ K) = +41.21\ kcal\ mol^{-1} \tag{II}$$

needed to reduce the Cu oxide to metal. It can also happen that the reaction of the resultant CO_2 with carbon, following the equation II, is kinetically retarded so that the actual CO concentrations are below the ones which could be expected in a theoretical model (Kronz 1997). This means that the redox conditions during crucible smelting may have been quite heavily oxidizing. It is thus very unlikely that a Boudouard equilibrium, which is based on a C surplus, will ever be reached. Much more probable is that CO will be oxidized to CO_2 through a surplus of O_2.

Five to six blowpipes are sufficient to reach the energy necessary for the first reaction, with temperatures of 1 100–1 200 °C. This was shown in our own experiments (Hauptmann et al., in prep.), which confirmed calculations by Rehder (1994). The temperatures were measured 3–4 cm away from the tips of the blowpipes in a narrowly defined space of the charcoal bed. Although during continuous blowing human breath may contain less oxygen than natural air (13.7% compared with 21% O_2 during normal breathing), there is always the immanent danger of a re-oxidation of the already-formed copper. The temperature can sink rapidly to 800 °C during fluctuations of the air supply. Additionally, one has to count in an enormous loss of heat through the open parts of the reaction vessel. A decrease of temperatures is also caused if the CO/CO_2-ratio in the fire increases due the endothermic reaction of CO_2 with charcoal, which would create a more reducing gas atmosphere. This is the reason, as the following energy balances show, that for example cuprite is easier to smelt than malachite (or other secondary copper ores):

$$Cu_2O + CO \rightarrow 2\,Cu + CO_2 \tag{I}$$
$$\Delta H\,(1\,073.15\ K) = -137.74\ KJ\ mol^{-1}$$

$$Cu_2[(OH)_2/CO_3] + CO \rightarrow Cu_2O + H_2O + 2\,CO_2 \tag{II}$$
$$\Delta H\,(1\,073.15\ K) = +153.54\ KJ\ mol^{-1}$$

Fig. 7.2.
Ranges of stability of copper and copper oxide as a function of temperature and oxygen partial pressure (after Elliott 1975). The process of reduction in the solid state is shown in (*1*), which leads to a copper sponge, as it is displayed in Fig. 7.3. The reduction in the shaded area of (*2*) occurs mainly via the liquid phase and has more in common with the actual conditions of the earliest copper production

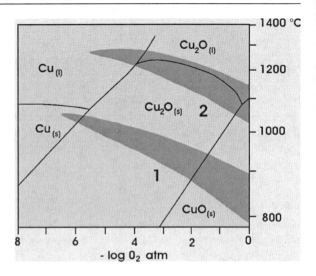

The reason for this is the fact that during the smelting of malachite its decomposition is preceded at low temperatures, which is connected with the releases of CO_2 and H_2O.

To reduce malachite as completely as possible to copper in the liquid state, the balance between two contrary tendencies has to be found: between high temperatures on one side and a reducing gas atmosphere on the other side. The diagram in Fig. 7.2 shows that for the reduction of copper in its liquid state at a temperature of 1 083 °C, an oxygen partial pressure of at least $10^{-5.5}$ atm is required. At higher temperatures the pressure can be slightly lower. At $T < 1 083$ °C, copper precipitates in the solid state and forms a porous sponge (Fig. 7.3), but this has never yet been found in an archaeological context. This might be because, with the spread of extractive metallurgy, i.e., the smelting of ores, the As content in the produced copper noticeably increases (Begemann et al. 1994), which leads to a slightly lower liquidus temperature. Pollard et al. (1990, 1991) showed that the uptake of As into copper increases at ca. 950 °C to nearly 10% and happens particularly fast in liquid metal. This has its main reason in the absorption of gaseous arsenic into copper, through which a liquid Cu-As alloy can be formed even at 685 °C. Pollard wanted to read from these deliberately used low-temperature processes in the Bronze Age in the British Isles. This theory contains two logical mistakes. The first one is the unfounded assumption of the existence of ores with a generally very high As content. If these did exist this would then mean according to his theory that the whole process had to be carried through at lower temperatures, so that the metal objects did not acquire As contents (>5%) that were too high. The second mistake lies in the fact that the produced metal had then to be melted anyhow for further production (see Pernicka 1995). Budd (1992) wanted to deal with this contradiction in forming the hypothesis that production and further treatment would have been spatially separated.

Rehder (1994) mentioned that with the existence of cuprite in copper a number of minor and trace elements are oxidized, such as Ni, Co, Zn, Pb, and Fe (see Fig. 2.4), meaning that during crucible smelting a refining effect might occur due to the tendency of copper to oxidize. Although this is essentially true, it cannot be confirmed

Fig. 7.3.
Porous copper sponge, which still shows the original texture of the ore. Developed by solid state reaction (line 1 in Fig. 7.2) in experimental crucible smelting. Not visible in the figure are fine cuprite layers, which cling to the surface of the copper particles (SEM picture, backscattered electron mode) (photo Karsten Hess)

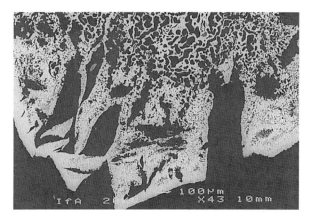

through our own smelting experiments in crucibles (Hauptmann et al., in prep.). The same concentrations of Co, Ni, As, Zn, and Fe have been measured in primary-produced copper lumps as in additionally remelted metal. This is in accordance with Merkel's (1990) observations that a noticeable change in the trace element pattern only occurs with a deliberate input of air into the copper bath. It seems thus a bit rash to postulate a technologically independent step based solely on the occurrence of cuprite. Therefore, the line of reasoning by Shalev and Northover (1987), claiming that the existence of cuprite and other Cu-Fe oxides in Chalcolithic slag in Shiqmim would thus indicate a refining process and with that an additional technological process independent of the actual smelting process, cannot be validated.

Reconstruction of Crucible Smelting Based on Archaeological Finds

It is quite unfortunate for the reconstruction of the copper smelting and melting processes used in the earliest times that the material most frequently found in archaeological contexts, the (s)melting crucibles, are so difficult to interpret. Thermodynamic parameters necessary for metallurgical operations are not evenly mirrored in all parts of the crucibles bodies. This is not surprising when one considers that ideally, the focus of the air supply coming from the tips of the blowpipes should be above the crucibles in the charcoal covering. The crucible rim would consequently be the most vitrified part of the vessel, while the base might not have been exposed to so high a temperature. The material that was processed in crucibles can also not always be precisely identified. Maggetti et al. (1991) could, for example, prove temperatures of around 1 100 °C only at some spots of Neolithic crucibles from Pfyn (Switzerland), but they were not able to determine if the vessels were used for the melting of metal or the smelting of ore. Yener and Vandiver (1993) were able to show that the crucibles possibly used for the melting of tin from Göltepe in eastern Anatolia (Turkey) were heated up to ca. 950–1 000 °C on the inside, but only up to ca. 700–800 °C on the exterior. The investigations of Kingery and Frierman (1974) into a Chalcolithic crucible fragment from Karanova (Bulgaria), which was supposedly used for the production of copper, revealed that it was only heated up to ca. 800 °C. They conclude that it was not even necessary to produce a special refractory ceramic for early copper smelting. The temperature they measured is rather low if the fragment was really used in metallurgical processes. The temperature must have been measured on the outside of the vessel.

Crucibles can have slag incrustations on their inner walls formed by reactions between ceramic lining, charged material and charcoal ash. When copper is melted, cuprite forms through oxidation on the upper rim and is incorporated into the slagged crucible wall. This adds to the typical red coloring of the slag (Fig. 7.4). The existence of Cu_2O-containing slag is not necessarily proof of metal melting; it also forms during the smelting of high-grade, secondary copper ores. The material from WF4 can be used here as an example. Figures 7.5 and 7.6 illustrate that the red hue of the slag crust from a crucible is explained by the finely dispersed distribution of cuprite in the melt of several Ca-Mg silicates. Plane scans by REM/EDX at the transition between crucible wall and slag show that the differences in concentration of SiO_2, K_2O, CaO and others were not caused by the melting of the crucible lining, but by the melting of sandy dolomite, a host rock of the copper ore in the DLS.

It is easy to recognize the smelting of Fe-containing copper ores. In that case, (Cu)-Fe oxides such as delafossite, magnetite and corresponding Fe silicates form

Fig. 7.4.
Tell Abu Matar, sample BAM 113. Slag incrustation from the upper part of a Chalcolithic smelting crucible for metal. Large drops of cuprite, in between a glassy matrix of Ca-Al silicate, from which occasional pyroxene and delafossite have crystallized. Border area towards the ceramic crucible wall (*W*) (SEM picture, backscattered electron mode) (from Hauptmann et al. 1993)

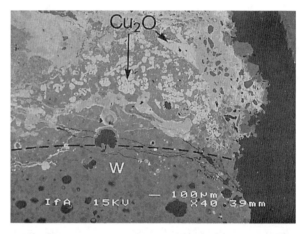

Fig. 7.5.
Wadi Fidan 4, sample 1482. Wall of an ore smelting crucible. The slag incrustation at the transition to the crucible wall (*W*) contains copper inclusions (*Cu*), which are embedded in a red slag of Ca-Mg silicates (*black bordered square*) and cuprite (SEM picture, backscattered electron mode). The chemical composition (SEM/EDS-analysis) is presented in Table 7.1

in the slag incrustation (Fig. 7.7). Normally, the Fe content strongly differs from that of the crucible wall; thus, a slag formation alone by means of ceramic overheating can be ruled out. But the absorption of the crucible lining due to the Fe silicate melts has of course to be expected, which is very uncommon in liquid copper or cuprite melts. Whether that permits the conclusion that casting crucibles and such, in which pure Cu ore has been smelted, are more common in the archaeological material, remains open.

Table 7.1.
Chemical composition (SEM/EDS-analysis) of sample 1482 shown in Fig. 7.5

Compound	Content (wt.-%)	
	Crucible	Slag
SiO$_2$	68.3	28.6
K$_2$O	10.4	1.9
CaO	2.7	33.4
MgO	–	7.7
FeO	8.0	6.7

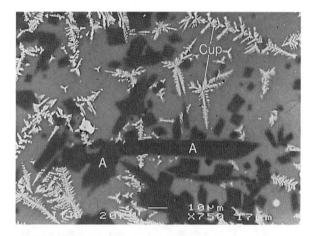

Fig. 7.6.
Wadi Fidan 4, sample 1482. Detail from Fig. 7.5; rhombohedal crystals of åkermanite (*A*). The glassy matrix is interspersed with cuprite dendrites (*Cup*), which give the red color to the slag incrustation (SEM picture, backscattered electron mode)

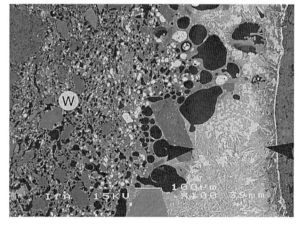

Fig. 7.7.
Nevali Çori, sample TR-4/17. Slag incrustation (between *arrows*) from the upper part of an Early Bronze Age crucible for the smelting of ore. In comparison with the copper crucible from Wadi Fidan 4 the main components crystallizing here are magnetite, delafossite and cuprite. Transition to the ceramic wall (*W*) (SEM picture, backscattered electron mode) (from Hauptmann et al. 1993)

Apart from slag incrustations in crucibles, slags themselves are very useful indicators of the reconstruction of the metallurgical processes from which they have been formed by the decomposition of the charged ore. Slag provides evidence about the two most important process parameters, temperature and gas atmosphere which – in contrast to modern technology – could not be controlled independently. Temperature and gas atmosphere were largely coupled and controlled directly through air supply. Slags dating into the Chalcolithic and the beginning of the Early Bronze Age have only been recently identified as such and have been analyzed more intensively (Shalev and Northover 1987; Moesta 1989a,b; Glumac and Todd 1991a; Yalçin et al. 1992; Hauptmann et al. 1993; Saez et al. 2003). This was mainly due to their mineralogical and chemical compositions which significantly differ from slags from later periods.

The phase content of the slags from Wadi Fidan 4 described in Sect. 6.1 allows an assessment of the redox conditions, if one compares it in a simplified form with the phase equilibria in the system Cu–Fe–O as a function of the oxygen partial pressure. This system was developed by Gadalla et al. (1962/1963) and Jacob et al. (1977) for temperatures of 1 000 °C and 1 200 °C (Fig. 7.8), which is largely within the parameters of our framework. Such an assessment is also valid for those slags that have not been (completely) liquefied due to their being exposed for too short a time to sufficiently high temperatures, caused, e.g., by an irregular distribution of the heat. As

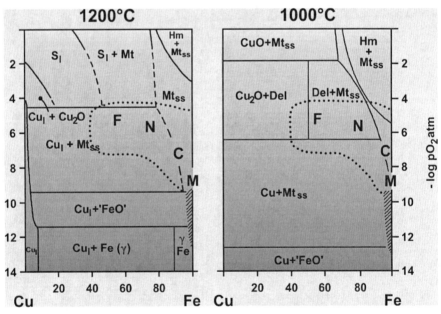

Fig. 7.8. Estimated formation area of Chalcolithic and Early Bronze Age slag, displayed in the p_{O2}/T-diagram of the systems Cu–Fe–O. In accordance with the estimated firing temperatures it was chosen to display them in two isothermal sections at 1 000 °C and at 1 200 °C. The figure also illustrates the approximate Cu content of the slag. Slag from the 1st millennium BCE, which was formed under much stronger reducing conditions, is also shown for reasons of comparison (author's data, *hatched field*). *F:* Wadi Fidan 4; *C:* Nevali Çori; *N:* Norsuntepe; *M:* Murgul; *S:* Slag; *Mt:* magnetite; *Hm:* hematite; *Del:* delafossite (Hauptmann et al. 1993)

well as the copper, the phases delafossite ($Cu^+Fe^{3+}O_2$) and cuprite are important and in the case of iron-richer compositions magnetite, whose associations show an oxygen partial pressure of $p_{O2} = 10^{-6}$ to 10^{-5} atm and above. The crystallization of magnetite-rich melts from Norsuntepe and Nevali Çori (Hauptmann et al. 1993) falls into the same category of formation. It is also consistent with the desulphurization of sulphides, which can be observed occasionally in the material of Tell Abu Matar and in Çayönü Tepesi (Hauptmann 1989; unpubl. data DBM).

This phase content seems to be typical of the slags from the Chalcolithic and beginning of the Early Bronze Age. It demonstrates that early smelting processes were performed under such strongly oxidizing conditions (Hauptmann et al. 1996) that they are even in parts comparable with converter slags in modern copper metallurgy (Trojer 1951). Moesta and Schlick (1989) observed a similar phase content in the Bronze Age slag from Mitterberg and are of the opinion that such strongly oxidizing conditions are only possible when the slag is not covered by charcoal. These notions might be true for smelting in larger shaft furnaces, but not for the kind of smelting processes performed in the crucibles that are being described here.

The extremely high Cu contents in the ancient slags can by no means be taken as an indication of a technologically immature, 'primitive' smelting. The analyses of ore finds from the Chalcolithic settlements in the Beersheba valley (Shalev and Northover 1987; Hauptmann 1989; Golden 1999; Shugar 2000) show that rich ores with 33–66% Cu (mostly of a cupritic composition) were directly smelted without any addition of fluxing agents. But the availability of such rich ores made a more complex technique most likely obsolete and unnecessary.

Modeling Other Possibilities

What kind of information do we obtain from experimental work, archaeological evidence and material investigations?

Although the space in a crucible for any reaction was very limited, it was obviously necessary to use large amounts of charcoal, in comparison to the small size of the crucibles, to successfully perform a smelting process. The crucibles had to be heated beforehand in a glowing charcoal bed to reach the required high temperatures and had then be covered by charcoal to create the indispensable reducing conditions. It is conceivable that charcoal or fire basins, as were found in Değirmentepe and other localities, had to be constructed to achieve this. The real technological challenge seems to have been keeping both the conditions of temperature and atmosphere in an ideal balance for the smelting process. This might be the reason for the high cuprite contents in the slag from WF4. The pie-sized inclusions substantiate the high smelting temperatures as well as the incomplete reduction of the ore.

There is no archaeological assemblage that has proved that in a crucible, in a single smelting process, a sufficiently large compact metal regulus would have been produced. The first results will most likely have been small metal lumps or prills, which were liquated from the gangue or the highly viscous slag and dripped through the charcoal bed into the crucible, where they agglomerated into smallish pieces of different sizes, dependent on the duration of the process (Hauptmann 2003). The metal entrapped in the slag would have been separated from it mechanically. The spatially limited area that had sufficiently high temperatures only partially enabled the formation of liquid slag melts and resulted in only parts of the charge being transformed. This is exactly what has been observed in the finds from Wadi Fidan 4. The formation of (liquid) slag was not – not at all as in later slag smelting processes (Craddock 1995) – a primary goal, but only a means to an end. The term 'slagless metallurgy' coined by Craddock (1990) might be correct. Keeping in mind the originally available high-percentage ores and the small quantities of slag that are generally found in Chalcolithic and Early Bronze Age contexts, the term has its merits.

There are other solutions to obtain copper in a two-step smelting process, and a different approach for the earliest smelting of copper is suggested by the archaeological record at Tell Hujayrat al-Ghuzlan, Aqaba. The site dates into the first half of the 4th millennium BCE (Görsdorf 2002). The first step of copper reduction from oxidic ores, collected at Timna and other nearby mineralizations (Hauptmann et al. forthcoming), is suggested to have taken place in a prototype of a wind-powered furnace, i.e., a simple stone construction. Slag formation was incomplete, i.e., the slag was regularly heterogeneously composed. It had an abundance of copper prills, undecomposed components of host rock and was only partly liquefied. Similar observations were made by Saez et al. (2003) in Cabezo Juré, Huelva Province, Southwestern Spain, where copper was produced in the 3rd millennium BCE. According to the archaeological evidence, a first-step smelting was proposed in furnaces using the strong prevailing winds of the region.

The second step would entail a melting of copper prills into larger pieces in crucibles, as suggested previously by Tylecote (1992a; Hauptmann et al. 1996). Such a division is also sensible for thermodynamic reasons, because the melting of metal requires far less energy than the smelting and decomposition of malachite to cuprite and its reduction to metal.

It is not easily understandable why metallurgy in Timna should have started with a comparatively complex technology. This includes the already mentioned, rather disputed Chalcolithic smelting furnace from Timna 39. It also includes the suggestion made by Rothenberg and Merkel (1995) and Merkel and Rothenberg (1999), who presume a deliberate use of fluxing agents already in the Qatifian (6th/5th millennium BCE). The whole construct is rather questionable, as it is not only the dating of the slag from site F2 into the Qatifian which cannot be upheld (see Adams and Genz 1995; Adams 1997; Kerner 2001), it is also the fact that Cu-Fe-mixed ores (with relatively high SiO_2 contents) occur in Timna, which could as easily explain the existence of Fe oxides in the glass-like solidified silicate matrix in slag (Hauptmann et al. 2005; see Fig. 4.8).

7.2 Early Bronze Age: Smelting in Wind-Powered Furnaces

The Beginning or Initial Early Bronze Age

The fact that so far no smelting furnaces have been found which date securely into the Chalcolithic or Early Bronze Age I is probably not an accident caused by a gap in archaeological knowledge, but the result of none having existed. The situation in the later Early Bronze Age and then the Middle Bronze Age is different.

7.2 · Early Bronze Age: Smelting in Wind-Powered Furnaces

In Anatolia, ore was still smelted in crucibles at the beginning of the Early Bronze Age at the end of the 4th millennium. This is demonstrated, for example, by the heavily slagged crucible fragments from Arslantepe and from Nevali Çori (Palmieri 1973, 1993; Hauptmann et al. 1993; Frangipane, pers. comm.). The crucibles utilized there were larger than the above-mentioned examples from the earlier stages of metal extraction. They had a diameter of ca. 20 cm and a height of at least 15 cm. At Arslantepe, a well-preserved cylindrically shaped ceramic crucible with a diameter of ca. 20 cm and a height of 30 cm, was excavated in level VI B2 which dates to 2900–2750 BCE. The crucible was half destroyed. Slag was stuck in the crucible, and it was thrown away.

The first palm-sized slag cakes appear here too, which show that ore with increased quantities of gangue were smelted. Such slag can be easily explained as being the result of smelting ores coming from the gossan of sulphide ore deposits, which occur frequently in those parts of Anatolia. The ore probably existed in a mixture consisting of Fe hydroxides, quartz and clay minerals, meaning mostly self-fluxing ores. It is typical that the slag now regularly contains fayalite.

The smelting of ore and gangue in a crucible was still being assessed skeptically by Tylecote (1974), who argued that the intergrowth of ore with host rock would prevent the formation of a metal regulus, because the copper would remain in the very viscous slag and thus be useless. However, archaeological field evidence proved that the production of slag rich in copper prills seems to have been the rule. This illustrates the fact that modern economic considerations cannot be transferred unilaterally onto the prehistory.

Smelting furnaces must have existed from the Early Bronze Age II, because several sites in the Aegean, Egypt and the Levant show slag heaps with a volume that cannot be explained by ore smelting in crucibles. But as the archaeological evidence is very limited, there has not been a chance to reconstruct the shape and design of those smelting furnaces. The Early Bronze Age smelting furnaces from Faynan are therefore of great importance.

Evidence of Wind-Powered and Natural Draught Furnaces, Respectively

The smelting furnaces excavated in Faynan 9 and 15 from the Early Bronze Age II–IV are the oldest known examples in the Near East. It has been pointed out in Sect. 5.2. that the smelting of ore in these furnaces was carried out using natural wind. The key arguments for this are the positioning of the furnaces at wind-exposed slopes or slope edges and the general lack of tuyères. Careful estimates show that 300–500 t of metal could have been extracted resulting in the roughly 5 000 t of slag, which had been produced during the 3rd millennium in Faynan, as is illustrated in Table 5.3. This seems to be a reasonable calculation regarding the large number of furnace locations at Faynan alone. Thirtyfour furnace locations have been counted as a minimum at two smelting sites out of seven, each having multiple backwalls that indicate repeated (10 to 20 times) firing at each place. Not counted were furnace locations at Ras en Naqb, Wadi Ghwair and Wadi Fidan. This means that many hundreds or even more than a thousand smelting processes have been carried out at these smelting sites over the centuries. This calculation seems reasonable and is in line with a calculation of Bachmann and Rothenberg (1980), who also calculated a thousand smelting processes at Iron Age Timna. It will be explained below that the type of smelting furnaces found at Faynan belong to so-called wind-powered furnaces, where natural wind drives through the tuyères into the inner part of the reaction vessel. In contrast, natural draught furnaces are operated using the draw from a chimney.

Early Bronze Age wind-powered or natural draught furnaces have been observed in Faynan, but they have not necessarily been recognized as such. One can propose with some justification that this kind of furnace existed outside the Arabah and played a far larger role in the development of metallurgy than has so far been believed.

The geographically closest sites, which would also have had the right conditions for such furnaces, are ca. 30 km away at the western margin of the Arabah. Yigal Israeli and David Alon informed us (March 1997) that three small smelting sites, on hills, are located between En Yahav and Hazeva (Fig. 3.1). The typical clay rods have been found there (pers. comm. Thomas Levy, March 1997). One of these sites had been described by McLeod (1962), who already suggested that the smelting furnace would have operated by natural draughts.

On the steeply rising table mountain above Timna 30 is another small smelting site, which Rothenberg (1980) dated into the transition from the 2nd to the 1st millennium BCE but made no more detailed study. It is thus rather interesting that Scharpenseel et al. (1976) published a ^{14}C-date from Timna 30 itself, which dates into the EBA III (HAM-215: 4020 ±100 BP = 2834–2459 BCE cal (see also Fig. 5.3). It is not known where exactly the sample was taken. The date has not been commented upon by the excavator. But it shows that in Timna 30 more ancient metallurgical activities took place than the ascribed dating of the smelting site suggests.

Rothenberg (1990: 71, fn. 18) also mentioned that numerous smelting sites in the Arabah are actually positioned on hilltops and therefore exposed to prevailing strong northern winds. He nonetheless doubts that the ancient smelting processes could have functioned with a natural wind supply due to the erratic nature of the winds. The same reason is given by Rothenberg and Shaw (1990) as to why they think it unlikely that the clay rods found in Timna 149, an Early Bronze Age IV smelting site, had anything to do with smelting. In their opinion, the partially slagged clay rods have not been exposed to high temperatures. They published only completely atypical, slightly bent clay rods, describing them as fragments from crucible-like refining vessels. These must be the exceptions, as the majority of the clay rods found at this site are as straight as the ones found in Faynan (Fig. 7.9), which makes the notions of the authors not very convincing. The conspicuous parallels to the Early Bronze Age smelting sites in Faynan hint much more firmly at a connection with ore smelting.

Only a short time later Rothenberg changed his mind completely. He now (Rothenberg and Glass 1992) draws attention to the fact that there is practically no hill between Wadi Amram and Yotvata on which a smelting has not been found. He then formed the opinion that the smelting sites were deliberately put on these hills, because of their exposed position, in order to use the winds of the Arabah for copper smelting.

Avner et al. (1994) mention remains of a smelting furnace, which was excavated north of Eilat on a hillside at Shehoret. A ^{14}C-sample dates the site into the Early Bronze Age III (RT-591: 4010 ±150 BP = 2771–2355 BCE cal (see also Fig. 5.3). The authors also assumed that the smelting furnace at the northern side of the hill was operated by natural winds. They add that remains of smelting furnaces can be

7.2 · Early Bronze Age: Smelting in Wind-Powered Furnaces

Fig. 7.9.
Timna 149. Clay rods ('Ladyfingers') from the smelting site dated by Rothenberg and Shaw (1990) in the Early Bronze Age IV. The clay rods depicted here are far more typical of the site's material than the ones published by the authors (see above). They are directly compatible to the ones from Faynan (see Fig. 5.15)

found on nearly all hills in the southern Arabah and also occasionally in the northern Arabah. This is further supported by the observation from Ricardo Eichmann (pers. comm. May 1997), who found a slag heap at a hill 25 km north of Aqaba on the eastern side of the Arabah.

Avner Goren (pers. comm., March 1997) and Hans-Gert Bachmann (pers. comm., September 1997) pointed out parallels in the southern Sinai, at least concerning the positioning of smelting sites on hilltops. Three such smelting sites were visited by Dr. Ali Abdelmotelib and the author at Bir Nasib, at the mouth of the Wadi Ba'Ba and in Wadi El-Humr. Castel et al. (1995) and Castel and Pouit (1997) describe "natural draught" furnaces for the smelting of copper from the 3rd millennium BCE in a mining district at Wadi Dara in the northern part of the eastern desert in Egypt. Their construction seems to be comparable with that of the furnaces in Faynan.

Similar smelting sites are known from the Aegean. Gale et al. (1985), Stos-Gale et al. (1988) and Bassiakos and Philaniotou-Hadjianastasiou (in press) all describe a copper smelting site on the island of Kythnos, which can be dated into the Early Cycladic I–II (3100–2400 BCE) due to associated finds. They presumed that the smelting furnaces, of which only fragments survive, were operated by natural winds, because all the slag heaps are positioned on a cliff above the coastline. A similar situation has been observed in Kephala on the Cycladic island of Kea (Pernicka 1987), where again slag was found at an exposed position on top of the cliffs. The pottery associated with the slag is in all probably also Early Cycladic (Maran 2000). On the island of Seriphos close to the village of Alevassos is a slag heap of later date but with pottery finds which indicate that the slag heap might go back into the 3rd millennium BCE (Weisgerber, pers. comm., September 1997), and a smelting site at Chrysokamino in northwestern Crete has become known, which dates into the Early to Middle Minoan period (ca. 2500–1500 BCE) (Betancourt 2006). Here copper has also been smelted in furnaces operated by natural wind and probably a pot bellow.

In summary, one can state that the smelting of ores in wind-powered furnaces in the Wadi Arabah and in the eastern Aegean during the Early Bronze Age developed very quickly into a very sophisticated technology. The origins of this development are still unknown. The surprisingly numerous finds and evidence of wind-powered furnaces in the Wadi Arabah (Avner 2002) make it possible that we see here a local development, which might be connected to this very early metal production, as has already been suggested. All these sites have in common that ore was transported from sources of different distances, sometimes dozens of kilometers away. The decisive point was the choice of a smelting site suitable for utilizing strong and continuous blowing winds.

Serious doubts have to be expressed concerning the suggestion from Moesta and Schlick (1989) that Bronze Age copper smelting at the Mitterberg near Salzburg would have been carried out in smelting furnaces with open front sides. The authors thus imply that the techniques for the process would have involved induced or natural draught furnaces or wind-powered furnaces. This is even more astonishing as Moesta (1986) himself commented upon the finds of tuyères at the Mitterberg. Tuyères and fragments of furnace walls have been found at several contemporary sites (Cierny et al. 1992; Cierny et al. 1995), all of which makes the use of natural wind operated furnaces unlikely. Apart from that, Kölschbach (1999) found based upon experimental work that furnaces with an open front would never reach a sufficiently high level of temperatures necessary for smelting copper.

Find-Based Reconstruction of Metallurgical Processes

The possibility of reconstructing smelting processes in natural draught or wind-powered furnaces from Faynan comes from analytical data first of all of slags, and of ores and metal, as well as from the excavation results and from experimental studies. Slag plays an important role here, because the later Early Bronze Age is the time period when the proper slagging process (Craddock 1995) proves the beginning of a metal smelting process sensu lato, which is named with the German key term "Rennfeuerverfahren." This term had so far only been used in connection with iron metallurgy and is there a synonym for the bloomery process. The term means the process of running of the slag out of the furnace when the firing has suitably liquefied the charged ore (Keesmann 1989; Hauptmann 2000). The examples of Arisman and Tepe Sialk show that the slagging process in Iran seems to have developed already in the 4th millennium (Schreiner 2002).

Copper smelting experiments have been carried out in Faynan in reconstructed wind-powered furnaces (Kölschbach 1999). Here Juleff's (1996) experiments need mentioning, in which steel was produced in such furnaces at Salamanawewa in Sri Lanka. This will be expanded on later. Rehder (1986, 1987) has published articles about the thermodynamic possibilities and limits of natural draught and wind-powered furnaces following theoretical considerations.

In Sect. 6.2 cuprite inclusions were mentioned, which are a characteristic of Early Bronze Age Mn silicate slags in Faynan and are, for a large part, responsible for their high Cu content (0.9–8%). They show that part of the ore was reduced only incompletely and rapidly incorporated into the liquid slag, where a further reduction was prevented. It is conspicuous that the Mn silicate slag itself often contains $Mn^{2+/3+}$ oxides, whose crystallization requires relatively high oxygen pressure (see Fig. 6.32, 7.11). The formation of slag melts in the wind-powered furnaces happened rapidly after the ore was charged. It was improved by the high reactivity of shale and sandy dolomite. The temperatures of slag formation measured during the experiments are ca. 1 150 to 1 200 °C. The range of the bulk compositions of the slags in the low melting part of the system CaO–MnO–SiO_2 implies that the ore had not been beneficiated, but had a self-fluxing composition. During the smelting only weakly reducing conditions have been reached with an oxygen partial pressure of $p_{O_2} \sim 10^{-6}$ atm and slightly below. This becomes clear in the trace element pattern of the copper, which shows only comparatively low contents of, e.g., iron and zinc (Fig. 6.34), because these elements react sensitively to oxygen and were thus, in this case, slagged to oxides.

This roughly determines curve A in Fig. 7.10, which displays schematically the reduction and slag formation from the point of charging to the cooling of the slag. This process cannot be compared with that of the 4th millennium BCE, because then Cu-Fe oxides dominated in the slag. But it is also different from the 'classic' bloomery process, where iron is produced along with the precipitation of fayalitic slag. This is represented by curve B in Fig. 7.10. The iron is reduced from the charge in the relatively cool part at the furnace top, while the formation of slag occurs generally afterwards in the lower part of the furnace, just above the tuyères where the atmosphere is more oxidizing because of the oxygen surplus (Tylecote et al. 1971). The upper limit of the p_{O_2} is reached during the cooling just as in a wind-powered furnace.

The course of the curves in Fig. 7.10 is idealized. The equilibrium conditions required for the smelting operations were in reality affected by various disturbances. This is particularly true for intermittent fuel supply in furnaces with such a small

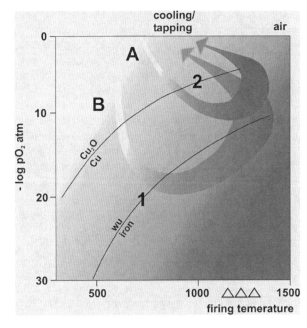

Fig. 7.10.
Schematic course of reduction and slag formation in natural draught or wind-powered furnaces (*A*) and shaft furnaces (*B*) caused by atmosphere and temperature variables, shown in the p_{O_2}/T- diagram of the systems Fe–Si–O and Cu–O. The reduction of ore to iron happens at *1*, while at *2* only the one to copper. *wu:* Wuestite (after data from Eugster and Wones 1962; Huebner and Sato 1970, and Abs-Wurmbach et al. 1983)

reaction space or as in the case of Faynan, pulsating oxygen intake caused by variable air supply due to natural wind variations. But all these factors had only limited negative influences on the process as a result of the Mn silicate immunity of the melts to oxygen fluctuations. This will be described in the next paragraph.

The Influence of Manganese-Rich Slag on Smelting Processes

The chemistry of manganese is in many ways similar to that of its neighbor in the periodic table of elements, iron. But manganese has, with valences between −3 and +7, more oxidation numbers than iron. It therefore reacts far more sensitively to fluctuations in the oxygen partial pressure. This leads to a relatively fast and quantitative thermal decomposition of Mn oxides under reducing conditions. The different behavior of Mn compounds during pyrotechnological processes has already been demonstrated in Fig. 6.32, where the ranges of stability of numerous oxides from Mn and Fe are shown as functions of temperature and oxygen partial pressure. One characteristic difference was immediately clear there: the Mn curves strew over a much larger area then those of Fe. The decisive point we wish to make is that slag forming MnO is still stable under a much higher p_{O2} pressure than 'FeO.' On the other hand, MnO can only be reduced to metal under a much lower oxygen partial pressure than 'FeO.' So the bivalent manganese has a much higher affinity to oxygen. The border of manganese reduction is far below that reached with the firing of charcoal (Fig. 2.4) being the thermodynamic possibility that smelters had at the time. The lower limit is given approximately by the equilibrium $C/CO/CO_2$, the Boudouard equilibrium.

Also, Mn silicates show important differences in the chemical-physical behavior of Fe compounds. Muan (1959a) discovered for example that Mn^{2+} is stable in silicates over a much larger p_{O2}/T-range than Fe^{2+}. This has the effect that tephroite, the most common slag-forming Mn silicate, as well as Mn metasilicates are stable even in air at liquidus temperatures (Muan 1959b). When the temperatures decrease, Mn^{2+} can be oxidized to higher-valent oxides and a SiO_2-rich phase.

The chemical-physical behavior of Mn and Mn oxides enables them to act, dependent on their valency and reactants, as oxidation as well as deoxidizing agents. The use of Mn^{4+}-compounds as oxidizers for the decoloration of glass melts has already been mentioned. Mn is also used as a deoxidizer in steel production, because the metal can be, compared with Fe, rapidly changed into bivalent oxides and then slagged.

The use of Mn ores had thus a number of advantages for early metal production. The precipitation of iron or manganese, respectively, by over-reduction is prevented. Studies by Körber and Oelsen (1940) showed that MnO is reduced only in traces, meaning the manganese reduction at temperatures up to 1 100 °C can be disregarded; the metal is thus nearly without any importance for the composition of copper. This is meant to have been the most common reason for the use of manganese-containing fluxing agents in early metallurgy (Steinberg and Koucky 1974). It has been postulated for Timna (Bachmann 1980), as well as Cyprus (Bachmann 1982) and Oman (Hauptmann 1985).

But slag formation under relatively strongly oxidizing conditions, even under free access of air (Abs-Wurmbach et al. 1983), seems to have been an important

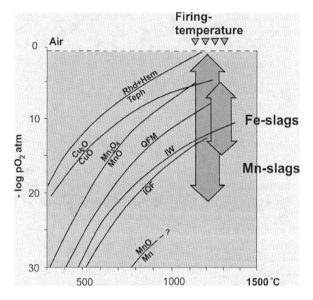

Fig. 7.11.
Ranges of stability of some oxides and silicates in the systems Fe–Si–O und Mn–Si–O. Possible areas of firing conditions for Mn-rich and Fe-rich slag are indicated. *Rhd:* Rhodonite; *HSM:* hausmannite; *Teph:* tephroite; *Q:* quartz; *F:* fayalite; *M:* magnetite; *I:* iron; *W:* wuestite (following data from Huebner and Sato 1970; Myers and Eugster 1983; Abs-Wurmbach et al. 1983)

factor in early metallurgy. This was already pointed out by Hauptmann (1985) when medieval smelting techniques in Oman were discussed (Fig. 7.11). Under these circumstances, Fayalitic melts would decompose rather rapidly into magnetite and a SiO_2-rich phase, and thus 'freeze'. The possibility of working under relatively strongly oxidizing as well as reducing conditions in Faynan must have helped along the rapid development of metal production in the Early Bronze Age, particularly in the wind-powered smelting furnaces used at the time.

Another effect should be mentioned here, which mainly concerns the extraction of iron but has also been observed in Cu metallurgy. The CO/CO_2 ratio has to be profoundly raised for the reduction of mixed FeO/MnO oxides, thus leading to a stronger carburization of the iron, which may cause a distinct improvement in the material. This is proven in Roman times by the widely appreciated Ferrum Noricum of the Southern Alps, which was actually steel. The MnO contents of the ore (up to over 4%) have favored the formation of steel (Straube 1996) just as in the Siegerland, Germany. Research by Gilles (1958) showed those slags from the Latène period and from the Middle Ages iron prod production to contain MnO values of up to 5% and 16%, respectively. These values can be explained by corresponding contents in the ores, which were highly valued for this reason.

The carburization of possible Fe contents has certainly not been approached deliberately at any time in Faynan, not even in the Mameluk period (see Sect. 6.3). Nevertheless, that the precipitation of high carburized iron happened during early copper metallurgy has been demonstrated by smelting experiments carried out, e.g., in Timna (Merkel 1983a; Tylecote and Boydell 1978), where even pig iron was produced (Craddock and Meeks 1987). Pig iron was also diagnosed in a Late Bronze Age copper ingot from Sardinia (Tylecote et al. 1983).

Using Mn silicate slag firing conditions ensures a rather insensitive reaction to oxygen oscillations as long as a T-range of 1 150–1 200 °C can be maintained. This must have been particularly good for the smelting in the Early Bronze Age wind-powered furnaces in Faynan, where the air supply could not yet be controlled with the help of bellows. Whether Mn slag has supported the development of the slag smelting process in a wider geographical area, as supposed by Craddock (1995), has to remain an open question, because the copper smelting, e.g., on the island of Kythnos (Gale et al. 1985), which is technologically and chronologically compatible with Faynan, is based on the use of Fe-rich ores.

Comparisons with Smelting Experiments

It is not clear in Faynan whether the entire charge has been fully liquefied and thus reached the theoretically possible, ideal condition for a separation of metal and slag. This is a very crucial question, which has to be asked for the early metal production in general – for the Early Bronze Age as well as the later periods. The experiments from Kölschbach (1999) make such a complete liquefaction seem unlikely. The simulation of medieval iron production in wind-powered or natural draught furnaces in Samanalawewa (Sri Lanka), which was carried out by Juleff (1996), also make it seem improbable. A direct comparison with Faynan is possible primarily due to the geometry of the smelting furnaces found in Sri Lanka, and also because of their position along the slope edges facing the prevailing western monsoon winds. The furnace from Samanalawewa displayed in Fig. 7.12 is, with over 2 m width, noticeably more broadly built, but there are also narrower furnaces with a frontal width of 0.5–0.95 m, which are comparable with those in Faynan. Furthermore, the air supply at the furnace's front is also similar: In Samanalawewa, as in Faynan, the wind is pressed through inbuilt tuyères set into the furnace walls. For this reason, Juleff (1996) proposed calling these furnaces wind-powered. The distance between the furnace front and back wall is 0.4 m and thus compares quite well with the archaeological evidence from Faynan. The same is probably true for the furnaces' limited height of 0.5 m in Samanalawewa. It needs to be highlighted that the reaction zone is limited to the front part of the furnace. The charge was only transformed into liquid slag in that front area, close to the tuyères, where at 1 500 °C the highest temperatures in the furnaces were reached. In this part, the ceramic furnace front was also vitrified. The back part of the furnace bottom and the rear wall were – just as in Faynan – not slagged and thus indicate that the area with very high temperatures was limited to a narrow strip along the furnace front. After the smelting process, which took slightly longer than seven hours, was finished, the furnace front was destroyed and the bloom removed. Juleff emphasized this was the reason why, in the archaeological field evidence, only the back walls of the furnaces have been found. This is again comparable with the evidence from Faynan. Part of the charge would not have transformed into liquid slag and would thus have had to be removed mechanically from the back part of the furnace.

The synthesis of the evidence from Samanalawewa as well as from Faynan 9 resulted in the reconstruction of a model furnace as shown in Fig. 7.13. This model was developed in the framework of a research study about process techniques of Early Bronze Age natural draught furnaces, which was carried out in the Lehr- und Forschungsgebiet Industrieofenbau und Wärmetechnik (Prof. G. Woelk, Prof. W. Bunk) at the Rheinisch-Westfälische Technische Hochschule Aachen. A comparison

7.2 · Early Bronze Age: Smelting in Wind-Powered Furnaces

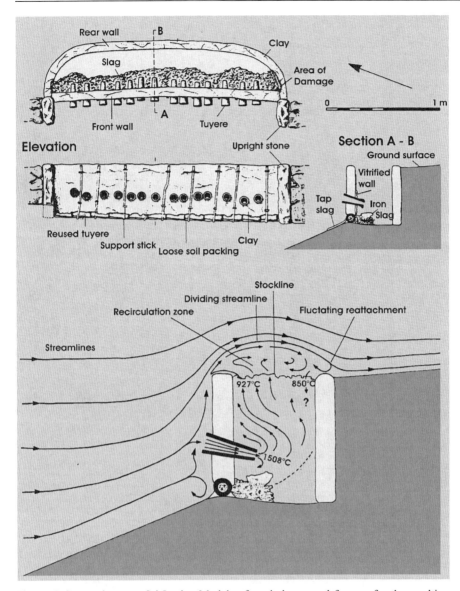

Fig. 7.12. Samanalawewa, Sri Lanka. Models of a wind-powered furnace for the smelting of iron ore, reconstructed after the archaeological evidence. The plan illustrates that the furnace is indeed wider than the furnace type from Faynan 9 (see Figs. 5.17, 5.18), but the basic arrangement is comparable. Notice the slag formation, which is restricted to the area around the tuyères (Juleff 1996)

of these drawings with Figs. 5.17 and 5.18 shows an integration of the most important archaeological facts from Faynan 9, namely the rounded furnace bottom, which rises towards the rear and the clay rods that had been incorporated into the ceramic furnace wall and were constructed as a panel to stabilize it.

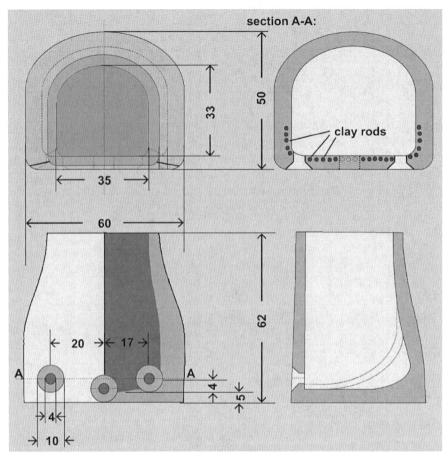

Fig. 7.13. Suggestion for the model of an Early Bronze Age wind-powered copper smelting furnace, resulting from a base of thermodynamic calculations as well as the archaeological evidence from Faynan 9 and from Samanalawewa (Sri Lanka). The furnace front (*lower left*) has two openings for the wind in its lower part. The wall is stabilized with vertically set clay rods (all measurements in cm) (Kölschbach 1999)

The production of metal in natural draught or wind-powered furnaces has never been seriously considered in archaeometallurgy, for copper even less so than for iron. Indeed it is not immediately recognizable where the advantages of natural draught furnaces are supposed to lie. The chief counterarguments state the possibly longer duration of the smelting process (30–100 hours compared with ca. 8–10 hours in furnaces with tuyères, Craddock 1995), which goes hand in hand with an enormously high fuel consumption (Killick 1992). Both arguments could have been of crucial importance in a semiarid area such as the Arabah. It has been assumed furthermore that the natural air supply would generally be too irregular and a calm would 'freeze' the furnace. It is also difficult to imagine that sufficiently reducing conditions could be created. These points apply primarily to the natural draught shaft furnaces, as they have been traditionally used over large areas in Africa for the production of iron. These furnaces mainly suck in air through tuyères or similarly constructed openings at the furnace bottom and utilize the chimney effect of the shaft. This is neither the case with the furnaces in Faynan nor in Samanalawewa, and so the counterarguments for the use of natural draught furnaces have been partially refuted by the experimental facts. Bachmann also concluded from his firing experiments in Timna 30 (Bachmann and Rothenberg 1980) that the application of additional means such as bellows was not a necessity in the beginning of copper metallurgy. The local winds would be sufficient for a simple process technique.

In summary, it can be stated that the Early Bronze Age wind-powered furnaces from Faynan hold a key position in the development of early metallurgy. They are, to begin with, the oldest documented smelting furnaces and they manifest a technological transition between the ore smelting in crucibles to that of smelting in furnaces run with an artificial air supply.

In comparison to the smelting processes in crucibles, the smelting processes described above in wind-powered furnaces can be compared with the technology of pottery kilns. In both cases, the firing is regulated by a natural air supply; in the earliest pottery kilns, firing chamber and stacking chamber were still identical. Since the first stage of smelting technology was not the Early Bronze Age metallurgy with wind-powered furnaces, but rather the use of crucible technology, which is based on completely different principles, it seriously brings into question the hypothesis that metallurgy might have developed out of the much older pottery craft. But Kayani and McDonnell (1997) recently took this idea up again. They compared the production of extremely high-fired Chalcolithic pottery ('Green Ubaid Ware') from northern Mesopotamia directly with the earliest metal technology. Regrettably, the authors have only discussed high-temperature processes in general and did not verify their theory with either thermodynamic considerations or with archaeological evidence.

Processing of Copper: Evidence from Khirbet Hamra Ifdan

The field evidence of reconstructing the development of smelting processes and the production of copper during the Early Bronze Age (but also the Late Bronze Age and Iron Age) is in general already difficult enough, but when it comes to the further working of the metal to its end products the difficulties become even larger. With the excavation and analyses of the metal workshop at Khirbet Hamra Ifdan (KHI), stratum II and III by Levy (2002) there is now, for the first time at Faynan, good evidence providing much better insights into the craft of copper processing during the 3rd millennium BCE.

Early Bronze Age metal workshops have been found across the Near East in Oman at Maysar 1 (Weisgerber 1981), in Iran at Shahr-i Sokhta (Heskel 1982), Tepe Hissar (Pigott et al. 1982), Shahdad (Hakemi 1997) and Arisman (Chegini et al. 2001), in Anatolia at Arslantepe (Palmieri et al. 1993), Kestel (and Göltepe, Yener 2000), Nor°untepe (level VIII – Müller-Karpe 1994; Pernicka et al. 2002), Tepecik, Cudeyde, Gözlu Kule/Tarsus (Goldman 1956), Değirmentepe (Esin 1988), and Hisarlik (Level II, Troy, Schliemann 1881). Interestingly, the nearby (ca. 100 km south) copper mines at Timna in southern Israel have produced no evidence of Early Bronze Age metal processing, and the island of Cyprus, the most important source of eastern Mediterranean metal during the Middle and Late Bronze Age (ca. 2000–1300 BCE), has virtually no evidence of contemporary Early Bronze Age metal production or processing. Until the recent excavations at KHI, the largest assemblage of EBA metallurgical remains came from Schliemann's excavations at Troy (Hisarlik) and from Arisman. At Troy 70 crucible fragments, 70 casting moulds and some other metal production remains were found. At Arisman, slag mounds, smelting furnaces, fragments of litharge cakes from silver extraction and casting moulds were found.

It is possible to identify an EBA *chaine d'operatoire* for the production of copper tools and ingots at the KHI settlement. As seen in the GIS map for the stratum III (EBA III, Levy et al. 2002) (Fig. 5.39), specialized metal processing activities were concentrated in a variety of the more than 80 rooms, courtyards and other spaces excavated at the site. The melting and casting of metal took place in the largest (southern) courtyard. Here, copper was melted in crucibles. Unfortunately, it was impossible to reconstruct their shape and size. Hundreds of moulds (see Fig. 5.38) were used to cast copper ingots and final products such as metal axes, chisels, pins and blades. The moulds appear to have been broken (probably with small grooved hammerstones) to retrieve the metal objects and then discarded in refuse areas surrounding the courtyard. Metal products were then moved north through the site to other rooms where hammering on anvils, grinding, polishing and other final production activities took place.

The finds from KHI show that a variety of metallurgical materials were delivered from outside to the settlement for further processing. These indicate that a number of different steps were carried out to finally produce metal artifacts. Among these finds were fist-sized pieces of heterogeneously composed slag with copper contents up to 10 wt.-% or more ("furnace conglomerate"), thousands of prills and small batches of copper with adhering slag. They are proof of a rather limited success of smelting processes carried out at the smelting sites in Faynan. Copper was not separated completely from the slag during smelting. The consequence was a systematic reworking of slag by crushing it to extract entrapped prills of metal. Some layers of crushed slag comparable with the one from Khirbet en-Nahas were found.

In addition, high-grade pieces of malachite, chrysocolla, and copper chlorides were transported from nearby Faynan mines, perhaps from Umm az-Zuhur, to KHI. At the KHI processing center, they were all repeatedly (s)melted in crucibles in order to remove slag and gain larger lumps of metal. Prills, small lumps of copper and flaw castings of ingots were collected to have sufficient quantities available to cast an ingot or a tool (Fig. 7.14). That this was also practiced elsewhere is exemplified at the late 3rd millennium BCE metal workshop at Shahdad, Iran (Hakemi 1997). Here, an identical collection of copper prills and metal lumps was excavated (Fig. 7.15) (Hauptmann 2005). These "metal lumps" resulted from the formation of malachite and other corrosion products through time.

The smelting of ores perhaps followed a traditional workshop recipe that goes back to the Chalcolithic period, when smelting activities were carried out inside the settlement. The hypothesis that ancient metal workshops had their own "recipes" – which might have been characteristic of certain regions – was suggested already by Junghans et al. (1960).

The multiple remelting and recycling of copper lumps and prills into larger units is probably identical with the production of washed or purified copper as described

Fig. 7.14.
Khirbet Hamra Ifdan, stratum III. Collection of copper prills, lumps of copper and flaw castings ("cup cake") for remelting and recycling to obtain larger portions of metal sufficient for casting an ingot or a final tool. The individual pieces are baked together by corrosion. Comparable finds were made for the late 3rd millennium BCE at Shahdad, Iran (Hauptmann 2005) (photo: O. Teßler, Staatliche Museen Preußischer Kulturbesitz Berlin)

Fig. 7.15.
Shahdad (Iran), building D, room 10, section D10. Late 3rd millennium BCE. "Cake" composed of copper prills and small lumps which were extracted after smelting by crushing slag. They were collected in a small pot or hollow in the ground to enlarge the mass of copper ready for casting. This find is comparable with that from Khirbet Hamra Ifdan. It was formerly addressed as an ingot (Stöllner 2004, p. 590) (photo: Astrid Opel, Deutsches Bergbau Museum)

in 3rd millennium BC cuneiform tablets from Ebla in Syria (Reiter 1997). This was called "urudu-luh-ha," in contrast to "urudu" which means unpurified copper. Indeed, relatively pure copper objects have been identified in Ebla (Palmieri and Hauptmann 2000). Previously it was suggested (Waetzold and Bachmann 1984) that such a purification would mean a deliberate refining of raw copper in the modern metallurgical sense. This would mean purification from impurities such as lead, arsenic, iron or nickel by blowing air into the liquid metal bath. Herewith, impurities get oxidized until the liquid copper itself is transformed to copper oxide. A typical by-product of the refining process is normally a surplus of cuprite (Cu$_2$O) in the

Chapter 7 · Copper Smelting Technology

Fig. 7.16. Khirbet Hamra Ifdan, stratum III (EBA III). Section of a crescent-shape bar ingot. The item is dotted by only tiny flaws and shows a low porosity. Scattered inclusions of copper sulphides are visible. In addition, the section shows a cooling rim following the incision of the casting mould. The copper contains ca. 1 wt.-% Pb and is low in As. No slag inclusions were observed. In comparison with LBA oxhide ingots from Uluburun (Hauptmann et al. 2002), KHI copper is of much higher quality! (Width of the sample: 2.3 cm)

crucible. As shown by the phase contents of slags and of copper, this was not the case at KHI, and there is no evidence as yet of refining in the modern sense at any other of the ancient Near Eastern EBA workshops. The skill and expertise of the KHI metallurgists were remarkable, as they produced high quality copper suitable for a nearly flawless casting (Fig. 7.16). This was the metal ready for export in the form of both ingots and finished tools.

It is important to note that alloying was not performed at KHI. Tin bronzes are extremely rare in the EBA III Southern Levant. It seems that their production was probably carried out at urban centers in the north and the northwest, e.g., at Bab edh-Dhrah (Rast 1979b) and at Zeiraqun (Bochum, unpublished results).

7.3 Technologically Controllable Smelting: Iron Age to Mameluk Period

The Archaeological Evidence

A number of observations at the smelting sites in the area of Faynan, which were operating in the Late Bronze Age but mainly in the Iron Age, point towards a much more highly developed technology in metal production. The extent of the slag heaps now comprises several tens of thousands of tons and thus indicates a metal production on a far greater scale than earlier production, which comprised only a few thousand tons(see Table 5.3). Iron Age metallurgy will be treated here in detail as it is the first archaeologically attestable phase of controlled smelting. The metallurgy of this time period has already been extensively dealt with by others (Koucky and Steinberg 1982a,b; Bachmann 1982; Tylecote 1992a; Craddock 1995), and numerous

7.3 · Technologically Controllable Smelting: Iron Age to Mameluk Period

smelting experiments have been carried out as well (see below). Metallurgical techniques from the Roman and Mamluk periods will only be described cursorily, because the analysis of other smelting sites has already shown that no decisively new developments are likely to have occurred (Hauptmann 1985; Rothenberg 1990; Mangin et al. 1992; Tylecote 1992a; Craddock 1995 and others).

The shape and size of the slag as well as the appearance of tuyères in the Faynan area indicate a far better controlled smelting process of ore, a much larger volume of smelting, i.e., a considerable technological improvement in metal production. In the Late Bronze and Iron Ages, Faynan (and Timna as well) join the general state of technology in the Old World, where copper and iron were produced in large quantities. It is unlikely that this technology appeared suddenly in the Wadi Arabah. One assumes instead that a developmental stage must have existed in the Middle Bronze Age, which has not yet been discovered in this region, or that a technological transfer from elsewhere must have taken place. Such a transfer could have come from one of the major copper producers in the Eastern Mediterranean, e.g., from the island of Cyprus. Stöllner (2005) rightly suggests that during the 2nd millennium BCE a concentration of copper production took place in the Eastern Mediterranean, which focused mainly on economically valuable ore deposits ("giants") such as Cyprus or Ergani Maden. At the same time, the production of copper from sulphide ores, through a number of steps, developed.

On Cyprus, as exemplified by the mid-second millennium BCE site of Politiko Phorades (Knapp et al. 2001), matte smelting was practiced. Copper sulphide was produced, which separated as a cake at the bottom of slag cakes tapped from the furnace. It is proposed that matte was further treated by roasting. Remains of metallic copper included in matte were identified by Hauptmann et al. (in prep.) in LBA slags from Kition. This intergrowth suggests a smelting either of mixed sulphidic and oxidic ores as practiced at EBA Shahr-i Sokhta / Iran (Hauptmann et al. 2003), or, alternatively, it may indicate to a smelting of partially roasted ores. Evidence of such a metallurgical procedure is still open.

Many LBA copper smelting sites are known in the eastern part of the Alps such as Trentino and Tyrol (Piel et al. 1992; Cierny 2005), Mitterberg (Eibner 1982) and the Paltental (overview in res montanarum 2004), where sulphidic copper ores have been roasted and smelted. Also high-quality steel (Ferrum Noricum) was produced in the Southern Alps at Magdalensberg (Straube 1996). The smelting was carried out in furnaces, which were run by forced draught through the use of jets, the so-called tuyères. They have been found in smelting sites from the Late Bronze Age / Iron Age periods in large numbers. The technique of forced draught was certainly an improvement on the natural draught or wind-powered furnaces or the crucible smelting and had already started in the Middle Bronze Age, when a variety of types of bellows were used (Weisgerber and Roden 1986; Craddock 1995). At Politiko Phorades, numerous tubular tuyères have been found, and most interestingly double tuyères were discovered, which show parallels to the types used in the Wadi Arabah (see Fig. 5.10 and Rothenberg 1990). In addition, tuyères with bent air channels were excavated, which are not only identical with those from the Late Bronze Age urban settlements at Enkomi and Kition, but also with those found at the Late Bronze Age metal factory of Qantir-Piramesse in the Nile Delta (Pusch 1990).

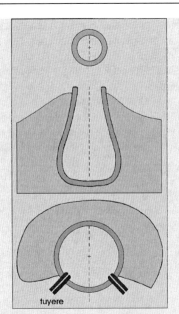
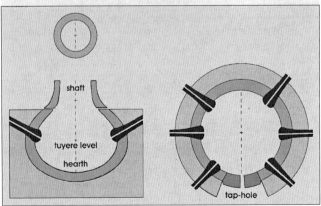

Fig. 7.17.
Suggestions for the reconstruction of copper smelting furnaces from the Late Bronze Age (*top*) and the Iron Age (*bottom*) in Timna 30. The reconstructions are based on finds of ceramic, slag tempered furnace wall fragments and slagged tuyères (Bachmann and Rothenberg 1980)

The number of smelting furnaces actually excavated is very limited, and most smelting sites produced only a few fragments of ceramic furnace lining or furnace wall. The only sites with furnaces sufficiently preserved are either on Cyprus (Muhly 1991; Fasnacht 1992; Knapp et al. 2001) or in the Alps (Hauser 1986; Cierny 2005), although even here there is disagreement about the shape and function of the furnaces. It is characteristic that Craddock (1995) questions the interpretation of the smelting furnaces found in Mitterberg as such, using their supposedly unpractical squarish shape and the lack of any signs of strong heat as an argument. His suggestion that the furnaces might be roasting installations needs to be considered. Probably, the best examples of Late Bronze and Iron Age smelting furnaces have been found in Timna 2 and Timna 30 (Rothenberg 1980). Based on these finds, the hypothesis was developed that the common type of smelting furnace since the end of the 2nd millennium BCE was a cylindrical shaft furnace with a height of between 0.5

and 1.5 m. The slag would have been tapped after the smelting process and would have cooled outside the furnace. It is not clear whether copper was (in parts) tapped with the slag, as suggested by experimental work (see below), or if it formed a regulus at the furnace bottom as suggested by Tylecote (1980). Bachmann and Rothenberg (1980) published reconstructions of the smelting furnaces from the Late Bronze and Iron Age in Timna 30, which are shown in Fig. 7.17. It is certain that the furnaces in Timna worked with forced draught just as the ones in Faynan did. This is confirmed by the numerous tuyères at both sites. The slag covering shows that they always led with an angle of 40–60° into the furnaces (Rothenberg 1990; Sect. 5.2).

It is not surprising that a large number of uncertainties about the exact functioning of these furnaces still exist. Merkel (1990) points out that so far no (secure) information about the number and positioning of the tuyères in the ancient furnaces has come up and – linked to this question – information about two of the most essential parameters for an assessment of the technique are missing, i.e., the atmosphere and the firing temperature. Information about the original height of ancient furnaces would be difficult, because they are normally destroyed, and the process technique could not be assessed without experimental reconstructions. He points towards the complete lack of bellows in Timna. The reconstructions of the smelting furnaces from Timna 30 by Bachmann and Rothenberg (1980) are not reconstructable at Faynan, although a surprising amount of parallel finds exist, based on the actual archaeological material there. For instance, it is not clear whether the original reaction chamber of the smelting furnace in Faynan 5 was really that large (see Figs. 5.11, 5.12). It is surprising that the installation uncovered by Rothenberg (1980) in Timna 30, which he has very tentatively interpreted as a smelting furnace, has not been further mentioned and has not been incorporated into the reconstruction suggestions by Bachmann and Rothenberg, and yet it is very similar to the dome-shaped fore furnace in Faynan 5.

Crushing Slags

The role of mechanically reworked/processed slag has not been sufficiently considered in the reconstruction of metallurgical processes. A typical characteristic of the Iron Age smelting sites at Faynan 5, Khirbet en-Nahas and Khirbet el-Jariye is enormous heaps of pounded and crushed slag, attesting to the reprocessing of slag for the extraction of metal prills as a mainstay in copper production (Bachmann and Hauptmann 1984). Tools used for that crushing were hammer stones, stone anvils and mortars cut into the sandstone (Figs. 5.34 and 7.18). These tools have also been found regularly at Early Bronze Age smelting sites in the Faynan area such as Barqa el-Hetiye, Ras en Naqb, Khirbet Hamra Ifdan, Faynan 8–13 etc. (see Chap. 5). This observation, together with the close vicinity and /or overlapping of Early Bronze and Iron Age smelting sites, clearly indicates that during the Iron Age ancient slags have been processed. We therefore can suggest that Iron Age people at Faynan took advantage of the copper left in the earlier slag heaps.

Reason for such reworking can be found in the already-mentioned assumption that a complete, quantitative separation of the charge into slag and metal (or matte) was not possible in the early periods of copper production. A perfect technical pro-

Fig. 7.18. Iron Age slag site at Aarja, Oman. The surface is scattered with hammerstones and anvil stones for crushing slags to extract metal prills (Costa and Wilikinson 1987). Their size and shape are identical with stone tools found at Khirbet en-Nahas and Faynan. This procedure seems to be a "worldwide," firmly established stage in ancient metallurgy and points to a very careful recovery of metal even in small quantities from slag

cess with a complete liquefaction of the charge in question has never been attested; it is an ideal expectation of modern technology (Hauptmann 2003). Logically, crushing slags to extract metal prills must have been followed by a subsequent (repeated) remelting of prills. As described in Chap. 6 this would lead to a chemical homogenization of the copper produced.

Similar evidence of reprocessed slag is well known from the Early Bronze Age copper smelting site at Chrysokamino (Bassiakos and Catapotis 2006), from Late Bronze Age copper smelting in the Alps, where the reprocessing is widespread (Metten 2004), in Anatolia (Wagenr et al. 1989), in Oman (Weisgerber 1981), in Spain (Rothenberg and Blanco-Freijeiro 1981 and others), from the Roman gold production in Três Minas in Portugal (Bachmann 1993), and at numerous smelting sites on Cyprus (Given and Knapp 2003). Crushed slag has also been reported from smelting sites in Timna (Rothenberg 1980, 1990). These observations clearly indicate that mechanical beneficiation of slag to recover metal inclusions was an overall firmly established step in metal production.

Reconstruction of Smelting Processes with Iron Age Finds

Merkel (1990) believes that tap slag in Timna is highly indicative of the level of process techniques in copper smelting during the Late Bronze and Iron Ages. This is as true, with certain exceptions, for Faynan. Bulk composition and phase content of

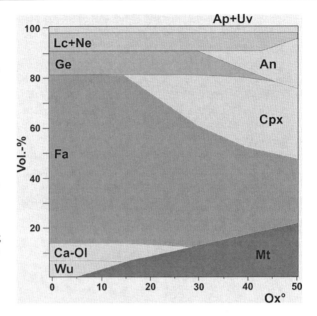

Fig. 7.19. Normative phase content of a Bronze Age slag as a function of the degree of oxidation. It becomes clear how the composition changes with increasing oxidation: fayalite decreases in quantity, while clinopyroxene and magnetite crystallize at its expense. *Ox:* degree of oxidation; *Lc:* leucite; *Ne:* nepheline; *Ge:* gehlenite; *Fa:* fayalite; *Ca-Ol:* calcio-olivine; *Wu:* wuestite; *Ap:* apatite; *Uv:* ulvite; *Kpx:* clinopyroxene (hedenbergite); *Mt:* magnetite (Hauptmann 1985)

tap slag would thus be the main sources of information for reconstructing the parameters of the smelting processes, as they are frozen during tapping. Iron Age furnace slags from Faynan, however, show slightly different conditions of just such a formation. They have not completely liquefied and contain sintered ore. And although their bulk compositions are compatible with fully liquefied tap slag, they include tephroite and pyroxenoids only in parts. A re-oxidation of these phases to Mn oxides indicates that the slag suffered oxygen intrusions when the temperature was still high as is possible when the charcoal bed is allowed to burn down after the end of the actual process. Such slag cooled down more slowly in comparison to tap slag.

Analyses of many ancient slag samples have shown that they may have a different mineralogical phase content even though their bulk composition varies little. This is particularly obvious in Fe-rich silicate slag with Cu(Fe) oxides and phases such as fayalite, hedenbergite, magnetite, and wuestite, because these can react during the smelting process very sensitively to changes in the oxygen concentration of the furnace's atmosphere. The normative phase content at differing degrees of oxidation can be calculated for slags with a known chemical composition by way of a calculation program developed by Keesmann et al. (1982). That has been done for Ottoman copper slag (Fig. 7.19) by Hauptmann (1985). But it has already been demonstrated that microscopic studies might also allow the conditions of formation to be defined for some phases or phase associations, at least approximately (Figs. 7.8, 7.10 and 7.11).

Ancient slag has already been described several times in this way (Craddock et al. 1985; Hauptmann and Roden 1988; Moesta et al. 1989; Moesta and Schlick 1989; Lutz 1990; Hauptmann et al. 1993; Metten 2003). A tendency towards changing from smelting processes carried out under higher oxidizing firing conditions during the Bronze Age to more strongly reducing conditions in the later periods becomes clear:

- Crucible slag, with cuprite, delafossite and magnetite as main components forms under relatively high oxidizing conditions at an oxygen partial pressure of 10^{-4} to 10^{-7} atm as an average. The slag samples all come from sites dating into the (end of the) Chalcolithic period in the Eastern Mediterranean and have been discussed in Sect. 6.1. Fayalite, as the 'typical' slag phase, occurs irregularly.

  Fayalite occurs more frequently in Early Bronze Age I slags (middle of the 4th millennium BCE) and has been found at Arslantepe, Çayönü Tepesi and Nevali Çori in southeast Anatolia (Hauptmann et al. 1993; Hauptmann 2003). This material is characterized by the crystallization of magnetite, followed by fayalite, clinopyroxene or SiO_2-rich glass. Fayalite has often oxidized into ferrifayalite or laihunite. It forms at a low temperature range and weakly reducing conditions, as are possible during a slow solidification of iron-rich silicate melts in the furnace or the crucible. Another typical phase found is iscorite ($Fe^{2+}_5 Fe^{3+}_2 SiO_{10}$) (Hauptmann et al. 1993; Ganzelewski 2000), which forms out of wuestite/magnetite through a reaction with the melt (Keesmann 1989; Rose et al. 1990), corresponding to

  $$\text{wuestite/magnetite} + \text{fayalite} \rightarrow \text{iscorite} \leftarrow \text{magnetite} \; (+SiO_2)$$

  Following Rose et al. (1990) iscorite crystallizes in an area of decreasing temperatures between 570 and 1 100 °C and increasing oxygen partial pressure. The formation range of such slag is along the buffer equilibrium

  $$\text{quartz} + \text{magnetite} \rightleftharpoons \text{fayalite (QFM)}$$

  at $p_{O2} \sim 10^{-8}$ to 10^{-10} atm, if one assumes a firing temperature of ca. 1 100–1 200 °C. This would mean in reality a charcoal covering of ca. 20–30 cm.

- In most smelting sites in the Old World, from Oman over Timna up to the Alps, slags have been produced regularly since the Late Bronze Age in the stability field of fayalite at an oxygen partial pressure of p_{O2} between 10^{-8} and 10^{-12} atm, which again illustrate stronger reducing conditions (Milton et al. 1976; Hauptmann 1985; Lutz 1990; Metten 2003). These seem to be the typical redox-conditions that could be achieved in shaft furnaces common at the time. We assume that much higher firing temperatures could be achieved in the Late Bronze Age in comparison with earlier times: temperatures above 1 200 °C possibly even up to 1 500 °C (see below).

It is more difficult to come to a conclusion about the slags from Faynan, because Mn silicate slags are stable over a larger p_{O2}/T-range than comparable iron-rich compositions (see Fig. 7.11). Only rising Fe concentrations in Iron Age copper indicate stronger reducing conditions in Faynan. While during the Bronze Age they remained in the 1/10 to 1/100%-range, they then increase up to >2%, as can be observed in contemporary copper and bronze objects from other regions (Craddock 1995). During copper production in the Mameluk period, precipitation of iron seems to have been a serious technological problem (see Sect. 6.3). It may be generally allowable to transfer the redox conditions observed in Timna, Oman or the Alps also to the Iron Age Faynan metallurgical processes.

Fig. 7.20.
Estimated ranges of slag formation from different periods in Faynan, depicted in a p_{O_2}/T-diagram of the systems Mn–Si–O, Fe–Si–O and Cu–O. *F:* fayalite; *Hsm:* hausmannite; *I:* iron; *M:* magnetite; *Q:* quartz; *Rhd:* rhodonite; *Teph:* tephroite; *W:* wuestite (after data by Eugster and Wones 1962; Abs-Wurmbach et al. 1983)

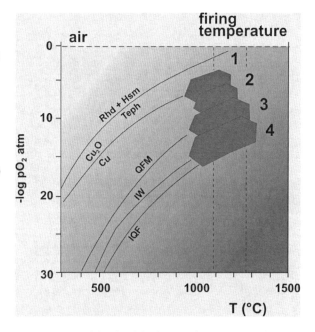

When the data discussed here are combined with observations concerning the metal technology of the 4th and 3rd millennium BCE in Faynan, then the following conclusion is inevitable: temperature and gas atmosphere, i.e., the oxygen partial pressure as a simplified version of the CO/CO_2 ratio, are two crucial parameters to define a smelting process. Both are determined by firing, preheating, postheating, and the amount of charcoal, and all of these criteria are accordingly determined by the shape and structure of the smelting furnaces or crucibles. In Faynan, the development of varying furnace shapes and other reaction vessels can be readily observed throughout the different periods. Following the model developed by Yazawa (1980) to classify modern copper smelting processes based on thermodynamic data obtained from changing oxygen partial pressures, a model for the individual stages of development based on the process technique can be constructed. It is illustrated in Fig. 7.20.

Fluxing Agents

Deliberate utilization of fluxing agents is, in modern metallurgy, essential for the controlling of smelting processes in order to produce low viscosity, easy flowing slag melts with high reactivities and low temperatures of solidification. According to their chemical composition, fluxing agents can be divided into basic (limestone, soda, and iron oxides), acidic (quartz, glass, shales) and neutral (fluorite, borax).

The modern importance of fluxing agents led in archaeometallurgy to the assumption that their intentional use had always been of importance. Wertime (1980) for example, thought that the utilization of fluxing agents had already started at the earliest stages of metallurgy. He believed the use of Fe-rich fluxing agents such as hematite and Ca-rich fluxes like limestone or shells was supposedly proven in Timna, while the Mn-ores also existing there would have only been used accidentally. Bachmann (1978) thought the iron-rich silicate slags from Timna 2 were typical products of SiO_2-rich ore with Fe-containing fluxing agents. This idea of manifold additions with completely different effects, to which Lupu et al. (1970) added dolomite and Zwicker et al. (1975) salt, might be in accordance with Wertime's concept that the earliest metallurgy was less based on the deliberate selection of rich ores but more on trial and error experiments using varying materials in the fire. Mostly is it possible to explain a 'fluxing agent component' as coming from the host rock of the ore itself. At Timna copper mineralizations in NaCl-bearing sandstone are as common as in carbonatic conglomerates or in dolomite; some copper ores are also intergrown with SiO_2 and Fe-hydroxide brecciae (Bartura et al. 1980; Bachmann and Rothenberg 1980), and Mn-ores exist, too (see Sect. 4.3).

There are also diverging ideas about the point in time when the effects of fluxing agents were first recognized and utilized deliberately as a technological innovation. Tylecote (1992a) suggests that it is contemporary with the appearance of the first larger masses of slag, without giving an actual date. In the case of Faynan, for example, this could be the Early Bronze Age. During the same period, the use of fluxing agents such as iron oxides and limestone were suggested for smelting activities at Chrysokamino, Crete (Bassiakos and Catapotis 2006). However, as long as the ore sources that supplied this smelting site are not known (Betancourt 2006; Stos-Gale and Gale 2006), this suggestion is vague. It is more likely that Tylecote had the enormous slag heaps that appear in the Old World roughly from the Late Bronze/Iron Age in mind, thus at the transition from the 2nd to the 1st millennium BCE. This is more convincing than the suggestion by Rothenberg and Merkel (1995) and Merkel and Rothenberg (1999), who propose the intentional utilization of iron ores as fluxing agents as early as the Late Neolithic. Bachmann (1980) and Bachmann and Rothenberg (1980) are more careful in their interpretation and indicate the possibility that in the earlier periods of copper production, self-fluxing ore might have been used. Archaeological evidence of the use of fluxing agents in Timna starts in the Iron Age (Timna 30, stratum I), where two storage containers filled with copper and manganese ore have been found.

The assessment of whether an ore deposit contains self-fluxing ore requires perspicacity, because highly reactive mixtures of ore and gangue or host rock, which form liquid slag melts at 1 200 °C without any problems, are quite common. In Faynan self-fluxing ores were available in the DLS formation with sandy dolomite and Mn-containing shale (see Sect. 4.3). Other (hydrothermal) ore deposits, such as those in Cyprus, Ergani Maden or Keban (see Sect. 4.2) contain mineral mixtures consisting of phyllosilicates (e.g., kaolinite, sericite, chlorite), of carbonates and Fe oxides, formed through the alteration of host rock (argillitization, sericitization, propylitization). Additionally, jarosite $((K,Na)Fe_3[(OH)_6 \cdot (SO_4)_2])$ is common in the oxidation zone of sulphidic mineralizations, which contains quite considerable metal concentrations. These minerals are intergrown with quartz and limonite occurring in the oxidation zones. This mixture thus forms a material that can be easily smelted under relatively widely varying conditions. The use of additional fluxing agents is hardly necessary here. If the SiO_2/FeO ratio is too high, quartz cannot react and remains in slag as an unmelted component. This can be observed in so-called 'free-silica slag' (see Sect. 6.1).

Bachmann (1982) suggests the manganese slag from Cyprus to prove a deliberate optimization of smelting processes. He assumes that Mn slag from Skouriotissa, Lythrodonda, Kornos and Polis was produced by the intentional addition of Mn oxides for fluxing and thus believes them to be the technologically most developed slag. Mn ores occur in Cyprus in the fine-grained sediments of the umber, which is stratigraphically separated from the sulphide ore body (Constantinou 1980). It is, however, most likely that transitional phases existed: Kortan (1970) demonstrated in his studies on the ore deposits of Skouriotissa that sulphide ores are partly intergrown with umber in alternating layers up to 15 m thick. Consequently, a fairly large quantity of self-fluxing ore has been present in ancient times.

It is not understandable why the use of fluxing agents is consistently postulated for the early metallurgy of copper, while at the same time it is not suggested for iron smelting (Tylecote 1987). It has already been explained above that the deliberate use of fluxing agents was not the only essential criterion influencing the development of metallurgy up to the Classical period.

Given the present state of knowledge, we believe that at Faynan fluxing agents were not utilized before the Iron Age. That is at least indicated by the chemical analyses of the slags. Figures 6.17 and 6.18 illustrate that the Early Bronze Age slag shows a relatively large range of variations in their bulk composition, which allows the conclusion that ore and host rock were used quite undeterminably. A certain focus on the kind of ore used can only be observed in the Iron Age, which indicates an intentionally controlled composition of the charge and would attest to a progress in technology. This would be in line with the findings from Timna 30. The chemical and mineralogical composition of the slag can, with all due care and bearing in mind the general chronological considerations, be used as a rough dating device just as the trace element pattern of the copper itself can.

Comparison with Smelting Experiments

Reconstructions of ancient smelting processes in shaft furnaces were already being tried out at the beginning of the last century, e.g., by Gowland (1899), who very consciously sought to compare ancient metallurgical techniques with still existing ethnographically techniques. He concentrated mostly on research into iron production, as the ancient techniques were much better documented. Smelting experiments concerning copper production were carried out much later, particularly by Tylecote and his team in the 1970s. They used, based on the finds in Timna, small shaft furnaces for their experiments. A summary of the results is published in Tylecote and Merkel (1985). The most comprehensive and still influential simulation of ancient smelting processes was carried out by Merkel (1983a, 1990) and Bamberger and Wincierz (1990). As far as possible, they tried to integrate Late Bronze and Iron Age archaeological evidence found in Timna into their experiments. Conclusions of direct importance for the understanding of contemporary smelting furnaces in Faynan will be discussed here.

The authors managed in their experiments to produce, after some hours, a low viscosity slag of eutectic composition, which was tapped out of the furnace and solidified in a forehearth. At the same time, some kilograms of the charge stuck inside the furnace as slag, which could only be removed by the destruction of the furnace. The furnace slag consisted of a mixture of solidified slag with inclusions of charcoal and ore and contained kilograms of metal lumps and prills. Bamberger and Wincierz (1980) thus concluded that after slag tapping it would have been necessary to reheat the furnace again up to >1 100 °C in order to achieve an agglomeration and separation of metal prills. It is not proven whether such a process would have been possible in the Late Bronze Age to liquefy ore charges totally.

The formation of furnace slag is caused by an irregular and varying distribution of temperature, which can be observed in shaft furnaces with tuyères in the lower part (Fig. 7.21). In African iron smelting furnaces and also in smelting experiments, top temperatures of 1 400–1 600 °C have been measured by optical pyrometers and by Pt/Rh-thermocouples, but only directly in front of the tuyères (Avery and Schmidt 1979; Tylecote et al. 1971; Rehder 1986; Merkel 1990; Juleff 1996). This part of a furnace is the reaction zone where slag is liquefied. Around these spots and in the upper part of furnaces, temperatures decrease to <700 °C. Therefore, parts of charged ores never were exposed to temperatures required to form a homogenous melt. The amount of partly or not liquefied furnace slag depends on the volume of the charcoal that was exposed to high temperatures, which in itself depends on the combustion caused by the air influx per time unit (Rehder 1987).

Merkel, Bamberger and Wincierz managed to tap roughly half of the produced copper together with slag; the other half remained in the furnace forming irregularly shaped lumps underneath the furnace slag and had to be removed with it by breaking

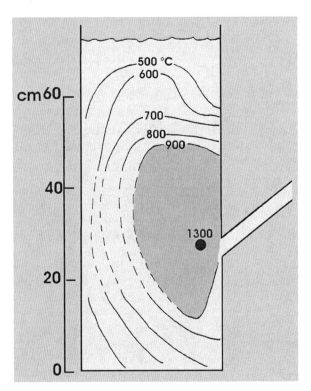

Fig. 7.21.
Distribution of temperature in a shaft furnace constructed for experimental copper production. The isotherms, given in °C, show a maximum close to the tuyères. This shows that the spatial zone for eutectic slag formation ('tap slag') is limited (from Tylecote and Boydell 1978)

the furnace. These lumps have nothing in common with the plano-convex ingots traded during the Bronze Age in the Near East (Weisgerber 1981), which had been interpreted previously as being formed in the furnace bottoms as reguli (Tylecote 1980). The copper lumps produced by Merkel and coworkers contained 20–50% iron, which results from Fe filaments or small globules in contact with charcoal. Charcoal pieces are often literally covered with iron. According to the phase equilibria, they are dissolved at further precipitation of copper and form two-phased Cu-Fe alloys. Simultaneously, copper with low Fe-contents is produced that is tapped with the slag. Possibly Late Bronze Age iron objects in Timna were made from such Fe filaments. Gale et al. (1990) analysis shows that the lead isotope composition of the iron objects is identical to that of the Timna copper. They see this as a good example of how iron production could have developed out of copper metallurgy.

These observations are essential for the understanding of the field evidence not only in the Faynan district. They support the impression that the efficiency of ancient smelting processes has possibly been overestimated, while the archaeological evidence was not adequately taken into account (Hauptmann 2003). The two types of slag, tap slag and furnace slag, found in Faynan 5, Khirbet en-Nahas and Khirbet el-Jariye, seem to be compatible with experimental results. Redoxidation of tephroite observed in Faynan can be explained by after heating. Tap slag is more conspicuous and seemingly present in larger quantities because furnace slag has been systematically reprocessed to extract copper prills. The remains left after this reprocessing have become no more than unobtrusive 'filling material', which was thrown, either together with the tap slag on slag heaps, or separately on other refuse spots. It must be assumed that the smelting furnaces, made from ceramic materials, were more or less completely destroyed directly after the smelting process to pull out (copper containing) slag remaining in the furnace. This shows parallels between the copper production and the bloomery process in iron metallurgy, where the furnace also had to be destroyed in order to break the bloom, in its solid state, outside the furnace. Such procedures have been described by *Theophilus Presbyter* for medieval metallurgy, who observed that the formation of solid copper inside the furnaces, as well as the breaking down of furnaces, was a general practice applied in metallurgy.

Experimental copper smelting support the assumption of Piel et al. (1992) that different slag types found at one and the same smelting site, e.g., in the Late Bronze Age copper production in the Alps are not necessarily the products of several different smelting processes. They believed it possible that slags all came from one smelting process, where a partial melt with low viscosity led to the formation of the well-known plate slag on one side, while the crude slag cakes are actually furnace slags whose formation and composition are similar to those of a part of a partially melted rock that is not mobilized : a so-called 'restite' (Hauptmann 2003). Systematically, they have been very finely crushed, just like in Faynan.

It is difficult to estimate how much slag and metal could have been produced in a single smelting event in the Iron Age furnaces in Faynan. It is not too optimistic to assume that ore with 10% copper was smelted, or in other words a metal:slag ratio of ~1:10. This suggests that at a tap slag weight of 40–60 kg, not more than 4–6 kg copper were produced. This, however, does not imply that the metal would have been produced during a single smelting. The fabric of the tap slag shows that they consist of separate layers alternating with cooling rims of Mn oxides. This fabric may point towards several tapping events, which might have happened in very short intervals and during one (continuous) smelting process (shaft furnace with a so-called open eye), but which could as well be the result of several events during much larger time intervals.

This phenomenon is also observed in oxhide-ingots from Cyprus, which are between 25 and 40 kg. It is not clear if they are the result of a single smelting event or if they were cast as the outcome of several smelting events, possibly even from recycled metal. This question has recently been touched upon by Budd et al. (1995). It is very difficult to imagine the production of such a quantity of metal being produced in a single smelting event in the Iron Age furnaces in Faynan. Even if sufficiently rich ores (with a Cu concentration of 30–40% that actually existed) were used, not more then 16 kg of metal could have been produced in one smelting process. Bachmann and Rothenberg (1980) assumed a metal production of 3.5–6 kg from the Iron Age furnaces of Timna 30 and 1–3 kg for the smaller Late Bronze Age furnaces (Rothenberg 1980), which is a far more realistic assumption. Larger quantities of metal have never been produced in experimental copper smelting (Merkel 1990). We thus tend to agree with the opinion voiced already by Maddin and Merkel (1990) that metal masses of 24–40 kg in the Late Bronze Age have to have been produced in several smelting events and possibly partly from recycled metal.

Chapter 8

Export of Ore and Copper: The Importance of Faynan in Prehistoric Palestine

8.1 Trade of Ore from Faynan in the Neolithic

The Importance of the Color Green in the Pre-Pottery Neolithic

With the domestication of plants and animals at the beginning of the 8th millennium BCE, the evolving husbandry became an economic basis of human societies for the first time. This change from food gathering to food production was an important cultural evolution. This 'Neolithic revolution' is the most obvious factor, which divides the hunter and gatherer society of the Palaeolithic from that of the early or Pre-Pottery Neolithic. The same time period witnesses new uses of color having importance and meaning throughout the entire Eastern Mediterranean. During the Palaeolithic, the colors used in symbolic contexts such as the well-known cave paintings, were mainly red and black, made from iron and manganese (hydr)oxides. During the Pre-Pottery Neolithic, green achieves more and more importance. Individual objects made from green minerals or stones had already been found in the Epipalaeo-lithic/Protoneolithic, for example a pendant from serpentinite with malachite in the Shanidar cave in northern Iraq (Solecki 1969) or pieces of secondary copper ore from the settlement of Hallan Çemi Tepesi in Anatolia (Rosenberg 1994). But the widely distributed use of green mineral pigments can be considered the third characteristic of the Pre-Pottery Neolithic, and here particularly for the PPNB. The use of 'greenstone,' as these materials are collectively known in archaeology, has been observed all over Palestine and Transjordan (Garfinkel 1987) as well as in the entire Eastern Mediterranean as far as the Balkans (Glumac 1985). It continues uninterrupted through the remainder of the Neolithic up to the beginning of the Bronze Age at the turn of the 4th to the 3rd millennium. The archaeological 'greenstone' includes mineralogically different materials, which comprise not only different rocks but virtually all green-colored minerals. Secondary copper ores, particularly malachite, form a large percentage. This is not surprising because PPNB settlements are often not far away from copper ore deposits.

In the Balkans, in Anatolia, over the Zagros Mountains up to Iran, the use of green mineral pigments gave direct stimulation to the development of metallurgical techniques. Geologically speaking, these areas, which are exceptionally rich in copper ore deposits, form a metallogenic belt along the fault line between the Eurasian plate in the north and the Afro-Arabian plate in the south (Jankoviè 1997). Secondary copper ore here is intergrown with cuprite and native copper (Maczek et al. 1952; Wagner et al. 1989).

A particularly noteworthy example is the excavation of Çayönü Tepesi in southeast Anatolia (Fig. 1.2). In this settlement, dating mainly into the 9th/8th millennium BCE (Özdoğan and Özdoğan 1999), several kilograms of secondary copper ores were found, from which hundreds of pendants and beads had been made. In addition, several dozens of simply shaped small tools such as awls, burins, hooks and fragments of copper sheets were also found, manufactured from native copper as shown by metallographic and chemical analyses (Maddin et al. 1999; unpubl. data Mainz/Heidelberg). The metal objects have been made by cold and hot working. This testifies to the use of fire to deliberately manipulate physical properties. The objects from Çayönü Tepesi can be seen as the earliest steps in the development of metallurgical techniques. Schoop (1994) showed in a compilation of Neolithic metal objects from the Near East that working of native copper and of secondary copper ores happened in a number of other places in Anatolia and in northwestern Iran.

In the Southern Levant, in Sinai and in the surroundings of the copper ore deposits in the Wadi Arabah, settlements dating into the PPNB have been excavated in which 'greenstone' has also been found (Bar-Yosef 1995; Fig. 8.1). At the eastern

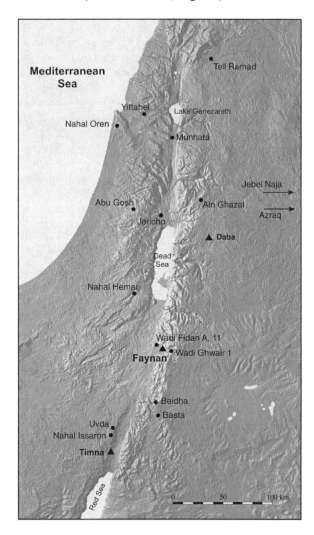

Fig. 8.1.
Settlements of the Pre-Pottery Neolithic in the Southern Levant and Sinai, in which 'greenstone' has been found. Also indicated is Tell Ramad, where a pendant made of native copper was excavated. The *triangles* mark some possible raw material sources of green mineral pigments

edge of the Arabah there are Basta, Baja, and Beidha, and in the Faynan region are the three settlements of Wadi Fidan A, Wadi Fidan 11, and Wadi Ghwair 1 (Hauptmann 1989; Adams 1991; see Sect. 5.2). The area in the direct vicinity of Timna, Nahal Issaron and some sites in the Uvda Valley date into the PPNB (Goring-Morris and Gopher 1983; Ronen 1980). Eshel (1990) found even in the Timna valley itself remains of this time period at the outcrop of the copper ore.

It is not surprising that excavations and test trenches in Wadi Fidan A and 11 as well as Wadi Ghwair 1 brought to light amounts of pure, intensively green-colored copper ores, which had been collected at the outcrops of the mineralizations of the DLS, partly also of the MBS. It has been mentioned before that these finds are the earliest attestations of a utilization of the Faynan deposit, which cannot be proven in any other way, for example by mining activities. The ores must have been collected in order to exchange them as 'exotic goods.' They undoubtedly played a significant role in the exchange or trade systems involving different raw materials. The ores were an important source of green mineral pigments next to a number of other "greenstones" from other sources in the region.

A number of examples will suffice to show the distribution of copper ores from Faynan during the Pre-Pottery Neolithic. Later it will be discussed if a similar development of metallurgy can be postulated for this region as it is to be observed in Anatolia.

Ores from Faynan in Settlements of the Southern Levant

'Greenstone' artifacts from different PPNB sites in the Southern Levant (Tubb 1985; Garfinkel 1987; Kingery 1988; Hauptmann 1989; Hauptmann 1997; Fig. 8.1) often originated from Faynan, as is shown by ore petrography and micro-analyses with SEM/EDS and X-ray diffractometry. Typical of this ore deposit is an association of dioptase, chrysocolla, ±blue-colored plancheite, malachite, Cu chlorides and/or Mn oxides (see Sect. 4.3). Particularly copper silicates, when occurring in a massive shape, are typical of mineralizations of the DLS. Blue-green quartz grains colored by inclusions of finely distributed Cu minerals are also typical. They occur in layers of arkosic sandstone above the DLS at Faynan, but they are not known from Timna (Hauptmann 1989). Cu silicates can originate from Timna too, but the Timna-formation, which is the western equivalent of the DLS, outcrops in a very limited area only at a single spot at Givat Sasgon (Fig. 4.4). Ore from this formation can therefore have only been available in a small amount. Typical of and unique to Timna are cuprified plant remains or the nodules of Cu sulphides, malachite and other secondary copper minerals, baked together with sandstone.

Trade connections between Faynan and the Jordan valley are proven by "greenstone" artifacts from PPNA levels at Jericho, where ca. 100 beads and pendants were found (Wheeler 1983; Talbot 1983). Regrettably, composition and origin of most finds were not analyzed and are just described as 'malachite'. Hauptmann (1989) identified amongst these finds unworked pieces of ore consisting of microcrystalline chrysocolla with malachite and atacamite from the Faynan DLS. One piece of ore possibly comes from Timna. The Pb isotope ratios of three of the analyzed ore samples from Jericho were later measured in Mainz. The analyses show (Table 8.1)

Table 8.1. Lead isotope ratios of three copper ores from Neolithic layers in Jericho (see Fig. 4.12). The chemical and mineralogical analyses of the ores are published in Hauptmann (1989)

Sample no. IfA	Inv. no.	Description	Pb (µg g^{-1})	^{208}Pb/^{206}Pb	^{207}Pb/^{206}Pb	^{204}Pb/^{206}Pb	^{208}Pb/^{204}Pb	^{207}Pb/^{204}Pb	^{206}Pb/^{204}Pb
IL-3/2a	JpB 4.6	Cu-ore, PPNB	12	2.1104	0.8657	0.05539	38.10	15.63	18.05
IL-3/2b	JpB 4.6	Cu-ore, PPNB	12	2.1137	0.8681	0.05553	38.07	15.63	18.01
IL-3/8b	JpB 5.50a	Cu-ore, PPNB	30	2.1149	0.8680	0.05555	38.07	15.62	18.00

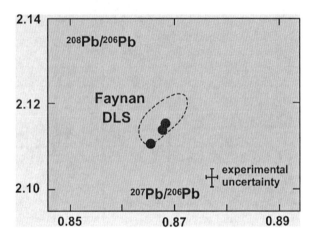

Fig. 8.2. Lead isotope ratios of three copper ores from PPNB layers in Jericho (see Table 8.1). For comparison, the isotope ratios of the Dolomite-Limestone-Shale Unit (DLS) from Faynan are also shown. The Jericho ores cannot be distinguished from them.

that their compositions are compatible with those from the DLS ores (Fig. 8.2), which further proves the supposed provenance.

Numerous disc- and butterfly-shaped beads have been found at Ain Ghazal, a Neolithic settlement close to the modern center of Amman (Rollefson and Simmons 1986). They are made from the intensively green-colored fluorapatite and calcite. The source of these objects is the Daba-Siwaqa marble located 50 km away (Fig. 8.1). The marble is a sequence of Cretaceous-Neogene limestone, marl and phosphorite affected by high-temperature contact metamorphism, whose coloring is caused by chromium (Nassir and Khoury 1982). Amongst the beads from Ain Ghazal are unprocessed pieces of ore consisting of fiber-like Cu silicates with malachite, originating from Faynan (unpubl. data DBM). Perhaps such pieces served as the basis for the production of green powder used for cosmetic purposes. By far the most awe inspiring finds showing the utilization of this product are 22 clay statues from Ain Ghazal (Fig. 8.3). The eye shadow and eyeliner around the eyes of some of these figurines have a foundation of green material: the X-ray diffractometry proved it to be powdered Cu silicate (Tubb 1985).

An interesting find from Beidha needs to be mentioned. The settlement is ca. 40 km southeast of Faynan on the Jordanian plateau. A deliberately shaped roll of powdered Cu silicates, Cu chlorides and malachite a few centimeters long (Figs. 8.4, 8.5) has

Fig. 8.3.
Ain Ghazal. The figurines from the PPNB settlement were shaped with lime plaster and clay around a core of reed. The eyes are made from white lime or chalk; pupil and eyeliner consist of bitumen. The eye shadow has a layer of dioptase underneath. The larger of the two statues, called 'Uriah' is 32 cm high (photo: Peter Durell, Stuart Laidlaw)

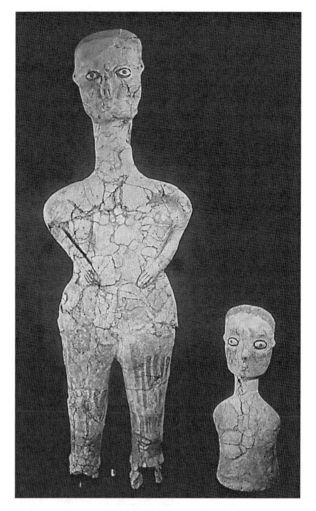

Fig. 8.4. Beidha, sample BSA 2725. Powdered copper ore, shaped into a roll, from the Pre-Pottery Neolithic settlement of Beidha. The material is particularly hardened; it cannot be ruled out that it was once mixed with an organic adhesive (glue) (photo: Brian Byrd, San Diego)

Fig. 8.5.
Beidha, sample BSA 2725. Microtexture of the sample shown in Fig. 8.4. It consists of angular fragments of Cu silicates, chlorides and carbonates very finely intergrown. White spots are barite and lead minerals. The matrix is mainly Cu carbonate (SEM picture, backscattered electron mode)

been uncovered there. We assume that it is actual cosmetic powder of the same kind as that which was used in Ain Ghazal. The shape could have been caused by it being kept in a small tube and through hardening during the long storage time. In all probability, the material originates from the DLS ores in Faynan.

Both settlements on the Jordanian plateau, Beidha and Baja, acquired most of their 'greenstone' from other locations. Most of the nearly 100 beads, mostly butterfly beads, consist of turquoise, which most probably originates from Sinai, which is geographically the closest source. Amazonite, from nearby pegmatites in the Precambrian basement of southwest Jordan (Sozzi et al. 1991), has also been identified. From Faynan, only a few pieces of copper ore from the DLS were excavated (Hauptmann 1998; unpubl. data DBM), while from Timna, just a few pieces of cuprified plant remains were found. The Daba-Siwaqa marble was obviously not of interest.

At the Pre-Pottery Neolithic settlement of Basta, numerous beads made of turquoise were identified (Hauptmann 2004). The nearest occurrence of this semiprecious stone is located in the Wadi Maghara, Gebel Adeida and Serabit el-Khadim in the western part of the Sinai Peninsula. The turquoise nodules are embedded in layers in the Nubian sandstone. Even if the main destination of the turquoise was probably Egypt (Beit-Arieh 1980), the origin of the material found at Basta can be assigned, with a high probability, to this region. This as more, as otherwise turquoise occurrences are only known at long distances from the Southern Levant, in Iran and Afghanistan (Moorey 1994). Beads made of turquoise were also identified among a suite of "greenstones" at Beidha.

Other examples of artifacts made of powdered Cu silicates come from the PPNB-cave of Nahal Hemar at the southern end of the Dead Sea. A green- and red-colored mask and more than 100 wooden beads were discovered here. The latter were covered with thin layers of lime plaster and green mineral powder and additionally decorated with red and black paint (Fe hydroxides, asphalt) (Bar-Yosef and Alon 1988). The green colored slip of the beads was identified by Kingery (1988) using X-ray diffraction as dioptase, which points to Faynan as the place of origin. It is not verified as yet, but the geographical proximity to Faynan makes it plausible that the color decoration of the mask is also made of Cu silicates from Faynan.

In summary it can be said that from Faynan during the PPNB mostly ores from the DLS were used. But they form only a relatively small amount of the 'greenstones' discovered in all the PPNB settlements. They were only used occasionally to manufacture beads, perhaps because their fine fiber intergrowth makes them quite soft and easily damageable (the hardness of chrysocolla, malachite and atacamite is ≤ 4 on the Mohs' scale, while turquoise, amazonite and apatite have a hardness of ≥ 5). Faynan ores are more amenable to pulverization intended for cosmetic purposes. An added advantage is the fact that Cu silicate, Cu chloride and malachite retain their color even in a powdered form, in contrast to amazonite, Cr-containing calcite and fluorapatite.

Ore from Timna has most likely also been used during the PPNB, which is indicated by the green beads and ore fragments from Nahal Issaron and the Uvda Valley (see above). The possibility that the raw material of these finds has its origin in Timna has regrettably never been studied. It would have proven the use of this deposit in the Neolithic and would have added in this way to the knowledge about the utilization history of the deposit.

8.2 Faynan as a Source of Copper in the Chalcolithic and Early Bronze Age I

Archaeological Overview

During the Neolithic, the overwhelming settlement pattern consisted of autonomous villages, while the increase in population during the second half of the 5th millennium BCE led to the formation of agriculturally based societies. Those had a more complex social structure, which was expressed through the existence of local centers with their own regional culture (Levy 1995). Weippert (1988) summarizes this under the titles *'Crafts, Village Society and Regionalism'*, and points out the growing functional differentiation of crafts and the wide span of different products. These new developments, in sociopolitical as well as economic matters, are taken to define the beginning of the Chalcolithic.

Among the Chalcolithic settlements spread from the Golan Heights to the Negev and Sinai, those connected with the distribution of metallurgy are particularly interesting. This is firstly the Beersheba basin with Tell Abu Matar, Bir Safadi, Neve Noy, Gilat, Shiqmim and others (Fig. 8.6), followed by the lower Jordan Valley with Jericho, Tuleilat Ghassul and Tell Abu Hamid. Some other sites are in the Judean desert, where the most famous metal hoard find of the entire Near East, consisting of 416 metal objects, was found at Nahal Mishmar (Bar-Adon 1980; Fig. 8.7). The ring-shaped gold ingots from Nahal Qanah (Gopher and Tsuk 1991) are the earliest in the Levant. Following the most recent data base of Burton and Levy (2001), the Chalcolithic in the Southern Levant can be dated into a period between ca. 4500–3600 BCE. The Nahal Mishmar hoard is dated between ca. 4600 and 3500 (see also Tadmor et al. 1995).

The cave of Peqi'in (Gal et al. 1997) and Tell Turmus (see Kerner 2001) northwest and north of Lake Tiberias shows how far the Chalcolithic culture of the Southern Levant reaches to the north. To the south a settlement gap so far exists in the central and southern Negev. Only at the northern end of the Gulf of Aqaba/Eilat the settlement density increases again, e.g., in the Uvda Valley, at Yotvata north of Timna

Fig. 8.6.
Distribution map of important Chalcolithic and Early Bronze Age I sites in the Southern Levant where metal objects and/or other evidence of metallurgical craftsmanship were found (after ILAN and Sebanne 1989; Shalev 1991; Avner et al. 1994; Khalil 1995; Levy 1995)

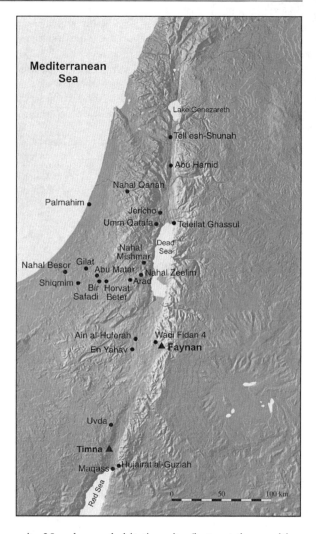

(Avner 2002), and at Timna site 39, where a habitation site (but not the smelting furnace found at a distance of ca. 150 m!) was dated by radiocarbon into the 5th/4th millennium BCE (Rothenberg 1998). Probably the largest settlements at the Gulf of Aqaba are Tell Maqass and Tell Hujayrat al-Ghuzlan (Khalil 1988, 1995; Khalil and Schmidt 2004), whose positions might have played an important role in the trading routes towards Sinai, Egypt and northwestern Arabia.

Although the transition to the Early Bronze Age I in the Southern Levant is suggested to be characterized by a change of cultures with new settlement structures, this time period will be treated here together with the Chalcolithic. Looked at from the viewpoint of the metallurgical craftsmanship, most traditions are still close to the Chalcolithic. Settlements from the Early Bronze Age I are Nahal Besor (Wadi Ghazzeh) site H, the nearby Ashqelon Afridar (see thematic volume Atiqot 45/2004), Tell Halif (near Gilat), and Tell esh-Shuna in the Jordan Valley.

8.2 · Faynan as a Source of Copper in the Chalcolithic and Early Bronze Age I

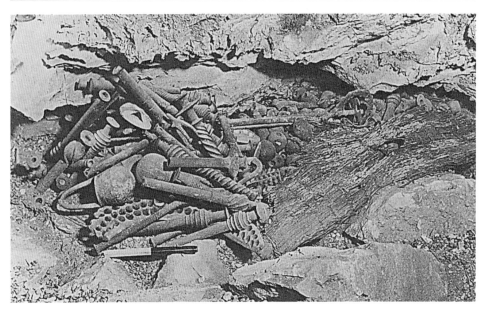

Fig. 8.7. The metal hoard from Nahal Mishmar (Judean Desert, Israel). One can recognize the standards made from Cu-As-Sb alloys and a perforated, sickle-shaped object made from hippopotamus ivory, allegedly all prestige items. In the background, a number of the many maceheads. The hoard find, when excavated, was covered in mats (from Bar-Adon 1980)

Varying quantities of metal objects were found in all these settlements, often with items used in metal workshops, thus proving the production of metal on the site itself, including pieces of ore, crucible fragments, slag, copper prills, and metal objects (Perrot 1955; Tylecote et al. 1974; Shalev and Northover 1987; Shalev 1991; Hauptmann 1989; Khalil 1988; Khalil 1995; Segal et al. 2004). Surprisingly, casting moulds are rare finds except at Tell Hujayrat al-Ghuzlan, where they were discovered in large quantities (Pfeiffer 2004). A few casting moulds were also found at Wadi Fidan 100. None of these settlements is in the direct proximity of ore deposits; the ore must thus have been imported and then worked on in those sites. Metal production and metal working, sometimes into objects of great artistic beauty, are quite obviously two of the most important innovations of the Chalcolithic period in Palestine. This development already led Gordon Childe (1951) to suppose that the Syro-Palestinian region would have been an independent center where early metallurgy had been developed. Other authors (Yakar 1985; ILAN and Sebanne 1989) preferred a diffusionistic model and thought it possible that early metallurgy came from the north of the Fertile Crescent to Palestine.

Metal Production in Settlements: Domestic Metallurgy

Finds from metallurgical activities in workshops of the mentioned sites not only came from inside the settlements, but were also concentrated within some of the houses. This form of production seems to be the typical organization of metallurgical activities adopted during the Chalcolithic and, as the example of Wadi Fidan 4 shows, also of the Early Bronze Age I in the Near East (Shalev 1994; Hauptmann and Weisgerber 1996). It seems therefore to be justified to speak of a 'workshop or domestic mode of production' and thus suggest a craft specialization on a modest level (see also Kerner 2001). The smelting of ores did not happen at the mines; rather, it occurred within the more secure domain of the settlements. This means for all the sites mentioned here, the ore had to be procured from rather more distant raw material sources. This would explain why it has not been possible to find any signs of smelting dating from the earliest stages of metallurgy in the direct vicinity of the deposits, despite very intensive research (Craddock 1989; Rothenberg and Merkel 1995; Hauptmann and Weisgerber 1996).

The transportation of ore over distances of 100 to 150 km is not that surprising as one might assume at first sight (Levy 1995), as it would form part of the tradition of trade and exchange carried on over thousands of years. It had already been demonstrated that green pigments, and that meant also secondary copper ores, were traded during the Pre-Pottery Neolithic over sometimes great distances in the ancient Near East. Levy's (1995) assumption that metal was produced inside the settlements because it was a well-kept secret due to the social stratification of the society does not need to be true. It had possibly the very pragmatic reason that metal, once smelted from the ore and remelted to larger batches, was a precious material. The value of this material in the Chalcolithic becomes obvious when the ratio between stone adzes and metal adzes (250:1) in Shiqmim is considered (pers. comm. Levy 1997). Except for the extraordinary amount of metal objects in the Nahal Mishmar hoard find, the production of metal objects must have been limited. Tools and other items for daily use such as adzes, sickles, knifes etc. were certainly overwhelmingly made from stone.

Only in a few of those settlements the finds from metal workshops have been more thoroughly analyzed, although the discussion about the possible raw material sources in the Arabah has been going on with varying arguments for some decades now. Perrot (1955, 1957) already quoted the deposits in Faynan as a possible source of origin for the metal finds in Tell Abu Matar. This was later verified by Shugar (2000) who intensively investigated metallurgical finds from this site. Bachmann (1981) assumed that Timna had provided the ore for the finds in Nahal Besor/Wadi Ghazzeh. Shalev and Northover (1987) thought again that Faynan would be the source of the ore discovered in Shiqmim, because it contained kaolinite, which was, in their opinion, a typical characteristic for the Faynan material. At Tell Maqass and Tell Hujayrat al-Ghuzlan it was ore from Timna, which was smelted there (Khalil 1988; Khalil and Riederer 1996; Hauptmann et al. 2007).

Chemical and Lead Isotope Analyses

Based on previous investigations (Hauptmann 1989) of ore, slag and metal artifacts from several different sites dating to the Chalcolithic or Early Bronze Age I in

Palestine, the question of the origin of copper ore from Faynan, or the Arabah region respectively, will be reassessed. It will be tested if ores found in settlements in the Negev and the Jordan Valley have been smelted to metal, or if some might just have been used, in the Neolithic tradition, only for cosmetic purposes. For these tests sixteen ore samples, slags and metal objects were chosen, of which some had already been analyzed (Table 8.2). The earlier chemical analyses are given here again, some in a revised version (Table 8.3). Additionally, lead isotope ratios of twelve further samples were analyzed (Table 8.4) in the Max-Planck-Institut für Chemie (Mainz, Germany).

Table 8.2. Summary of workshop finds and metal objects from different Chalcolithic period and the Early Bronze Age I sites (Wadi Ghazzeh/Nahal Besor) in Palestine. Inventory-numbers from Abu Matar and Bir Safadi are based on Perrot's (pers. comm. 6/1988) information. The two adzes BAM 117 and AO 22905 are thus from Bir Safadi. Previously published differing information by Hauptmann (1989), Tylecote et al. (1974) and Miron (1992) are probably due to unclear data. *nn:* No number

Sample no. IfA	Inv. no.	Description	Reference
Tell Abu Matar			
IL-5/1	BAM 387/2	Bead (zylinder)	Perrot (1955)
IL-5/3	BAM 888c	Cu-ore	Perrot (1955)
IL-5/4	BAM 940	Cu-slag	Perrot (1955)
Bir Safadi			
IL-5/2	BAM 814(BES 117?)	Adze	Perrot (1955)
IL-8/1	BAM 117(BES 117)	Adze	Tylecote et al. (1974)
IL-8/2	AO 22905	Adze	Perrot (1972)
IL-8/3	AO 22906	Macehead	Perrot (1957)
Teleilat Ghassul			
IL-10/1a	A	Chisel	Mallon et al. (1934)
IL-10/1b	B	Chisel	Mallon et al. (1934)
nn	nn	Chisel	Miron (unpubl.)
Tell Abu Hamid			
JD-43/1	AH 89.3676	Sheet of metal	Dollfuß et al. (in press)
JD-43/2	AH 89.3434	Awl	Dollfuß et al. (in press)
JD-43/3	AH 86.639.1	Awl	Dollfuß et al. (in press)
Wadi Ghazzeh/NahalNahal Besor			
IL-2/8	nn	Cu-droplet	MacDonald (1932)
IL-2/9	3	Awl	MacDonald (1932)

Table 8.3. Chemical analysis of finds described in Table 8.2. Some of the samples are nearly completely corroded and therefore have a total substantially below 100 wt.-%. All data are given in µg g^{-1}, except when weight percent are given. *nn*: No number; *nd*: not detected; *na*: not analyzed; <: below detection limit

Sample no. IfA	Inv. no.	Cu (%)	Fe (%)	Pb	Zn	Sn	As	Sb	Bi	Co	Ni	Ag
Tell Abu Matar												
IL-5/1	BAM 387/2	86.8	0.76	0.69%	<400	<100	3.98%	5.45%	0.34%	<13	0.96%	3300
IL-5/3	BAM 888c	55.3	1.61	230	90	<100	40	3	4	100	<7	45
IL-5/4	BAM 940	25.9	30.1	200	500	na	na	na	na	na	700	na
Bir Safadi												
IL-5/2	BAM 814 (BES 117?)	99.0	0.14	45	10	<30	2200	140	<10	20	4400	110
IL-8/1	BAM 117 (BES 117)	96.9	<0.025	39	<30	<100	2700	20	<10	<30	3300	90
IL-8/2	AO 22905	95.6	<0.025	37	<30	<100	20	17	<10	<30	<30	54
IL-8/3	AO 22906	87.5	0.34	4.46%	<30	<100	6330	7900	770	<30	270	530
Tuleilat Ghassul												
IL-10/1a	A	70.0	0.21	15	75	<100	1260	25	<10	10	3400	10
IL-10/1b	B	86.0	0.04	10	25	<100	880	135	20	4	3600	15
nn		89.3	0.03	90	100	nd	1900	nd	nd	nd	3400	270
Tell Abu Hamid												
JD-43/1	AH 89.3676	96.5	<0.01	20	<20	<50	12	<5	<5	<10	<20	74
JD-43/2	AH 89.3434	96.4	<0.01	<10	<20	<50	30	<5	<5	<10	<20	64
JD-43/3	AH 86.639.1	93.7	<0.01	<10	<20	<50	<10	<5	<5	<10	1080	1270
Wadi Ghazzeh/Nahal Besor												
IL-2/8	nn	82.6	5.11	0.24%	250	<50	1200	72	<10	92	400	165
IL-2/9	3	99.1	<0.025	140	<30	<50	<10	27	12	<10	46	56
IL-2/10	nn	99.0	<0.025	120	<30	<50	200	19	<10	200	120	105

8.2 · Faynan as a Source of Copper in the Chalcolithic and Early Bronze Age I

Table 8.4. Lead isotope ratios of the Chalcolithic metal objects mentioned in Table 8.2. Pb-concentrations in µg g⁻¹ except IL-8/3. *nn*: No number

Sample no. IfA	Inv. no.	Pb	$^{208}Pb/^{206}Pb$	$^{207}Pb/^{206}Pb$	$^{204}Pb/^{206}Pb$	$^{208}Pb/^{204}Pb$	$^{207}Pb/^{204}Pb$	$^{206}Pb/^{204}Pb$
Tell Abu Matar								
IL-5/1	BAM 387/2	590	2.0729	0.8357	0.05324	38.93	15.70	18.78
IL-5/3	BAM 888c	230	2.0910	0.8496	0.05429	38.51	15.65	18.42
IL-5/4	BAM 940	130	2.1103	0.8624	0.05515	38.26	15.64	18.13
Bir Safadi								
IL-5/2	BAM 814 (BES 117?)	20	2.0704	0.8351	0.05320	38.92	15.70	18.80
IL-8/1	BAM 117 (BES 117)	25	2.0737	0.8357	0.05315	39.02	15.72	18.81
IL-8/2	AO 22905	10	2.1142	0.8635	0.05513	38.35	15.66	18.14
IL-8/3	AO 22906	2.10%	2.0661	0.8234	0.05230	39.50	15.74	19.12
Tuleilat Ghassul								
IL-10/1b	B	4	2.0537	0.8262	0.0527	38.97	15.67	18.97
Tell Abu Hamid								
JD-43/3	AH 86.639.1	10	2.0640	0.8350	0.05338	38.67	15.64	18.73
Wadi Ghazzeh/Nahal Besor								
IL-2/8	nn	4150	2.1126	0.8648	0.05534	38.18	15.63	18.07
IL-2/9	3	137	2.1185	0.8674	0.05540	38.24	15.66	18.05
IL-2/10	nn	70	2.0994	0.8584	0.05494	38.21	15.62	18.2

Tell Abu Matar

Finds from the excavations in the 1950s are published by Perrot (1955, 1957, 1972) and Golden (1999). Further (unpublished) material is stored in the Centre du Recherche Français in Jerusalem. Perrot uncovered several kilograms of ore and slag, fragments of crucible as well as furnace lining, and some metal objects (three maceheads, one standard, one cylinder bead and two awls). A second excavation in 1990/91 unearthed some kilograms of ore and slag as well as crucible and furnace fragments (Gilead and Rosen 1992; Shugar 2000). The quantity of the entire material makes Abu Matar the largest metal working site of all the Chalcolithic settlements in the Beersheba valley, possibly of the whole Southern Levant.

Tylecote et al. (1974) analyzed an adze with the inventory-number BAM 117, giving as the find spot 'Safadi near Abu Matar'. Merkel (1977) analyzed obviously the same sample but gave Abu Matar as the place of origin. The author gave it the same nomination and analyzed it again. The results are given in the tables here. The adze fragment BAM 814, which the author got from Prof. Jean Perrot, also supposedly came from Abu Matar but was very similar to the adze Tylecote had originally analyzed. Dr. Commenge looked into the matter in 1997, and following her information both samples came from one adze with the original inventory number BES 117, excavated in Bir Safadi and will therefore be dealt with there.

Figure 8.8 illustrates that ore sample BAM 888c from Tell Abu Matar falls into group B of the isotope compositions, the same group that contains the MBS ore samples from Faynan and also slag and metal droplets from Timna and Faynan. The composition of the sample composition is indeed quite close to that of an ore sample from the MBS in Faynan. Its origin from this particular formation is underlined by

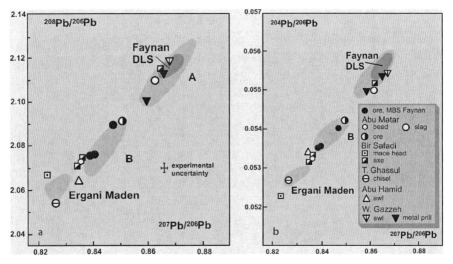

Fig. 8.8. Pb-isotope abundance ratios from artifacts, slag and ore from different Chalcolithic and the Early Bronze Age I sites in Palestine (Wadi Ghazzeh/Nahal Besor). For comparison, the range of composition of ore and metal samples from the DLS and three ore samples from the MBS in Faynan (●) is also given. *A* and *B* mark the groups of ore and metal from Faynan and Timna shown in Fig. 6.46

mineralogical studies undertaken by Hauptmann (1989), which demonstrated that all four ore samples analyzed so far from the excavation in Abu Matar originated in Faynan.

The isotope ratios of slag BAM 940, plotting into group A are worth mentioning. The ore samples in this group could have originated from Timna as well as from the DLS in Faynan, where Cu-Mn ore occur. However, the slag consists mainly of Fe oxides + SiO_2 and contains MnO only as a minor component (MnO = 0.15%) and can consequently not have been smelted from Faynan ore but must originate from Timna. This is not very surprising, considering that in Abu Matar a few ore samples from Timna were identified by petrographic studies (unpubl. data DBM). Concerning the archaeological interpretation, this fact is of importance, because Perrot has so far assumed (pers. comm. November 1996) that this raw material source, located 100 km further south, played no role in the trade connections of Tell Abu Matar.

The cylinder bead BAM 387/2 is of foreign and unknown origin. Although its lead has an isotopic composition identical to those of the adzes, the chemical composition proves that this artifact cannot have been smelted from ore coming from the Arabah region. It contains over 5% Sb and 4% As, thus showing the typical Cu-As-Sb alloy that can be found in the material of the Levantine Chalcolithic. The isotopic composition is nearly identical to that of the macehead 61-297 from Nahal Mishmar (Tadmor et al. 1995).

Bir Safadi

Miron (1992) gives the information that three adzes were found in Bir Safadi close to Tell Abu Matar.

The adze AO 22905 has comparable lead isotope ratios and a similar chemical composition as ore samples of group A from the Arabah (Fig. 8.8). The artifact is made of very pure copper, particularly As and Ni with 20 and <30 µg g^{-1} are drastically lower then in the other two adzes, which will be mentioned later. It is not possible to clarify the origin of the raw material, as both Timna and Faynan deposits appear equally in group A, and a single sample can not be used as a basis of a statistical statement, as will be shown in the discussion of the Early Bronze Age finds from Arad and Bab edh-Dhra. Ores from Safadi could be related to the Arabah region, but not to a specific deposit (Hauptmann 1989).

Samples BAM 117 and BAM 814 will be treated separately, due to the uncertainty about the inventory numbers, although they might actually originate from one and the same adze. They plot into group B. Lead isotope ratios of both adzes are close to those of two ore samples from the MBS in Faynan, which might be taken to indicate their origin from the ore of this formation. But high As and Ni concentrations (0.2% As and 0.33–0.44% Ni) speak against it. Even if absolute concentrations of single elements are not reliable indications of provenance studies, it is a matter of fact that As and Ni are by one and even two decades higher than those of the ore samples from Tell Abu Matar and Bir Safadi (Fig. 8.9), which have been analyzed from Hauptmann (1989). In addition, copper prills from crucible or furnace lining in Abu Matar contained only <10 µg g^{-1} As and only between 350 and 480 µg g^{-1} Ni (Hauptmann 1989). In all the ore samples from the MBS, in the Chalcolithic raw copper from Faynan (Fig. 4.6) and in over 200 ore and metal samples

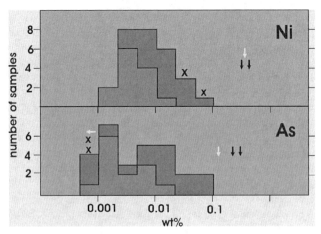

Fig. 8.9. As and Ni contents of ore from the Massive Brown Sandstone (MBS) in Faynan and from ore samples found in Abu Matar, whose origin from Faynan has been determined mineralogically (*medium grey*). The com-position of Early Bronze Age I raw copper from Faynan (*white*) and of two metal prills from crucibles in Abu Matar (**X**) are shown for comparison. The composition of the two adzes from Abu Matar (*dark arrow*) and from Tuleilat Ghassul (*light-colored arrow*) is not comparable with these

from Timna (Leese et al. 1985/86) there is not one, which shows a comparable concentration. It is therefore possible that the Ni concentrations of ores from Timna 39 (0.1–0.5%) given by Rothenberg (1978: 8) have to be taken more as the result of a semi-quantitative measurement. Such high As and Ni contents have only been measured in the Faynan copper of much later periods (Hauptmann et al. 1992), when the ore from the DLS is mined. However, the DLS ore is in its isotopic composition, not comparable to the adzes. This leads to the assumption that both adzes have probably not been made of the ore coming from the Arabah region.

The search for appropriate ore or metal comparisons reveals that the As- and Ni-rich copper objects occur mostly in eastern Anatolia, such as in Arslantepe (Caneva and Palmieri 1983) and Hassek Höyük (Schmitt-Strecker et al. 1992). In the latter site at least one object (Hsk. W. 81-20/HDM 1184) dating into the Early Bronze Age I has been excavated, whose lead isotope composition is nearly identical with that of the adzes. It is also of interest that ore samples analyzed so far from the copper deposit at Ergani Maden show a lead-isotopic composition (Fig. 8.8), which is, although not identical, at least similar to that of the artifact(s) from Safadi. An unequivocal connection between the artifacts and the deposit at Ergani Maden cannot be established, as the adze(s) have a much lower Co/Ni ratio (0.004–0.009) than any of the analyzed ores. A comparable problem has already been pointed out by Schmitt-Strecker et al. (1992) for the artifacts from Hassek Höyük. The artifacts possess, similar to the metal object(s) from Safadi, a Co/Ni ratio of ca. 0.001, while the sulphide ore from the modern open cast mine as well as the oxidic copper ore from the surface (Hess 1998) at Ergani have a ratio of 1–10. The comparison of the Co/Ni ratios from ore samples and raw copper from Faynan (Fig. 6.38) makes it very unlikely that the Co/Ni ratio would have changed that much during smelting.

In Bir Safadi, as in Abu Matar, metal of foreign origin has been found. The macehead (AO 22906) is neither isotopically nor chemically (0.8% Sb, 0.6% As, 0.08% Bi) comparable with the ore and metal samples from the Arabah.

Tuleilat Ghassul

Mallon et al. (1934) described, for level 4 of the settlement, two complete and one fragmentary chisel and several fragments of awls. Furthermore, they reported 'unidentifiable' pieces, which might be the pieces of slag that are stored in the Pontifical Biblical Institute, Jerusalem. For one of the awls Mallon et al. (1934) reported 7% Sn, which in his opinion should correspond with the composition of an awl from Umm Qatafa west of the Jordan Valley. This is more than unusual, because level 4 of Tuleilat Ghassul dates into the turn of the 5th to the 4th millennium BCE (Bourke 2001), and tin bronzes appear in the Southern Levant only from the middle of the 3rd millennium on. Three allegedly Chalcolithic tin bronzes from the deep sounding carried out in 1929 at Tell Halaf (northern Syria) have been published by Otto and Witter 1952, but Buchholz (1967) assesses this dating very critically.

The two complete chisels were analyzed again and consist of relatively pure copper. In both objects, no Sn was found above the detection limit at 100 µg g^{-1}. As and Ni concentrations are unusual since they are in the range of 1/10%. A third analysis, which was carried out in 1983 by Miron (unpubl.), is comparable. The analyzed object was probably the fragmentary chisel, because both complete items, when analyzed by us, showed no sign of an earlier sampling. The comparison with ore and raw copper from the Arabah produces the same problem that has already been described for the two adzes from Bir Safadi: they contain As and Ni each in the range of 1/10%. This irregularity is further underlined when lead isotope ratios of the chisel IL-10/1b are considered. They are unusually low and very close to the composition of an ore sample from Wadi Abu Khusheibah, which nevertheless cannot be the source of the artifact, because the As and Ni concentrations of the ore sample are two and three decades, respectively, lower than that of the chisel. It can be observed, however, that the lead isotope composition of the chisel comes close to that of the so-far defined field of Ergani Maden. A connection with this deposit can therefore not be excluded.

Tell Abu Hamid

At Tell Abu Hamid, three small awls and a centimeter-sized metal sheet have been found so far. Slag or crucible fragments, which could prove a local production or processing of metal, have not been uncovered in excavations (Dollfuss, in prep.). This is very sparse material compared with the finds in the Beersheba valley. Three objects were analyzed chemically, and for one (AH 86.6391), lead isotope ratios have been measured (Table 8.4; Hauptmann et al. 1997). Trace element concentrations are very low in all samples; only awl AH 86/6391 showed, with 0.1% Ni and 0.12% Ag, concentrations in a quantity that has not been found in the ore of the Arabah region (see Fig. 4.9 and Leese et al. 1985/86; Hauptmann et al. 1992). It is again difficult to explain how these finds could have originated from Arabah ore, even though their lead isotope ratio falls within group B (Fig. 8.8).

Nahal Besor/Wadi Ghazzeh

There are a number of sites along the Wadi Ghazzeh, and from one of these, site H, come some archaeometallurgical finds. The site dates into the Early Bronze Age Ia and is therefore the latest of the localities treated here.

MacDonald (1932) and Roshwalb (1981) reported a dagger, two hooks and several awls, with a square section, as well as Cu ore, slag and copper prills. The studies by Hauptmann (1989) have shown that the copper ores stored in the collection of the Institute of Archaeology in London (ca. 150 g) consist of Cu silicates and Cu-bearing conglomerates containing blue quartz grains typical of Faynan. They thus, in all probability, do not originate from Timna as Bachmann (1981) assumed, but from the hanging wall of the DLS in Faynan.

The analyses of two copper prills (IL-2/8, 2/10) and an awl (IL-2/9) showed that pure copper was utilized, which is compatible to the ore from Faynan in its trace element content. Isotope ratios $^{208/206}$Pb vs. $^{207/206}$Pb of all three objects plot into group A of the Arabah data (Fig. 8.8). In combination with the mineralogical ore analyses, one can thus conclude that ore from Faynan has been smelted to metal here, during which small quantities of slag seem also to have been produced (MacDonald 1932).

8.3 Metal Trade in the Early Bronze Age II–IV

Archaeological Aspects

The urbanization in Palestine, which led to the foundation of cities, starts at the beginning of the Early Bronze Age II, i.e., half a millennium later than in Mesopotamia. This process affected the entire countryside up to the marginal zones along the desert. The already existing, as yet unfortified settlements develop into cities and towns with city walls and clear signs of architectural city planning. In several of these towns, metal objects and/or signs of metal workshops have been found (Fig. 8.10). Genz and Hauptmann (2002) studied possible connections between urbanization and metallurgy in the Early Bronze Age. They concluded, after an analysis of the distribution pattern of metal objects and evidence of metal processing, that only marginal connections existed, if any.

Three of these cities, Arad, Bab edh-Dhra, and Jericho (as well as the smaller settlement of Numeira) are of particular interest due to their geographical proximity to Faynan.

The southernmost city, Arad, is located in an arid climatic zone at the northern fringe of the Negev desert. The development of this heavily fortified town during EBA II was decisively influenced by the copper trade with Sinai, or such is the opinion of Amiran et al. (1973), Amiran (1978) and Ilan and Sebanne (1989), and they also consider Arad as the trade center for copper in southern Palestine. This hypothesis is not, however, supported by the latest research from Hauptmann et al. (1999). The authors demonstrate rather that the largest percentage of the copper

Fig. 8.10.
Important sites with metal objects and remains of metal workshops of the Early Bronze Age II–IV in Palestine. The sites in which 'crescent-shaped ingots' from the Early Bronze Age IV/Middle Bronze Age I were found are marked with an asterisk. The numerous sites in the Uvda Valley as well as the one south of Timna ('Arabah-sites') are collectively shown (modified after data from Ilan and Sebanne 1989)

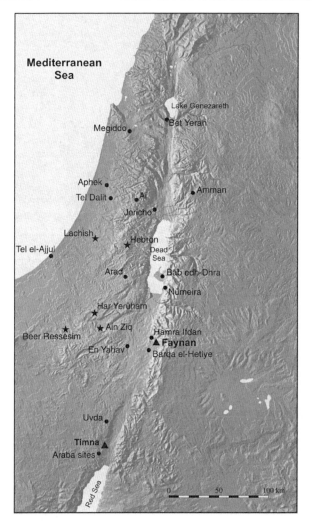

objects from Arad were smelted from ore obtained from the Wadi Arabah, most likely from Faynan (see Fig. 8.16), but not from ore deposits in Sinai. Arad's sphere of influence nevertheless must have reached far to the south, up to the Sinai Peninsula, because the pottery found there is very similar to that of Arad (Amiran et al. 1973). Arad was deserted in the late Early Bronze Age II and with it a number of other towns in the region, some of which had been destroyed. Other towns flourished during the Early Bronze Age III, for example Bab edh-Dhra and Numeira, where massive fortification walls were constructed at this time (Coogan 1984).

Bab edh-Dhra was the most prominent urban settlement along the eastern shore of the Dead Sea. The geographical position of the town meant that it held a key position between the Jordan Valley and the Wadi Arabah and controlled, at the same time, one of the few routes to the Jordanian plateau. Close to the town of Bab edh-Dhra is the largest Early Bronze Age necropolis of the entire Near East, whose tremendous size remains enigmatic: it is assumed that ca. 1 000 inhabitants lived in Bab edh-Dhra, but ca. 20 000 tombs with roughly 500 000 interments have been estimated (Rast and Schaub 1978).

The site of Numeira, ca. 13 km south of Bab edh-Dhra, was much smaller (Rast 1979a). It was only some 70 km away from Faynan and thus the site closest to its ore deposits. Numeira and Bab edh-Dhra were both destroyed at the end of the Early Bronze Age III, at ca. 2350 BCE (Rast 1979b), and only Bab edh-Dhra was resettled during EBA IV, although on a smaller scale.

Jericho, situated on the northern bank of the Dead Sea, was already an important site in the Pre-Pottery Neolithic. During the Early Bronze Age, its defensive walls were renovated not less than seventeen times. It has been assumed that Jericho's trade connections, unlike Arad's, were concentrated to the north due to its position at the southern end of the Jordan Valley (Weippert 1988). This orientation towards the north, within the context of the Kura-Araxes culture, becomes obvious when the so-called Khirbet el-Kerak ware (Beth Yerah-Ware) appears in larger quantities in the Early Bronze Age III. The idea and also the ability to produce this, black- and red-colored pottery, using changing reducing and oxidizing firing techniques, is often seen in connection with newcomers, who might have come originally from eastern Anatolia and Transcaucasia (Fritz 1985; Ben-Tor 1992).

In addition to the larger cities, there existed a number of smaller, undefended settlements. They lay in the direct vicinity of the larger cities or along the trading routes that connected them, for example along the northern coast of the Sinai (Oren 1989). Several smaller settlements in the southern Sinai (Beit-Arieh 1983), in the Uvda Valley at Timna (Ilan and Sebanne 1989) and close to Faynan demonstrate the links to the ore deposits.

The opinions about the importance of metallurgy and the provenance of metals in the Early Bronze Age differ. In several EBA towns and settlements metal objects and production remains have been found, as is shown in a compilation by Genz and Hauptmann (2002). But the actual number is low, particularly when the sudden rise in copper production at Faynan is considered. In Bab edh-Dhra, ca. 35 metal objects have been excavated, 21 of which come from tombs and fourteen from the settlement (Maddin et al. 2003). In Numeira, altogether only fourteen objects were found, but from Arad over 200 (Amiran 1978; Amiran et al. 1997). The hoard find from Kfar Monash (Hestrin and Tadmor 1963; Ben-Tor 1992) dated into the EBA II/III is more impressive as it consists of over 800 small metal plates and over 30 weapons and tools. This selection of sites certainly shows the large distribution of metal tools and weapons. But it is difficult to imagine that metal was already, in the Early Bronze Age II/III, an everyday item, certainly not in the shape of bronze objects as had been proposed by Fritz (1985). Amongst all the finds so far discovered there is only one object that contains a large amount of tin (see below). More convincing is the opinion of Kenyon (1955), who saw the real onset of metal crafts only towards the end of the Early Bronze Age, when the process of urbanization had already peaked.

Fritz (1985) suggested that Early Bronze Age metal objects were imported from Anatolia. Kempinski (1989), however, assumed that Palestine itself was, at the beginning of the EBA II, the main source of copper for Egypt. Around 2700 BCE, this trading network appears to have broken down, which he explains with the beginning of mining in the Sinai and the Egyptian economic interest in Lebanon and Syria.

In fact, a visible increase in the distribution of metal in Palestine can only be shown, in archaeological material, during the period that has been called Middle Bronze Age I and is now generally known under its Anglo-Saxon term as Early Bronze Age IV or Intermediate Bronze Age (2200–2000), depending on the author. The difficulties with the terminology illustrate the problems still existing in defining this phase properly. It is generally seen as an 'in-between period', between the urban cultures of the Early and the later Middle Bronze Age (Gophna 1992). But in this period too, metal objects have been found mainly in tombs; they appear only seldom in settlements. On a few occasions, hoard finds with ingots have been excavated (Dever and Tadmor 1976), which will be described later. The metal was worked most frequently into weapons; typical items are a riveted dagger, throwing-spears, spearheads, fenestrated axes and crescent-shaped axes. Metal tools are rather seldom found. The analysis of numerous EBA IV metal objects showed (Branigan et al. 1976; Eaton and McKerell 1976) that tin bronze slowly becomes more widespread along with the already traditional As copper. This actually means that in technical terms in Palestine the real Bronze Age only begins at the end of the 3rd millennium CE.

Ilan and Sebanne (1989) observed that during the Early Bronze Age the distribution pattern of metal objects is very different from that of the remains of metal workshops, i.e., crucibles, slag, ore, etc. They thus inferred that the production of metal must have taken place in the vicinity of the raw material sources – quite the opposite of the situation during the Chalcolithic and the Early Bronze Age I. This hypothesis corresponds well with the results of our fieldwork, which determined the earliest, clearly provable highpoint in copper production near the ore deposits of Faynan to be in the late Early Bronze Age.

In this section, examples will be given to illustrate Faynan's role in the Early Bronze Age II–IV (2950–2000) up to the transition towards the Middle Bronze Age and to show whether locally produced copper was the only metal traded in the region. In archaeological and cultural terms, these periods are normally treated separately, although the idea of the EBA IV as a time without cultural innovations has recently been abandoned. In archaeometallurgical terms, it is sensible to treat these periods as one and give due weight to the fact that more remote areas, like Faynan, might have shown a cultural development differing from that of areas in the limelight of the Fertile Crescent (Rothenberg and Glass 1992; Avner 2002). The field evidence so far has not allowed us to differentiate the technology of mining and smelting between these periods. Of great assistance in the provenance studies is the fact that from this point on, the metal extraction was only carried out in mines exploiting the Dolomite-Limestone-Shale Unit (DLS). The isotopic composition of this ore is far more homogenous than that of the ore in the sandstone (see Fig. 6.45).

Chemical and Lead-Isotope Analyses

Thirty metal objects and one ore sample have been analyzed from the three Early Bronze Age sites of Bab edh-Dhra, Numeira and Jericho. An overview of these objects and their chronological position is given in Table 8.5. Chemical analyses are compiled in Table 8.6. They have been partly made by atomic absorption spectrometry (AAS) in the Deutsches Bergbau-Museum Bochum and partly by neutron acti-

Table 8.5. Catalogue of analyzed Early Bronze Age metal objects. As most finds are not published, they are given with the (sometimes preliminary) dating of the museums. Both adzes with the numbers JD-47/1 and 2 (inventory numbers 1960a,b) were originally corroded together. They must have been sampled in this form by Maddin. The needle JD-46/10 is now securely dated to EBA I (pers. comm. Schaub January 2004). *Loc.:* Locus; *EBA:* Early Bronze Age; *–:* not published; *nn:* no number; *MASCA:* The Museum Applied Science Center for Archaeology, University Museum, University of Pennsylvania. Given is the present place of storage of the objects

Sample-no. IfA Heidelberg/Mainz	Inv. no.	Description	Reference
Bab edh – Dhra			
JD-46/1 / HDM 702	nn	Dagger (EBA ?)	Museum Amman; –
JD-46/2 / HDM 703	nn	Crescent-shaped axe, tomb A 44 (EBA III)	Museum Amman; –
JD-46/3 / HDM 707	017	Dagger; tomb D 1 (EBA III/IV)	Museum Amman; Rast and Schaub (1978)
JD-46/4 / HDM 708	A 1256	Spearhead; Areal XVI.1, Loc.2 (EBA III/IV)	Museum Kerak; Rast (1979b)
JD-46/5 / HDM 711	A 1257	Riveted dagger; Areal XVI.1, Loc.2 (EBA III/IV)	Museum Kerak; Rast (1979b)
JD-46/6 / HDM 712	A 1995	Riveted dagger; Grab A 22, Loc.16 (EBA II/III)	Museum Kerak; Maddin et al. (2003)
JD-46/7	1955	Awl; tomb A 22 (EBA II/III)	Museum Kerak; Maddin et al. (2003)
JD-46/8	2880	Awl; square XIII.1, Loc. 202 (EBA III)	MASCA, Philadelphia; –
JD-46/9	2901	Awl; square XVI.4, Loc. 10 (EBA III)	MASCA, Philadelphia; –
JD 46/10	2923	Awl; tomb J 2, Loc. 15 (EBA I)	MASCA, Philadelphia; –
JD-46/11	2926	Tube; surface (?)	MASCA, Philadelphia; –
JD-46/12	3108	Chisel; square XVI.2, Loc. 107 (EBA III)	MASCA, Philadelphia; –
JD-46/13	3125	Lump of metal; square XIX.1, Loc. 93 (EBA III)	MASCA, Philadelphia; –
JD-46/14	3141	Lump of metal; square XIII.1, Loc.2 (EBA III)	MASCA, Philadelphia; –
Numeira			
JD-47/1 / HDM 709	1960a	Adze; square SE 4/1, Loc. 50 (EBA II/III)	Museum Kerak; Maddin et al. (2003)
JD-47/2 / HDM 710	1960b	Adze; square SE 4/1, Loc. 50 (EBA II/III)	Museum Kerak; –
JD-47/3	2820	Awl; square NE 4/2, Loc. 2 (?)	MASCA, Philadelphia; –
JD-47/4	2909	Metal fragment; square NE 5/1, Loc. 3	MASCA, Philadelphia; –
JD-47/5	2910	Awl; waste dump	MASCA, Philadelphia; –
JD-47/6	2914/1	Ore; square NE 5/1, Loc. 5	MASCA, Philadelphia; –
JD-47/7	2914/2	Metal droplet; square NE 5/1, Loc. 5	MASCA, Philadelphia; –
JD-47/8	2918	Sheet of metal; square NE 4/4, Loc. 33	MASCA, Philadelphia; –
JD-47/9	2956	Awl; square NE 8/1, Loc. 33	MASCA, Philadelphia; –
JD-47/10	3127	Needle; square J2, Loc. 15 (EBA I)	MASCA, Philadelphia; –
Jericho			
IL-3/x / HDM 704	4125	Dagger; tomb A 95 (EBA IV)	Museum Amman; Kenyon (1960)
IL-3/y / HDM 705	5747	Crescent-shaped axe; tomb A 144 (EBA III)	Museum Amman; Kenyon (1960)
IL-3/12	SA 1077	Metal droplet; JpH 113.26 (EBA IV); –	Institute of Archaeology; London; Khalil (1980)
IL-3/14	SA 987	Metal droplet; JpH 300.5 (EBA IV?); –	Institute of Archaeology; London; Khalil (1980)
IL-3/15	nn	Metal droplet; JpH (?) (EBA IV?); –	Institute of Archaeology; London; –
IL-3/20	SA 814	Metal droplet; JpH 8.5 (EBA)	Institute of Archaeology; London; –
IL-3/25	3077	Chisel; JpO 112.44 (EBA)	Institute of Archaeology; London; Holland (1982)

8.3 · Metal Trade in the Early Bronze Age II–IV

Table 8.6. Chemical analyses of the artefacts described in Table 8.5. All analyses are given in $\mu g\,g^{-1}$ or in weight percent. The samples having a 'HDM' number in the first column were analysed by Prof. Pernicka, Tübingen/Mannheim by neutron activation, all other samples by AAS in Bochum. Of the samples HDM 704 and HDM 705 the iridium contents have been additionally measured; these are <0.008 and 0.016 $\mu g\,g^{-1}$. *na*: Not analyzed; *nd*: not detected; *nn*: no number; <: below detection limit

Sample-Nr. IfA/ Heidelberg/Mainz	Inv. no.	Cu (%)	Fe (%)	Pb (%)	Zn	Sn	As	Sb	Bi	Co	Ni	Ag	Au	Se
Bab edh – Dhra													
JD-46/1 / HDM 702	nn	92.5	1.56	0.012	20	30	1.71%	130	na	20	55	15	3.3	9
JD-46/2 / HDM 703	nn	107	0.19	0.086	40	90	0.98%	150	na	65	240	160	8.6	6.5
JD-46/3 / HDM 707	017	97.8	0.023	0.047	<15	30	4.70%	50	na	5	32	180	4.19	32
JD-46/4 / HDM 708	A 1256	99.3	0.014	0.002	<8	60	2.57%	15	na	4	10	220	55.6	9.9
JD-46/5 / HDM 711	A 1257	88.1	0.032	0.132	<16	11.1%	2600	140	na	40	1920	1870	21	63
JD-46/6 / HDM 712	A 1995	87.0	0.39	9 $\mu g\,g^{-1}$	<7	420	3.73%	60	na	35	75	25	3.86	8.8
JD-46/7	1955	96.4	0.307	0.202	30	<50	5200	36	<5	40	580	40	na	na
JD-46/8	2880	94.3	0.026	0.152	30	<50	4600	195	<5	<20	260	75	na	na
JD-46/9	2901	80.3	<0.025	0.005	80	<50	35	7	<5	<20	75	10	na	na
JD 46/10	2923	65.7	0.028	0.007	120	<50	85	7	<5	40	220	<5	na	na
JD-46/11	2926	95.7	<0.025	0.251	<30	<50	65	<5	<5	<20	200	35	na	na
JD-46/12	3108	94.7	<0.025	0.796	35	<50	110	10	<5	<20	265	10	na	na
JD-46/13	3125	77.7	0.118	0.045	55	<50	330	5	<5	40	200	195	na	na
JD-46/14	3141	92.8	0.209	0.035	<30	<50	1.82%	4400	<5	<20	235	555	na	na
Numeira														
JD-47/1 / HDM 709	1960a	106	0.018	0.077	<8	50	220	10	na	5	320	120	1.25	5.1

Table 8.6. Continued

Sample-Nr. IfA/ Heidelberg/Mainz	Inv.no.	Cu (%)	Fe (%)	Pb (%)	Zn	Sn	As	Sb	Bi	Co	Ni	Ag	Au	Se
Numeira (continued)														
JD-47/2 / HDM 710	1960b	96.4	0.19	0.048	<8	60	1.19%	70	na	110	nd	200	6.5	8.9
JD-47/3	2820	81.7	0.37	0.70	16.4%	0.53%	770	0.24%	35	<20	495	840	na	na
JD-47/4	2909	71	0.25	0.10	<30	<50	2300	<5	<5	25	610	50	na	na
JD-47/5	2910	66.8	0.036	0.018	<30	<50	25	<5	<5	<20	85	10	na	na
JD-47/6	2914/1	40.4	0.19	0.02	45	<50	10	<5	<5	<20	10	55	na	na
JD-47/7	2914/2	64.3	0.44	0.043	<30	<50	7300	20	<5	70	1400	25	na	na
JD-47/8	2918	63.0	0.051	0.094	<30	<50	35	<5	<5	<20	55	215	na	na
JD-47/9	2956	93.7	<0.025	0.11	<30	<50	50	<5	<5	<20	110	<5	na	na
JD-47/10	3127	91.3	0.45	0.032	<30	<50	1.43%	60	<5	70	905	130	na	na
Jericho														
IL-3/x / HDM 704	4125	95.8	<0.04	1.78	<20	<50	1.56%	90	<2	165	2.21%	625	7.7	11.4
IL-3/y / HDM 705	5747	100.0	1.09	0.221	43	290	9100	47	<3	104	2700	57	6.5	5.6
IL-3/12	SA 1077	91.3	<0.025	0.017	60	<50	78	<10	<10	<10	<10	530	na	na
IL-3/14	SA 987	97.1	<0.025	0.036	<30	<50	9500	390	<10	<10	190	150	na	na
IL-3/15	nn	98.8	<0.025	0.013	<30	<50	40	27	15	<10	110	100	na	na
IL-3/20	SA 814	95.1	<0.025	0.43	<30	<50	1.73%	66	<10	54	1.63%	2100	na	na
IL-3/25	3077	99.4	<0.025	0.003	<30	<50	130	<10	<10	10	185	62	na	na

Table 8.7. Lead-isotope ratios of the Early Bronze Age metal objects shown in Table 8.5. The differentiation in Cu and Cu-AS follows the bimodal curve in Fig. 8.10, meaning Cu-As contains at least 0.98% As. *nn*: No number

Sample-Nr. IfA / Heidelberg/Mainz	Inv. no.	Metal	Pb (µg g^{-1})	^{208}Pb/^{206}Pb	^{207}Pb/^{206}Pb	^{204}Pb/^{206}Pb	^{208}Pb/^{204}Pb	^{207}Pb/^{204}Pb	^{206}Pb/^{204}Pb
Bab edh – Dhra									
JD-46/1 / HDM 702	nn	Cu-As	33	2.1550	0.8975	0.05771	37.34	15.55	17.33
JD-46/2 / HDM 703	nn	Cu-As	360	2.1145	0.8708	0.05583	37.87	15.60	17.91
JD-46/3 / HDM 707	017	Cu-As	84	2.1130	0.8779	0.05659	37.34	15.51	17.67
JD-46/4 / HDM 708	A 1256	Cu-As	4	2.1146	0.8725	0.05610	37.69	15.55	17.83
JD-46/5 / HDM 711	A 1257	Cu-Sn	800	2.1155	0.8665	0.05535	38.22	15.66	18.07
JD-46/6 / HDM 712	A 1995	Cu-As	7	2.0940	0.8530	0.05444	38.46	15.67	18.37
JD-46/7	1955	Cu	320	2.1195	0.8697	0.05565	38.09	15.63	17.97
JD-46/8	2880	Cu	420	2.1168	0.8675	0.05550	38.14	15.63	18.02
JD-46/12	3108	Cu	3520	2.1199	0.8700	0.05569	38.06	15.62	17.96
JD-46/14	3141	Cu	130	2.0993	0.8630	0.05534	37.94	15.6	18.07
Numeira									
JD-47/1 / HDM 709	1960a	Cu	180	2.1185	0.8693	0.05563	38.09	15.63	17.98
JD-47/2 / HDM 710	1960b	Cu-As	240	2.1183	0.8691	0.05561	38.09	15.63	17.98
JD-47/3	2820	Cu-Zn	1400	2.0910	0.8523	0.05453	38.35	15.63	18.34
JD-47/10	3127	Cu-As	160	2.1138	0.8652	0.05537	38.18	15.63	18.06
Jericho									
IL-3/x / HDM 704	4125	Cu-As-Ni	1770	2.0647	0.8289	0.05287	39.05	15.68	18.92
IL-3/y / HDM 705	5747	Cu	650	2.1193	0.8694	0.05562	38.10	15.63	17.98
IL-3/12	SA 1077	Cu	20	2.1186	0.8685	0.05550	38.17	15.65	18.02
IL-3/14	SA 987	Cu	140	2.1050	0.8590	0.05504	38.24	15.61	18.17
IL-3/15	nn	Cu	70	2.1131	0.8656	0.05530	38.21	15.65	18.08

vation analysis (NAA) at the Max-Planck-Institut für Kernphysik in Heidelberg. Six of the artifacts from Bab edh-Dhra and Numeira had already been chemically analyzed using different methods (AAS, PIXE, XRF) at the University Museum of Archaeology and Anthropology (Philadelphia, USA) (Maddin et al. 2003). Details about the chemical analyses are given in the Appendix.

Lead-isotope ratios of nineteen objects were measured at the Max-Planck-Institut für Chemie in Mainz. They are presented in Table 8.7.

Bab edh-Dhra

Nine of the fourteen analyzed artifacts contain, as the most important minor component, arsenic in a concentration of up to several percent. It shows a bimodal distribution (Fig. 8.11), where one group is between 35 and 330 $\mu g\ g^{-1}$ and the other between 0.98 and 4.70%. This second group thus again shows a much higher concentration (up to the factor of 10) than that measured in ore samples from Faynan, although the field evidence in Faynan has now proved the deliberate mining of As-richer ores from the DLS. The figure also illustrates that the artifacts from Jericho and Numeira fall into the same distribution. The comparison with the material from Arad makes it clear that copper objects from Bab edh-Dhra, Numeira and Jericho are higher in arsenic. In the Arad material are only two awls with As > 1%, dating into the Early Bronze Age II and III. Although the arsenical copper in Bab edh-Dhra has been mainly used for daggers and axes, this does not mean that it has been used deliberately for a certain group of objects, as one of the riveted daggers (JD-46/5) has only very little arsenic. The four crescent-shaped axes from there each have a different metal composition. This is made clear in the analyses published by Maddin et al. (2003) and by Miron (1992). Maddin and his team even assume that one of these (Nr. 7336) was melted together from recycled metal as it had 0.2% tin.

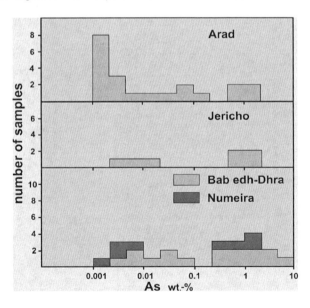

Fig. 8.11.
Distribution of arsenic concentrations in Early Bronze Age artifacts from Palestine. It becomes clear that in Bab edh-Dhra, Numeira and Jericho, copper higher in arsenic was utilized than had been used in Arad (Hauptmann et al. 1999)

These observations fit well with the opinion voiced by many authors that arsenical copper may not be necessarily the result of a deliberate addition of arsenic ores to copper (Charles 1980) but are much more due to natural impurities in copper ores (Gale et al. 1985; Pernicka 1995).

The As vs. Ni diagram (Fig. 8.12) illustrates that a number of artifacts from varying sites do not conform, due to their high As content, to the composition of the raw copper from Faynan and cannot have been produced from it.

Of these artifacts, the five As-rich objects from Bab edh-Dhra (JD-46/1,2,3,4,6) show deviations in their ^{208}Pb/^{206}Pb-, ^{207}Pb/^{206}Pb- and ^{204}Pb/^{206}Pb-ratios from the above-defined field valid for Faynan and Timna (Fig. 8.13). The hypothesis already mentioned is thus supported by isotope analysis: the copper did not originate in the Arabah. And it is hardly probable that arsenical copper would come from deposits in the Sinai. In Wadi Tar, however, a deposit with Cu arsenides has recently been found (Ilani and Rosenfeld 1994) showing signs of 'ancient' mining, but the Pb and Zn concentrations, with a quantity of several percent, of this ore mineralization and the isotopic pattern are different (see Fig. 4.14). That the Sinai is not the source of at least the crescent-shaped axe (JD-46/2) is not surprising, because the origin of this type of axe is generally not sought in the south but in the Syro-Mesopotamian area (Miron 1992). This does not necessarily mean that metal comes from the same area, as that region has hardly any ore deposits. But it is possible that arsenical copper comes from neighboring Iran or Anatolia. At least the riveted dagger (A 1995) is, with all three lead isotope ratios, still just within the field defined by Wagner et al. (1989) for Anatolia. The other three objects come from ore deposits, which must be of a similarly high age to those in the Arabah.

One dagger (sample number: JD-46/1) has a completely different isotopic pattern: ^{208}Pb/^{206}Pb = 2.1550; ^{207}Pb/^{206}Pb = 0.8975; ^{204}Pb/^{206}Pb = 0.05771. This composition indicates an origin for the ore from geologically very old deposits, probably of Palaeozoic or Precambrian age, which excludes all raw material sources in the

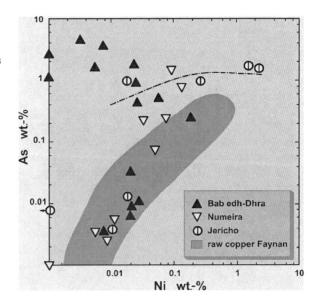

Fig. 8.12.
As and Ni concentrations of Early Bronze Age artifacts from Faynan. All data plotting above the dotted line also have lead-isotope ratios not compatible with those in Faynan ore. All given in wt.-%

Chapter 8 · Export of Ore and Copper: The Importance of Faynan in Prehistoric Palestine

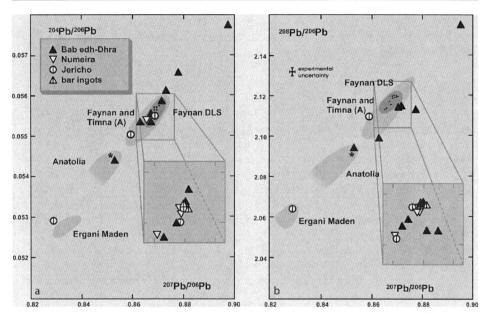

Fig. 8.13. Lead isotope ratios of Early Bronze Age copper artifacts from Bab edh-Dhra, Numeira and Jericho. The messing awl from Numeira is marked with a *, because it stems from an unclear stratigraphic context. Lead isotope ratios of ore and metal objects from the Dolomite-Lime-Shale Unit (DLS) at Faynan; other samples from Faynan and Timna (group 'A', see Fig. 6.44) as well as from Anatolia and Ergani Maden (after data from Hauptmann et al. 1992; Wagner et al. 1989; Seeliger et al. 1985) are displayed for comparison

Aegean and in Anatolia. Only ore deposits in Saudi Arabia have yielded similar isotope ratios (see Fig. 4.14). At this point in time, it is not possible to judge which role these ore deposits played in the prehistoric trade connections in the Eastern Mediterranean, respectively the Levant. Cultural connections between Palestine and Midian in northwestern Arabia are well known from the Late Bronze and Iron Age (Bawden 1989), and Midianite and Qurayyah ware polychrome pottery, found both in the mining districts of Timna (Rothenberg 1998) and Faynan (Fritz 1994b; Levy et al. 2004), suggests an involvement of the Midianites in mining and metal production in the Arabah (Rothenberg 1998, Levy et al. 2004). It is noteworthy that Sillitoe (1979) lists 150 "ancient" mines and slag heaps ("prospects") in northwestern Arabia.

Isotopic data from the Early Bronze Age tin bronzes from Troy and Poliochni (Pernicka et al. 1984, 1990) may be enlightening too; they show extreme compositions similar to the dagger. These bronzes might be comparable with the dagger from Bab edh-Dhra, because it is suggested that the isotopic composition of the lead in tin bronzes is determined by the copper ore and not by tin. The authors argue that cassiterite, which was the most probable source of tin in the Early Bronze Age, did not normally have more than 100 $\mu g\ g^{-1}$ lead. This suggests that not only tin, but also copper must have been imported from foreign mining regions. Muhly et al. (1991) and Pernicka (1995) assume that such ore might have come from central Asia, e.g., Kazakhstan, Uzbekistan and Afghanistan, where such ancient rock is extant. Recent research has proved Bronze Age tin mining in Middle Asia (Alimov et al. 1998).

Fig. 8.14.
Crescent-shaped bar ingots were probably the most prominent shape in which copper was traded all over the Southern Levant during the Early Bronze Age. Examples shown from Yeruham (from Dever and Tadmor 1976)

The existence of a piece of unalloyed copper in Bab edh-Dhra with a similar exotic isotope pattern shows that not only ready-mixed tin bronzes have been imported from these sources. On the other hand, the isotopic composition of the dagger JD-46/5 does not exclude the possibility that locally produced copper might have been mixed with imported tin. Figure 8.13 shows that the isotopic pattern of this dagger lies well within the area formed by the ores and metals from the DLS in Faynan. The dagger has been dated into the Early Bronze Age III and is thus the earliest tin bronze in the Southern Levant, if one ignores for the moment the awls from Tuleilat Ghassul. Maddin et al. (2003) express doubts about the correct dating of the dagger into the EBA III in Palestine. They argue that its import from greater Syria cannot be excluded. Two other tin bronzes, a flat adze and a fenestrated axe from the Early Bronze Age II/III have been found in Tell ez-Zeiraqun (Mittmann pers. comm. 1996; unpubl. data DBM).

Numeira

At Numeira, fourteen metal artifacts and four crucible fragments were found. The latter indicate the treatment of copper in the settlement. Together with the ore sample JD-47/6 one could even assume smelting occurred at the site.

The most important finds are two adzes (JD-47/1, 2), which have been investigated metallographically and chemically using PIXE-analyses by Maddin et al. (2003). The results of their chemical analyses are compatible to ours as one sample (independent from the usage of a and b by Maddin et al. and as in the list quoted here) is an adze with 1% As and the other sample is an adze with very pure copper. A needle from Numeira (JD-47/10) also consists of As copper, while four other objects are from comparatively pure copper.

The lead isotope ratios of the two adzes and the needle are identical with each other and with the pattern of Faynan copper, which has been produced since the Early Bronze Age from the ore of the DLS. It is also possible that at least some of the artifacts found in Numeira have even been produced there.

The awl JD-47/3 is a rather unusual find. It contains, next to Sn and Sb (one tenth of a percent), 16.4% Zn and can thus be classified as brass. Artifacts made of brass are known from other Early Bronze Age II/III sites, such as for example from Thermi III and IV (Begemann et al. 1992) and from Amorgos, where an Early Cycladic dagger with zinc was found (Renfrew 1967). The objects from Thermi have a zinc content of over 1%, in one case of 16.9%. The artifact from Numeira does not come from a secure archaeological context (W. Rast, pers. comm. February 1989), but it has been found in an assemblage with blue-green glazed pottery, the same as what is known from the smelting site at El-Furn in the Faynan area (Hauptmann et al. 1985). A dating into the 13th century CE can thus not be excluded. The chemism of the artifact does not determine nor exclude a dating into either the Early Bronze Age or the medieval period. It is thus another brass find in the eastern Mediterranean, which cannot be dated with any security (see overview Begemann et al. 1992).

Jericho

Only three of seven samples analyzed here show a composition that might indicate a provenance of the metal from the Arabah deposits. These are a crescent-shaped axe (IL-3/y) and two metal prills (IL-3/12; IL-3/15).

The axe dates into the Early Bronze Age III and can consequently be compared chronologically with the one from Bab edh-Dhra, but it has different lead isotope ratios. It cannot be excluded that this axe might have been made from local copper, thus differing from the one from Bab edh-Dhra. The isotope pattern is practically identical with the copper produced from the ore mineralizations of the DLS in Faynan. The usage of this particular ore also explains easily, in a marked difference to the artifacts from the Chalcolithic period, the relatively high As and Ni contents. Otherwise, trace elements are low except for lead, which is in the range of a tenth of a percent.

It is probable that copper from Faynan was still used in Jericho during the Early Bronze Age IV. This is not only confirmed by numerous 'crescent-shaped bar ingots,' which were used during this time period (see below). Chemical and lead isotope composition of the metal prills IL-3/12 and IL-3/15 imply the same; the results lie inside the Faynan DLS-field. The sample IL-3/14, on the other hand, has an unknown provenance. The sample is isotopically isolated and also has a much too high Sb content (390 μg g^{-1}) for any of the deposits in the Arabah.

Next to the evidence of the use of local resources, Cu-As-Ni alloys have been found at Jericho, which do not appear among the artifacts from Bab edh-Dhra, Numeira or Arad. They cannot be derived from ores in the Arabah. These finds include the dagger IL-3/x with over 2% Ni, 1% As and 1.8% Pb. The dagger was probably manufactured in Jericho, as it is typical of the late Early Bronze Age of the Southern Levant (Gophna 1992). A metal prill (IL-3/20) with a compatible composition also argues for the production of the weapon in Jericho itself. The lead isotope pattern of the dagger matches the field defined for Ergani Maden (Fig. 8.13). But it is not clear whether or not the metal originated from there, because the low Co/Ni ratios of the metal objects do not match the ore from Ergani Maden. This has already been discussed in the last chapter.

Crescent-Shaped Ingots from Faynan

More than 100 crescent-shaped ingots, ca. 12–15 cm long with a more or less T-shaped section have been excavated in settlements of the Early Bronze Age IV/Middle Bronze Age I in the Southern Levant west, north and southwest of the Dead Sea, in Har Yeruham, Ain Ziq, Beer Resisim, Hebron, Lachish, Har Zayyad, and Jericho (Tufnell 1958; Dever and Tadmor 1976; Gophna 1992; Segal et al. 1999b) (Figs. 8.7 and 8.14). The large number of these ingots indicates that this must have been the most typical shape in which copper was traded in the region at that time period. The Israel Museum in Jerusalem has recently bought a few dozens of such ingots from the art market. They are supposedly from a refuge in the mountains south of Hebron. All sites in which these ingots have been found are west and northwest of Faynan. Crescent-shaped ingots have not been found in Transjordan.

Maddin and Stech-Wheeler (1976) analyzed three ingots from Har Yeruham and four from Hebron for their chemical and metallographical composition. They discovered that all of them consist of unalloyed, lead-containing copper and had been recast. Dever and Tadmor (1976) based their assumption that the crescent-shaped ingots from Hebron and Har Yeruham all come from one source on this similar chemical make-up. The find context implies that the ingots might have belonged to itinerant smiths.

The first lead isotope analyses were carried out by Brill and Barnes (1988) on an ingot from Har Yeruham. Stos-Gale (1991) published further analyses results from an ingot found in Ain Ziq. The isotope ratios of both ingots are nearly identical (Har Yeruham: $^{208}Pb/^{206}Pb = 2.1199$; $^{207}Pb/^{206}Pb = 0.87072$; $^{204}Pb/^{206}Pb = 0.055648$; Ain Ziq: $^{208}Pb/^{206}Pb = 2.11906$; $^{207}Pb/^{206}Pb = 0.86996$; $^{204}Pb/^{206}Pb = 0.05683$) and plot into group 'A' of the compositions from Faynan and Timna. Both authors infer quite correctly that the isotopic composition of the ingots is compatible with that of the ores from Timna. In spite of that, it is open to question whether Timna was the copper source for the settlements in the Negev during the Early Bronze Age IV, as had been suggested by Merkel and Dever (1989) as well as Rothenberg (1991). The archaeological evidence and the present dating do not exclude such an assumption. It has already been mentioned (Sect. 4.3) that a small ore mineralization has been discovered at Givat Sasgon in the Timna valley, which was mined in the transitional period between the Early and Middle Bronze Age, and a small smelting site

(Rothenberg and Shaw 1990) is nearby. Also, the earliest ^{14}C-dates for mines in Timna are within the EBA III so that from the analytical data, Timna could have been the source of the metal from which the ingots were made. Segal et al. (1999b) analyzed 23 crescent-shaped bar ingots from Ain Ziq and Beer Resisim for their chemical and lead isotope composition and concluded that they would have been manufactured using copper from Faynan.

Since then, however, there has been clear evidence that the source of the ingots is the copper district of Faynan. In the copper mines, nos. 11 and 18 in Wadi Khalid pottery from the Early and Middle Bronze Age have been found (Hauptmann et al. 1985), and ^{14}C-dates of smelting sites clearly point to an enormous metal production in the EBA (see Tables 5.1 and 5.3). The decisive evidence comes from excavations at Khirbet Hamra Ifdan, where the largest metal processing workshop of the Early Bronze Age in the Near and Middle East (Levy et al. 2002) was found (see also Sect. 7.2). The dimensions of this factory (dated to ca. 2700–2200 BCE) are only paralleled by the dimensions of the EBA smelting and metal processing site of Arisman in Iran (Chegini et al. 2001). At Khirbet Hamra Ifdan, many hundred fragments of open casting moulds show that crescent-shaped bar ingots were cast on site. This form of casting mould was also found during our survey over a wide area around the EBA smelting site of Barqa el-Hetiye. In total, in stratum III at Khirbet Hamra Ifdan, 58 ingots were found, which were for the most part complete; fifteen of which had been buried as a hoard.

In order to find out where the source of the crescent-shaped bar ingots in the Southern Levant was, twenty of the ingots from Khirbet Hamra Ifdan (KHI) were

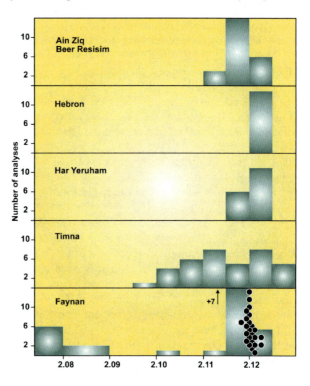

Fig. 8.15.
Histogram of ^{208}Pb/^{206}Pb ratios in ores and raw copper from Faynan, in ores from Timna 4, and in crescent-shaped bar ingots from several localities in the southern Negev. *Black dots* indicate first results of ingots from KHI, those from Ain Ziq and Beer Resisim are taken from Segal et al. (1999). In principle, the origin of the ingots could be Timna too, but the overwhelming compositions point to Faynan – and is supported by the archaeological evidence of mining and smelting in this area

analyzed for their chemical and lead isotope composition. Lead-isotope abundance ratios were analyzed at the Geochronological Laboratory of the University of Münster (Hauptmann et al., forthcoming). In addition, more than 70 South Levantine crescent-shaped bar ingots were analyzed at the Max-Planck-Institut für Chemie in Mainz. Shown in Fig. 8.15 are all compositions of the Khirbet Hamra Ifdan ingots, and all other compositions of ingots which were analyzed previously (Stos-Gale 1991; Segal et al. 1999). The lead isotope ratios of the ingots are identical to those from the DLS ores and the raw copper produced in Faynan. In addition, they match the composition of artifacts from Arad (Fig. 8.16). There is thus a high statistical probability that these artifacts come from the widely outcropping but isotopically uniform ore mineralizations of the DLS in Faynan.

The isotopic clustering of the ingots is underlined by the chemical composition. It was already shown in Sect. 6.3 that minor and trace elements of these ingots, along with other finished objects, in comparison with ores from the DLS, and raw copper from numerous smelting sites in Faynan, show a pronounced bunching. The ingots thus show – like other finished products from KHI – the "typical" average composition of EBA II–IV copper from Faynan. At the current state of research we have to assume that copper was brought from several different smelting sites in the Faynan area to this central place for further production and distribution: KHI. It is not possible to establish the exact mine or smelting site from which the material for the numerous copper items in KHI had been brought. It is therefore not possible to discuss the provenance in as much detail as was possible for the oxhide ingots from Cyprus. There, after performing more than 1 000 lead isotope analyses of ingots,

Fig. 8.16.
Lead isotope ratios of copper artifacts from Arad and Sinai. With two exceptions, all artifacts are compatible with copper from Faynan. Ore samples analyzed from Sinai make evident that these deposits can indeed be eliminated as possible sources. The *arrows* point towards the median value of thirteen objects (Hauptmann et al. 1999)

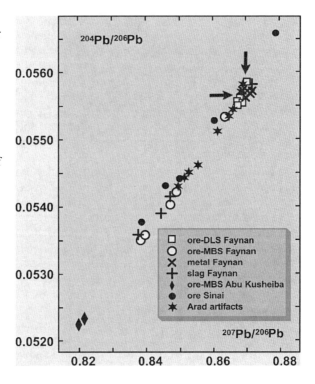

ores and slag, Gale and Stos-Gale (Gale 1999, 2005) are convinced that the mine at Apliki in the northwestern part of the island must have played an important role in the Late Bronze Age copper production. By using analytical data they are thus able to prove mining for this period, for which otherwise mining and archaeological evidence was missing.

An origin of crescent-shaped bar ingots from Timna lacks convincing archaeological evidence. First of all, the geographical position would be a strong factor *pro* Faynan. It is unimaginable that the trading route from the southern Arabah towards the settlements (south)west of the Dead Sea and further north would have left Faynan untouched. Archaeological evidence with the concentration of Early Bronze Age settlements, extensive copper mining and smelting and copper processing at Khirbet Hamra Ifdan and Barqa el-Hetiye points to the Faynan district as the focal point. The bar ingot trade from Faynan supports previous observations of a pronounced "metallurgical drift" of Faynan copper to the north, northwest and southwest from the Chalcolithic period on.

8.4 Conclusions from the Distribution Pattern of Ore and Metal Obtained from Faynan

Ore

The green hue of the ores from the mineralizations in Faynan has been one of the reasons for their use in producing adornments and cosmetic products since the Pre-Pottery Neolithic. Copper silicates such as chrysocolla, dioptase, malachite and other secondary copper minerals, the main ores available here, were distributed in the settlements of Southern Palestine and on the Jordanian Plateau in a circle of more then 250 km. But Faynan was not the only source of green-colored minerals. Other minerals of the same color such as turquoise, amazonite, fluorapatite etc. from different sources in the region were also used. These indicate, on the one hand, a general exchange of goods within a limited geographical/cultural region, and on the other hand, already in the Pre-Pottery Neolithic, a trade link between the Jordanian Plateau and copper deposits on the west side of the Sinai peninsula.

In the Southern Levant, no indications exist (in contrast to Anatolia) that the utilization of copper ore might have led to metallurgical experiments or even initiated the production of metal. The Southern Levant remains in the sequence of metallurgical stages of development, as suggested by Strahm (1994), until and during most of the Chalcolithic period at a precursory stage of metallurgy. At this stage, copper minerals and other 'greenstones' were worked with the same traditional techniques as stone material. Here metallurgy begins only at the $5^{th}/4^{th}$ millennium BCE, when, in the entire Near East, metal was produced by smelting processes. This fact is not changed by the find of a pendant made from (native) copper, which was found at Tell Ramad (Fig. 8.1) close to Damascus (France-Lanord and de Contenson 1969), dating thus into the Pre-Pottery Neolithic. The association with obsidian finds makes it plausible that it was an Anatolian import. Rothenberg's (1995, 1999) suggestion of a Neolithic copper production in Timna can also not be taken as proof due to the unconvincing dating of the site in question (see Chap. 5). This was also commented upon by Adams (1997).

8.4 · Conclusions from the Distribution Pattern of Ore and Metal Obtained from Faynan

The reason cannot have been insufficient knowledge of the usage of fire. It can be assumed that this insight was well known in the entire Near East during the Early Neolithic. Other raw materials were purposefully manipulated with fire, e.g., during the working of flint tools or the production of lime plaster, which was then the first product achieved by a chemical transformation of material through the use of fire and water. Lime plaster was used for walls and floors in buildings, but also for the production of sculptures, figures, plastered skulls and beads. The study of this material has been the focus of several publications during the last twenty years (summations for example by Gourdin and Kingery 1975; Kingery et al. 1988). This not least because the so-called 'white ware' or 'vaisselle blanche' made from lime plaster has often been seen as a precursor of pottery production. Research by Affonso (1997) showed, however, that the percentage of burnt lime was smaller than formerly presumed in the production of the 'white ware'. Much more often locally available material such as marble, limestone, shale, chalk or ash was used, which led to the light color of the vessels, without the need for firing (Goren 1991).

That native copper was used 3 000 years later in the Levant than in Anatolia has nothing to do with a cultural north-south discrepancy. The excavations in Jericho, Beidha, Ain Ghazal, Nahal Hemar and other sites show clearly that the Pre-Pottery Neolithic was a well-developed culture in the Levant, which lacked nothing in comparison to the northern cultures (Schoop 1994). The Southern Levant was embedded in an extensive long-distance trade: Obsidian was imported from regions as far as eastern Anatolia, and elephant ivory and hippopotamus tusk were imported from Egypt. This is in contrast to the Aegean, which, in the fourth millennium BCE, was characterized by the existence of relatively isolated regional groups of the Final Neolithic.

The reasons for the comparatively late beginning of metal extraction and production must be sought in the nature of the two main ore deposits in the Arabah, in which native copper does not occur. In the ore deposits in Anatolia, native copper intergrown with secondary ores was the basis for the usage and exploitation of the new material, causing a number of metallurgical activities during the Neolithic period. This developmental stage is thus missing in the Arabah.

Ores from Faynan – and less frequently from Timna due to the geographical situation – were transported mainly into the settlements in the Beersheba valley, but also to the Jordan valley. The ore and copper objects found in Maadi, in the Nile delta, might perhaps also come from Faynan (Pernicka and Hauptmann 1989). With the exception of turquoise, used for jewelry, no other ores were traded. This is an indication that the beginning of metallurgy in the Southern Levant was based predominantly on local raw materials. There is no evidence yet of any transportation of copper ore for metallurgical activities in the $5^{th}/4^{th}$ millennium BCE in Transjordan.

It is highly interesting that the largest export of the ores went into the Beersheba valley, because it clearly indicates that the earliest metallurgical developments did not happen at the deposits themselves but rather far away from them. It is also an indication of a certain amount of control of metallurgical activities from one or several 'monopolies' in the Negev where Tell Abu Matar played an outstanding role among numerous other settlements (Levy and Adams 1999; Shugar 2000; Kerner 2001). As can be demonstrated by metallurgical finds from the settlements Wadi Fidan 4 and Wadi Faynan 100, local control of mining and metal production in the Faynan district began in the Early Bronze Age I. Based upon observations available at present it is not clear whether or not there was any 'quality control', or specific choice made, for the export of ores from Faynan. It is remarkable that among the ore exported to the Beersheba valley, Fe-containing cupritic ores, with SiO_2 concentrations, were particularly frequent (Tylecote 1974; Shalev and Northover 1987; Hauptmann 1989; unpubl. data DBM, Shugar 2000). This distinct kind of ore is easier to smelt than malachite and contains components necessary for slag formation. It was thus ideally suited for the first steps of extractive metallurgy by smelting ores in crucibles. The ore was smelted to metal inside the settlements.

Any indirect evidence of the use of ore from Faynan and Timna is of great importance because mining activities most likely occurred at both places during this time period but proof is (still) missing. The lack of proof even led to the fundamental questioning of Chalcolithic metal production from locally available ores (Ilan and Sebanne 1989; Sebanne and Avner 1993). The oldest mines in Faynan show, from their entire technological level, a probable Chalcolithic date. But the only means to date a mine are pottery finds and stone hammers, and the earliest such material is identical to those found in Wadi Fidan 4. They date into the EBA I between 3500 and 3100 BCE and are consequently much younger than the oldest evidence of metal production in the Beersheba valley.

The situation in Timna is similar. A smelting site has been found in Timna 39, which appears as 'Chalcolithic' in publications (Rothenberg 1978). But the radiocarbon dating of the smelting furnace there showed only an age of 1 945 ±309 BP, i.e., AD 50 ±309 (Burleigh and Hewson 1979). This dating is not altered by repeated publications by Rothenberg (1998, 1999) and Merkel and Rothenberg (1999), because the recently presented ^{14}C-date from the middle of the 5th millennium BCE (5 485 ±45 BP = 4355–4258 cal. BCE; see Fig. 5.3) is not from the smelting furnace itself but from a nearby stove (for detailed discussion of the dating problem see Adams 1999). The mines in Timna, area G, also labeled Chalcolithic by Ordentlich and Rothenberg (1980) are most likely more recent. They have been dated indirectly via finds from the smelting site F2, which itself dates into the Late Bronze Age according to a ^{14}C-date (Burleigh and Matthews 1982). This date was confirmed by thermoluminescence dating of slag from site F2 (Hauptmann et al. 2007). The earliest mines in Timna (mine S28, Ordentlich and Rothenberg 1980) date from 3 890 ±70 BP (= 2468–2211 cal. BCE) to 4000 ±90 (= 2620–2412 cal. BCE) and are thus from the Early Bronze Age III.

But there are strong indications of the use of ores from Timna during the 4th millennium, which were found in Tell Magass and nearby Tell Hujayrat al-Ghuzlan at Aqaba. The settlements date into the first half of the 4th millennium BCE (Görsdorf 2002), and thus in the Late Chalcolithic period. Many kilograms of slag, crucible fragments, copper ore, a large number of casting moulds, and copper objects were retrieved from the sites, which are ample evidence of extensive copper smelting activities (Khalil 1988; Khalil and Riederer 1998; Brückner et al. 2002; Hauptmann et al., forthcoming). Next to rounded stone hammers with a shaft hole found at Timna and at Hujayrat (Brückner et al. 2002), it was the texture of copper ores that made it possible to pinpoint their origin as the Timna district and therefore proved mining activities during that time period.

While the Beersheba basin, with Tell Abu Matar and Shiqmim as the most prominent settlements, was the area where copper ore from Faynan was at first smelted

8.4 · Conclusions from the Distribution Pattern of Ore and Metal Obtained from Faynan

and distributed, Tell Magass and Tell Hujayrat al-Ghuzlan near Aqaba served as the centers for the Late Chalcolithic/Early Bronze Age I copper production for the Timna area. They are proof that the earlier copper trading from Faynan, where the social pattern of metal production showed that the actual production occurred inside settlements that were at a large distance from the ore deposit itself, is also true for this later trade from Timna.

'Pure' Copper

The distribution of ore from Faynan continued until the 4th millennium BCE. From the middle of this millennium onwards, copper was smelted at Faynan itself, as is evidenced by the settlements of Wadi Fidan 4 and of Wadi Faynan 100. It can thus be assumed, with all probability, that a large part of the artifacts produced from "pure" copper during this period and the subsequent Early Bronze Age, which have been found in the Southern Levant, were smelted from Wadi Arabah copper, mainly from Faynan ore, and that this sort of copper constituted the "metallurgical province" of the Beersheba valley in the 5th, 4th and 3rd millennia BCE. The copper district of Faynan, for many hundreds of years, was a major copper supplier. The finds from other alloys at several sites show that at no time was Faynan the only supplier of ore and/or metal to the Southern Levant. Because copper at that time was oftentimes alloyed with lead, what happened in later periods remains open, while only sparse analytical data are available.

In the Chalcolithic period, judging from the shape of the artifacts, simple tools such as adzes, chisels and awls were produced from pure copper in open cast moulds and subsequently treated by hammering and annealing. Tadmor et al. (1995) were able to use chemical and lead isotope analyses to show that at least some, if not all of the tools from the Nahal Mishmar hoard made from pure copper, were very likely produced from Faynan ore (Fig. 8.17). This implies that the other tools and even some of the other objects could have come from there, too.

No artifacts such as daggers or spearheads have been found. A socketed hammer with a shaft hole from the Nahal Mishmar hoard (no. 61-150) shows a very bad casting with large gas bubbles. The same is true for another three chisels – at least according to the photographs published by Bar-Adon (1980). It is especially true for the chisel 61-179, which is almost rectangular in shape and shows pronounced shrinkages. Such casting defects might be typical features of casting pure copper, not only as it occurs in the Wadi Arabah, with a very low concentration of arsenic. The same phenomena were also observed in many of the Late Bronze Age oxhide ingots in the Mediterranean (Hauptmann et al. 2002): they often are characterized by numerous gas bubbles and interdendritic porosity, which led in part to enormously high volumes of porosity. Oxhide ingots were cast from pure copper that originated in Cyprus.

It is well known that As concentrations, even <1%, may positively affect the casting properties of copper, and this is the reason for the widespread distribution of arsenical copper at the beginning of the Early Bronze Age (see below).

A similarly low concentration of trace elements such as in the artifacts from Nahal Mishmar is known from the adzes in Bir Safadi, two of the three objects in Tell Abu Hamid and the adzes and awls from Shiqmim analyzed by Shalev and Northover

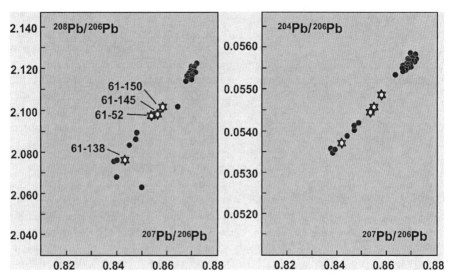

Fig. 8.17. Lead isotope analyses of artifacts made from pure copper in the Nahal Mishmar hoard (*stars*). They are on one mixing line between the extreme data for the compositions of ore, slag and Cu prills, which occurred in Faynan (*solid circles*), making their origin from there more than probable (after Tadmor et al. 1995)

(1987). The rather inconvenient shape of some of the adzes and chisels leads to the question if these artifacts from pure copper were really meant as tools, as has been pointed out by Tadmor et al. (1995). These items could also be ingots. That would explain why some of the adzes after the casting and cold working were annealed again. Annealing as the last step in the production process would have eliminated the previously achieved results from hardening by cold hammering – not a sensible decision if the items were meant to be used as actual tools. Two examples of such a treatment exist, one of them from the Nahal Mishmar hoard. The edge of one adze is nearly totally recrystallized (Potaszkin and Bar-Avi 1980; contra Shalev and Northover 1993). The adze BAM 814 from Bir Safadi is slightly recrystallized and shows isolated twinning (Fig. 8.18), which indicates tempering as being the last production step. This supports the opinions of Golden et al. (2000), who deduced from the context in Shiqmim that these 'tools' were not used as such partly due to their very valuable material, and of Kerner (2001), who suggested these artifacts to be "prestige items" of lower quality but not tools in the same sense as stone tools. Another example of a similar problem comes from Budd (1992). He observed that the most conspicuous result from a metallurgical study of Late Neolithic/Chalcolithic copper axes from several sites in Austria and the Czech Republic was that the last production step was not cold working but tempering/annealing.

We can not exclude that copper from Faynan, and from Timna as well, found its way to Lower Egypt. Archaeologically, cultural interactions between the Nile Delta and the Southern Levant are well established (Van den Brinck and Levy 2002). But the present state of knowledge cannot provide us with a clear picture when it comes to metallurgy. Finds of copper ore at Maadi show striking petrographic similarities to a sedimentary copper mineralization such as that from Faynan (Pernicka and

8.4 · Conclusions from the Distribution Pattern of Ore and Metal Obtained from Faynan

Fig. 8.18.
Bir Safadi, adze BAM 814 (BES 117). Segment of the adze edge showing deformation caused by hammering (cold working). Partial recrystallization and twinning indicate a final tempering, which eradicated at least part of the hardening effect of the hammering (*arrows*) (reflected light, width of the displayed: 3.5 cm)

Hauptmann 1989), and rectangular-shaped copper ingots from the same locality clearly point to connections with Tell Magass and Tell Hujayrat al-Ghuzlan near Aqaba, where ore from Timna was smelted and cast into rectangular-shaped moulds during the Late Chalcolithic period. Lead isotope analyses of artifacts from Maadi match those from the Wadi Arabah (unpublished results Bochum/Münster). However, up until now the mineralogical, geochemical and lead isotopic signature of many of the copper ore deposits in the western part of Sinai and in the eastern desert of Egypt are not known.

It is conspicuous that Arad yielded nearly no As and no As-Ni copper at all, but all objects were made of pure copper. This shows that following the tradition of the Chalcolithic trade connections, the main distribution for the Faynan copper still seemed to have been in the northwest. The study by Hauptmann et al. (1999) confirmed Faynan as the main source of the copper in Arad and thus supported the argument from Finkelstein (1990) and Avner et al. (1994), that Arad was not the trading center for copper from Sinai and hence did not control the copper production on the Sinai Peninsula. In the EBA II, copper was produced for the first time in large quantities in the Faynan region (see Chap. 5), in an 'industrial stage', as defined by Strahm (1994), in the development of metallurgy. The information from Barqa el-Hetiye and Khirbet Hamra Ifdan demonstrate that the control of the metal production is at this point in time inside the Faynan region, not outside as was the case during the Chalcolithic period.

It is even possible that copper from Faynan, or, alternatively, from Timna may have reached Mesopotamia. At the transition between the 4th and the 3rd millennium BCE (Jamdat Nasr), copper objects were found that are characterized by extraordinary high lead isotope ratios ($^{208}Pb/^{206}Pb > 2.10$; $^{207}Pb/^{206}Pb > 0.85$; Begemann and Schmitt-Strecker 2006). Such isotope ratios are not known from any copper deposit in Anatolia, but they were observed in the two copper districts in the Wadi Arabah. Field evidence also proves large-scale copper mining and metallurgy for the middle of the 3rd millennium BCE at Faynan and Timna. At Faynan there is ample evidence of mass production, which obviously started as early as the Early Bronze Age II, when the metal was traded to Arad. Trade connections between the (Southern) Levant and Mesopotamia existed since the 4th millennium BCE (Begemann and Schmitt-Strecker 2006; Pilip 2003).

After evaluating copper objects and copper ore finds from Chalcolithic and Early Bronze Age sites, Ilan and Sebanne (1989) stated that the trade with local ore stopped at the beginning of the EBA I and metal was traded instead. This might have been the tendency. But the ore finds from Wadi Ghazzeh/Nahal Besor and Nahal Tillah (unpubl. data DBM) show that it was not the rule. Excavations at the EBA I settlement of Ashqelon-Afridar at the coast of the Mediterranean Sea brought to light evidence of metal production using ores from Faynan and perhaps from Timna, as well (Segal et al. 2004). Admittedly, the authors did not report on ores and assumed that also metal, next to copper ores, could have been traded to the sites. On the other hand, copper ore smelting is proved by numerous finds of crucibles and smelting slags containing fayalite.

No evidence exists at Faynan itself of such an early change from exporting ore to exporting metal, and the field evidence shows sizeable metal production only from the EBA II on. As this observation is based only on surface material, it might be altered in the future by new excavations. But generally, the trade in metal seems to have started in the Chalcolithic period in the eastern Mediterranean. This is indicated by finds of artifacts made of "foreign" metal in a local style, but not of any "foreign" ore.

In the Early Bronze Age in Palestine, a change happened not only in the production technology, but also in the available sorts of metal and alloys, which are in part different from those in the Chalcolithic period: pure copper, As copper, As-Ni copper and tin bronze (see below) – though there is no contradiction that the unalloyed, very pure copper axes from EBA I Yiftahel should not have originated from Arabah copper (Shalev and Braun 1997).

Arsenic-Antimony-Copper-Alloys

During the last decades, several objects of Cu-As-Sb alloys with varying As and Sb concentrations have been found in a number of sites in Palestine. The macehead from Bir Safadi and the cylinder from Abu Matar, which have been described above, are two of them. By far the largest number of finds that are made from such an alloy comes from the Nahal Mishmar hoard. They are usually referred to as prestige objects (Tylecote 1976; Key 1980; Shalev 1991; Shalev and Northover 1993; Tadmor et al. 1995). The alloys that were used result from smelting fahl ores or their weathering products. Cu-As-Sb alloys exclusively appear during the Chalcolithic period in Palestine and form, besides "pure" copper, the widest distributed copper alloys at this time. It is telling that the prestige objects and not the "tools" were made from such (natural) alloys. It is a good example that such early alloys have also been chosen for their aesthetic value and not only for their improved physical abilities.

It remains open to question whether ingots or tools made from pure Arabah copper would have been used to dilute highly concentrated parent Cu-As-Sb alloys. The As and the Sb concentrations of the prestige objects from Nahal Mishmar, as well as those from other sites, vary from <1 until >15%, but a deliberate (?) mixture can only be proven in individual cases and not as a general rule (Tadmor et al. 1995).

There are no raw material sources known in the Southern Levant and in the Sinai, which could possibly have yielded the material for the Cu-As-Sb alloys, and it is

8.4 · Conclusions from the Distribution Pattern of Ore and Metal Obtained from Faynan

highly unlikely that any might be found in the future. The alloys are definitely of foreign origin. But archaeological and stylistic criteria (Moorey 1988; Tadmor et al. 1995) and the composition of the stone cores included in many of the maceheads that were investigated for their petrographic composition (Goren 1995) show that these 'prestige objects' were manufactured in Palestine. The analysis of stone cores from some of the maceheads from Nahal Mishmar, Shiqmim and Nahal Zeelim points also to a local production. The petrographic analyses show the cores to be made from glauconitic chalk, marl and clay (Goren 1995; Shalev et al. 1992), which all occur in Negev and the Arabah. An unformed piece of metal made from a Cu alloy with 5% Sb, 2% As and 1% Pb, found in the Nahal Qanah cave (Gopher and Tsuk 1991; Fig. 8.6), is very important for this discussion as it could possibly be a semi-finished product. The study of these finds has demonstrated again that the As-Sb copper production was completely separate from the Arabah copper production (see Tadmor et al. 1995). No workshop in the region has yielded any indication that ore had been smelted during the 4th millennium, which fits the description of these alloys. There are also no signs of their being further worked.

At this moment in time, it is only possible to speculate on the origin of the As-Sb copper. It has been proposed that it could come from deposits in Anatolia, Armenia, Azerbaijan or from the (Trans)Caucasus (Tadmor et al. 1995). Newly found metal objects made from arsenical copper and copper arsenic-nickel alloys (see below) make an origin of the Levantine copper finds from Anatolia at least more plausible. A trade in metal between Anatolia and Palestine would fit well into the general picture of the relations between both regions as for example the trade in obsidian has long been known (Yellin et al. 1996). The find of double spirals or flat violin-shaped stone figurines (Levy and Alon 1985) would also fit. In isolated cases, even ores with the appropriately high As and Sb concentrations have been discovered (unpubl. data Rome), which correspond, in their isotopic composition, to the objects from the Nahal Mishmar hoard.

The high-quality Chalcolithic metal-work, which reached a high point with the utilization of lost wax casting for the production of prestige goods, was not continued into the Early Bronze Age. The characteristic method for the manufacture of artifacts in this time period is instead rather simple, open casting with subsequent cold working (Shalev 1994). One reason for this change might be found in an interruption of the trade in As-Sb-containing copper, because this kind of metal, being particularly suitable for the lost wax cast method, might not have been available anymore.

Arsenical Copper

In the 4th millennium BCE, arsenical copper (with As between 0.X and several percent) was in use all over the eastern Mediterranean and beyond. It was a sort of metal that was, in comparison to pure copper, more suitable for smelting and casting, and it reached a greater strength by cold and hot working. Arsenical copper was distributed in the Transcaucasus and was the dominant copper alloy in the Aegean and in western Anatolia during the Early Bronze Age II (Muhly and Pernicka 1992). The spread of arsenical copper is also observed in the Southern Levant. It led Selimkhanov

(1977) to suggest the beginning of the Early Bronze Age to be named "Arsenkupferzeit" predating the period when tin bronze appeared.

As early as the 4th and the 3rd millennium BCE, numerous objects were found, which in comparison with the pure copper from the local sources in the Wadi Arabah, had elevated As concentrations (and also partially elevated Ni concentrations, see below). This arsenical copper was used for making tools that were found in the Chalcolithic settlements of Tuleilat Ghassul, Tell Abu Hamid and Tell Abu Matar. These objects pose a problem: either the deposits in the Arabah once contained As- (and Ni-) richer ores than today, which is not likely, judging from all available information, or yet another sort of metal with a different provenance was at hand. This model might be supported by investigations carried out by Shugar (2000), who analyzed slags, ores and other metallurgical remains from recently conducted excavations at Tell Abu Matar. He analyzed As-rich copper prills embedded in slags. These prills were found to be incompatible with local copper from the Arabah. He suggested an import either of ores, or, alternatively of metal from ore deposits in Anatolia. From a large variety of possible ore sources he suggested as the most probable candidate, next to ore deposits at the Black Sea coast, the deposits at Kaman-Kalehöyük in Middle Anatolia, southeast of Ankara. Due to the isotopic evidence it is not possible to contradict this hypothesis. But it would be a surprise if copper (ore) from a source in the middle of Anatolia would have found its way to the Southern Levant, when many other ore deposits in the east and north of this area are geographically closer (see overview in Wagner and Öztunali 2000; Hauptmann et al. 2002). In the developed Early Bronze Age, the Jordan Valley with Pella, Tell esh-Shuna, Jericho, and Bab edh-Dhra and Numeira, all sites where arsenical copper was found among "pure" copper seem to have been part of this trading network with the north. This is in contrast to Arad, where no As and no As-Ni copper objects were found.

In fact, there are numerous ore deposits in (eastern) Anatolia from where arsenical copper could have originated (Çukur and Kunç 1989; Wagner et al. 1989; unpubl. data Rome/DBM), as well as in Iran (Heskel and Lamberg-Karlovsky 1980) and even, but in very scarce amounts, on Cyprus (Panayiotou 1979). In Arslantepe As- and As-Ni-rich copper ores have been worked into corresponding metal mixtures (Palmieri et al. 1993). We also point to objects found at Chalcolithic Norsuntepe, which probably originated from the ore deposit of Ergani Maden (Pernicka and Schmidt 2002). Apart from Cyprus, this is the largest copper deposit in the whole region of southeast Anatolia and the Levant. From the end of the 4th millennium BCE, arsenical copper was found, e.g., at Arslantepe and Hassek Höyük on the Upper Euphrates, which could also have originated from this source (Schmitt-Strecker et al. 1998; Hauptmann et al. 2002). At Arslantepe, arsenical copper was also imported from the huge ancient mining district of Murgul and other deposits on the Black Sea coast close to the border of Georgia.

A spread of arsenical copper from Anatolia to the south in the 4th and 3rd millennium BCE would not come as a surprise. Cultural interactions between Anatolia and the Southern Levant were established as early as the Neolithic period, evidenced, e.g., by finds of obsidian, which occurs in Anatolia but not in the Southern Levant. These trade connections may have been instrumental in bringing about the beginnings of the later Khirbet-Kerak culture (Kura-Araxes culture), which may date back to the third millennium BCE. Fritz's (1985) assumption that metal objects were imported from Anatolia has thus some credibility. Philip's (1989) opinion that all daggers in the Levant were imported from Anatolia is, however, not supported by the isotope analyses. It is quite clear, in view of the extensive remains of mining and metallurgy in Faynan, that the copper production in Palestine in the 4th and 3rd millennia BCE is not a tradition coming from Anatolia, but has local roots (Gophna 1992). It is obvious that along with the use of local resources a trading net, running parallel to the eastern coast of the Mediterranean Sea, existed and was in use (Philip et al. 2003).

Copper-Arsenic-Nickel Alloys

Next to arsenical copper we can distinguish another very characteristic sort of metal that is frequently and exclusively distributed in the ancient Near and Middle East, mainly in the 4th millennium BCE, but more frequently in the 3rd millennium BCE. It has not been observed in finds from Bronze Age Europe. It shows a pronounced correlation between arsenic and nickel (Fig. 8.12). The highest concentrations of the two elements are in the lower percentage level. In two artifacts from Jericho, As and Ni are more than 1% each. Again, copper high in arsenic and nickel do not occur in the ore deposits of the Wadi Arabah, and it seems that another foreign metal was distributed in the Southern Levant. There is some indication of this e.g., in the Nahal Mishmar hoard. It contains some artifacts with high Ni and As concentrations, some of which also have a high Sb concentration as well and some with Sb only in traces (Key 1980; Shalev and Northover 1993; Tadmor et al. 1995). An even more important indication is a collection of 31 maceheads and an adze probably of Late Chalcolithic age, which have appeared during the last years on the art market (Mish-Brandl pers. comm. 1996). Fourteen of the maceheads were made of copper high in arsenic and nickel (unpubl. data Bochum).

Figure 8.19 shows the distribution of copper-arsenic-nickel alloys with each As > 1% and Ni > 1%. They are common in the area between northern Syria, southern Anatolia and along the Upper Euphrates. It is particularly conspicuous that Cu-As-Ni alloys have been found on several sites, which stretch like pearls on a string along the Jordan River into Palestine. The idea of a trade route comes immediately to mind: The finds from Bet Yerah (Miron 1992), Tell esh-Shuna (Rehren et al. 1997) and Jericho are of importance, as they date into the Early Bronze Age. In addition, Cu-As-Ni alloys were found at Kfar Monash, which probably date to the EBA I (Tadmor 2002), too.

Both kinds of alloys, the As-Ni as well as the As-Sb copper, show that their distribution was tightly controlled. Although the sources of the As-Ni copper are not yet known, it is likely that they also come from the north (see Sect. 8.3). It is thus imaginable that this kind of metal was considered more precious than the local product.

On the other hand, Cu-As-Ni objects occur in five sites in Mesopotamia, on the island of Bahrain, in Umm an-Nar and in Oman. Tallon (1987) analyzed 71 comparable objects from Late Chalcolithic/EBA levels at Susa. Cuneiform texts and archaeological evidence (Weisgerber 1981) have established that during the Bronze Age the copper trade went from modern Oman – the 'Magan' of the Sumerians – through the Persian Gulf to Mesopotamia. One of the shipping harbors was Umm en-Nar (Frifelt 1995), and an important trading station was 'Dilmun', in all probability modern Bahrain. The Mesopotamian finds, at least, can be accounted for by the ophiolite-hosted copper ore deposits in Oman, in opposition to Pernicka's (1995) opinion, because in the Bronze and Iron Age Cu-As-Ni alloys were produced there in large quantities. Geochemical and lead isotope analyses of ores and artifacts proved that this As- and Ni-rich copper was smelted from local ores (Prange 2001). It shows that As-Ni-containing copper is not necessarily a deliberately produced artificial alloy, but it can result from the smelting of copper ores, which are intergrown with nickel-arsenide minerals such as nickeline (NiAs) or similarly composed secondary minerals formed by weathering.

As the origin of the Cu-As-Ni objects in the Near East is not always clear, one should look at the geological conditions, because ores with such a composition are relatively rare. Generally, arsenic- and nickel-containing copper ores are typical of deposits embedded in rocks of the earth's mantle and lower crustal formations such as peridotites and related rocks as they occur in the lower part of ophiolite complexes. Such rocks are exposed, for example, in the Samail Ophiolite Complex in Oman. They can be considered as possible sources of Cu-As-Ni artifacts. Aside from Oman, there are ophiolite-hosted ore mineralizations in the eastern Mediterranean. These ophiolites are located along the Zagros tectonic line (Thalhammer et al. 1986; Çatagay 1987), where the Arabian Plate is subducted under the Tethyan Eurasian metallogenic belt. This belt includes the mountain range of the Taurides in eastern

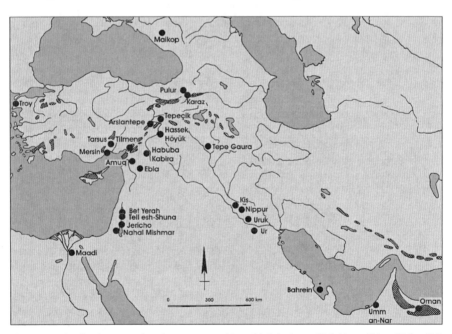

Fig. 8.19. Distribution map showing copper artifacts high in As and Ni (As > 1% and Ni > 1%) and also the occurrence of ophiolites (*dotted area*). Notice the distribution of the artifacts along a line from northern Syria along the Jordan valley towards the south (added to after a compilation by Pernicka 1995). *1:* Jericho; *2:* Tell esh-Shuna (Rehren et al. 1997); *3:* Bet Yerah (Miron 1992); *4:* Ebla (Palmieri and Hauptmann 2000); *5:* Oman (Prange 1998); *6:* Umm an-Nar (Hauptmann 1995); *7:* Bahrain (Prange 1998); *8:* Hacinebi (Özbal et al. 1999)

8.4 · Conclusions from the Distribution Pattern of Ore and Metal Obtained from Faynan

Anatolia, where the most prominent ore deposit of Ergani Maden is located, and it continues far into the southern part of Iran, which is close to the Samail Ophiolite Complex of Oman (Fig. 8.19). In eastern Anatolia, a rock formation with the name listwaenite that is geologically connected with ultrabasic rocks is characterized by enrichments of arsenic, nickel, and other elements such as antimony and gold (Ucurum 2000). At the present state of knowledge, it must remain open whether or not these ores were exploited in ancient times, because they were not explored under archaeometallurgical aspects. In addition, it is conspicuous that Syrian-Anatolian artifacts were found in the direct vicinity of the ultrabasic rocks of the Baer-Bassit- and Hatay-ophiolitic complex. Also here, no research for appropriate ore mineralizations has been carried out so far in these ophiolites.

In Cyprus, part of the ophiolite outcrops along the Zagros tectonic line can be excluded as a source of Cu-As-Ni alloys, because most of the copper deposits are embedded in volcanic rocks in the upper part of the ophiolitic rock sequence in the Troodos Mountains. Ores from here are low in As and Ni, i.e., far below 1%. Only one small mineralization is known from Limassol Forest, which contains polymetallic ores rich in As, Ni, and also in Co (Panayiotou 1980; Foose et al. 1985). Yet there is no evidence of the use of these ores in ancient times.

If ore mineralizations in ultrabasic rocks can possibly be raw material sources of Cu-As-Ni alloys, then artifacts in Syria/Anatolia and Palestine do not necessarily have to originate in the Caucasus, as has been suggested by Pernicka (1995). A comparison with the main geological units in the region (Fig. 3.9), as explained, could offer a much simpler solution. Cultural exchange between eastern Anatolia and the Southern Levant – as a possible basis for copper trade – existed since the Neolithic and is well documented by the distribution of the Kura-Araxes culture.

In Fig. 8.20, lead isotope data are compiled for a number of Cu-As-Ni artifacts from different localities in the Near East. These are a collection of fifteen maceheads probably of Chalcolithic Age from Israel (unpublished data Bochum/Mainz/Münster), five adzes from the Kfar Monash hoard, Israel (unpublished data Bochum/Mainz), and one fragment of a Chalcolithic adze from Maadi in Lower Egypt (see Pernicka and Hauptmann 1989). From Anatolia we included seven objects from the "royal" tomb at Arslantepe from the Late Chalcolithic/Early Bronze Age (Hauptmann et al. 2002), from Hassek Höyük four needles from the EBA (Schmitt-Strecker et al. 1992), and from Norsuntepe two needles from the Late Bronze Age (Pernicka et al. 2002).

Despite their different geographic and chronological origin, they show a pronounced cluster in the lower left-hand corners of the diagrams, close to the "field" of the ophiolite-hosted copper deposit of Ergani Maden as it was defined by Seeliger et al. (1986). In the cases of Hassek Höyük and Norsuntepe at the Upper Euphrates, it has been considered quite probable that the Early Bronze Age metal artifacts originated from the ores of Ergani Maden or other geologically compatible ore deposits such as, e.g., Kisabekir (TG 177; Seeliger et al. 1985). Metal finds from the "royal" tomb in Arslantepe originate most likely from southeastern Anatolia too, if not from Ergani Maden itself. That constitutes a high probability that the As- and Ni-rich Early Bronze Age metal objects from Kfar Monash, some of the Chalcolithic (?) maceheads from Israel and the artifact from Jericho have also been manufactured from material of that region. They point thus to centuries of metal imports from southeastern Anatolia, which is also evidenced archaeologically.

Chapter 8 · Export of Ore and Copper: The Importance of Faynan in Prehistoric Palestine

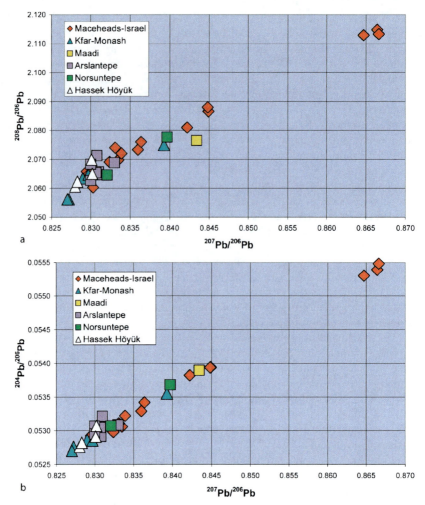

Fig. 8.20. $^{208}Pb/^{206}Pb$ vs. $^{207}Pb/^{206}Pb$ and $^{204}Pb/^{206}Pb$ vs. $^{207}Pb/^{206}Pb$ isotope abundance ratios of Chalcolithic and EBA Cu-As-Ni artifacts in the Near and Middle East. *Shaded areas:* ophiolite-hosted copper ore deposits of Ergani Maden and of Oman and for comparison the "field" of Anatolia. The lead isotope data of most of the objects show a pronounced cluster close to the ore "field" of Ergani Maden. Three out of ten maceheads from Israel are compatible with ores (and Cu-As-Ni objects) from Oman (sources: Maceheads, probably of Chalcolithic Age: unpublished data Bochum/Mainz/Münster; Kfar Monash: unpublished data Bochum/Mainz (see also references!); Maadi unpublished data Bochum/Mainz (see also Pernicka and Hauptmann 1989; Arslantepe: Hauptmann et al. 2002; Norsuntepe: Pernicka and Schmidt 2002; Hassek Höyük: Schmitt-Strecker et al. 1992)

Among the maceheads, there are samples that have unusual lead-isotope compositions. Their isotopic composition differs from all so far known analyses from Anatolia; in three cases they even reach $^{208}Pb/^{206}Pb$ = ca. 2.115. The maceheads were smelted from geologically very old ore, as is known from the Samail Ophiolite Complex of Oman.

8.4 · Conclusions from the Distribution Pattern of Ore and Metal Obtained from Faynan

If the archaeological dating is correct, the connection to Oman would be surprising. The Cu-As-Ni objects of the southeast Arabian Peninsula, with the exception of a few, and the copper ores from the Samail Ophiolith Complex are all between ^{208}Pb/^{206}Pb = ca. 2.075 und 2.12 (Prange 2001; Weeks 2003). The copper objects, though, date without exception into the first peak of copper production in the area, the Umm an-Nar period of the 3rd millennium BCE, which corresponds with the more developed Early Bronze Age of the Levant. This does not exclude the possibility that copper might have been smelted earlier in Oman. Copper-base objects from Hafit graves with multiple burials (3 100–2 700) suggest that local copper production had already begun by this period (Weeks 2003). But there is practically no evidence available yet that copper from Oman could have come to the Levant in the wake of the starting Uruk-expansion in the 4th millennium BCE. This would be surprising as there is no contact between the Uruk culture and the Southern Levantine area (Philips 2003). But an EBA II chisel from Pella needs to be mentioned (sample no. 180048), which, according to its lead isotope composition (^{208}Pb/^{206}Pb = 2.1000, ^{207}Pb/^{206}Pb = 0.85529, ^{206}Pb/^{204}Pb = 18.330; STOS in Philips et al. 2003), might be derived from Faynan copper. The chemical composition of the chisel, however, seems to be arbitrary. If an As- and Ni-rich composition is correct (cited in Table 2.1: >1% As, >1% Ni), then the metal would not be compatible with Faynan copper but with copper from Oman.

Tin Bronzes

Eaton and McKerell (1976) share the opinion that the transition from arsenical copper to deliberately produced tin bronzes during the third millennium BCE was complex in the Near East. But although both alloys seemingly exclude each other, it is still questionable when and if at all arsenical copper was produced intentionally. The only possible indication is a piece of roasted arsenopyrite found in Early Bronze Age Jericho (Khalil and Bachmann 1981). An alloy might also have been used intentionally for the dagger JD-46/3, which has nearly 5% arsenic. It dates into the Early Bronze Age III/IV and would thus be in accordance with Philip's (1991) observations, who suggests alloying techniques started before the beginning of the Middle Bronze Age in Palestine. But it needs to be considered how this scenario fits, e.g., with the fact that arsenical copper objects appear in (Trans)-Caucasia in large numbers (Chernykh 1992). It is not likely that those are all made from purposefully alloyed copper.

As indicated by the artifacts from Bab edh-Dhra and from Tell ez-Zeiraqun (see above), "classical" tin bronze appears in the Southern Levant for the first time in the Early Bronze Age II/III. This is a secure indication of the beginning of a deliberate alloying. The technique spreads in the second half of the 3rd millennium BCE, but at the latest with the beginning of the Middle Bronze Age it is widely distributed in the area (Branigan et al. 1976; Eaton and McKerell 1976; Maddin et al. in press). This is the same time period as the hoard finds with crescent-shaped ingots. At the beginning of the 2nd millennium BCE, there must have thus been large quantities of pure copper from Faynan, and possibly Timna, available. No tin or tin ores in archaeological contexts have been found in the Levant from this time period. The earliest finds of tin come from several localities off the coast of Haifa, Israel, where circular-shaped and other types of tin ingots with a weight of ca. 100 kg were found (Galili and Shmueli 1982). They date into the late 2nd millennium BCE. From here, an Iron Age tin ingot was also found (Artzy 1983). These finds are an indication of maritime trade routes rather than routes on land.

Faynan and Timna

The geographical area for which this study tries to define a general metallurgical development in the 4rd and 3rd millennium BCE comprises the Nile Delta and Upper Egypt in the West, the Sinai Peninsula, Palestine and Syria in the center, Anatolia in the north and Mesopotamia in the east. The ore deposits of Faynan and Timna are at the periphery of a corridor, which forms, due to its climatically favorable position, a connection between the larger political units of Egypt on one side and Anatolia and Mesopotamia on the other side. The question arises if the metallurgical development of the two ore deposits can be treated separately, or if both deposits have a common history. The latter seems to be more likely as the technological development, reconstructed from survey data and material analysis, is very similar. E.g., the clay rods found at all the smelting sites of the Early Bronze Age, the technical ceramics, the slag typology and the carefully constructed nozzles from the Iron Age shall be sufficient examples of such a development.

The so-far described examples show that the copper deposit in the Faynan district was mined from the Neolithic to the end of the Early Bronze Age. Faynan ore made its way not only into the Chalcolithic and EB I settlements of the Southern Levant, but also along the route to the northwest to the Beersheba valley, to Ashkelon, Afridar (at the coast of the Mediterranean Sea), and further along the northern coast of the Sinai to the Nile Delta. Archaeometallurgical finds from Ashkelon, Afridar indicate that metal possibly was also traded further by sea (Segal et al. 2004). The size of the slag heaps in Faynan demonstrates that the production in the later Early Bronze Age was far above the local needs. It becomes clear that a regular 'flow of metal' must have existed into the northwest, to the north and again to the Old Kingdom and Intermediate Period in Egypt as well. Faynan copper, hence, was not only of regional importance; the copper district was connected to international trade routes.

But copper from Faynan was never the sole raw material available in the Southern Levant. Probably from the Early Bronze Age on and likely before this point in time, it was only part of the materials which were traded from the north, i.e., from southeast Anatolia and with the Kura-Araxas culture in the Transcaucasus. In the middle of the 4th millennium BCE before the Uruk intrusion, local Late Chalcolithic complex societies (chiefdoms) obtained metal and Canaan lithic tools through long distance trade (Stein et al. 1998). Indigenous in-settlement metallurgy was practiced in the Altinova region (Esin 1979; Hauptmann 1992) and at the Taurus piedmont far to the north, e.g., at Hacinebi (Stein 1998; Özbal 1999). Handmade red-black pottery indicates intensive contacts with Transcaucasia in Anatolia at the end of the 4th millennium BCE (level VIB2). From Anatolia, Cu-As-Ni objects found their way to the Southern Levant and the Nile Delta. Even if connections between Mesopotamia and the Levant during the 4th millennium BCE are difficult to detect (Philip 2003),

8.4 · Conclusions from the Distribution Pattern of Ore and Metal Obtained from Faynan

recent lead isotope analyses by Begemann and Schmitt-Strecker (2006) of Jemdet Nasr and Uruk-period metal objects found in Mesopotamia opened the possibility that they originated from Faynan copper.

It is unknown where the main trading partners for the metal were in the Late Bronze and Iron Age. Foreign influence from the 14th to the 12th century BCE is proved by Midianite pottery at Barqa el-Hetiye and Khirbet en-Nahas (Levy et al. 2004) and especially by two scarabs from the 12th/10th centuries, which indicate Egyptian contacts at Khirbet en-Nahas. To answer this, more studies of provenance are necessary, which have not fallen into the framework of this study.

In contrast to Faynan, the copper district of Timna is in a relatively isolated position in the south of Wadi Arabah. Surprisingly enough, Timna is surrounded by a number of prehistoric sites (Fig. 7.18). Ilan and Sebanne (1989) and Avner (2002) report on sites in the Uvda Valley, on several sites south of Timna (called Arabah-sites in Fig. 7.18), as well as on sites at Yotvata. Although some evidence of mining and smelting in this early time period seems to come from Timna too, the relevance of these sites is not sufficiently known yet. It is, however, conspicuous that these sites, which date into the entire 3rd millennium BCE and are in the close vicinity of ore deposits, show hardly any evidence of the execution of metallurgical craft activities. The data known so far indicate only a limited production of copper during the Early Bronze Age in Timna itself. Rothenberg and Glass (1992) allow the production of copper, on a small scale, to only start at the end of the 3rd millennium BCE with the change from the Early to the Middle Bronze Age, at a point in time when the main activities in Faynan decrease for the first time. But there are some indications that mining was carried out (Ordentlich and Rothenberg 1980), and at Givat Sasgon (site 250) Rothenberg and Shaw (1990) describe mining into the Timna Formation as well as a slag heap (site 149) from the Early Bronze Age IV. The EBA II settlements in the Uvda Valley close to Timna give additional indirect proof of metal extraction at this time (Sebanne and Avner 1989).

The export of ore and later copper from Timna does not seem to have been directed towards the north. The ore in Abu Matar cannot be connected to Timna with certainty (Hauptmann 1989), and the production at Timna of crescent-shaped bar ingots found in Palestine as assumed by Merkel and Dever (1989) is relatively unlikely, when the situation of finds in Faynan is considered.

In contrast, Timna seems to have early connections to the western trade network towards Egypt. This is demonstrated by the two Late Chalcolithic settlements of Tall Magass (Khalil 1992) and Tall Hujayrat al-Ghuzlan (Brückner et al. 2002). In both sites there are numerous signs of copper production, and both acquired the ore from the relatively close deposits at Timna and its surroundings (Hauptmann et al., forthcoming). The two sites underline, as possible gateways for the trade to the Red Sea, the larger than regional importance of the deposits. The connections between Wadi Arabah and the Nile Delta concerning copper production at the very end of the Chalcolithic is demonstrated by the numerous rectangular and oval casting moulds from Tell Hujayrat al-Ghuzlan, which show a stunning similarity to the copper ingots from Maadi (Rizkana and Seeher 1989). At the moment, the question remains open, if and to what extent the copper finds from Maadi have been produced from Faynan ore as had been suggested by Pernicka and Hauptmann (1989) or if they are really made from Timna ore. The archaeological evidence points now, with the surprising number of casting moulds from Hujayrat, more to Timna. The cultural connections between Timna and Egypt are, at least in the Late Bronze Age with the Egyptian temple in Timna and also later in the Iron Age, much stronger than those between Faynan and Egypt.

The role of the ore deposits in the southern and southwestern part of the Sinai is not sufficiently studied, as many of the mining activities and slag heaps there have not been dated. They were most likely, for large time periods, under the influence of the Egyptian empires. Turquoise from Serabit el-Khadim, the only deposit of this particular semi-precious stone in the entire area of the Near East, has been found in Basta and Beidha (Hauptmann 2004, unpubl. results Bochum) and thus proves the use and exchange of this stone already in the PPNB. Rothenberg (1970), Beit-Arieh (2003) and Avner (2002) found indications of the use of copper ore during the Chalcolithic and the EBA in the Sinai, when copper also was produced in the region of Wadi Dara, Ras Gharib and Umm Balad in the eastern Egyptian desert (Castel et al. 1997, 1998). The site of Bir Nasb showed an extensive production of metal during the New Kingdom; nevertheless, this smelting site has only been cursorily surveyed (Rothenberg 1987) and thus earlier activities cannot be ruled out.

Chapter 9

Summary

This book deals with the early metal extraction and production of copper from the ore deposits in Faynan, at the eastern edge of the Wadi Arabah between the Dead Sea and the Red Sea. The metal production in this area can clearly be used as an example of the archaeometallurgy of copper in the entire Southern Levant, as Faynan is the most important source of raw material in the region. The topics of research included questions concerning the geology of mineral deposits in order to form the base for further understanding of early mining and smelting, as well as geochemistry and the lead isotope analyses of ore, metal and slag with the aim of reconstructing technical processes and studies of provenance.

Faynan is a sedimentary mineralization, which originally had formed one joint ore district with Timna at the western side of the Wadi Arabah. This initial mineralization was parted into two districts by the diverging lateral movement of the African and the Arabian shield along the Jordan rift valley. The main ore deposit in Faynan consists of a 1–2 m thick layer of oxidic copper ores embedded in the shale of the 'Dolomite-Limestone-Shale-Unit' (DLS) of the lower Cambrian. The most important ores are Cu silicates, Cu chlorides and malachite, which are intergrown with Mn oxides and Mn hydroxides. The percentage of trace elements coming into the metal during smelting is very low, with the exception of lead. The ore in vein fillings in the 'Massive-Brown-Sandstone unit' (MBS) is of a lesser extent. They consist of cuprite + limonite (tile-ore), malachite, Cu chlorides, and sparse Cu sulphides. The trace element concentration in the MBS ore is lower by a factor of ten than in the DLS ore. The mineralogy and ore petrography of Faynan are highly characteristic and makes it possible to differentiate it from other deposits in the Levant, even from Timna. The lead isotope ratios of the Faynan share a large range of variation. The lead isotope ratios normalized to ^{206}Pb ratios are higher than those of any other ore mineralizations in the Eastern Mediterranean, except Timna. But on the other hand, some lead isotope ratios are the lowest ever measured in ore from the Mediterranean. The Cu ores from the DLS have a much smaller range in the same field of isotope ratios. They are not different from the MBS ores in this respect. The utilization of Mn-rich ore as fluxing agents for a better quality metal-slag division has, in consequence, not changed the isotopic composition of the lead in the copper. The DLS ore gives another opportunity to differentiate between the Faynan and Timna deposits, as the same formation in Timna was not accessible in ancient times.

Ore from Faynan has been used since the Pre-Pottery-Neolithic period (9th/8th millennium BCE), but at that point only for cosmetic purposes, in the form of powder or as raw material for beads. The actual production of metal from Faynan ores is confirmed for the 5th/4th millennium BCE. It can be proven in the Late Chalcolithic settlements in the Beersheba valley, where small-scale metal production took place at the household level. This points towards a control over metallurgical activities by a 'monopoly' in the Negev. No evidence has been found of the production of metal in Jordan in this period. This is of sociopolitical importance as it shows that the earliest signs of metal production do not come from the area of the deposits but from villages further away. This model seems to work for the Levant but might also be valid for other regions. The production of copper in Faynan and therefore the first really local control of the production chain can only be proven in the second half of the 4th millennium BCE (Early Bronze Age) in Wadi Fidan 4. This site evidenced the first workshop with all necessary steps in the metallurgical production chain. The ore there was mostly mined from the MBS and smelted in small crucibles at the household level. A more industrial copper production became full-fledged for the first time in the Early Bronze Age II/III (mid 3rd millennium BCE), when ore from the DLS was mined using the room and pillar method and smelted in natural draught furnaces. These are the earliest smelting furnaces known so far. The copper was mostly traded to the northwest of Faynan and the main trading posts there were Barqa el-Hetiyeh and Khirbet Hamra Ifdan. These sites were local control posts along the way to the Arabah, and at the same time they were also specialized villages, where crude copper was cast into ingots as well as tools. There are only a few indications of copper production during the Middle Bronze Age. But in the Late Bronze Age and particularly in the Iron Age, copper was produced in truly industrial quantities – this time under the control of the Edomite Kingdom. Mining reaches, with the help of new technologies in shaft construction, hitherto unknown depths (over 70 m) and into new layers of the DLS. By far the largest quantity of the 150 000–200 000 tons of slag found in Faynan stems from this time and is concentrated around only three production centers (Khirbet en-Nahas, Faynan, and Khirbet el-Jariye). This indicates a rather centralized organization of the metal production. In Roman times from around the turn of the era to the 3rd century CE another boom in copper production took place in which, very typically, the old mines were reused again. Faynan presented us with the largest, completely preserved mine, Umm el-Amad, in the entire Roman empire. The mining and smelting of metal during the Mameluk period is rather limited; it is not even clear whether iron or copper were produced.

The spatial distribution and isolation of the slag heaps from different periods allowed not only a very secure dating, it also enabled detailed chemical and mineralogical studies. These studies showed interesting differences in the slag from different time periods, which also allowed further interpretations. The Early Bronze Age I slag has a low melting point and a rather inhomogeneous matrix. It consists chemically either of Fe oxide, SiO_2 and CaO or of CaO, MgO and SiO_2. The precise composition can be directly explained by the choice of ore and the connected host rock (MBS or DLS). Fluxing agents were not used, and the crucible lining hardly influenced the chemistry either. Extremely high concentrations of Cu oxides and copper (up to 60% Cu_2O) demonstrate that high-percentage Cu ores were available at the time. The main components of the slag are Cu/Fe oxides (delafossite, cuprite, magnetite) with richer clinopyroxenes or åkermanite, merwinite, and diopside. Fayalite, normally the main phase in archaeometallurgical slag, has not been found.

The reason for this can be found in the only weakly reducing firing atmosphere, as is typical of the earliest smelting over a wide geographical distribution (it has also been reported from Anatolia). Contrary to the usual theories about early metallurgy, the results of the studies in Faynan showed that the smelting technology of the Early Bronze Age I was still quite simple. The aim of the smelting was not the formation of liquefied slag to gain an easier separation of gangue from the precipitating/developing copper, but only the reduction of copper (to a liquid state). Independent of the degree with which the charged ore was slagged with gangue, the copper was separated by crushing and the single prills and pieces were melted together in a second work process.

In later periods, the large-scale use of DLS ores produced Mn-rich silicate slag with varying percentages of CaO, MgO, etc. The main components of the slag are tephroite, knebelite, Mn-rich pyroxenoids and to a lesser degree, Mn oxides. The chemistry of the slag shows that during the EBA II/III most self-fluxing ores have been mined without any further examination as long as they contained sufficient quantities of Mn oxides (ca. 20–40%). High CaO concentrations indicate a heavy absorption of the ceramic furnace lining. Only during the Iron Age were Mn ores used deliberately as fluxing agents. The Roman slag shows a much higher SiO_2 content than earlier slag. This verifies the field evidence, which already indicated that in Roman times ore was mined again from the MBS. The high range of stability of silicate slag under varying gas atmospheres had positive repercussions for the development of technology. In comparison with Fe-rich compositions, the ore from Faynan was in little danger of reaching temperatures and redox conditions that would lead to a reduction of Mg oxide into metal, which would have resulted in unwanted alloy components in the copper. It also allows the formation of melts even under strongly oxidizing conditions, when Fe-richer slag would have long precipitated magnetite and frozen. This becomes clear by the association of tephroite + cuprite in Early Bronze Age slag.

The Faynan copper is characterized, with the exception of lead and sometimes phosphorus, by a relatively low concentration of trace elements, which sets it apart from the other ore deposits in the Near East. They show variations through time just like the slag and are influenced by other additions; they therefore allow questions of provenance to be discussed. Elements sensitive to oxygen such as Fe, Zn and sometimes lead are also used as indicators of technological processes. The stronger reducing gas atmosphere in Iron Age smelting processes led to a higher concentration of these trace elements in raw copper then in earlier periods. The partition coefficients of trace elements between copper and slag shows that independently of the oxygen partial pressure, As and Ni most strongly precipitate into the metal. Co, on the other hand, coalesces with the slag, which leads to a lower Co-Ni ratio in the metal than in the ore. The lead isotope ratios, however, are identical in ore and metal. Early Bronze Age I copper shows the largest range of variation in these concentrations and forms, together with some of the metal finds from Timna, a separate group. In the Early Bronze Age II and III new depositional layers are opened, and chemically and isotopically 'new kinds of metal' are thus produced, which have a very homogenous composition and at the same time a much higher concentration of trace elements. The copper of the Bronze and Iron Ages is identical with DLS ore; they form one isotopic cluster.

Faynan played an essential part in Levantine metallurgy, which started, compared to Iran and Anatolia, rather late. A 'metal connection' is recognizable, which runs from Faynan westwards up to the modern Gaza strip, possibly even through the Sinai along the Mediterranean coast to the Nile Delta and in the other direction towards Galilee. Faynan was, from the beginning, not the only copper source in the region. The far-reaching connections of the Kura-Araxes culture, which reached from the Transcaucasus far to the south to Bilad el-Sham, made an exchange of metal from the deposits in Anatolia, Armenia and the Caucasus possible.

Examples of this trade can be found in the Cu-As-Sb and Cu-As-Ni alloys, which were used for the skillfully produced copper objects in the hoard find at Nahal Mishmar and several other Late Chalcolithic settlements in the Southern Levant. The alloys are certainly not of local origin although their provenance from the above-mentioned regions can only be assumed. In the Early Bronze Age, other sorts of copper and other metals and alloys appear, whose source might again be in Anatolia and possibly even Central Asia. Towards the Middle Bronze Age, crescent-shaped ingots made from Faynan copper came into many settlements in the Southern Levant.

References

Abdel Motelib A (1987) The Cupriferous Sediments in West Central Sinai. MSc thesis, Cairo University

Abdel Motelib, A (1996) Geological and Mineralogical Studies of Some Manganese Occurrences of Egypt. PhD thesis, Cairo University

Abecht J, Peters TJ (1975) Hydrothermal Synthesis of Pyroxenoids in the System $MnSiO_3$–$CaSiO_3$ at Pf = 2 kb. Contrib Mineral Petrol 50:241–246

Abrecht J, Peters TJ (1980) The Miscibility Gap Between Rhodonite and Bustamite Along the Join $MnSiO_3$–$Ca_{0.60}Mn_{0.40}SiO_3$. Contrib Mineral Petrol 74:261–269

Abs-Wurmbach I, Peters T, Langer K, Schreyer W (1983) Phase Relations in the System Mn–Si–O: an Experimental and Petrological Study. N Jb Miner Abh 146(3):258–279

Abu-Ajamieh MM, Bender F, Eicher RN (1988) Natural Resources in Jordan. National Resources Authority, Amman

Adams R (1991) The Wadi Fidan Project, Jordan 1989. Levant XXII:181–183

Adams R (1992) Romancing the Stones: New Light on Glueck's 1934 Survey of Eastern Palestine as a Result of Recent Work by the Wadi Fidan Project. In: Bienkowski P (ed) Early Edom and Moab. Sheffield Archaeol Monogr 7:177–186

Adams R (1997) On Early Copper Metallurgy in the Levant: A Response to Claims of Neolithic Metallurgy. In: Gebel HG, Kafafi Z, Rollefson G (eds) The Prehistory of Jordan II. Ex Oriente, Berlin, pp 651–656

Adams R (1998) The Development of Copper Metallurgy During the Early Bronze Age of the Southern Levant: Evidence from the Feinan Region, Southern Jordan. Dissertation, University of Sheffield

Adams R (2003) External Influences at Faynan during the Early Bronze Age: A Re-Analysis of Building I at Barqa el-Hetiye, Jordan. Palestine Exploration Quarterly 135(1):6–21

Adams R, Genz H (1995) Excavations at Wadi Fidan 4: A Chalcolithic Village Complex in the Copper Ore District of Feinan, Southern Jordan. Palestine Explor Quart 127:8–20

Affonso MTC (1997) Mineralogische und geochemische Untersuchungen von Kleinplastik und Baumaterialien aus dem akeramischen Neolithikum des Nahen Ostens. Dissertation, Naturwiss.-Mathemat. Fakultät, University of Heidelberg

Agricola G (1556) De Re Metallica Libri XII. Translated and annotated by Frauenstadt G, Prescher H, Deutscher Verlag der Wissenschaften Berlin 1974

Aharoni Y, Evenari M, Shanan L, Tadmor NH (1960) The Ancient Desert Agriculture of the Negev, V: An Israelite Settlement at Ramad Matred. Israel Explor J 10:23–36, 97–111

Al-Shanti AMS, Mitchell AHG (1976) Late Precambrian Subduction and Collision in the Al-Amar-Idsa Region, Arabian Shield, Kingdom of Saudi Arabia. Tectonophysics 30(3,4):T41–43

Alimov K, Boroffka N, Bubnova M, Burjakov J, Cierny J, Jakubov J, Lutz J, Parzinger H, Pernicka E, Radililovskij V, Ruzanov V, Širinov T, Staršinin D, Weisgerber G (1998) Zinnbergbau in Mittelasien. Eurasia Antiqua 4:137–199

Amar Z (1997) Gold Production in the 'Arabah Valley in the Tenth Century. Israel Exploration J 47:1–2, 100–103

Amiran R (ed) (1978) Early Arad: The Chalcolithic Settlement and Early Bronze Age City. Israel Explor Soc, Jerusalem

References

Amiran DHK (1991) The Climate of the Ancient Near East: The Early Third Millennium B.C. in the Northern Negev of Israel. Erdkunde 45(3):153–162

Amiran R, Beith-Arieh I, Glass J (1973) The Interrelationship between Arad and Sites in Southern Sinai in the Early Bronze Age II. Israel Explor J 23:194–197

Amiran R, Ilan O, Sebanne M (forthcoming) Early Arad, Vol. III, Finds from the 6th to the 18th Seasons, 1971–1978 and 1980–1984, Jerusalem

Anfinset N (1996) Social and technological aspects of mining, smelting and casting copper. An ethnoarchaeological study from Nepal. MA thesis, Dept Archaeology, University of Bergen, Norway

Artzy M (1983) Arethusa of the Tin Ingot. Bull. Amer School of Oriental Research 250:51–55

Aurenche O, Évin J (1987) Radiocarbon Dates List. In: Aurenche O, Évin J, Hours F (eds) Chronologies in the Near East. Brit Archaeol Records, Intern Ser, Oxford 379:691–736

Avery DH, Schmidt P (1979) A Metallurgical Study of the Iron Bloomery, Particularly as Practised in Buhaya. J Metals 31:14–20

Avner U, Carmi I, Segal D (1994) Neolithic to Bronze Age Settlement of the Negev and Sinai in Light of Radiocarbon Dating: A View from the Southern Negev. In: Bar-Yosef O, Kra R (eds) Late Quaternary Chronology and Paleoclimates of the Eastern Mediterranean. Radiocarbon, Tucson and Cambridge, pp 265–300

Ayalon A, Beyth M, Vulkan U (1985) Geochemistry of Radioactive Mineralization Occurrences in the Timna Valley. Trans Inst Mining Metallurgy B 94:197–201

Bachmann HG (1978a) Schlacken: Indikatoren archäometallurgischer Prozesse. In: Hennicke HW (ed) Mineralische Rohstoffe als kulturhistorische Informationsquelle. Hagen, Verlag Deutscher Emailfachleute, pp 66–103

Bachmann HG (1978b) The Phase Composition of Slags from Timna Site 39. In: Rothenberg B (ed) Archaeometallurgy: Chalcolithic Copper Smelting. Inst Archaeo-Metall Studies, Monograph 1, London, pp 21–23

Bachmann HG (1980) Early Copper Smelting Techniques in Sinai and in the Negev as Deduced from Slag Investigations. In: Craddock PT (ed) Scientific Studies in Early Mining and Extractive Metallurgy. Brit Museum Occ Papers 20, London, pp 103–134

Bachmann HG (1981) Identification of a Copper Mineral from Wadi Ghazzeh, Israel. In: Roshwalb A (ed) Protohistory in the Wadi Ghazzeh: A Typological and Technological Study Based on the Macdonald Excavations. Dissertation, Institute of Archaeology, University of London

Bachmann HG (1982) Copper Smelting Slags from Cyprus: Review and Classification of Analytical Data. In: Muhly JD, Maddin R, Karageorghis V (eds) Early Metallurgy in Cyprus, 4000–500 BC. Acta Intern Archaeolog Symp Larnaca 1981, Nicosia, pp 143–152

Bachmann HG (1993) Zur Metallurgie der römischen Goldgewinnung in Três Minas und Campo de Jales in Nordportugal. In: Steuer H, Zimmermann U (ed) Montanarchäologie in Europa: Archäologie und Geschichte – Freiburger Forschungen zum ersten Jahrtausend in Südwestdeutschland, Bd. 4, Thorbecke, Sigmaringen, pp 153–160

Bachmann HG, Hauptmann A (1984) Zur alten Kupfergewinnung in Fenan und Hirbet en-Nahas im Wadi Arabah in Südjordanien. Ein Vorbericht. Der Anschnitt 35(4):110–123

Bachmann HG, Rothenberg B (1980) Die Verhüttungsverfahren von Site 30. In: Conrad HG, Rothenberg B (eds) Antikes Kupfer im Timna-Tal. Der Anschnitt Beiheft 1, Deutsches Bergbau-Museum Bochum, pp 215–236

Bachmann HG, Lupu A, Rothenberg B, Tylecote RF (forthcoming) Early Copper Metallurgy in the Sinai Peninsula. In: Naim G, Rothenberg B (eds) New Researches in Sinai: Archaeology and Archaeometallurgy. London

Baierle HU (1993) Vegetation und Flora im südwestlichen Jordanien. Dissertationes Botanicae 200, Borntraeger, Berlin, Stuttgart

Baierle HU, Frey W, Jagiella C, Kürschner H (1989) Die Brennstoffressourcen im Raum Fenan (Wadi Arabah, Jordanien) und die bei der Kupfererzverhüttung verwendeten Brennstoffe. In: Hauptmann A, Pernicka E, Wagner GA (eds) Archäometallurgie der Alten Welt/Old World Archaeometallurgy. Der Anschnitt, Beiheft 7, Deutsches Bergbau-Museum Bochum, pp 213–222

Ball MW, Ball D (1953) Oil prospects of Israel. Bull Amer Ass Petroleum Geol 37:1–113

Bamberger M, Wincierz P (1990) Ancient Smelting of Oxide Copper Ore. In: Rothenberg B (ed) Researches in the Arabah 1959–1984, Vol. II: The Ancient Metallurgy of Copper. Inst. Archaeo-Metall. Studies, London, pp 123–157

Bamberger M, Wincierz P, Bachmann HG, Rothenberg B (1986) Ancient Smelting of Oxide Copper Ore: Archaeological Evidence at Timna and Experimental Approach. Metallwissenschaft + Technik 40:1166–1174

Bar-Adon P (1980) The Cave of the Treasure. Israel Explor Soc, Jerusalem

Bar-Matthews M (1987) The Genesis of Uranium in Manganese and Phophorite Assemblages, Timna Basin, Israel. Geol Magazin 124:211–229

Bar-Yosef O (1995a) Earliest Food Producers – Pre Pottery Neolithic (8000–5500). In: Levy TE (ed) The Archaeology of Society in the Holy Land. Facts on File, New York, pp 190–204

Bar-Yosef O (1995b) The Origins of Modern Humans. In: Levy TE (ed) The Archaeology of Society in the Holy Land. New York, Facts on File, pp 110–123

Bar-Yosef O, Alon D (1988) Nahal Hemar Cave, the Excavations. Atiqot XVIII:1–30

Barker GW, Creighton OH, Gilbertson DD, Hunt CO, Mattingly DJ, McLaren SJ, Thomas DC (1997) The Wadi Faynan Project, Southern Jordan: a Preliminary Report on Geomorphology and Landscape Archaeology. Levant XXIX:19–40

Barker GW, Adams R, Creighton OH, Gilbertson DD, Grattan JP, Hunt CO, Mattingly DJ, McLaren SJ, Mohamed HA, Newson P, Reynolds TEG, Thomas DC (1998) Environment and Land Use in the Wadi Faynan, Southern Jordan: The Second Season of Geoarchaeology and Landscape Archaeology (1997). Levant XXX:5–25

Bartels C, Bingener A, Slotta R (2006) Das Schwazer Bergbuch. Veröffentlichungen aus dem Deutschen Bergbau-Museum Bochum 142

Bartels C, Fessner M, Klappauf L, Linke FA (2007) Kupfer, Blei und Silber aus dem Goslarer Rammelsberg von den Anfängen bis 1620. Veröffentlichungen aus dem Deutschen Bergbau-Museum Bochum 151

Bartov J (1974) A Structural and Paleogeographic Study of the Central Sinai Faults and Domes. Dissertation, Dept of Geology, Hebrew University, Jerusalem

Bartura Y (1966) Type Sections of Paleozoic Formations in the Timna Area. Geol Surv Israel 3/66

Bartura Y, Würzburger U (1974) The Timna Copper Deposit. In: Bartholome P (ed) Gise-ment Stratiformes et Provinces Cuproferes. Societe Geologique de Belgique, Liege, pp 277–285

Bartura Y, Hauptmann A, Schöne-Warnefeld G (1980) Zur Mineralogie und Geologie der antik genutzten Kupferlagerstätte im Timna-Tal. In: Conrad HG, Rothenberg B (eds) Antikes Kupfer im Timna-Tal. Der Anschnitt, Beiheft 1, Deutsches Bergbau-Museum Bochum, pp 41–56

Bassiakos Y, Catapotis M (2006) Reconstruction of the Copper Smelting Process at Chrysokamino Based upon the Analysis of Ore and Slag Samples. In: Betancourt P (ed) The Chrysokamino Metallurgy Workshop and its Territory. Hesperia Suppl 36:329–353

Bassiakos Y, Philaniotou-Hadjianastasiou O (in press) Early copper production on Kythnos: archaeological evidence, material and analytical reconstruction of metallurgical processes. In: Day PM, Doonan R (eds) Sheffield Studies in Aegean Archaeology. Oxbow Press, Oxford

Basta EZ, Sunna BF (1971) Mineralogy and Mode of Occurrence of Copper Ores in Wadi Araba, Jordan. Extr Bull Inst Egypt, Tome LII:199–224

Basta EZ, Sunna BF (1972a) Petrological Studies on Some Precambrian Igneous Rocks from Feinan District, Wadi Araba, Jordan. Bull Fac Science Univ Cairo 44:175–193

Basta EZ, Sunna BF (1972b) The Manganese Mineralisation at Feinan District, Jordan. Bull Fac Science Univ Cairo 44:111–126

Baumgarten Y, Eldar I (1984) Neve Noy – A Chalcolithic Site Near Beer-Sheba. Qadmoniot 17:51–56 (Hebrew)

Bawden G (1989) Midian, Moab and Edom. The History and Archaeology of Late Bronze Age and Iron Age Jordan and Northwest Arabia. Sheffield JSOT Suppl. Series 24:37–52

Begemann F, Schmitt-Strecker S (forthcoming) Die Metallindustrie Mesopotamiens von den Anfängen bis zum 2. Jahrtausend v. Chr. 5. Materialanalysen. Untersuchungen zur Herkunft der Rohstoffe

Begemann F, Pernicka E, Schmitt-Strecker S (1994) Metal Finds from Ilipinar and the Advent of Arsenical Copper. Anatolica XX:203–219

References

Begemann F, Pernicka E, Schmitt-Strecker S (1995) Thermi on Lesbos: a case study of changing trade patterns. Oxf J Archaeology 14:123–136

Begemann F, Schmitt-Strecker S, Pernicka E (1992) The Metal Finds from Thermi III–IV: a Chemical and Lead Isotope Study. Studia Troica II:219–239

Beit-Arieh I (1983) Central-Southern Sinai in the Early Bronze Age II and its Relationship with Palestine. Levant XV:39–48

Beit-Arieh I (1985) Serabit el-Khadim: New Metallurgical and Chronological Aspects. Levant XVII:89–116

Belgiorno MR (2000) Project "Pyrame" 1998–1999: Archaeological, metallurgical and historical evidence at Pyrgos (Limassol). Report Dept Antiquities Cyprus, pp 1–17

Belli O (1991) Ore Deposits and Mining in Eastern Anatolia in the Urartian Period: Silver, Copper and Iron in Urartu. In: Merhav R (ed) Urartu – A Metalworking Center in the First Millenium B.C., Jerusalem, The Israel Museum Cat no 324, pp 14–41

Ben-Tor A (1992) The Early Bronze Age. In: Ben-Tor A (ed) The Archaeology of Ancient Israel. The Open Univ Israel, pp 81–125

Bender F (1965) Zur Geologie der Kupfererzvorkommen am Ostrand des Wadi Araba, Jordanien. Geol Jb 83:181–203

Bender F (1968) Geologie von Jordanien. Beiträge zur Regionalen Geologie der Erde, Bd. 7, Borntraeger, Berlin, Stuttgart

Bender F (1974) Explanatory Notes on the Geological Map of the Wadi Araba, Jordan (Scale 1:100000, 3 sheets). Geol Jb B10:1–62

Bentor YK (1956) The Manganese Occurrence of Timna (Southern Israel), a Lagoonal Deposit. Symp del Manganeso XX Congr Geol Int Mexico, pp 159–172

Bentor YK (1985) The Crustal Evolution of the Arabo-Nubian Massif with Special Reference to the Sinai Peninsula. Precambrian Res 28:1–74

Berezhnoi AS, Karyakin LI, Dudavskii IE (1952) Doklady Akad. Nauk. S.S.R., 83, 339 cited in Gadalla AMM 1962/63

Betancourt P (2006) Other Metallurgical Materials. In: Betancourt P (ed) The Chrysokamino Metallurgy Workshop and its Territory. Hesperia Suppl 36:137–147

Beyschlag F, Vogt JH, Krusch P (1914–1916) The Deposits of Useful Minerals and Rocks. MacMillan, London

Beyth M (1987) Mineralization Related to Rift Systems: Examples from the Gulf of Suez and the Dead Sea Rift. Tectonophysics 141:191–197

Beyth M, Segev A (1983) Lower Cretaceous Basaltic Plug in the Timna Valley. Israel J Earth Science 32:165–166

Beyth M, Longstaff FJ, Ayalon A, Matthew A (1997) Epigenetic Alteration of the Precambrian Igneous Complex of Mount Timna, Southern Israel: Oxygen-Isotope Studies. Israel J Earth Science 46:1–11

Bigot M (1975) Geology and Ore Characterization. Wadi Araba Copper Project – Final Report vol IVA. Unpubl report Bureau de Recherches Geologique et Minières (BRGM), Orléans

Bintliff JL (1982) Paleoclimate Modelling of Environmental Changes in the East Mediterranean Region since the LastGlaciation. In: Bintcliff J, van Zeist W (eds) Paleoclimates, Paleoenvironments and Human Communities in the Eastern Mediterranean Region in Later Prehistory. BAR Intern Ser, Oxford, 133.1:485–527

Biringuccio V (1540) Pirotechnia. Translated by H.S. Mudd 1942. Reissued 1959, Massachusetts Institute of Technology, Cambridge, Mass

Bisson MS (2000) Precolonial Copper Metallurgy: Sociopolitical Context. In: Vogel J (ed) Ancient African Metallurgy, pp 83–145. Altamira-Press, Walnut-Creek

Blake GS (1930) The Mineral Resources of Palestine and Transjordan. Palestine, Vol. 2. Printing & Stat. Off., Jerusalem

Blanckenhorn M (1912) Naturwissenschaftliche Studien am Toten Meer und im Jordantal. Friedländer, Berlin

Bokhari FY, Kramers JD (1982) Lead Isotope Data from Massive Sulfide Deposits in the Saudi Arabian Shield. Scientific Comm, pp 1766–1769

Boni M, Koeppel V (1985) Ore-Lead Isotope Pattern from the Iglesiente-Sulcis Area (SW Sardinia) and the Problem of Remobilization of Metals. Mineral Deposita 20:185–193

Boom G van den, Rösch H (1969) Modalbestand und Petrochemie der Granite im Gebiet von Aqaba-Quweira, Südjordanien. Beih Geol Jb 81:113–148

Bottinga Y, Weil D (1972) The Viscosity of Magmatic Silicate Liquids: a Model for Calculation. Amer J Science 272:438–475

Bowen N L, Schairer JF, Posnjak E (1933) The System $CaO-FeO-SiO_2$. Amer J Science 5th Ser. 26:193–284

Branigan K, McKerell H, Tylecote RF (1976) An Examination of Some Palestinian Bronzes. J Hist Metall Soc 10(1):5–23

Brill RH, Barnes IL (1988) The Examination of some Egyptian Glass Objects. In: Rothenberg B (ed) The Egyptian Mining Temple at Timna. Researches in the Arabah 1959–1984, I. Inst Archaeo-Metall Studies, London, pp 217–222

Brill RH, Cahill N (1988) A Red Opaque Glass from Sardis and Some Thoughts on Red Opaques in General. J Glass Studies 30:16–27

Brown GE (1982) Olivines and Silicate Spinels. In: Ribbe P (ed) Orthosilicates. Reviews in Mineralogy, Vol. 5, Mineralog Soc America, pp 275–381

Brownell WE (1976) Structural Clay Products. Applied Mineralogy, Vol. 9. Springer-Verlag, Wien/New York

Brückner H, Eichmann R, Herling L, Kallweit H, Kerner S, Khalil L, Miqdadi R (2002) Chalcolithic and Early Bronze Age Sites near Aqaba, Jordan. In: Eichmann R (ed) Ausgrabungen und Surveys im Vorderen Orient I. Orient-Archäologie, Bd. 5:217–331

Bruins H (1994) Comparative Chronology of Climate and Human History in the Southern Levant from the Late Chalcolithic to Early Arabic Period. In: Bar-Yosef O, Kra R (ed) Late Quaternary Chronology and Paleoclimates of the Eastern Mediterranean. Radiocarbon, Tucson & Cambridge, pp 301–314

Buchholz HG (1967) Analysen prähistorischer Metallfunde aus Zypern und den Nachbarländern. Berliner Jb Vor- u Frühgeschichte 7:189–XXXX

Buchwald F, Leisner P (1990) A Metallurgical Study of 12 Prehistoric Bronze Objects from Denmark. J Danish Archaeology 9:64–102

Budd P (1991) The Properties of Arsenical Copper Alloys: Implications for the Development of Eneolithic Metallurgy. In: Budd P, Chapman B, Jackson C, Janaway R, Ottaway B (eds) Archaeological Sciences '89. Oxbow Monographs 9, pp 132–142

Budd P (1992) Alloying and Metalworking in the Copper Age of Central Europe. Bull Metals Museum, Sendai 17:3–14

Budd P, Gale N, Pollard M, Thomas RG, Williams PA (1992) The Early Development of Metallurgy in the British Isles. Antiquity 66:677–686

Budd P, Gale D, Pollard M, Thomas RG, Williams PA (1993) Evaluating Lead Isotope Data: Further Observations. Archaeometry 35:225–240

Budd P, Pollard M, Scaife B, Thomas RG (1995) Oxhide Ingots, Recycling and the Mediterranean Metals Trade. J Mediterran Archaeol 8(1):1–32

Burgath KP, Hagen D, Siewers U (1984) Geochemistry, Geology, and Primary Copper Mineralization in Wadi Araba, Jordan. Geol Jb B53:3–53

Burleigh R, Hewson A (1979) British Museum Radiocarbon Measurements XI. Radiocarbon 21(3):349

Burleigh R, Matthews K (1982) British Museum Radiocarbon Measurements XIII. Radiocarbon 24(2):165

Burton RF (1878/1979) The Gold-mines of Midian and the Ruined Midianite Cities. Reprint Falcon-Oleander, London, 238 p

Burton RF, Drake CF (1872) Unexplored Syria 2. London

Çagatay A (1978) Genetische Ergebnisse einer geologisch-mineralogischen Untersuchung der Kupfererzlagerstätten und -vorkommen in Südostanatolien. Bull Mineral Research Explor Inst Turkey 89, pp 48–74

Çagatay A (1987) The Pancarli Nickel-Copper Sulfide Mineralization, Eastern Turkey. Miner Deposita 22:163–171

References

Caneva CMF, Palmieri AM (1983) Metalwork at Arslantepe in Late Chalcolithic and Early Bronze I: The Evidence from Metal Analyses. Origini XII:637–653

Caneva CMF, Palmieri AM (1985) I metalli di Arslantepe nel quadro dei più antichi sviluppi della metallurgia vicino-orientale. Quad "La ricerca scientifica", pp 112–137

Castel G, Pouit G (1997) Anciennes mines métalliques dans la partie Nord désert oriental d'Egypte. Archéo-Nil 7:101–112

Castel G, Köhler C, Mathieu B, Pouit G (1998) Les mines du ouadi Um Balad. Bull de L'Institute Francais d'Archéologie Orientale 98:57–87

Castel G, Mathieu B, Pouit G, El Hawari MA, Shaaban G, Hellal H, Abdallah T, Ossama A (1995) Wadi Dara Copper Mines. In: Feisal A (ed) Proc 1st Intern Conf Ancient Egyptian Mining and Metallurgy and Conservation of Metallic Artifacts. Ministry Cult Cairo, April 1995, pp 15–31

Charles JA (1967) Early Arsenical Bronzes – A Metallurgical View. Amer J Archaeol 71:21–26

Charles JA (1980) The Coming of Copper and Copper-Base Alloys and Iron: A Metallurgical Sequence. In: Wertime TA, Muhly JD (eds) The Coming of the Age of Iron. Yale University Press, New Haven, London, pp 151–181

Chernykh E (1992) Ancient Metallurgy in the USSR. Cambridge University Press, Cambridge

Childe VG (1951) Man makes himself. New Amer. Library, New York

Cierny J, Weisgerber G, Perini R (1992) Ein spätbronzezeitlicher Hüttenplatz in Bedello/Trentino. Universitätsforschungen zur Prähistorischen Archäologie, Bd. 8, pp 97–105

Cierny J, Hauptmann A, Hohlmann B, Marzatico F, Schröder B, Weisgerber G (1995) Endbronzezeitliche Kupferproduktion im Trentino. Der Anschnitt 47(3):82–92

Cina A (1988) Sulfide and Arsenide Mineralization in Ophiolite Ultramafic and Gabbroic Rocks of Albanides. Ofioliti 13:85–87

Coghlan HH (1939/40) Prehistoric Copper and Some Experiments in Smelting. Trans Newcomen Soc XX:49–65

Conrad HG, Rothenberg B (eds) (1980) Antikes Kupfer im Timna-Tal. Der Anschnitt, Beiheft 1, Deutsches Bergbau-Museum Bochum

Constantinou G (1980) Metallogenesis Associated with the Troodos Ophiolite. In: Panayiotou A (ed) Ophiolites. Proc Intern Ophiolite Symp Cyprus 1979. Nicosia, pp 663–674

Constantinou G (1982) Geological Features and Ancient Exploitation of the Cupriferous Sulphide Orebodies of Cyprus. In: Muhly JD, Maddin R, Karageorghis V (eds) Early Metallurgy in Cyprus, 4000–500 BC. Acta Intern Archaeolog Symp, Larnaca 1981. Nicosia, pp 13–24

Coogan MD (1984) Numeira 1981. Bull Amer School Oriental Res 255:75–81

Cooke SRB, Aschenbrenner S (1975) The Occurrence of Metallic Iron in Ancient Copper. J Field Archaeology 2:251–266

Cooke SRB, Nielsen BV (1978) Slags and Other Metallurgical Products. In: Rapp G, Aschenbrenner SE (eds) Excavations at Nichoria in Southwest Greece I. Univ Minnesota Press, Minneapolis, pp 182–326

Costa PM (1978) The Copper Mining Settlement of Arja: a Preliminary Survey. J Oman Studies 4:9–14

Costa PM, Wilkinson TJ (1987) The hinterland of Sohar. Archaeological surveys and excavations within the region of an Omani seafaring city. J Oman Studies 9:10–238

Costin CL (1991) Craft specialization: issues in defining, documenting, and explaining the organization of production. In: Schiffer MB (ed) Archaeological Method and Theory 3. University Press Tucson, Arizona, pp 1–56

Craddock PT (1976) The Composition of the Copper Alloys Used by the Greek, Etruscan and Roman Civilization. J Archaeol Science 3:93–113

Craddock PT (1986) The Metallurgy and Composition of Etruscan Bronze. Studi Etruschi 52:211–271

Craddock PT (1989) The Scientific Investigation of Early Mining and Metallurgy. In: Henderson J (ed) Scientific Analysis in Archaeology. Oxf Univ Comm Archaeol Monograph 19, pp 178–212

Craddock PT (1990) Copper Smelting in Bronze Age Britain: Problems and Possibilities. In: Crew P, Crew S (eds) Early Mining in the British Isles. Plas Tan y Bwlch Occ Papers 1:69–71

Craddock PT (1995) Early Metal Mining and Production. Edinburgh University Press, Edinburgh

Craddock PT (2000) From Hearth to Furnace: Evidence for the Earliest Metal Smelting Technologies in the Eastern Mediterranean. In: Hauptmann A (coord.) Early Pyrotechnology. The Evolution of the First Fire-Using Industries. Paléorient XXVI:151–165

Craddock PT, Freestone I (1988) Debris from Metallurgical Activities at Site 200. In: Rothenberg B (ed) Researches in the Arabah 1959–1984, I: The Egyptian Mining Temple at Timna. Inst Archaeo-Metallurg Studies, London, pp 192–203

Craddock PT, Meeks N (1987) Iron in Ancient Copper. Archaeometry 29:187–204

Craddock PT, Freestone IC, Gale N, Meeks N, Rothenberg B, Tite M (1985) The Investigation of a Small Heap of Silver Smelting Debris from Rio Tinto, Huelva, Spain. In: Craddock PT, Hughes MJ (eds) Furnaces and Smelting Technology in Antiquity. Brit Museum Occ Papers 48:199–214

Crew P (1991) The Iron and Copper Slags at Baratti, Populonia, Italy. J Hist Metall Soc 25:109–115

Crew P, Crew S (Ed.) (1990) Early Mining in the British Isles. Plas Tan y Bwlch Occ Papers 1

Çukur A, Kunç S (1989) Analyses of Tepeçik and Tülintepe Metal Artifacts. Anatolian Studies XXXIX:113–120

D'Anville (1732) Patriarchatus Hierosolymitanus. Paris 1987 (map)

Davies O (1979) Roman Mines in Europe. Arno Press, New York

De Jesus PS (1980) The Development of Prehistoric Mining and Metallurgy in Anatolia. BAR Intern Ser, Oxford, 74

Deer WA, Howie RA, Zussman J (1982) Rock-Forming Minerals, vol 1A: Orthosilicates. Longman: London, New York

Delaloye M, De Souza H, Wagner JJ, Hedley I (1980) Isotopic Ages on Ophiolites from the Eastern Mediterranean. In: Panayiotou A (ed) Ophiolites. Proc Intern Ophiolite Symp Cyprus 1979. Nicosia, pp 292–295

Delibes G, Fernandez M, Fernandez Posse MD, Martin C, Montero I, Rovira S (1991) Almizaraque, Alméria, Spain. In: Mohen JP, Éluère C (eds) Découverte du Metal. Picard, Paris, pp 303–316

Dever WG, Tadmor M (1976) A Copper Hoard from the Middle Bronze Age. Israel Explor J 26:163–173

Doe BR, Zartmann RE (1979) Plumbotectonics, The Phanerozoic. In: Barnes HL (ed) Geochemistry of Hydrothermal Ore Deposits. John Wiley, New York, pp 22–70

Doelter C, Leitmeier H (1926) Handbuch der Mineralchemie III, 2. Steinkopff, Dresden und Leipzig

Dollfus G, Kafafi Z (1988) Abu Hamid: Village du 4e Mill. de la Vallee du Jourdain. Centre Culturel Francais et Department des Antiquites de Jordanie, Paris

Donnan CB (1973) A Precolumbian Smelter from Northern Peru. Archaeology 26:289–297

Doonan RCP, Day PM, Dimopoulou N, Kilikoglou V (forthcoming) Lame excuses for Emerging Complexity in Early Bronze Age Crete: The Metallurgical Finds from Poros Katsambas. Sheffield Studies in Aegean Archaeology

Dothan M (1959) Excavations at Meser. Israel Explor J 9:13–29

Dougherty RC, Caldwell JR (1966) Evidence of Early Pyrometallurgy in the Kerman Range. Iran. Science 153:27–40

Driehaus J (1965) "Fürstengräber" und Eisenerze zwischen Mittelrhein, Mosel und Saar. Germania 43:32–49

Dubertret L (1962) Carte Géologique Liban, Syrie et bordure des pays voisins. Muséum Nat d'Hist, Paris

Duell P (1938) The Mastaba of Mereruka. Chicago

Eaton ER, McKerrell H (1976) Near Eastern Alloying and some Textural Evidence for the Early Use of Arsenical Copper. World Archaeology 8:169–192

Eckstein K, Hauptmann A, Rehren T, Richter U, Schwabenicky W (1994) Hochmittelalterliches Montanwesen im Sächsischen Erzgebirge und seinem Vorland. Der Anschnitt 46(4,5):114–132

El Sharkawi MA, El Aref MM, Abdel Motelib A (1990) Syngenetic and Paleokarstic Copper Mineralization in the Paleozoic Platform Sediments of West Central Sinai, Egypt. Spec Publ Int Ass Sediment 11:159–172

El-Shazley EM (1979) Formation of Old Plates at the Precambrian-Phanerozoic Boundary in Northern Africa and South-Western Asia. In: Al Shanti AMS (ed) Proc Symp Evolution and Mineralization of the Arabian-Nubian Shield, Jeddah 1978, 2. Pergamon Press, Oxford, pp 173–186

El-Shazley EM, Abdel Naser S, Shukri B (1955) Contribution to the Mineralogy of the Copper Ore Deposits in Sinai. Pap Geol Survey Egypt 1:1–13

Elliott JF (1976) Phase Relationships in the Pyrometallurgy of Copper. Metall Trans 7B:17–33

Engel T (1992) Nouvelles recherches sur les charbons de bois du Chalcolithique, de l'age du Bronze et de l'age du Fer de Fenan, Wadi Araba, Jordanie. Bull Soc Bot Fr 139, Actual. bot. 2/3/4, pp 553–563

Engel T (1993) Charcoal Remains from an Iron Age Copper Smelting Slag Heap at Feinan, Wadi Arabah (Jordan). Vegetation, History and Archaeobotany 2:205–211

Engel T (1995) Holzkohlen aus dem Vorderen Orient: Jahrtausend-alte Reste einstiger Wälder. Das Altertum 40:311–320

Engel T, Frey W (1996) Fuel Resources for Copper Smelting in Antiquity in Selected Woodlands in the Edom Highlands to the Wadi Araba/Jordan. Flora 191:29–39

Engel T, Frey W, Kürschner H (in prep) Botanische Untersuchungen des Brennstoffvorrats für die antike Kupfergewinnung in Fenan

Epstein S, Shimada I (1983) Metalurgia de Sicán. Una reconstrucción de la producción de la aleación de Cobre en el Cerro de los Cementerios, Perú. Beitr Allg u Vergl. Archäologie 5:380–430

Eshel I (1990) Timna. Excavations and Surveys in Israel 9:171–172

Esin U (1986) Dogu Anadolu'ya ait Bazi Prehistorik Cüruf ve Filiz Analizleri. Anadolu Arastirmalari X (Jb Kleinasiatische Forschung in Memoriam V. Bahadir Alkim, vol 10), Istanbul, pp 143–160

Esin U (1988) Degirmentepe (Malatya) Kutarma Kazisi, IX Kazi Sonuclari Toplantisi I, pp 79–125

Esin U (1996) Asikli, Ten Thousand Years Ago: a Habitation Model from Central Anatolia. In: Housing and Settlement in Anatolia: A Historical Perspective. Habitat, Istanbul, pp 31–42

Espelund A (1991) A Retrospective View of DirectIron Production. In: Espelund A (ed) Bloomery Ironmaking During 2000 Years, vol I. Trondheim, pp 71–99

Eugster HP, Wones DR (1962) Stability Relations of the Ferruginous Biotite, Annite. J Petrology 3:82–125

Evans AM (1992) Erzlagerstättenkunde. Enke, Stuttgart

Fall P (1990) Deforestation in Southern Jordan: Evidence from Fossil Hyrax Middens. In: Bottema S, Entjes-Nieborg G, van Zeist W (eds) Man's Role in the Shaping of the Eastern Mediterranean Landscape. Balkema, Rotterdam, pp 271–281

Fasnacht W (1992) Excavations at Ayia Varvara-Almyras. Fourth Preliminary Report. Reports Dept. Antiquities, Cyprus, pp 59–74

Fasnacht W, Peege C, Hedley I (2000) Agia Varvara–Almyras – Final Excavation Report. Report of the Department of Antiquities, Cyprus, pp 101–115

Finkelstein I (1990) Early Arad – Urbanism of the Nomads. Zeitschr Deutsch Palästina-Verein 106:34–50

Flörke OW (1959) Über Kieselsäurekristalle in Gläsern. Glastechnische Berichte 32(1):1–9

Foose MF, Economou M, Panayiotou A (1985) Compositional and Mineralogical Constraints on the Genesis of Ophiolithe Hosted Nickel Mineralizations in the Pevkos Area, Limassol Forest, Cyprus. Miner. Deposita 20:234–240

Forbes RJ (1971) Studies in Ancient Technology 8. Brill, Leiden

France-Lanord A, De Contenson H (1969) Une pendeloque en cuivre natif de Ramad. Paleorient 1:107–115

Frank F (1934) Aus der Araba I. Zeitschr Deutsch Palästina-Verein 57:191–280

Freestone I (1987) Compositon and Microstructure of Early Opaque Red Glass. In: Bimson M, Freestone I (eds) Early Vitreous Materials. Brit Museum Occ Paper 57:173–191

French DH (1962) Excavations at Can Hasan. First Preliminary Report, 1961. Anatolian Studies XII:27–40

Frifelt K (1995) The Island of Umm an-Nar 2: The Third Millennium Settlement. Jutland Archaeol Soc Publ, Århus

Fritz V (1985) Einführung in die Biblische Archäologie. Wiss. Buchgesellschaft, Darmstadt

Fritz V (1994a) Eine neue Bauform der Frühbronzezeit in Palästina. In: Dietrich M, Loretz O (eds) Beschreiben und Deuten in der Archäologie des Alten Orients. Altertumskunde Vord. Orient 4 (Festschr. R. Mayer-Opificius), Ugarit-Verlag, Münster, pp 85–92

Fritz V (1994b) Vorbericht über die Grabungen in Barqa el-Hetiye im Gebiet von Fenan, Wadi el-Araba (Jordanien) 1990. Zeitschr Deutsch Palästina-Verein 110(2):125–150

Fritz V (1996) Ergebnisse einer Sondage in Hirbet en-Nahas, Wadi el-Araba (Jordanien). Zeitschr Deutsch Palästina-Verein 112(1):1–9

Fröhlich K (1953) Goslarer Bergrechtsquellen des früheren Mittelalters, insbesondere das Bergrecht des Rammelsberges aus der Mitte des 14. Jahrhunderts. Giessen

Fu P, Kong Y, Zhang L (1982) Domain Twinning of Laihunite and Refinement of ist Crystal structure. Geochemistry (Beijing) 1, 115–133. In: Kan X, Coey JMD (eds) Mössbauer Spectra, Magnetic and Electrical Properties of Laihunite, a mixed Valence Iron Olivine Mineral. Amer Miner 70, 1985:576–580

Gadalla AMM, Ford WF, White J (1962/63) Equilibrium Relationships in the System CuO–Cu$_2$O–SiO$_2$. Trans British Ceramic Soc 62:45–66

Gal Z, Smithline H, Shalem D (1997) A Chalcolithic Burial Cave in Peqi'in, Upper Galilee. Israel Explor J 47:145–154

Gale NH (1989) Archaeometallurgical Studies of Late Bronze Age Ox-Hide Copper Ingots from the Mediterranean Region. In: Hauptmann A, Pernicka E, Wagner GA (eds) Old World Archaeometallurgy. Der Anschnitt, Beiheft 7, Deutsches Bergbau-Museum Bochum, pp 247–268

Gale NH (1991a) Copper Oxhide Ingots: Their Origin and Their Place in the Bronze Age Metals Trade in the Mediterranean. In: Gale NH (ed) Bronze Age Trade in the Mediterranean. Studies Mediterran Archaeology XC:197–239

Gale NH (1991b) Metals and Metallurgy in the Chalcolithic Period. Bull Americ School Oriental Res 282/283:37–61

Gale NH (1999) Lead isotope characterisation of the ore deposits of Cyprus and Sardinia and its application to the discovery of the sources of copper for Late Bronze Age oxhide ingots. In: Young S, Pollard M, Budd P, Ixer RA (eds) Metals in Antiquity. BAR Int Series S 792, Archaeopress, Oxford, pp 110–121

Gale NH (2005) Die Kupferbarren von Uluburun, II: Bleiisotopenanalysen von Bohrkernen aus den Barren. In: Yalcin Ü, Pulak C, Slotta R (eds) Das Schiff von Uluburun – Welthandel vor 3000 Jahren. Exhibition catalogue (= Veröffentlichungen aus dem Deutschen Bergbau-Museum Bochum 138), pp 141–147

Gale NH, Stos-Gale ZA (1995) Comments on "Oxhide Ingots, Recycling, and the Mediterranen Metals Trade". J Mediterranean Archaeology 8:33–41

Gale NH, Papastamataki A, Stos-Gale ZA, Leonis K (1985) Copper Sources and Copper Metallurgy in the Aegean BronzeAge. In: Craddock PT, Hughes MJ (eds) Furnaces and Smelting Technology in Antiquity. Brit Museum Occ Papers 48:81–101

Gale NH, Bachmann HG, Rothenberg B, Stos-Gale ZA, Tylecote RF (1990) The Adventitious Production of Iron in the Smelting of Copper. In: Rothenberg B (ed) Researches in the Arabah 1959–1984, vol II: The Ancient Metallurgy of Copper. Inst Archaeo-Metall Studies, London, pp 182–191

Galili E, Shmueli N (1982) A Cargo of Copper and Tin Ingots from the Sea of Haifa. Center for Maritime Studies News 8

Ganor E (1975) Atmospheric Dust in Israel – Sedimentological and Meteorological Analysis of Dust Deposition. Unpublished Ph.D. Thesis, Dept of Geology, Hebrew University, Jerusalem

Garfinkel Y (1987) Bead Manufacture on the Pre-Pottery Neolithic B Site of Yiftahel. J Israel Prehist Soc 20:79–90
Garfunkel Z (1970) The Tectonics of the Western Margins of the Southern Araba. Dissertation, Dept of Geology, Hebrew University, Jerusalem
Garfunkel Z (1981) Internal Structure of the Dead Sea Leaky Transform (Rift) in Relation to Plate Kinematics. Tectonophysics 80:81–108
Geerlings W (1983) Ins Bergwerk verurteilt – die "damnatio ad metalla". Der Anschnitt 35(4,5):130–136
Geerlings W (1985) Zum biblischen und historischen Hintergrund der Bergwerke von Fenan in Jordanien. Der Anschnitt 37(5,6):158–162
Gelhoit P (2002) Zur Verfahrenstechnik bronzezeitlicher Kupfergewinnungsanalgen. Dissertation, Rheinisch-Westfälische Technische Hochschule, Aachen
Genz H (1997) Problems in Defining a Chalcolithic for Southern Jordan. In: Gebel HG, Kafafi Z, Rollefson G (eds) Prehistory of Jordan II. Berlin, ex oriente, pp 441–448
Genz H (2001) The Organisation of Early Bronze Age Metalworking in the Southern Levant. Paléorient 26:55–65
Genz H, Hauptmann A (2002) Chalcolithic and EBA Metallurgy in the Southern Levant. In: Yalcin Ü (ed) Anatolian Metal II. Der Anschnitt, Beiheft 15, Deutsches Bergbau-Museum Bochum, pp 149–157
Gilead I (1990) The Neolithic-Chalcolithic Transition and the Qatufian of the Northern Negev and Sinai. Levant 22:47–63
Gilead I (1994) The History of the Chalcolithic Settlement in the Nahal Beer Sheva Area: The Radiocarbon Aspect. Bull Amer Soc Oriental Res 296:1–13
Gilead I, Rosen S (1992) New Archaeometallurgical Evidence for the Beginnings of Metallurgy in the Southern Levant: Excavation at Tell Abu Matar, Beersheba (Israel) 1990/1. Inst Archaeo-Metall Studies Newsletter 18:11–14
Gilead I, Rosen S, Fabian P (1991) Excavations at Abu Matar (the Hatzerim Neighborhood), Beer Sheva. J Israel Prehist Soc 24:173–179
Gilles JW (1958) Versuchsschmelze in einem vorgeschichtlichen Rennofen. Stahl & Eisen 78:1690–1695
Ginzbourg D (1963) Petrography of the Loess in the Beer Sheva Basin. Dissertation, Dept. of Geology, Hebrew University, Jerusalem
Given M, Knapp A B (2003) The Sydney Cyprus Survey Project. Social Approaches to Regional Archaeological Survey. Monumenta Archaeologica 21, University of California, Los Angeles
Glasser VP (1962) The Ternary System CaO–MnO–SiO$_2$. J Amer Ceramic Soc 45:242–249
Glueck N (1935) Explorations in Eastern Palestine, II. Ann Amer School Oriental Res 15/1935. (continued in Bull Amer School Oriental Res 71, 75, 79, 80, 82, 1935, 1938, 1941)
Glumac PD (1985) Earliest Known Copper Ornaments from Prehistoric Europe. Ornament 8:15–17
Glumac PD, Todd JA (1991a) Early Metallurgy in Southeast Europe: The Evidence for Production. In: Glumac PD (ed) Recent Trends in Archaeometallurgical Research. MASCA Research Papers 8:9–19
Glumac PD, Todd JA (1991b) Eneolithic Copper Smelting Slags from the Middle Danube. In: Pernicka E, Wagner GA (eds) Archaeometry '90, Birkhäuser, Basel, pp 155–164
Golden J (1998) The Dawn of the Metal Age: Social Complexity and the Rise of Copper Metallurgy during the Chalcolithic in the Southern Levant, ca. 4500–3500 BC. Dissertation, Univ. of Pennsylvania
Golden J, Levy TE, Hauptmann A (2000) Recent discoveries concerning Chalcolithic metallurgy at Shiqmim, Israel. J Archaeological Science 28:951–963
Goldenberg G (1990) Die montanarchäologische Prospektion. In: Erze, Schlacken und Metalle. Früher Bergbau im Südschwarzwald. Freiburger Universitätsblätter 109:85–113
Goldenberg G (1993) Frühe Blei-, Silber- und Kupfergewinnung im Südschwarzwald: Hüttenplätze und Bergschmieden. In: Steuer H, Zimmermann U (eds) Montanarchäologie in Europa. Archäologie und Geschichte, 4, Thorbecke, Sigmaringen, pp 231–248

Goldenberg G (1996) Archäometallurgische Untersuchungen zur Entwicklung des Metallhüttenwesens im Schwarzwald. Blei-, Silber- und Kupfergewinnung von der Frühgeschichte bis zum 19. Jahrhundert. In: Goldenberg G, Otto J, Steuer H (eds) Archäometallurgische Untersuchungen zum Metallhüttenwesen im Schwarzwald. Archäologie u. Geschichte: Freiburger Forschungen zum ersten Jahrtausend in Südwestdeutschland, Bd. 8, pp 1–274

Goldman H (1956) Excavations at Gözlü Kule, Tharsus II. From Neolithic to the Bronze Age. Princeton University Press, Princeton (NJ)

Gopher A, Tsuk T (1991) Ancient Gold. Rare Finds from the Nahal Qanah Cave. Catalogue 321, The Israel Museum, Jerusalem

Gophna R (1992) The Intermediate Bronze Age. In: Ben-Tor A (ed) The Archaeology of Ancient Israel. The Open Univ Israel, pp 126–158

Gophna R, Lipschitz N, Lev-Yadun, S (1986/87) Man's Impact on the Natural Vegetation of the Central Coastal Plain of Israel During the Chalcolithic Period and the Bronze Age. Tel-Aviv 13/14:71–84

Gordon Jr. RL, Knauf EA, Hauptmann A, Roden C (1986) Antiker Eisenbergbau und alte Eisenverhüttung im Aglun. Archiv für Orientforschung 33:231–233

Goren Y (1991) The Beginnings of Pottery Production in Israel: Technology and Typology of Proto-Historic Ceramic Assemblages in Eretz-Israel (6th–4th mill. BC). Dissertation, Hebrew University, Jerusalem (Hebrew)

Goren Y (1995) Shrines and Ceramics in Chalcolithic Israel: The View through the Petrographic Microscope. Archaeometry 37:287–305

Görg M (1982) Punon – ein weiterer Distrikt der Beduinen? Biblische Notizen 19:15–21

Goring-Morris AN, Gopher A (1983) Nahal Issaron: A Neolithic Settlement in the Southern Negev. Israel Explor J 33:149–162

Gourdin WH, Kingery WD (1975) The Beginnings of Pyrotechnology: Neolithic and Egyptian Lime Plaster. J Field Archaeology 2:133–150

Gow NN, Lozej GP (1986) Bulk mineable micro gold deposits of the Carlin type: a review with reference to the exploration potential of the Red Sea – Jordan Valley Rift Area. Proc Second Jord Geol Conference, Amman, pp 95–111

Gowland W (1899) The Early Metallurgy of Copper, Tin and Iron in Europe, as Illustrated by Ancient Remains and the Primitive Processes Survived in Japan. Archaeologia 56:267–322

Gulson BL (1986) Lead Isotopes in Mineral Exploration. Developments in Economic Geology 23. Elsevier, Amsterdam

Gutt W (1967) Crystallization of Merwinite from Melilite Compositions. J Iron & Steel Inst 206:840–841

Hakemi A (1997) Shahdad. Archaeological Excavations of a Bronze Age Center in Iran. ISMEO, Centro Scavi e recherche archaeogiche. Reports and Memoirs 27. Rom

Hallimond AF (1919) The Crystallography of Vogtite, an Anorthic, Metasilicate of Iron, Calcium, Manganese and Magnesium from Acid Steel-Furnace Slags. Mineral Magazine 18:368–372

Hanbury-Tenison JW (1986) The Late Chalcolithic to Early Bronze Age I Transition in Palestine and Transjordan. BAR Intern Ser 311, Oxford

Hansen M, Anderko K (1958) The Constitution of Binary Alloys. McGraw Hill, New York

Harding GL (1971) An Index and Concordance of Pre-Islamic Arabian Names and Inscriptions. Toronto

Hart S (1992) Iron Age Settlement in the Land of Edom. In: Bienkowski P (ed) Early Edom and Moab: The Beginning of the Iron Age in Southern Jordan. Sheffield Archaeol Monogr 7:93–98

Hauptmann A (1980) Zur frühbronzezeitlichen Metallurgie von Shahr-i Sokhta (Iran). Der Anschnitt 32 (2,3):55–61

Hauptmann H (1982) Die Grabungen auf dem Norsuntepe, 1974. In: Keban Projekt, 1974–1975 Activities. Middle East Technical University Keban Projekt Publications, Series, No. I, 7, Ankara, pp 41–70

Hauptmann A (1985) 5000 Jahre Kupfer in Oman, I: Die Entwicklung der Kupfermetallurgie vom 3. Jahrtausend bis zur Neuzeit. Der Anschnitt, Beiheft 4, Deutsches Bergbau-Museum Bochum

Hauptmann A (1989a) The Earliest Periods of Copper Metallurgy in Feinan/Jordan. In: Hauptmann A, Pernicka E, Wagner GA (eds) Archäometallurgie der Alten Welt/Old World Archaeometallurgy. Der Anschnitt, Beiheft 7, Deutsches Bergbau-Museum Bochum, pp 119–135

Hauptmann A (1989b) Chemical Analyses of Prehistoric Metal Artefacts from the Indian Subcontinent. Jb Röm German Zentralmuseum 36(1):261–267

Hauptmann A (1991) From the Use of Ore to the Production of Metal. The Discovery of Copper Metallurgy at Feinan, Wadi Arabah/Jordan. In: Mohen JP, Éluère C (eds) Découverte du Métal. Millénaires 2, pp 397–412

Hauptmann A (2000) Reconstructing Ancient Smelting Processes: Applied Mineralogy in Archaeometallurgy. In: Rammlmair D, Mederer J, Oberthür Th, Heimann Rb, Pentinghaus H (eds) Applied Mineralogy. Balkema, Rotterdam, pp 29–31

Hauptmann A (2002) Rationales of liquefaction and metal separation in earliest copper smelting: basics for reconstructing chalcolithic and Early Bronze Age smelting processes. In: Archaeometallurgy in Europe. Proc Internat Conference, Milano 2003, pp 459–468

Hauptmann A (2003) Developments in Copper Metallurgy During the Fourth and Third Millennia BC at Feinan, Jordan. In: Craddock PT, Lang J (eds) Mining and Metal Production Through the Ages. British Museum Press, London, pp 90–100

Hauptmann A (2004) "Greenstones" from Basta. Their Mineralogical Composition and Possible Provenance. In: Nissen HJ, Muheisen M, Gebel HG (eds) Basta I. The Human Ecology. Bibliotheca neolithica Asiae meridionalis et occidentalis & Yarmouk University, Monograph of the Faculty of Archaeology and Anthropology 4, pp 169–176

Hauptmann A, Horowitz A (1980) Zur Geomorphologie und Paläomorphologie des Modellgebietes. In: Conrad HG, Rothenberg B (eds) Antikes Kupfer im Timna-Tal. Der Anschnitt, Beiheft 1, Deutsches Bergbau-Museum Bochum, pp 57–67

Hauptmann A, Klein S (1999) Über die Bildung von Glasuren in alten Kupferschmelzöfen. In: Busz R, Gercke P (eds) Türkis und Azur: Quarzkeramik in Orient und Okzident. Staatl. Museen Kassel, pp 114–121

Hauptmann A, Roden C (1988) Archäometallurgische Untersuchungen zur Kupferverhüttung der Frühen Bronzezeit in Feinan, Wadi Arabah, Jordanien. Jb Röm German Zentralmuseum Mainz 35:510–516

Hauptmann A, Wagner I (2007) Prehistoric Copper Production at Timna: TL-dating and Evidence from the East. forthcoming

Hauptmann A, Weisgerber G (1985) Vom Kupfer zur Bronze: Beiträge zum frühesten Berg- und Hüttenwesen. In: Born H (ed) Archäologische Bronzen – antike Kunst, moderne Technik. Staatl Museen Preuß Kulturbes, Berlin, pp 16–36

Hauptmann A, Weisgerber G (1987) Archaeometallurgical and Mining-Archaeological Investigations in the Area of Feinan, Wadi Arabah (Jordan). Ann Dept Antiquities Jordan XXXI:419–437

Hauptmann A, Weisgerber G (1996) The Early Production of Metal in the Near East. In: Bagolini B, Lo Schiavo F (eds) The Copper Age in the Near East and Europe. Proc Intern Union Prehist and Protohist Science 10:95–101

Hauptmann A, Weisgerber G, Knauf EA (1985) Archäometallurgische und bergbauarchäologische Untersuchungen im Gebiet von Fenan, Wadi Arabah (Jordanien). Der Anschnitt 37:163–195

Hauptmann A, Pernicka E, Wagner GA (1988) Untersuchungen zur Prozeßtechnik und zum Alter der frühen Blei-Silbergewinnung auf Thasos. In: Weisgerber G, Wagner GA (eds) Antike Edel- und Buntmetallgewinnung auf Thasos. Der Anschnitt, Beiheft 6, Deutsches Bergbau-Museum Bochum, pp 88–112

Hauptmann A, Weisgerber G, Bachmann HG (1989) Frühe Kupfergewinnung im Gebiet von Fenan, Wadi Arabah, Jordanien. In: Domergue C (ed) Mineria y Metallurgia en las Antiguas Civilizaciones Mediterraneas y Europeas. Ministerio de Cultura Madrid, pp 39–56

Hauptmann A, Maddin R, Weisgerber G (1991) Metallographic Investigation of Two Miner's Tools from Wadi Khalid, Feinan. Ann Dept Antiquities Jordan XXXV:195–201

Hauptmann A, Begemann F, Heitkemper E, Pernicka E, Schmitt-Strecker S (1992) Early Copper Produced at Feinan, Wadi Araba, Jordan: The Composition of Ores and Copper. Archeomaterials 6:1–33

Hauptmann A, Lutz J, Pernicka E, Yalcin Ü (1993) Zur Technologie der frühesten Kupferverhüttung im Vorderen Orient. In: Frangipane M, Hauptmann H, Liverani M, Matthiae P, Mellink M (eds) Between the Rivers and Over the Mountains – Archaeologica Anatolica et Mesopotamica Alba Palmieri Dedicata. Universita La Sapienza, Rome, pp 541–572

Hauptmann A, Bachmann HG, Maddin R (1996) Chalcolithic Copper Smelting: New Evidence from Excavations at Wadi Fidan 4. In: Demirci S, Özer AM, Summers GD (eds) Archaeometry '94. Tübitak, Ankara, pp 3–10

Hauptmann A, Begemann F, Schmitt-Strecker S (1998) Copper Objects from Arad – Their Composition and Provenance. Bull Amer Schools Oriental Res 314:1–17

Hauptmann A, Maddin R, Prange M (2002) On the texture and composition of copper and tin ingots excavated from the shipwreck of Uluburun. Bull Amer School Oriental Res 328:1–30

Hauptmann A, Rehren Th, Schmitt-Strecker S (2003a) Early Bronze Age Copper Metallurgy at Shahr-i Sokhta (Iran), reconsidered. In: Stöllner T, Körlin G, Steffens G, Cierny J (eds) Man and Mining – Mensch und Bergbau. Studies in honour of Gerd Weisgerber on occasion of his 65th birthday. Der Anschnitt, Beiheft 16, Deutsches Bergbau-Museum Bochum, pp 197–213

Hauptmann A, Schmitt-Strecker S, Begemann F, Palmieri M (2003b) Chemical Composition and Lead Isotopy of Metal Objects from the "Royal" Tomb and Other Related Finds at Arslantepe, Eastern Anatolia. Paléorient 28:43–70

Hauptmann A, Khalil L, Schmitt-Strecker S (2007) Evidence for Late Chalcolithic / Early Bronze Age I copper production from Timna ores at Tall Magass, Aqaba. In: Eichmann R, Khalil L, Schmidt K (eds) Prehistoric Aqaba. Orient Archäologie. Marie Leidorf, Rahden, Westfalen

Hauser L (1986) Die bronzezeitlichen Kupferschmelzöfen in "Fennhals" über Kurtatsch. Der Schlern 60:75–87

Healy JF (1978) Mining and Metallurgy in the Greek and Roman World. Thames & Hudson, London

Heimann A, Steinitz G, Mor D Shaliv G (1996) The Cover Basalt Formation, its Age and its Regional Setting: Implications from K-Ar and ^{40}Ar/^{39}Ar Geochronology. Israel J Earth Science 45(2):55–71

Heitkemper E (1988) Kupfervererzungen in kambrischen Sedimenten des Fenan-Gebietes, SW-Jordanien. Diplomarbeit, Institut für Geologie, Ruhr-Universität Bochum

Helke A (1964) Die Kupfererzlagerstätte Ergani Maden in der Türkei. N Jb Mineral, Abh 101:233–270

Helmig D (1986) Versuche zur analytisch-chemischen Charakterisierung frühbronzezeitlicher Techniken der Kupferverhüttung in Shahr-i-Sokhta / Iran. Diplomarbeit, Institut für Chemie, Ruhr-Universität Bochum

Heskel D, Lamberg-Karlovsky CC (1980) An Alternative Sequence for the Development of Metallurgy: Tepe Yahya, Iran. In: Wertime TA, Muhly JD (eds) The Coming of the Age of Iron. Yale Univ. Press, New Haven & London, pp 229–265

Hess K (1998) Zur frühen Metallurgie am oberen Euphrat: Untersuchungen an archäometallurgischen Funden vom Arslantepe aus dem 4. und 3. Jahrtausend v. Chr. PhD thesis, Institut für Mineralogie, University of Frankfurt/Main

Hess K, Hauptmann A, Wright H, Whallon R (1999) Evidence of fourth millennium BC silver production at Fatmali-Kalecik. In: Rehren T, Hauptmann A, Muhly JD (eds) Metallurgica Antiqua. Der Anschnitt, Beiheft 8, Deutsches Bergbau-Museum Bochum, pp 57–67

Hestrin R, Tadmor M (1963) A Hoard of Tools and Weapons from Kfar Monash. Israel Explor J 13(4):265–290

Hezarkhani Z, Keesmann I (1996) Archäometallurgische Untersuchungen an Kupferschlacken im Zentraliran. Metalla (Bochum) 3:101–125

Hilmy HE, Mohsen LA (1965) Secondary Copper Minerals from West-Central Sinai. J Geol U.A.R., 9:1–12

Hlawatsch C (1907) Eine trikline, rhodonitähnliche Schlackenphase. Zeitschr Kristallographie 42:590–593

Hook DR, Freestone IC, Meeks N, Craddock, PT, Moreno A (1991) The Early Production of Copper-Alloys in Southeast Spain. In: Pernicka E, Wagner GA (eds) Archaeometry '90. Birkhäuser, Basel, pp 65–76

Horowitz A (1979) The Quaternary of Israel. McGrawhill, New York

Houtermans FG (1960) Die Bleimethoden der geologischen Altersbestimmung. Geol Rundschau 49:168–196

Huebner JS (1969) Stability Relations of Rodochrosite in the System Manganese-Carbon-Oxygen. Amer Mineral 54:457–481

Huebner JS, Sato M (1970) The Oxygen Fugacity-Temperature Relationships of Manganese Oxide and Nickel Oxide Buffers. Amer Mineral 55:934–952

Ilan O, Sebanne M (1989) Copper Metallurgy, Trade and the Urbanization of Southern Canaan in the Chalcolithic and the Early Bronze Age. In: de Miroschedji P (ed) L'ur-banisation de la Palestine à L'age du Bronze ancien. BAR Intern Ser, Oxford, 527(1):139–162

Ilani S, Rosenfeld A (1994) Ore Source of Arsenic Copper Tools from Israel during Chalcolithic and Early Bronze Ages. Terra Nova 6:177–179

Ilani S, Flexer A, Kronfeld J (1987) Copper Mineralization in Sedimentary Cover Associated with Tectonic Elements and Volcanism in Israel. Miner Deposita 22:269–277

Itamar A (1988) Polymetallic Mineralization in Makhtesh Ramon and Har Arif, Central Negev, Israel. Geol Surv Israel Bull 80:1–56

Jacob KT, Fitzner K, Alcock CB (1977) Activities in the Spinel Solid Solution, Phase Equilibria and Thermodynamics in the System Cu–Fe–O. Metall Trans B 8:451–460

Jankoviè S (1997) The Carpatho-Balkanides and adjacent area: a sector of the Tethyan Eurasian metallogenic belt. Miner Deposita 32:426–433

Jarrar GH (1984) Late Proterozoic Crustal Evolution of the Arabian-Nubian Shield in the Wadi Araba, SW-Jordan. Braunschweiger Geol Paläont Diss 2

Jarrige JF (1984) Towns and Villages of Hill and Plain. In: Lal BB, Gupta SP (eds) Frontiers of the Indus Civilization. New Delhi, pp 289–300

Joffe AH, Dessel JP (1990) Redefining Chronology and Terminology for the Chalcolithic of the Southern Levant. Current Anthropol 36:507–518

Juleff G (1996) An Ancient Wind-Powered Iron Smelting Technology in Sri Lanka. Nature 379(4):60–63

Junghans S, Sangmeister E, Schröder M (1960) Metallanalysen kupferzeitlicher und frühbronzezeitlicher Bodenfunde aus Europa. Studien zu den Anfängen der Metallurgie, Bd. 1. Mann, Berlin

Junghans S, Sangmeister E, Schröder M (1968) Kupfer und Bronze in der frühen Metallzeit Europas, Bd. 1–3. Mann, Berlin

Junghans S, Sangmeister E, Schröder M (1974) Kupfer und Bronze in der frühen Metallzeit Europas, Bd. 4. Mann, Berlin

Kafafi Z (1986) White Objects from Ain Ghazal, near Amman. Bull Amer School Oriental Res 261:51–55

Kan X, Coey JMD (1985) Moessbauer-Spectra, Magnetic and Electric Properties of Laihunite, a Mixed Valence Iron Olivine Mineral. Amer Miner 70:576–580

Karcz I, Key CA (1966) Note on the Pre-Paleozoic Morphology of the Basement in the Timna Area (Southern Israel). Israel J Earth Science 15:47–56

Kassianidou V (2003) Archaeometallurgy: Data, Analyses and Discussion. In: Given M, Knapp B (eds) The Sydney Cyprus Survey Project. Social Approaches to Regional Archaeological Survey. Monumenta Archaeologica 21:214–221

Kassianidou V, Rothenberg B, Andrews P (1995) Silver Production in the Tartessian Period. The Evidence from Monte Romero. ARX 1(1):17–34

Kayani PI, McDonnell G (1997) Formative Pyrotechnology in Northern Mesopotamia. Paléorient XXII:133–141

Keesmann I (1985) Verfahrensbedingungen aus Phasenbestand und chemischer Analyse von alten Kupferschlacken (Afunfun, Agadez/Niger). Analyt Chemie 320:660

Keesmann I (1989) Chemische und mineralogische Detailuntersuchungen zur Interpretation eisenreicher Schlacken. In: Pleiner R (ed) Archaeometallurgy of Iron, 1967–1987. Symposium Liblice 1987, Prague, pp 17–34

Keesmann I (1991) Rio Tinto: Die Technik der Silbergewinnung zu Beginn des Mittelalters. In: Argent, plomb et cuivre dans l'histoire. Lyon 1991, pp 1–13

Keesmann I (1993) Naturwissenschaftliche Untersuchungen zur antiken Kupfer- und Silberverhüttung in Südwestspanien. In: Steuer H, Zimmermann U (eds) Montanarchäologie in Europa. Archäologie und Geschichte: Freiburger Forschungen zum ersten Jahrtausend in Südwestdeutschland, Bd. 4. Thorbecke, Sigmaringen, pp 105–122

Keesmann I, Heege A (1990) Archäometallurgische Untersuchungen an Material der Ausgrabung 1984 am "Steinbühl" bei Nörten – Hardenberg, Ldkr. Northeim. Nachr Niedersachsens Urgesch 59:87–109

Keesmann I, Hilgart T (1992) Chemische und mineralogische Untersuchungen der Schlacken von Manching. Manching 15:391–413

Keesmann I, Onorato M (1998) Los Millares (Provinz Almeria, Süd-Spanien): Naturwissenschaftliche Untersuchungen zur frühen Technologie von Kupfer und Kupfer-Arsen-Bronze. In: Hauptmann A, Pernicka E, Rehren T, Yalçin Ü (eds) The Beginnings of Metallurgy. Der Anschnitt, Beiheft 9, Deutsches Bergbau-Museum Bochum, pp 317–332

Keesmann I, Bachmann HG, Hauptmann A (1982) Norm-Berechnungsverfahren für eisenreiche Schlacken. Fortschr Miner 60, Beih. 1:110–111

Keesmann I, Kronz A, Maier P (1994) Zur Kupfertechnologie von Los Millares (Provinz Almeria, Südwest-Spanien). Ber Deutsche Miner Gesellsch 1:134

Keidar Y (1984) Mineralogy and Petrography of Copper Nodules in the Upper White Sandstone in the Timna Valley. M.Sc. Thesis, Ben Gurion Univ. Negev (Hebrew.)

Kempinski A (1989) Urbanization and Metallurgy in Southern Canaan. In: de Miroschedji P (ed) L'urbanisation de la Palestine à L'age du Bronze ancien. BAR Intern Ser, Oxford, 527:163–168

Kenyon K (1955) A Crescentic Axehead from Jericho, and a Group of Weapons from Tell el Hesi. Ann Rep Inst Archaeol Univ London 11:1–9

Kenyon K (1960) Excavations at Jericho, vol I: The Tombs Excavated in 1952–4. Harrison & Sons, London

Kerner S (2001) Das Chalkolithikum in der südlichen Levante. Orient-Archäologie, Bd. 8. Leidorf, Rahden, Westfalen

Key CA (1980) The Trace Element Composition of the Copper and Copper Alloys Artifacts of the Nahal Mishmar Hoard. In: Bar-Adon P (ed) The Cave of the Treasure. Israel Explor Soc, Jerusalem, pp 238–243

Khalil L (1988) Excavation at Magass-Aqaba, 1985. Dirasat XV(7):71–109

Khalil L (1992) Some Technological Features from a Chalcolithic Site at Magass-Aqaba. In: Hadidi A (ed) Studies in the History and Archaeology of Jordan, IV. Amman, pp 143–148

Khalil L (1995) The Second Season of Excavation at Al-Magass-Aqaba. Ann Dept Antiquities Amman XXXIX:65–79

Khalil L, Bachmann HG (1981) Evidence of Copper Smelting in Bronze Age Jericho. J Hist Metall Soc 15:103–106

Khalil L, Riederer J (1998) Examination of Copper Metallurgical Remains from a Chalcolithic Site at el-Magass, Jordan. Damaszener Mitt 10:1–9

Khoury H (1986) On the Origin of Stratabound Copper-Manganese Deposits in Wadi Araba, Jordan. Dirasat XIII(7):227–247

Kienow S, Seeger M (1965) Mikroskopie von Schamotte- und Silikaerzeugnissen. In: Freund H (ed) Handbuch der Mikroskopie in der Technik, Bd. IV, Teil 3: Mikroskopie in der Sintertechnik, insbesondere der keramischen und pulvermetallurgischen Produkte. Umschau, Frankfurt/M., pp 101–154

Killick D, van der Merwe N, Gordon R, Grébénart D (1988) Reassessment of the Evidence for Early Metallurgy in Niger, West Africa. J Archaeol Science 15:367–394

Killick D, van der Merwe N, Gordon R, Grébénart D (1992) The Relevance of Recent African Iron-Smelting Practice to Reconstructions of Prehistoric Smelting Technology. MASCA Res Papers Science Archaeology 8:8–20

References

Kind HD (1965) Antike Kupfergewinnung zwischen Rotem und Totem Meer. Zeitschr Deutsch Palästina-Verein 81:56–73

Kind HD, Gilles KJ, Hauptmann A, Weisgerber G (2005) Coins from Faynan, Jordan. Levant 37:169–195

Kingery WD (1988) Plaster Beads. In: Bar-Yosef O, Alon D (eds) Nahal Hemar Cave. Atiqot XVIII:45–46

Kingery WD, Frierman J (1975) The Firing Temperature of a Karanovo Sherd and Inferences about South-East European Chalcolithic Refractory Technology. Proc Prehist Soc 42:204

Kingery WD, Vandiver P, Prickett M (1988) The Beginnings of Pyrotechnology, part II: Production and Use of Lime and Gypsum Plaster in the Pre-Pottery Neolithic Near East. J Field Archaeology 15:219–244

Kisnawi A, De Jesus P, Rihani B (1983) 2. Preliminary Report on the Mining Survey, Northwest Hidjaz, 1982. Atlal 7:76–83

Kitamura M, Buming S, Shohei B, Morimoto N (1984) Fine Textures of Laihunite, a Non-Stoichiometric Distorted Olivine-Type Mineral. Amer Mineral 69:54–160

Klein S, Hauptmann A (1999) Iron Age Leaded Tin Bronze form Khirbet Edh-Darih, Jordanien, J Archaeolog Science 26:1075–1082

Klein S, Lahaye Y, Brey GP, Von Kaenel HM (2004) The Early Roman Imperial Aes Coinage II: Tracing Copper Sources by Lead- and Copper-Isotope Analysis – Copper Coins of Augustus and Tiberius. Archaeometry 46:469–480

Knapp B, Kassianidou V, Donnelly M (1998) Excavations at Politiko-*Phorades* 1997. Rep Dept Antiquities, Cyprus, pp 247–268

Knapp B, Kassianidou V, Donnelly M (1999) Excavations at Politiko-*Phorades* 1998. Rep Dept Antiquities, Cyprus, pp 125–146

Knapp B, Kassianidou V, Donnelly M (2002) Excavations at Politiko-*Phorades*: A Bronze Age copper smelting site on Cyprus. Antiquity 76:319–320

Knauf EA (1992) The Cultural Impact of Secondary State Formation: The Case of the Edomites and Moabites. In: Bienkowski P (ed) Early Edom and Moab: The Beginning of the Iron Age in Southern Jordan. Sheffield Archaeol Monogr 7:47–54

Knauf EA, Lenzen C (1987) Edomite Copper Industry. In: Hadidi A (ed) Studies in the History and Archaeology of Jordan, III. Amman, pp 83–88

Kölschbach S (1999) Experimente zur Simulation prähistorischer Kupfergewinnung: Zur Verfahrenstechnik von Windöfen. Dissertation, Rhein Westfäl Techn Hochschule Aachen

Kondoh S, Kitamura M, Morimoto N (1985) Synthetic Laihunite, an Oxidation Product of Olivine. Amer Mineral 70:737–746

Körber F, Oelsen W (1940) Die Reduktionsgleichgewichte von Oxyden und Oxydgemengen als Grundlage wichtiger Probleme der Eisenerzeugung. Zeitschr Elektrotechnik 46:188–194

Kortan O (1970) Zur Bildung der Schwefelkies-Kupferkies-Vorkommen Cyperns unter besonderer Berücksichtigung der Lagerstätte Skouriotissa. Dissertation TU Clausthal

Koucky F, Steinberg A (1982a) Ancient Mining and Mineral Dressing on Cyprus. In: Wertime SF (ed) Early Pyrotechnology. Smithsonian Institution Press, Washington, pp 149–180

Koucky F, Steinberg A (1982b) The Ancient Slags of Cyprus. In: Muhly JD, Maddin R, Karageorghis V (eds) Early Metallurgy at Cyprus, 4000–500 B.C. Acta Intern Archaeol Symp, Larnaca 1981. Nicosia, pp 117–141

Krawczyk E (1986) Petrologische und geochemische Untersuchungen an Kupferverhüttungsschlacken aus dem südlichen Teil des Wadi Arabah (Israel). Diplomarbeit, Fachb. Geowissenschaften, University of Mainz

Krawczyk E, Keesmann I (1988) Ergänzende Untersuchungen zur Kupfermetallurgie von Timna im Wadi Arabah, Israel. Jb Röm German Zentralmuseum 35:516–522

Kresten P (1984) The Mineralogy and Chemistry of Selected Ancient Iron Slags from Dalarna, Sweden. Jernkontorets Forskning H29

Kronz A (1997) Phasenbeziehungen und Kristallisationsmechanismen in fayalitischen Schlackenschmelzen – Untersuchungen an Eisen- und Buntmetallschlacken. Dissertation, Fachb Geowissenschaften, University of Mainz

Kruse R, Dörfler W, Jöns H (1997) Angewandte Prospektionsmethoden. In: Jöns H (ed) Frühe Eisengewinnung in Joldelund, Kr. Nordfriesland. Ein Beitrag zur Siedlungs- und Technikgeschichte Schleswig-Holsteins, Teil 1. Universitätsforschungen zur Prähistorischen Archäologie, Habelt, Bonn, Bd. 40, pp 12–44

Kubaschewski O, Alcock CB (1979) Metallurgical Thermochemistry. Intern Ser Materials Technology. Pergamon-Press, Oxford

Kuniholm PI (1996) Long Tree-Ring Chronologies for the Eastern Mediterranean. In: Demirci S, Özer AM, Summers GD (eds) Archaeometry '94. Tübitak, Ankara, pp 401–409

Kürschner H (1986) A Physiognomical-Ecological Classification of the Vegetation of Southern Jordan. In: Kürschner H (ed) Contributions to the Vegetation of Southeast Asia. Reichert, Berlin, pp 45–77

Kuxmann U, Benecke T (1966) Untersuchungen zur Löslichkeit von Sauerstoff in Kupfersulfid-Schmelzen. Erzmetall 19:215–221

Kuxmann U, Kurre K, (1968) Die Mischungslücke im System Kupfer-Sauerstoff und ihre Beeinflussung durch die Oxide CaO, SiO_2, Al_2O_3, MgO, Al_2O_3, ZrO_2. Erzmetall 21:199–209

Lechtman H (1996) Arsenic Bronze: Dirty Copper or Chosen Alloy? A View from the Americans. J Field Archaeology 23:477–514

Leese M, Craddock P, Freestone IC, Rothenberg B (1985/1986) The Composition of Ores and Metal Objects from Timna, Israel. Wiener Ber Naturwiss Kunst 2/3:90–120

Lenz H, Bender F, Besang C, Harre W, Kreuzer H, Müller P, Wendt I (1972) The Age of Early Tectonic Events in the Zone of the Jordan Geosuture, Based on Radiometric Data. 24th Intern Geol Congres, Montreal 3:371–379

Letsch J (1982) Neolithische und chalkolithische Keramik Thessaliens: Material, Rohstoffe und Herstellungstechnik. Dissertation. University of Cologne

Levy T (1995) Cult, Metallurgy and Rank Societies – Chalcolithic Period (ca. 4500–3500 BCE). In: Levy T (ed) Archaeology of Society in the Holy Land. Facts on File, New York, pp 226–244

Levy TE, Alon D (1985) An Anthropomorphic Statuette from Shiqmim. Atiqot XVII:187–189

Levy TE, Shalev S (1989) Prehistoric Metalworking in the Southern Levant: Archaeometallurgical and Social Perspectives. World Archaeology 20:350–372

Levy TE, Alo D, Smith Y, Yekutieli Y, Rowan Y, Goldberg P, Porat N, van den Brinck ECM, Witten A, Golden J, Grigson C, Kansa E, Dawson L, Holl A, Moreno J, Kersel M (1997) Egyptian-Canaanite interaction at Nahal Tillah, Israel (ca. 4500–3000 BCE): an interim report on the 1994–95 excavations. Bull Amer Schools Oriental Res 307:1–51

Levy TE, Adams R, Shafiq R (1999) The Jabal Hamrat Fidn Project: Excavations at Wadi Fidan 40 Cemetery, Jordan (1997). Levant 31:293–308

Levy T, Adams R, Hauptmann A, Prange M, Schmitt-Strecker S, Najjar M (2002) Early Bronze Age metallurgy: a newly discovered copper manufactory in southern Jordan. Antiquity 76:425–437

Levy TE, Adams R, Witten A, Anderson J, Arbel Y, Kuah S, Moreno J, Lo A, Wagonner M (2001) Early Metallurgy, Interaction, and Social Change: The Jabal Hamrat Fidn (Jordan) Research Design and 1998 Archaeological Survey: Preliminary Report. Ann Dept Antiquities Jordan 45:159–187

Levy T, Adams RB, Anderson JD, Najjar N, Smith N, Arbel Y, Soderbaum L, Muniz A (2003) An Iron Age Landscape in the Edomite Lowlands: Archaeological Surveys along the Wadi al-Guwayb and Wadi al-Jaryie, Jabal Hamrat Fidan, Jordan, 2002. Ann Dept Antiquities Jordan 47:247–277

Levy T, Adams R, Najjar M, Hauptmann A, Anderson JD, Brandl B, Robinson MA, Higham T (2004) Reassessing the Iron Age Chronology of Biblical Edom: New Excavations and [14]C Dates from Khirbat en Nahas (Jordan). Antiquity 78:302, 865–879

Lillich W (1963) Report on the Detailed Geological Mapping of the Copper Occurrences at Wadi Abu Kusheiba/Wadi Araba. Unveröff Ber Deutsche Geol Mission Jordanien. Arch. Bundesanstalt Geow. & Rohstoffe, Hannover

Lippert A (1993) Frühe Zeugnisse von Kupfermetallurgie im Raum Mühlbach am Hochkönig-Bischoshofen. In: Günther W, Eibner C, Lippert A, Paar W (eds) 5 000 Jahre Kupferbergbau Mühlbach am Hochkönig – Bischofshofen. Mühlbach/Hochkönig, pp 27–40

Locke A (1926) Leached Outcrops as Guides to Copper Ore. Williams & Wilkins, Baltimore
Lorenz I (1988) Thermolumineszenz-Datierung an alten Kupferschlacken. Dissertation Naturwiss.-Mathemat. Fakultät University of Heidelberg
Lupu A (1970) Metallurgical Aspects of Chalcolithic Copper Working at Timna (Israel). Bull Hist Metall Soc 4:21–23
Lupu A, Rothenberg B (1970) The Extractive Metallurgy of the Early Iron Age Copper Industry in the Arabah, Israel. Archaeol. Austriaca 47:91–129
Luraschi A, Elliott JF (1976) Phase Relationship in the Cu–Fe–O–System, 1 100 to 1 350 °C. In: Yannopoulos JC, Agarwal JC (eds) Extractive Metallurgy of Copper, vol I: Pyrometallurgy and Electrolytic Refining. Metall Soc Amer Inst Metall Eng, New York, pp 90–114
Lutz J (1990) Geochemische und mineralogische Aspekte der frühen Kupferverhüttung in Murgul/Nordost-Türkei. Dissertation Nat.-Math. Gesamtfakultät, University of Heidelberg
Lutz J, Wagner G, Pernicka E (1994) Chalkolithische Kupferverhüttung in Murgul, Ostanatolien. In: Wartke RB (ed) Handwerk und Technologie im Alten Orient. Philip v. Zabern, Mainz, pp 60–66
Macdonald E (1932) Beth Pelet II. Praehistorica Fara, London
Macdonald B (1992) The Southern Ghors and Northeast Arabah Archaeological Survey. Sheffield Archaeol Monogr 5
Maczek M, Preuschen E, Pittioni R (1952) Beiträge zum Problem der Kupfererzverwertung in der Alten Welt, Teil I. Archaeologia Austriaca 10:120–144
Maddin R, Merkel JF (1990) Metallographic and Statistical Analyses. In: LoSchiavo F, Maddin R, Merkel JF, Muhly JD, Stech T (eds) Metallographic and Statistical Analyses of Copper Ingots from Sardinia. Qaderni 17:43–199
Maddin R, Stech-Wheeler T (1976) Metallurgical Study of Seven Bar Ingots. Israel Explor J 26:170–173
Maddin R, Muhly JD, Stech-Wheeler T (1980) Distinguishing Artifacts Made of Native Copper. J Archaeol Science 7:221–225
Maddin R, Muhly JD, Stech T (1999) Early Metalworking at Çayönü. In: Hauptmann A, Pernicka E, Rehren T, Yalçin Ü (eds) The Beginnings of Metallurgy. Der Anschnitt, Beiheft 9, Deutsches Bergbau-Museum Bochum, pp 37–44
Maddin R, Muhly JD, Stech T (2003) Metallurgical Studies on Copper Artefacts from Bab edh-Dhra. In: Rast WE, Schaub RT (eds) Bab edh-Dhra: Excavations at the Town Site (1975–1981). Winonan Lake, Ind., Eisenbrauns
Maggetti M, Baumgartner D, Galetti G (1991) Mineralogical and Chemical Studies on Swiss Neolithic Crucibles. In: Pernicka E, Wagner GA (eds) Archaeometry '90. Birkhäuser, Basel, pp 95–104
Majidzadeh Y (1976) An Early Prehistoric Coppersmith Workshop at Tepe Ghabristan. Archäol Mitt Iran 6:82–92
Mallon A, Koeppel R, Neuville R (1934) Teleilat Ghassul. Rome
Mangin M, Keesmann I, Birke W, Ploqin A (1992) Mines et Métallurgie chez les Éduens: Le District Sidérurgique Antique et Médiéval du Morvan-Auxois. Annales Litter Univ Besançon
Maran J (2000) Das ägäische Chalkolithikum und das erste Silber Europas. In: I°ık C (ed) Studien zur Religion und Kultur Kleinasiens und des ägäischen Raums. Asia Minor Studien 39:179–193
Martinek KP (1996) Archäometallurgische Untersuchungen zur frühbronzezeitlichen Kupferproduktion und -verarbeitung auf dem Buchberg bei Wiesing, Tirol. Fundber Österreich 34, 1995:575–584
McAndrew J (1957) Crystallography and Composition of Crednerite. Acta Crystallogr 10: 276–287
McKerell H, Tylecote RF (1972) The Working of Copper-Arsenic Alloys in the Early Bronze Age and the Effect on the Determination of Provenance. Proc Prehist Soc 38:209–218
McLeod BH (1962) The Metallurgy of King Solomon's Copper Smelters. Palestine Explor Quart 94:68–71
Mellart J (1967) Çatal Hüyük, A Neolithic Town in Anatolia. Thames & Hudson, London

Merkel JF (1977) Neutron Activation Analysis of Copper to Examine Timna as an Ore Source during the Chalcolithic and Early Bronze Ages in Israel. Unpubl MSc Thesis, Center for Ancient Studies, University of Minnesota

Merkel JF (1983a) Reconstruction of Bronze Age Copper Smelting: Experiments Based on Archaeological Evidence from Timna, Israel. Dissertation, University of London

Merkel JF (1983b) Summary of Experimental Results for Late Bronze Age Copper Smelting and Refining. MASCA J 2(6):173–178

Merkel JF (1985) Ore Beneficiation during the Late Bronze/Early Iron Age at Timna, Israel. MASCA J 3(5):164–169

Merkel JF (1990) Experimental Reconstruction of Bronze Age Copper Smelting Based on Archaeological Evidence from Timna. In: Rothenberg B (ed) Researches in the Arabah 1959–1984, vol. II: The Ancient Metallurgy of Copper. Inst Archaeo-Metall Studies, London, pp 78–122

Merkel JF, Dever WG (1989) Metalworking Technology at the End of the Early Bronze Age in the Southern Levant. Inst Archaeo-Metall Studies Newsletter 14:1–4

Merkel JF, Rothenberg B (1999) The Earliest Steps to Copper Metallurgy in the western Araba. In: Hauptmann A, Pernicka E, Rehren T, Yalçin Ü (eds) The Beginnings of Metallurgy. Der Anschnitt, Beiheft 9, Deutsches Bergbau-Museum Bochum, pp 149–165

Merkel JF, Shimada I (1988) Arsenical Copper Smelting at Batán Grande, Peru. Inst Archaeo-Metall Studies Newsletter 12:4–7

Meshel Z (2006) Were There Gold Mines in the Eastern Arabah? In: Bienkowski P, Galor K (eds) Crossing the Rift: Resources, Routes, Settlement Patterns and Interactions in the Wadi Arabah. Levant Supplementary Series 3. Oxbow Books, Oxford, pp 231–238

Metten B (2003) Beitrag zur spätbronzezeitlichen Kupfermetallurgie im Trentino (Südalpen) im Vergleich mit anderen prähistorischen Kupferschlacken aus dem Alpenraum. Metalla 10(1,2):1–122

Milton C, Dwornik E, Finkelman RB, Toulmin III P (1976) Slag from an Ancient Copper Smelter at Timna, Israel. J Hist Metall Soc 10:24–33

Miron E (1992) Axes and Adzes from Canaan. Prähistorische Bronzefunde, Abt. IX, Bd. 19. Steiner, Stuttgart

Moesta H (1983) Erze und Metalle – Ihre Kulturgeschichte im Experiment. Springer-Verlag, Berlin Heidelberg New York

Moesta H (1986) Bronze Age Copper Smelting. Interdiscipl Science Rev 11(1):73–87

Moesta H, Kopcewicz B (1982) Bronzezeitliche Verhüttungsprozesse in den Ostalpen. I. Naturwissenschaften 69:493–494

Moesta H, Schlick G (1989b) The Furnaces of Mitterberg: An Oxidising Bronze Age Copper Process. Bull Metals Mus, Sendai 14:5–16

Moesta H, Schnau G (1983) Bronzezeitliche Hüttenprozesse in den Ostalpen, III: Die Abscheidung des metallischen Kupfers. Naturwissenschaften 70, 142–143

Moesta H, Rüffler R, Schnau-Roth G (1989a) Zur Verfahrenstechnik der bronzezeitlichen Kupferhütten am Mitterberg: Mößbauer- und mikroskopische Studien. In: Hauptmann A, Pernicka E, Wagner GA (eds) Archäometallurgie der Alten Welt / Old World Archaeometallurgy. Der Anschnitt, Beiheft 7, Deutsches Bergbau-Museum Bochum, pp 141–153

Moorey PRS (1985) Materials and Manufacture in Ancient Mesopotamia: The Evidence of Archaeology and Art, Metals and Metalwork, Glazed Materials and Glass, BAR Intern Ser, Oxford, 237

Moorey PRS (1988) The Chalcolithic Hoard from Nahal Mishmar, Israel, in Context. World Archaeology 20(2):171–189

Moorey PRS, Schweizer F (1972) Copper and Copper Alloys in Ancient Iraq, Syria and Palestine: Some New Analyses. Archaeometry 14:177–198

Morimoto N et al. (1988) Nomenclature of Pyroxenes. Amer Miner 73(9,10):1123–1133

Morton H, Wingrove J (1969) Constitution of Bloomery Slags. Part I: Roman. J Iron & Steel Inst 207:1556–1564

Morton H, Wingrove J (1972) Constitution of Bloomery Slags. Part II: Medieval. J Iron & Steel Inst 210:478–488

Muan A (1959a) Stability Relations among some Manganese Minerals. Amer Mineral 44:946–960

References

Muan A (1959b) Phase Equilibria in the System Manganese Oxide–SiO$_2$ in Air. Amer J Science 257:297–315

Muan A, Osborn EF (1965) Phase Equilibria among Oxides in Steelmaking. Addison-Wesley, Reading, Mass

Muhly JD (1980) The Bronze Age Setting. In: Wertime TA, Muhly JD (eds) The Coming of the Age of Iron. Yale University Press, New Haven/London, pp 25–67

Muhly JD (1984) Timna and King Solomon. Bibliotheca Orientalis XLI(3–4):276–292

Muhly JD (1986) Prehistoric Background Leading to the First Use of Metals in Asia. Bull Metals Mus, Sendai 11:21–42

Muhly JD (1988) The Beginnings of Metallurgy in the Old World. In: Maddin R (ed) The Beginnings of the Use of Metals and Alloys. MIT Press, Cambridge/Mass, pp 2–20

Muhly JD (1989) Çayönü Tepesi and the Beginnings of Metallurgy in the Ancient World. In: Hauptmann A, Pernicka E, Wagner GA (eds) Archäometallurgie der Alten Welt / Old World Archaeometallurgy. Der Anschnitt, Beiheft 7, Deutsches Bergbau-Museum Bochum, pp 1–11

Muhly JD (1991) The Development of Copper Metallurgy in Late Bronze Age Cyprus. In: Gale NH (ed) Bronze Age Trade in the Mediterranean. Studies Mediterran Archaeology XC:180–196

Muhly JD (1995) Lead Isotope Analysis and the Archaeologist. J Mediterran Archaeol 8(1):54–58

Muhly JD, Pernicka E (1992) Early Trojan Metallurgy and Metals Trade. In: Herrmann J (ed) Heinrich Schliemann: Grundlagen und Ergebnisse moderner Archäologie 100 Jahre nach Schliemanns Tod. Akademie, Berlin, pp 309–318

Muhly JD, Maddin R, Stech T (1988) Cyprus, Crete and Sardinia: Copper Oxhide Ingots and the Bronze Age Metals Trade. Rep Dept Antiquities Cyprus, pp 218–298

Muhly JD, Begemann F, Öztunali Ö, Pernicka E, Schmitt-Strecker S, Wagner G A (1991) The Bronze Age Metallurgy of Anatolia and the Question of Local Tin Sources. In: Pernicka E, Wagner GA (eds) Archaeometry '90. Birkhäuser, Basel, pp 209–220

Müller-Karpe A (1994) Altanatolisches Metallhandwerk. Offa-Bücher, Bd. 75. Wachholtz, Neumünster

Müller-Karpe M (1989) Neue Forschungen zur frühen Metallverarbeitung in Mesopotamien. Jb Röm-German Zentralmuseum Mainz 36:179–192

Müller-Neuhof B, Schmidt K, Khalil L, Eichmann R (2003) Warenproduktion und Fernhandel vor 6000 Jahren. Tall Hujayrat al Ghuzlan bei Aqaba. Alter Orient 4:22–25

Musil A (1907/8) Arabia Petrea. 2 Bände, Hölder, Wien

Naim GM, Rothenberg B (in prep) New Researches in Sinai: Archaeology and Archaeometallurgy. London

Najjar M, Abu Dayya A e S, Suleiman E, Weisgerber G, Hauptmann A (1990) Tell Wadi Feinan: The First Pottery Neolithic Tell in Southern Jordan. Ann Dept Antiquities Jordan XXXIV:27–56

Nassir SG, Khoury HN (1982) Geology, Mineralogy and Petrology of Daba Marble, Jordan. Dirasat 9:107–130

Neev D (1975) Tectonic Evolution of the Middle East and the Levantine Basin (Eastermost Mediterranean). Geology 3:683–686

Neuninger H, Pittioni R, Siegl W (1964) Frühkeramikzeitliche Kupfergewinnung in Anatolien. Archaeol Austriaca 35:98–110

Nissen HJ (1986) The Occurrence of Dilmun in the Oldest Texts of Mesopotamia. In: Al Khalifa Shaika H, Rice M (eds) Bahrein Through the Ages. KPI, London, pp 335–339

Noll W (1991) Alte Keramiken und ihre Pigmente. Schweizerbart, Stuttgart

Noth M (1966) Das 4. Buch Mose: Numeri. Göttingen, pp 210–213

Nurse RW, Midgley HG (1953) Studies on the Melilite Solid Solutions. J Iron & Steel Inst 174:121–131

Obenauer K (1954) Mikroskopische Untersuchungstechnik der Eisenhüttenschlacken. In: Freund H (ed) Handbuch der Mikroskopie in der Technik. Band 2, Teil 2, Mikroskopie der Bodenschätze. Umschau Frankfurt/Main, pp 459–518

Oelsen W (1952) Die Desoxydation von Kupferschmelzen mit Eisen, mit Phosphor und mit Schwefel, Giesserei, Techn.-Wiss. Beih. 618, pp 383–387

Ordentlich I, Rothenberg B (1980) Die Funde aus den freigelegten Bergbaurelikten. In: Conrad HG, Rothenberg B (eds) Antikes Kupfer im Timna-Tal. Der Anschnitt, Beiheft 1, Deutsches Bergbau-Museum Bochum, pp 169–180

Oren E (1989) The Overland Route between Egypt and Canaan in the Early Bronze Age, Preliminary Report. Israel Explor J 23:198–205

Osborn EF (1943) The Compound Merwinite ($3\,CaO \cdot MgO \cdot 2\,SiO_2$) and its Stability Relations within the System $CaO–MgO–SiO_2$ (Preliminary Report). J Amer Ceram Soc 26(10):321–332

Osborn EF, DeVries RC, Gee KH, Kraner HM (1954) Optimum Composition of Blast Furnace Slags as Deduced from Liquidus Data for the Quaternary System $CaO–MgO–Al_2O_3–SiO_2$. Trans Amer Inst Metall Eng 200:33–45

Otto H, Witter W (1952) Handbuch der ältesten vorgeschichtlichen Metallurgie in Mitteleuropa. Barth, Leipzig

Özdoğan M, Özdoğan A (1999) Archaeological Evidence on the Early Metallurgy at Çayönü Tepesi. In: Hauptmann A, Pernicka E, Rehren T, Yalçin Ü (eds) The Beginnings of Metallurgy. Der Anschnitt, Beiheft 9, Deutsches Bergbau-Museum Bochum, pp 13–22

Öztunali Ö (1989) Lagerstättenkundliche Probleme in der Archäometallurgie. In: Hauptmann A, Pernicka E, Wagner GA (eds) Archäometallurgie der Alten Welt/Old World Archaeometallurgy. Der Anschnitt, Beiheft 7, Deutsches Bergbau-Museum Bochum, pp 293–297

Palmieri A (1973) Scavi nell'Area sud-occidentale di Arslantepe. Ritrovamento di una Struttura Templare dell' Antica Età del Bronzo. Origini 7:55–228

Palmieri A, Hauptmann A (2000) Metals from Ebla: Chemical Analyses of Metal Artefacts from the Bronze and Iron Ages. In: Proc. I. Intern Congr Archaeol Ancient Near East, Rome 1998, pp 1259–1272

Palmieri A, Sertok K, Chernykh E (1993) From Arslantepe Metalwork to Arsenical Copper Technology in Eastern Anatolia. In: Frangipane M, Hauptmann H, Liverani M, Matthiae P, Mellink M (eds) Between the Rivers and Over the Mountains – Archaeologica Anatolica et Mesopotamica Alba Palmieri Dedicata. Universita La Sapienza, Rom, pp 573–599

Panayiotou A (1980) Cu-Ni-Co-Fe Sulphide Mineralization, Limassol Forest, Cyprus. In: Panayiotou A (ed) Ophiolites. Proc Intern Ophiolite Symp, Cyprus 1979. Nicosia, pp 102–116

Parker TW, Nurse RW (1943) Merwinite in the System $CaO–MgO–SiO_2$. J Iron & Steel Inst. CXLVIII:475–488

Patterson CC (1971) Native Copper, Silver and Gold Accessible to Early Metallurgists. Amer Antiquity 36:286–321

Pauly E (1964) The Wadi Araba Copper Exploration, Report on Phase A. Unveröff. Bericht Fa. Otto Gold, Köln

Percy J (1870) The Metallurgy of Lead. John Murray, London

Pernicka E (1987) Erzlagerstätten in der Ägäis und ihre Ausbeutung im Altertum. Jb Röm-German Zentralmuseum 34(2):607–714

Pernicka E (1992) Evaluating Lead Isotope Data: Comments on E.V. Sayre, K.A. Yener, E.C. Joel and I.L. Barnes Statistical Evaluation of the Presently Accumulated Lead Isotope Data from Anatolia and Surrounding Regions. Archaeometry 34(2):322–326

Pernicka E (1993) Evaluating Lead Isotope Data: Further Observations. Comments on P. Budd, D. Gale, A.M. Pollard, R.G. Thomas & P.A. Williams. Archaeometry 35(2):259–261

Pernicka E (1995) Gewinnung und Verbreitung der Metalle in prähistorischer Zeit. Jb Röm-German Zentralmuseum 37(1), 1990:21–129

Pernicka E (1999) Trace Element Fingerprinting of Ancient Copper: A Guide to Technology or Provenance? In: Young SMM, Pollard M, Budd P, Ixer R (eds) Metals in Antiquities. BAR Intern Series 792:163–171

Pernicka E (2004) Kupfer und Silber in Arisman und Tappeh Sialk und die frühe Metallurgie im Iran. In: Stöllner T, Slotta R, Vatandoust A (eds) Persiens Antike Pracht. Exhibition catalogue Bochum (= Veröffentlichungen aus dem Deutschen Bergbau-Museum Bochum 128), pp 232–239

Pernicka E, Hauptmann A (1989) Chemische und mineralogische Analyse einiger Erz- und Kupferfunde von Maadi. In: Rizkana I, Seeher J (eds) Maadi III, The Non-Lithic Small Finds and the Structural Remains of the Predynastic Settlement. Veröff Deutsches Archäologisches Institut Kairo 80:137–140

Pernicka E, Seeliger T, Wagner GA, Begemann F, Schmitt-Strecker S, Eibner C, Öztunali Ö, Baranyi I (1984) Archäometallurgische Untersuchungen in Nordwestanatolien. Jb Röm-German Zentralmuseum 31(2):533–599

Pernicka E, Begemann F, Schmitt-Strecker S, Grimanis AP (1990) On the Composition and Provenance of Metal Artefacts from Poliochni on Lemnos. Oxford J Archaeol 9(3): 263–298

Pernicka E, Begemann F, Schmitt-Strecker S, Wagner GA (1993) Eneolithic and Early Bronze Age Copper Artefacts from the Balkans and their Relation to Serbian Copper Ores. Prähist Zeitschr 68(1):1–54

Pernicka E, Schmidt K, Schmitt-Strecker S (2002) Zum Metallhandwerk. In: Schmidt K (ed) Norşuntepe. Kleinfunde II. Philip von Zabern, Mainz, pp 115–137

Pernicka E, Eibner C, Öztunali Ö, Wagner G (2003) Early Bronze Age Metallurgy in the Northeast Aegean. In: Wagner GA, Pernicka E, Uerpmann HP (eds) Troia and the Troad. Springer-Verlag, Berlin Heidelberg New York, pp 143–172

Perrot J (1955) The Excavations at Tell Abu Matar near Beer Sheba. Israel Explor J 5:17–40, 73–84, 167–189

Perrot J (1957) Les Fouilles d'Abou Matar près de Beersheba. Syria 34:1–38

Perrot J (1972) La Prehistoire Palestinienne. Supplement de la Bible, VI, Letouzey & Ané, Paris, pp 286–446

Perrot J (1984) Structures d'habitat, mode de vie et environment. Les Villages souterrains des pasteurs de Beer Sheva dans le sud d'Israel, au IVe millénaire avant l'ère Chretienne. Paleorient 10(1):75–96

Petrie F (1906) Researches in Sinai. London

Pfeiffer K (2004) Keramisches Gerät in der Südlevante – Gussformen und Gusstiegel aus der ersten Hälfte des 4. Jt. v. Chr. in der südlichen Levante. MA thesis, Freie Universität Berlin

Philip G (1989) Metal Weapons of the Early and Middle Bronze Ages in Syro-Palestine. BAR Intern Ser, vol I, II. 526. Oxford

Philip G (1991) Tin, Arsenic, Lead: Alloying Practices in Syria-Palestine around 2000 B.C. Levant XXIII:93–104

Philip G (2003) Contacts between the "Uruk" World and the Levant during the Fourth Millennium BC: Evidence and Interpretation. In: Artefacts of Complexity. Tracking the Uruk in the Near East. Brit School Archaeol Near East. Iraq Archaeological Rep 5:207–235

Philip G, Clogg P W, Dungworth D (2003) Copper Metallurgy in the Jordan Valley from the Third to the First Millennium BC: Chemical, Metallographic and Lead Isotope Analyses from Pella. Levant 35:71–100

Piel M, Hauptmann A, Schröder B (1992) Naturwissenschaftliche Untersuchungen an bronzezeitlichen Kupferverhüttungsschlacken von Acqua Fredda/Trentino. Universitätsforschungen zur prähistorischen Archäologie, Bd. 8, Habelt, Bonn, pp 463–472

Pigott VC, Natapintu S (1988) Archaeological Investigations into Prehistoric Copper Production. In: Maddin R (ed) The Beginnnings of the Use of Metals and Alloys. MIT Press, Cambridge/Mass., pp 156–162

Pigott V, Howard S, Epstein SM (1982) Pyrotechnology and cultural change at Bronze Age Tepe Hissar. In: Wertime TA (ed) Early pyrotechnology. The evolution of the first fire-using industries. Smithsonian Inst. Press, Washington, pp 215–236

Pilz R (1917) Kupfererzlagerstätten in der Gegend von Arghana Maden. Zeitschr Prakt Geologie 25:191–198

Pittioni R (1957) Urzeitlicher Bergbau auf Kupfererz und Spurenanalyse: Beiträge zum Problem der Relation Lagerstätte-Fertigobjekt. Archaeol Austriaca Beiheft 1

Pleiner R (2000) Iron in Archaeology. Archaeologický Ústav, Praha

Pollard AM, Thomas RG, Williams PA (1990) Experimental Smelting of Arsenical Copper: Implications for Early Bronze Age Copper Production. In: Crew P, Crew S (eds) Early Mining in the British Isles. Plas Tan y Bwlch Occ Papers 1:72–74

Pollard AM, Thomas RG, Ware DP, Williams PA (1991) Experimental Smelting of Secondary Copper Minerals: Implications for Early Bronze Age Metallurgy in Britain. In: Pernicka E, Wagner GA (eds) Archaeometry '90, Birkhäuser, Basel, pp 127–136

Potaszkin R, Bar-Avi K (1980) A Material Investigation of Metal Objects from the Nahal Mishmar Treasure. In: Bar-Adon P (ed) The Cave of the Treasure. Israel ExplorSoc, Jerusalem, pp 235–237

Pouit G, Castel G (1999) Les exploitations pharaoniques, romaines et arabes de cuivre, fer et or. L'exemple des ouadi Dara (désert oriental d'Egypte). In: Cauuet B (ed) L'or dans l'Antique, Aquitania suppl. 9, pp 113–144

Prange M (1998) Vergleichende Untersuchungen zur Charakterisierung des omanischen Kupfers mittels chemischer und isotopischer Analysenmethoden. Dissertation, Fakultät für Chemie, Ruhr-Universität Bochum

Prange M (2001) 5 000 Jahre Kupfer in Oman, Band II: Vergleichende Untersuchungen zur Charakterisierung des omanischen Kupfers mittels chemischer und isotopischer Analysemethoden. Metalla 8,1/2:1–126

Prange M, Götze HJ (1995) Analyse früher Kupfer- und Eisenschlacken mittels oprischer Atomspektrometrie (ICP-OES). Metalla 2:29–41

Presslinger H, Walach G, Eibner C (1988) Bronzezeitliche Verhüttungsanlagen zur Kupfererzeugung in den Ostalpen. Berg- u. Hüttenmänn Mh 133:338–344

Prokop FW, Schmidt-Eisenlohr WF (1966) Erfahrungen bei der Untersuchung von Kupfererzvorkommen im Wadi Araba, Jordanien. Zeitschr Erzbergbau Metallhüttenwesen XIX:111–120

Pusch E (1990) Metallverarbeitende Werkstätten der frühen Ramessidenzeit in Qantir-Piramesse/Nord. Ägypten u. Levante 1:75–113

Quenell AM (1951) The Geology and Mineral Resources of (former) Transjordan. Colon Geol Miner Res 2:85–115

Quintero L (1996) Flint-Mining in the Pre-Pottery Neolithic: Preliminary Report on the Exploitation of Flint at Neolithic Ain Ghazal in Highland Jordan. In: Kozlowski SK, Gebel HG (eds) Neolithic Chipped Stone Industries of the Fertile Crescent and their Contemporaries in Adjacent Regions. Stud Near Eastern Production, Subsistence and Environment 2:233–242

Raghavan V (1976) Ancient Sites in the Wadi Arabah and Nearby. Unpubl Rep Brit Inst Archaeology Amman

Raghavan V (1980) Notes on some Neolithic and Later Sites in Wadi Arabah and the Dead Sea Valley. Levant XII:40–60

Raghavan V (1985) The Charakter of the Wadi Arabah. In: Hadidi A (ed) Studies in the History and Archaeology of Jordan, II. Dept Antiquities Jordan, Amman, pp 95–102

Raghavan V (1988) Phase Diagrams of Ternary Iron Alloys, Part 3: Ternary Systems Containing Iron & Phosphorus. Indian Inst Metals Calcutta, pp 68–73

Ramdohr P (1975) Die Erzmineralien und ihre Verwachsungen. Akademie, Berlin

Rast WE (1979a) Settlement in Numeira. In: Rast WE, Schaub RT (eds) The Southeastern Dead Sea Plain Expedition: An Interim Report of the 1977 Season. Ann Amer Soc Oriental Res 46:35–44

Rast WE (1979b) Patterns of Settlement at Bab edh-Dhra. In: Rast WE, Schaub RT (eds) The Southeastern Dead Sea Plain Expedition: An Interim Report of the 1977 Season. Ann Amer Soc Oriental Res 46:7–34

Rast WE, Schaub RT (1978) A Preliminary Report of Excavation at Bab edh-Dhra, 1975. Ann Amer Soc Oriental Res 43:1–32

Rast WE, Schaub RT (1980) Preliminary Report of the 1979 Expedition to the Dead Sea Plain, Jordan. Bull Amer Soc Oriental Res 240:21–61

Rehder JE (1986) Primitive Furnaces and the Development of Metallurgy. J Hist Metall Soc 20(2):87–92

Rehder JE (1987) Natural Draft Furnaces. Archeomaterials 2:47–58

Rehder JE (1994) Blowpipes versus Bellows in Ancient Metallurgy. J Field Archaeol 21: 345–350

Rehren T (1997a) Tiegelmetallurgie: Tiegelprozesse und ihre Stellung in der Archäometallurgie. Habilitationsschrift, Fakultät f. Werkstoffwissenschaften und werkstofftechnologie, TU Bergakademie Freiberg

Rehren T (1997b) Die Kupfersulfid-Krusten der Pfyner Schmelztiegel. Unveröff. Manuskript, Deutsches Bergbau-Museum Bochum

References

Rehren T (1998) Rationales in Old World Base Glass Compositions. J Archaeol Science 27: 1225–1234

Rehren T, Hess K, Philip G (1997) Fourth Millennium BC Copper Metallurgy in Northern Jordan: Evidence from Tell esh-Shuna. In: Gebel HG, Kafafi Z, Rollefson G (eds) Prehistory of Jordan II. Berlin, pp 625–640

Reiter K (1997) Die Metalle im Alten Orient unter besonderer Berücksichtigung altbabylonischer Quellen. Ugarit-Verlag, Münster

Renfrew C (1967) Cycladic Metallurgy and the Aegean Early Bronze Age. Proc Prehistoric Soc 35:12–47

Rickard TA (1932) Man and Metals. New York/London

Rizkana I, Seeher J (1989) Maadi III. The Non-Lithic Small Finds and the Structural Remains of the Predynastic Settlement. Veröff Deutsches Archäologisches Inst Kairo 80. Mainz, Pilipp von Zabern

Roden C (1988) Blasrohrdüsen. Ein archäologischer Exkurs zur Pyrotechnologie des Chalkolithikums und der Bronzezeit. Der Anschnitt 40(3):62–82

Rollefson G, Simmons A (1986) The 1985 Season at 'Ain Ghazal: Preliminary Report. Ann Dept Antiquities Jordan XXX:41–56

Roman I (1990) Copper Ingots. In: Rothenberg B (ed) Researches in the Arabah 1959–1984, vol II: The Ancient Metallurgy of Copper. Inst Archaeo-Metall Studies, London, pp 176–181

Ronen A (1980) Biqat Uvda 20 (Nahal Reuel). Hadashot Archaeologiot 74/75:47–48

Rose D, Endlicher G, Mucke A (1990) The Occurrence of "Iscorite" in Medieval Iron Slags. J Hist Metall Soc 24(1):27–32

Rosenberg M (1994) Hallan Çemi Tepesi: Some Further Observations Concerning Stratigraphy and Material Culture. Anatolica 20:121–140

Roshwalb A (1981) Protohistory in the Wadi Ghazzeh: A Typological and Technological Study Based on the MacDonald Excavations. Dissertation, Instit Archaeol, University of London

Rostoker W, McNattan M, Gebhard E (1983) Melting / Smelting of Bronze at Isthmia in Greece. J Hist Metall Soc 17,1:23–27

Rostoker W, Pigott VC, Dvorak JR (1989) Direct Reduction to Copper Metal by Oxide-Sulfide Mineral Interaction. Archeomaterials 3(1):69–87

Rothenberg B (1973) Das Tal der biblischen Kupferminen. Lübbe, Bergisch-Gladbach

Rothenberg B (1978) Excavations at Timna Site 39, a Chalcolithic Copper Smelting Site and Furnace and its Metallurgy. In: Rothenberg B (ed) Archaeometallurgy: Chalcolithic Copper Smelting. Inst Archaeo-Metall Studies, Monograph 1, London, pp 1–15

Rothenberg B (1980) Einleitung zu: Conrad HG, Rothenberg B (eds) Antikes Kupfer im Timna-Tal. Der Anschnitt, Beiheft 1, Deutsches Bergbau-Museum Bochum, pp 17–32

Rothenberg B (1983) Ancient Jordan City may Rival Timna's Place in Copper History. Inst Archaeo-Metall Studies Newsletter 4:4–5

Rothenberg B (1985) Copper Smelting Furnaces in the Arabah, Israel: the Archaeological Evidence. In: Craddock PT, Hughes MJ (eds) Furnaces and Smelting Technology in Antiquity. Brit Museum Occ Papers 48:123–150

Rothenberg B (1986) Radiocarbon (C14) Dating Helps to Solve Riddle of Timna's Late Bronze Age Smelting Furnaces. Inst Archaeo-Metall Studies Newsletter 9:7

Rothenberg B (1987) Pharaonic Copper Mines in Southern Sinai. Inst Archaeo-Metall Studies Newsletter 10/11:1–7

Rothenberg B (1988a) Researches in the Arabah 1959–1984, vol 1: The Egyptian Mining Temple. Inst Archaeo-Metall Studies, London

Rothenberg B (1988b) Early Islamic Copper Smelting – and Worship – at Beer Ora, Southern Arabah (Israel). Inst Archaeo-Metall Studies 12:1–4

Rothenberg B (1990) Copper Smelting Furnaces, Tuyeres, Slags, Ingot-Moulds and Ingots in the Arabah: The Archaeological Data. In: Rothenberg B (ed) Researches in the Arabah 1959–1984, vol II: The Ancient Metallurgy of Copper. Inst Archaeo-Metall Studies, London, pp 1–77

Rothenberg B (1991) Archaeo-Metallurgy on the Trail of History. Inst. Archaeo-Metall. Studies Newsletter 17, 1–7. The Ghassulian-Beersheba Chalcolithic Enigma. Inst. Archaeo-Metall Studies Newsletter 17:6–7

Rothenberg B (1998) Who were the "Midianite" copper miners of the Arabah? About the Midianite enigma. In: Rehren T, Hauptmann A, Pernicka E, Yalçin Ü (eds) Metallurgica Antiqua. In honour of Hans-Gert Bachmann and Robert Maddin. Der Anschnitt, Beiheft 8, Deutsches Bergbau-Museum Bochum, pp 197–212

Rothenberg B (1999) Archaeo-Metallurgical Researches in the Southern Arabah 1959–1990, part I: Late Pottery Neolithic to Early Bronze Age IV. Palestine Explor Quart 131:68–89

Rothenberg B, Blanco-Freijeiro A (1981) Studies in Ancient Mining and Metallurgy in South-West Spain. Metal in History I. Inst Archaeo-Metall Studies, London

Rothenberg B, Glass J (1992) The Beginnings and the Development of Early Metallurgy and the Settlement and Chronology of the Western Arabah, from the Chalcolithic Period to Early Bronze Age IV. Levant XXIV:141–157

Rothenberg B, Merkel JF (1995) Late Neolithic Copper Smelting in the Arabah. Inst Archaeo-Metall Studies Newsletter 15/16:1–8

Rothenberg B, Shaw CT (1990) The Discovery of a Copper Mine and Smelter from the End of the Early Bronze Age (EB IV) in the Timna Valley. Inst Archaeo-Metall Studies Newsletter 15/16:1–8

Roy S (1968) Mineralogy of Different Genetic Types of Manganese Deposits. Econ Geol 63:760–786

Ryndina N, Indenbaum G, Kolosova V (1999) Copper Production from Polymetallic Sulphide Ores in the Northeastern Balkan Eneolithic Culture. J Archaeolog Science 26:1059–1068

Saez R, Nocete F, Nieto J, Capitán À, Rovira S (2003) The extractive metallurgy of copper from Cabezo Jurè, Huelva, Spain: chemical and mineralogical study of slags dated to the third millennium BC. Canadian Mineral 41:627–638

Sangmeister E (1971) Aufkommen der Arsenbronze in SO-Europa. Actes du VIII Congrès International des Sciences Préhistoriques et Protohistoriques, Bd. I. Beograd, pp 131–138

Sayre EV, Yener KA, Joel EC, Barnes IL (1992) Statistical Evaluation of the Presently Accumulated Lead Isotope Data from Anatolia and Surrounding Regions. Archaeometry 34(1):73–105

Scharpenseel HW, Pietig F, Schiffmann H (1976) Hamburg University Radiocarbon Dates. Radiocarbon 18(3):286–287

Schaub RT, Rast WE (1984) Preliminary Report of the 1981 Expedition to the Dead Sea Plain. Bull Amer Soc Oriental Res XX:35–60

Schlegel H, Schüller A (1952) Die Schmelz- und Kristallisationsgleichgewichte im System Kupfer–Eisen–Schwefel und ihre Bedeutung für die Kupfergewinnung. Freiberger Forschungsh Reihe B, 2:3–32

Schlick-Nolte B (1999) Ägyptische Fayence und Ägyptisch Blau im alten Ägypten. In: Busz R, Gercke P (ed) Türkis und Azur: Quarzkeramik in Orient und Okzident. Staatl. Museen Kassel, pp 12–51

Schliemann H (1881) Ilios. Stadt und Land der Trojaner. Forschungen und Entdeckungen in der Troas und besonders auf der Baustelle von Troja. Leipzig, Brockhaus

Schlimbach J (1992) Versuche zur Kupferverhüttung in der Vor- und Frühzeit. Unveröffentlichte Studienarbeit, Institut für Metallhüttenkunde u. Elektrometallurgie, RWTH Aachen

Schmidt R, Eichmann R, Müller-Neuhof B, Khalil L (2003) Tall Hujayrat al-Ghuzlan 2002. Occident & Orient 8(1):9–11

Schmiedl J, Stofko M, Repcak V (1971) Gleichgewichtsuntersuchungen im System Cu–Fe–S–O. Neue Hütte 7:390–395

Schmitt-Strecker S, Begemann F, Pernicka E (1992) Chemische Zusammensetzung und Bleiisotopenverhältnisse der Metallfunde vom Hassek Höyük. In: Behm-Blancke M (ed) Hassek Höyük – Naturwissenschaftliche Beiträge. Istanbuler Forsch 38:108–123

Schmitt-Strecker S, Begemann F, Pernicka E (1994) Untersuchungen zur Metallurgie der Späten Uruk- und Frühen Bronzezeit am oberen Euphrat – Resumé. In: Wartke B (ed) Handwerk und Technologie im Alten Orient. Ph. v. Zabern, Mainz, pp 99–109

Schneider G, Zimmer G (1984) Technische Keramik aus antiken Bronzegußwerkstätten in Olympia und Athen. Berliner Beitr Archäometrie 9:17–60

Schoop UD (1995) Die Geburt des Hephaistos. Technologie und Kulturgeschichte neolithischer Metallverwendung im Vorderen Orient. Intern Archäologie, Bd. 24. Marie Leidorf, Rahden, Westfalen

Schreiner M (2002) Mineralogical and Geochemical Investigations into Prehistoric Smelting Slags from Tepe Sialk/Central Iran. Diploma thesis, Technical University Bergakademie Freiberg

Schröder B, Yalcin Ü (1991) Milet 1990: Geologische Begleituntersuchungen. Istanbuler Mitt 41:149–156

Schubert E (1981) Zur Frage der Arsenlegierungen in der Kupfer- und Frühbronzezeit Südosteuropas. In: Lorenz L (ed) Studien zur Bronzezeit, Festschrift für Wilhelm Albert v. Brunn. Ph. v. Zabern, Mainz, pp, 447–459

Schürmann E (1958) Die Reduktion des Eisens im Rennfeuer. Stahl & Eisen 19:1297–1308

Schwab MJ, Neumann F, Litt T, Negendank JFW, Stein M (2004) Holocene palaeoecology of the Golan Heights (Near East): investigation of lacustrine sediments from Birkat Ram crater lake. Quaternary Science Rev 23:16–17, 1723–1731

Schwarz G (1980) Die Verteilung von Cu und anderen Elementen auf die während der Verhüttung von Kupfererzen im Flammofen entstehenden Mineralphasen der Schlacke. Dissertation FB Geowissenschaften, University of Hamburg

Scott GS (1912) Copper and its Alloys in Early Times. J Inst Metals VII(1):23–49

Seaton P (1995) A Note on Possible Chemical Industries at Teleilat Ghassul. In: Bourke S, Descendres JP (eds) Trade, Contact and the Move of Peoples in the Eastern Mediterranean. Mediterran Archaeology, Suppl. 3:27–30

Sebanne M, Avner U (1993) Biq'at Nimra: A Tomb of the Early Bronze Age I. Atiqot 22:33–40

Seeliger T, Pernicka E, Wagner GA, Begemann F, Schmitt-Strecker S, Eibner C, Öztunali Ö (1985) Archäometallurgische Untersuchungen in Nord- und Ostanatolien. Jb Röm-German Zentralmuseum 32:597–659

Segal I, Halicz L, Cohen R (1999) A study of ingots and metallurgical remains from 'Ein Ziq and Beer Resisim, Central Negev, Israel. In: Young SMM, Pollard AM, Budd P, Ixer RA (eds) Metals in Antiquity. BAR Intern Series S 792, Oxford, pp 179–186

Segal I, Ilani S, Rosenfeld A (2000) Wadi Tar Copper-Arsenic Ore – Lead Isotope Study: Was it Used in Canaan during the Chalcolithic, EB and MBI Periods? Geol Survey Israel 12:244–246

Segal I, Halicz L, Kamenski A (2004) The Metallurgical Remains from Asquelon, Afridar – Areas E, G and H. Atiqot 45:311–330

Segev A (1986) Lithofacies Relations and Mineralization Occurrences in the Timna-Formation, Timna Valley. Ph.D. Thesis, Hebrew University, Jerusalem (Hebrew)

Segev A (1987) The Age of the Latest Precambrian Volcanism in Southern Israel, Northeastern Sinai and Southwestern Jordan – A Re-Evaluation. Precambrian Res 36:277–285

Segev A, Sass E (1989) Copper-Enriched Syngenetic Dolostones as a Source for Epigenetic Copper Mineralization in Sandstones and Shales (Timna, Israel). Geol Assoc Canada Spec Paper 36:647–658

Segev A, Beyth M, Bar-Matthews M (1992) The Geology of the Timna Valley with Emphasis on Copper and Manganese Mineralization – Updating and Correlation with the Eastern Margins of the Dead Sea Rift. Geol Survey Israel Rep No GSI 14:1–31

Segev A, Goldshmidt V, Itamar A, Rybakov M (1996) Effects of Mesozoic Magmatism on Composition, Structure, and Metallic Mineralization in the Ramon Area (Southern Israel): Magnetometric and Gravimetric Evidence. Israel J Earth Sc 45(2):89–112

Selimkhanov I R (1972) Ergebnisse von spektralanalytischen Untersuchungen an Metallgegenständen des 4. und 3. Jahrtausends aus Transkaukasien. Germania 44:221–233

Shalev S (1991) Two Different Copper Industries in the Chalcolithic Culture of Israel. In: Mohen JP, Éluère C (eds) Découverte du Métal. Millénnaire 2:413–424

Shalev S (1994) The Change in Metal Production from the Chalcolithic Period to the Early Bronze Age in Israel and Jordan. Antiquity 68:630–637

Shalev S, Braun E (1997) The Metal Objects from Yiftahel II. In: Braun E (ed) Yiftah'el. Israel Antiquities Authority Rep 2:92–96
Shalev S, Northover PJ (1987) Chalcolithic Metal and Metalworking from Shiqmim. In: Levy TE (ed) Shiqmim I. BAR Intern Ser 356:357–371
Shalev S, Northover PJ (1993) The Metallurgy of the Nahal Mishmar Hoard Reconsidered Archaeometry 35(1):35–47
Shalev S, Northover PJ (1999) Recasting the Nahal Mishmar Hoard: Experimental archaeology and metallurgy. In: Hauptmann A, Pernicka E, Rehren T, Yalçin (eds) The Beginnings of Metallurgy. Der Anschnitt, Beiheft 9, Deutsches Bergbau-Museum Bochum, pp 295–299
Shanks RI (1936) Shism al Tasa 26/37D. DGMR Rep No. 3/204/6023, Jeddah
Shaw SH (1948) The Geology and Mineral Resources of Palestine. Bull Imp Inst 46:87–103
Shimada I, Merkel JF (1991) Copper-Alloy Metallurgy in Ancient Peru. Scientific Amer 265(1): 62–75
Shimron AE (1972) The Precambrian Structural and Metamorphic History of the Elat Area. Dissertation, Dept. of Geology, Hebrew University, Jerusalem
Shpitzer M, Beyth M, Matthews A (1989) Evidence for Magmatic Differentiation in the Timna Intrusive Complex. Israel Geolog Soc Ann Meeting, Ramot, pp 52
Shugar AN (2000) Archaeometallurgical investigation of the chalcolithic site of Abu Matar, Israel: a reassessment of technology and its implications for the Ghassulian culture. PhD thesis, Institute of Archaeology, University of London
Siedner G, Horowitz (1974) Radiometric Ages of Late Caenozoic Basalts from Northern Israel: Chronostratigraphic Implications. Nature 250:23–26
Sillitoe RH (1972) Formation of Certain Massive Sulphide Deposits at Sites of Sea-Floor Spreading. Trans Inst Min Met 81,B:141–148
Sillitoe RH (1979) Metallogenic Consequences of Late Precambrian Suturing in Arabia, Egypt, Sudan and Iran. In: Al-Shanti AMS (ed) Proc. Symp. Evolution and Mineralization of the Arabian-Nubian Shield, Jeddah 1978, vol 1., Pergamon Press, Oxford, pp 109–120
Simon K (1993) Zum ältesten Erzbergbau in Ostthüringen und Sachsen. Argumente und Hypothesen. In: Steuer H, Zimmermann U (ed) Montanarchäologie in Europa. Archäologie und Geschichte: Freiburger Forschungen zum ersten Jahrtausend in Südwestdeutschland, Bd. 4, Thorbecke, Sigmaringen, pp 89–104
Slatkine A (1961) Nodules Cupriferes du Negev Meridional (Israel). Bull Res Counc Israel Sect. G10:292–301
Smirnov SS (1954) Die Oxydationszone sulfidischer Lagerstätten. Akademie, Berlin
Smith CS (1967) Tal-i Iblis: Metallurgical Archaeology. Iran 5:146–147
Smuts J, Steyn J G D, Boeyens J C A (1969) The Crystal Structure of an Iron Silicate, Iscorite. Acta Crystallographica B25:1251–1255
Solecki RS (1969) A Copper Mineral Pendant from Northern Iraq. Antiquity 43:311–314
Sozzi M, Vannucci S, Vaselli O, Vianello Fabio (1991) Indagini sulle amazzoniti provenienti dai siti neolitici della Giordania Meridionale. Stud. Ecologia Quaternario 13:35–42
Sperl G (1979) Eisen im alten Kupfer. Freiberger Forschungsh. Reihe B217:17–25
Sperl G (1980) Über die Typologie urzeitlicher, frühgeschichtlicher und mittelalterlicher Eisenhüttenschlacken. Stud Industrie – Archäol VII. Öster. Akad. Wiss., Wien
Spindler K (1971) Zur Herstellung der Zinnbronze in der frühen Metallurgie Mitteleuropas. Acta Prehistorica et Archaeologica 2:199–253
Stacey JS, Kramers JD (1975) Approximation of Terrestrial Lead Isotope Evolution by a Two-Stage Model. Earth Planetary Sc Letters 26:207–221
Stacey JS, Doe RB, Robertson RJ, Delevaux MH, Gramlich JW (1980) A lead isotope study of mineralization in the Saudi Arabian Shield. Contrib Miner Petrol 74:175–188
Stein G, Edens C, Edens JP, Boden K, Laneri N, Özbal H, Earl B, Adriaens AM, Pittman H (1998) Southeast Anatolia before the Uruk Expansion: Preliminary Report on the 1997 Excavation at Hacınebi, Turkey. Anatolica XXIV:143–193
Steinberg AR, Koucky FL (1974) Preliminary Metallurgical Research on the Ancient Cypriot Copper Industry. In: Stager L, Walker A, Wright EG (eds) Amer Expedition to Odalion, Cyprus. Suppl. Bull Amer School Oriental Res 18:149–178

References

Steinhof M (1994) Untersuchungen an Holzkohlen aus eisenzeitlichen Schlackenhalden in Khirbet en-Nahas (Wadi Arabah, Jordanien). Diplomarbeit, Inst. Syst. Botanik, FU Berlin

Stöllner Thomas (2003) Mining and Economy – A Discussion of Spatial Organisation and Structures. In: Stöllner T, Körlin G, Steffens G, Cierny J (eds) Man and Mining – Mensch und Bergbau. Der Anschnitt, Beiheft 16, Deutsches Bergbau-Museum Bochum, pp 415–446

Stos-Gale ZA (1991) Lead Isotope Studies – A Bar Ingot and a Macehead from the Negev. In: Rothenberg, B., Archaeo-Metallurgy on the Trail of History. Inst Archaeo-Metall Studies Newsletter 17:5–6

Stos-Gale ZA (1993) Isotopic Analyses of Ores, Slags and Artefacts: The Contribution to Archaeometallurgy. In: Francovich R (ed) Archaeologia delle attività estrattive e metallurgiche. Firenze, pp 593–627

Stos-Gale ZA, Gale NH (1994) Metals. In: Knapp B, Cherry JF (ed) Provenience Studies and Bronze Age Cyprus. Monographs in World Archaeology 21. Prehistory Press, Madison/Wisc., pp 92–121

Stos-Gale ZA, Gale NH (2006) Lead Isotope and Chemical Analyses of Slags from Chrysokamino. In: Betancourt P (ed) The Chrysokamino Metallurgy Workshop and its Territory. Hesperia Suppl. 36:299–319

Stos-Gale ZA, MacDonald CF (1991) Sources of Metals and Trade in the Bronze Age Aegean. In: Gale NH (ed) Bronze Age Trade in the Mediterranean. Stud Mediterran Archaeology XC:249–284

Stos-Gale ZA, Gale NH, Zwicker U (1986) The Copper Trade in the Southwest Mediterranean Region. Preliminary Scientific Evidence. Rep Dept Antiquities Cyprus Nicosia, pp 122–144

Stos-Gale ZA, Gale NH, Papastamataki A (1988) An Early Bronze Age Copper Smelting Site on the Aegean Island of Kythnos. In: Jones E (ed) Aspects of Ancient Mining and Metallurgy. Bangor, pp 23–30

Stos-Gale ZA, Gale NH, Houghton J (1995) The Origin of Egyptian Copper. Lead Isotope Analysis of Metals from Amarna. In: Davies WV, Schofield L (ed) Egypt, the Aegean and the Levant. Brit Museum, London, pp 127–135

Stos-Gale Z, Maliotis GG, Noel H, Annetts N (1997) Lead Isotope Characteristics of the Cyprus Copper Ore Deposits Applied to Provenance Studies of Copper Oxhide Ingots. Archaeometry 39(1):83–123

Strahm C (1994) Die Anfänge der Metallurgie in Mitteleuropa. Helvetia Archaeologica 25(97): 2–39

Straube H (1996) Ferrum Noricum und die Stadt auf dem Magdalensberg. Springer-Verlag, Wien

Stuiver M, Reimer P (1993) Extended ^{14}C Data Base and Revised CALIB 3.0 ^{14}C Age Calibration Program. Radiocarbon 35(1):215–230

Tadmor M (2002) The Kfar Monash Hoard Again – A View from Egypt and Nubia. In: van den Brinck, E, Levy TE (eds) Egypt and the Levant – Interrelations from the 4th through the Early 3rd millenium B.C.E. Continuum, London and New York, pp 239–251

Tadmor M, Kedem D, Begemann F, Hauptmann A, Pernicka E, Schmitt-Strecker S (1995) The Nahal Mishmar Hoard from the Judean Desert: Technology, Composition, and Provenance. Atiqot XXVII:95–148

Tafel V (1951) Lehrbuch der Metallhüttenkunde Bd. 1. Hirzel, Leipzig

Talbot G (1983) Beads and Pendants from the Tells and Tombs. In: Kenyon KM, Holland TA (ed) Excavations at Jericho, vol V, London, pp 788–801

Taute W (1994) Pre-Pottery Neolithic Flint Mining and Flint Workshop Activities Southwest of the Dead Sea, Israel (Ramat Tamar and Mesad Mazzal). In: Gebel HG, Kozlowski SK (eds) Chipped Stone Industries of the Fertile Crescent. Studies in Early Near Eastern Production, Subsistance and Environment 1:495–509

Taylor NW, Williams FJ (1935) Reactions between Solids in the System $CaO–MgO–SiO_2$ in the Temperature Range 600–1200 °C. Bull Geological Soc 46:281f

Thalhammer O, Stumpfl EF, Panayiotou A (1986) Postmagmatic, Hydrothermal Origin of Sulfide and Arsenide Mineralizations at Limassol Forest, Cyprus. Miner Deposita 21:95–105

Theophilus P (1984) Diversarum Artium Schedula (Technik des Kunsthandwerks im 12. Jahrhundert). Translated and annotated by W. Theobald. VDI Verlag, Düsseldorf

Thompson TL (1975) The Settlements of Sinai and the Negev in the Bronze Age. Beihefte Tübinger Atlas Vord. Orient B8, Wiesbaden

Tilley CE (1952) Some Trends of Basaltic Magma in Limestone Syntexis. Amer J Science (Bowen Volume), pp 529–545

Tite MS, Freestone IC, Bimson M (1983) Egyptian Fayence: An Investigation of their Methods of Production. Archaeometry 25(1):17–27

Trojer F (1951) Delafossit in einer Kupferverblaseschlacke. Mikroskopie 7(8):232–236

Trojer F (1963) Die oxydischen Kristallphasen der anorganischen Industrieprodukte. Schweizerbart, Stuttgart

Trömel G, Schwerdtfeger K (1963) Untersuchungen im System Eisen–Phosphor–Sauerstoff. Archiv Eisenhüttenwesen 34:55–59

Tubb KW (1985) Preliminary Report on the Ain Ghazal Statues. Mitt. Deutsche Orientges 117:117–134

Tufnell O (1958) Lachish IV (Tell Ed-Duweir), The Bronze Age. Oxford University Press, Oxford

Turner WES (1958) Report on Crucibles. In: Tufnell O (ed) Lachish IV (Tell Ed-Duweir), The Bronze Age. Oxford, pp 145–146

Twaltschrelidze A (2001) Erzlagerstätten in Georgien. In: Gambaschidze I, Hauptmann A, Slotta R, Yalçın Ü (eds) Georgien – Schätze aus dem Land des Goldenen Vlies. Exhibition Catalogue (= Veröffentlichungen aus dem Deutschen Bergbau-Museum Bochum 100), pp 78–89

Tylecote RF (1970) Early Metallurgy in the Near East. Metals and Materials 7:285–293

Tylecote RF (1974) Can Copper be Smelted in a Crucible? J Hist Metall Soc 8(1):54

Tylecote RF (1976) A History of Metallurgy. The Metals Society, London

Tylecote RF (1982a) Early Copper Slags and Copper-Base Metal from the Agadez Region of Niger. J Hist Metall Soc 16 (2):58–64

Tylecote RF (1982b) Metallurgical Crucibles and Crucible Slags. In: Odin JS, Franklin AD (eds) Archaeological Ceramics. Smithsonian Institution Press, Washington D.C., pp 231–242

Tylecote RF (1987) The Early History of Metallurgy in Europe. Longman Archaeology Series, London

Tylecote RF (1992a) A History of Metallurgy. Inst Materials, London

Tylecote RF (1992b) Extraction Metallurgy; Historical Development and Evolution of the Processes. In: Antonnacci Sanpaolo E (ed) Archeometallurgia Ricerche e Prospettive. Bologna, pp 25–51

Tylecote RF, Boydell PJ (1978) Experiments on Copper Smelting Based upon Early Furnaces Found at Timna. In: Rothenberg B (ed) Archaeometallurgy: Chalcolithic Copper Smelting. Inst Archaeo-Metallurg Studies, Monograph 1, London, pp 27–49

Tylecote RF, Merkel JF (1985) Experimental Smelting Techniques: Achievments and Future. In: Craddock PT, Hughes MJ (eds) Furnaces and Smelting Technology in Antiquity. Brit Museum Occ Papers 48:13–20

Tylecote RF, Austin JN, Wraith AE (1971) The Mechanism of the Bloomery Process. J Iron Steel Inst 5:342–363

Tylecote RF, Rothenberg B, Lupu A (1974) The Examination of Metallurgical Material from Abu Matar, Israel. J Hist Metall Soc 8(1):32–34

Tylecote RF, Ghaznavi HA, Boydell PJ (1977) Partitioning of Trace Elements between the Ores, Fluxes, Slags and Metal during the Smelting of Copper. J Archaeol Science 4(4):305–333

Tylecote RF, Balmuth MS, Massoli-Novelli R (1983a) Copper and Bronze Metallurgy in Sardinia. J Hist Metall Soc 17:63–78

Tylecote RF, Balmuth MS, Massoli-Novelli R (1983b) Copper and Bronze Metallurgy in Sardinia. In: Balmuth MS, Rowland RJ Jr (eds) Studies in Sardinian Archaeology. University of Michigan, pp 115–162

Uçurum A (2000) Listwaenites in Turkey: Perspectives on Formation and Precious Metal Concentration with Reference to Occurrences in East-Central Anatolia. Ofioliti 25(1):15–29

Van den Boom G, Lahloub M (1962) The Iron-Ore Deposit "Warda" in the Southern Ajlun District. Unveröff. Ber. Deutsche Geol. Mission Jordanien, Arch. Bundesanst. f. Geowissenschaften & Rohstoffe, Hannover

Van den Boom G, Rösch H (1969) Modalbestand und Petrochemie der Granite im Gebiet von Aqaba-Quweira, Südjordanien. Beih Geol Jahrb 81:113–148

Vieweger D (2003) Archäologie der biblischen Welt. UTB Göttingen

Vogel R, Berak J (1950) Das System Eisen–Eisenphosphid–Kupferphosphid–Kupfer. Archiv Eisenhüttenwesen E 1181:327–336

Voigt M (1985) Village on the Euphrates: Excavations at Neolithic Gritille in Turkey. Expedition 27(1):10–24

Voss O (1988) The Iron Production in Populonia. In: Sperl G (ed) The First Iron in the Mediterranean. PACT 21:91–100

Waetzold H, Bachmann HG (1984) Zinn- und Arsenbronzen in den Texten aus Ebla und aus dem Mesopotamien des 3. Jahrtausends. Oriens Antiquus XXIII:1–17

Wagner GA (1988) Die Anfänge der Kupfermetallurgie in Kleinasien. Die Geowissenschaften 6(11):323–329

Wagner GA (1998) Age Determination of Young Rocks and Artifacts. Springer-Verlag, Berlin, Heidelberg New York

Wagner GA, Öztunali Ö (2000) Prehistoric copper sources in Turkey In: Yalçin Ü (ed) Anatolian Metal I. Der Anschnitt, Beiheft 13, Deutsches Bergbau-Museum Bochum, pp 31–67

Wagner GA, Weisgerber G (1985) Silber, Blei und Gold auf Sifnos. Prähistorische und antike Metallproduktion. Der Anschnitt, Beiheft 3, Deutsches Bergbau-Museum Bochum

Wagner GA, Weisgerber G (1988) Antike Edel- und Buntmetallurgie auf Thasos. Der Anschnitt, Beiheft 6, Deutsches Bergbau-Museum Bochum

Wagner GA, Pernicka E, Seeliger T, Lorenz I, Begemann F, Schmitt-Strecker S, Eibner C, Öztunali Ö (1986) Geochemische und isotopische Charakteristika früher Rohstoffquellen für Kupfer, Blei, Silber und Gold in der Türkei. Jb Röm-German Zentralmuseum 33:723–752

Wagner GA, Begemann F, Eibner C, Lutz J, Öztunali Ö, Pernicka E, Schmitt-Strecker S (1989) Archäometallurgische Untersuchungen an Rohstoffquellen des frühen Kupfers Ostanatoliens. Jb Röm-German Zentralmuseum, 36(2):637–686

Wainwright A (1934) The Occurrence of Tin and Copper near Byblos. J Egyptian Archaeology 20:29–32

Waldbaum J (1989) Copper, Iron, Tin, Wood: The Start of the Iron Age in the Eastern Mediterranean. Archeomaterials 3(2):111–122

Weinstein JM (1984) Radiocarbon Dating in the Southern Levant. Radiocarbon 26:297–366

Weisgerber G (1978) Evidence of Ancient Mining Sites in Oman: a Preliminary Report. J Oman Studies 4:15–28

Weisgerber G (1981) Mehr als Kupfer in Oman – Ergebnisse der Expedition 1981. Der Anschnitt 33(5–6):174–263

Weisgerber G (1988) Oman: A Bronze-Producing Centre during the 1st Half of the 1st Millennium BC. In: Curtis J (ed) Bronzeworking Centres of Western Asia, c. 1000–539 B.C. Kegan Paul, London, pp 285–295

Weisgerber G (1989) Montanarchäologie. Grundzüge einer systematischen Bergbaukunde für Vor- und Frühgeschichte und Antike. In: Hauptmann A, Pernicka E, Wagner GA (eds) Archäometallurgie der Alten Welt. Der Anschnitt, Beiheft 7, Deutsches Bergbau-Museum Bochum, pp 79–98

Weisgerber G (1990) Montanarchäologische Forschungen in Nordwestiran 1978. Archäologische Mitt Iran 23:73–84

Weisgerber G (1991) Bergbau im alten Ägypten. Das Altertum 37:140–157

Weisgerber G (1996) Montanarchäologie – mehr als Technikgeschichte: Das Beispiel Fenan (Jordanien). Schriftenreihe Georg-Agricola-Gesellschaft 20:19–34

Weisgerber G (1997) Zur Geschichte der Bergbauarchäologie. In: Der Harz als frühmittelalterliche Industrielandschaft. Nachr Niedersachsens Urgesch 66(1):7–19

Weisgerber G (2006) The mineral wealth of ancient Arabia and ist use I: Copper mining and smelting at Feinan and Timna – comparison and evaluation of techniques, production, and strategies. Arabian archaeology and epigraphy 17:1–30

Weisgerber G, Hauptmann A (1988) Early Copper Mining and Smelting in Palestine. In: Maddin R (ed) The Beginning of the Use of Metal and Alloys. MIT-Press, Cambridge/Mass, pp 52–62

Weisgerber G, Pernicka E (1995) Ore Mining in Prehistoric Europe. In: Morteani G, Northover JP (eds) Prehistoric Gold in Europe. Kluwer, Dordrecht/Boston/London, pp 159–182

Weisgerber G, Roden C (1986) Griechische Metallhandwerker und ihre Gebläse. Der Anschnitt 38(1):2–26

Weissbrod T (1987) The Mineral Resources of Sinai. In: Gvirtzman G, Shmueli A, Gardus Y, Beit-Arieh I, Haz-El M (eds) Sinai, part 1. Tel Aviv University Publ., pp 287–296 (Hebrew)

Wenning R (1987) Die Nabatäer – Denkmäler und Geschichte. Novum Testamentum et Orbis Antiquus, Bd. 3

Wertime TA (1964) Man's First Encounters with Metallurgy. Science 146:1257–1267

Wertime TA (1967) A Metallurgical Expedition through the Persian Desert. In: Caldwell JR (ed) Investigations at Tal-i Iblis. Springfield/Illinois, pp 327–339

Wertime TA (1973) The Beginnings of Metallurgy: A New Look. Science 182:875–887

Wertime TA (1980) The Pyrotechnological Background. In: Wertime TA, Muhly JD (eds) The Coming of the Age of Iron. Yale Univ. Press, New Haven/London, pp 1–24

Weyl WA (1959) Coloured Glasses. London

Wheeler M (1983) Greenstone Amulets. In: Kenyon KM, Holland TA (eds) Excavations at Jericho, vol V. The British Academy, London, pp 781–788

Willies L (1990) An Early Bronze Age Tin Mine in Anatolia, Turkey. Bull Peak Distr Mines Hist Soc 11(2):91–96

Willies L (1991) Ancient copper mining at Wadi Amram, Israel. Bull Peak Distr Mines Hist Soc 11(2):109–138

Woelk D (1966) Agatharchides von Knidos: Über das Rote Meer. Bamberg

Wright K (1997) Wadi Faynan 4th–3rd millennia. Levant XXIX:253–255

Wright K (1998) The Wadi Faynan Fourth and Third Millennia Project, 1997: Report on the First Season of Test Excavations at Wadi Faynan 100. Levant XXX:33–60

Würzburger U (1969) Sulphides and Oxides in Igneous Rocks of the Timna Massif. Israel J Earth Science 18:3–4

Würzburger U (1970) Copper Silicates in the Timna Ore Deposit. Israel J Chemistry 8:443–457

Yakar J (1985) The Later Prehistory of Anatolia, part II: The Late Chalcolithic and Early Bronze Age. BAR Intern Ser, Oxford, 268(2)

Yalçin Ü, Hauptmann A (1995) Archäometallurgie des Eisens auf der Schwäbischen Alb. In: Beiträge zur Eisenverhüttung auf der Schwäbischen Alb. Forschungen und Berichte zur Vor- und Frühgeschichte in Baden-Württemberg 55, pp 269–309

Yalçin Ü, Pernicka E (1999) Zur Technologie der frühneolithischen Kupferverwendung in Aşıklı Höyük. In: Hauptmann A, Pernicka E, Rehren T, Yalçin Ü (eds) The Beginnings of Metallurgy. Der Anschnitt, Beiheft 9, Deutsches Bergbau-Museum Bochum, pp 45–54

Yalçin Ü, Hauptmann H, Hauptmann A, Pernicka E (1992) Norsuntepe'de Geç Kalkolitik Çagi Bakir Madenciligi Üzerine Arkeometallurjik Arastirmalar. VIII. Arkeometri Sonuçlari Toplantisi. Ankara, pp 381–389

Yamaguchi T, Shirashi T (1971) Kinetic Studies of Eutectoid Decompostion of $CuFe_5O_8$. J Amer Ceram Soc 54(11):556–558

Yazawa A (1980) Distribution of Various Elements between Copper, Matte and Slag. Erzmetall 33:377–382

Yazawa A, Eguchi M (1976) Equilibrium Studies in Copper Slags Used in Continuous Converting. In: Yannopoulos JC, Agarwal JC (eds) Extractive Metallurgy of Copper, I. Amer Inst Metall Engineering, New York, pp 3–20

Yekutieli Y, Shalev S, Shilstein S (2005) 'Ein Yahav – a copper smelting site in the 'Arava. Bull Amer Soc Oriental Res 340:1–21

Yellin J, Levy TE, Rowan YM (1996) New Evidence on Prehistoric Trade Routes: The Obsidian Evidence from Gilat, Israel. J. Field Archaeol 23:361–380

Yener A (1983) The Production, Exchange and Utilization of Silver and Lead Metals in Ancient Anatolia: A Source Identification Project. Anatolica 10:1–15

Yener A (2000) The Domestication of Metals. The Rise of Complex Metal Industries in Anatolia. Brill, Leiden-Boston-Köln

Yener A, Vandiver P (1993) Tin Processing at Göltepe, an Early Bronze Age Site in Anatolia. Amer J Archaeol 97:207–238

Yener A, Sayre EV, Joel EC, Özbal H, Barnes IL (1991) Stable Lead Isotope Studies of Central Taurus Ore Sources and Related Artifacts from Eastern Mediterranean Chalcolithic and Bronze Age Sites. J Archaeol Sc 18:1–37

Yilmaz Y (1993) New Evidence and Model on the Evolution of the Southeast Anatolian Orogen. Geol Soc Amer Bull 105:251–271

Zak I, Freund R (1981) Assymmetry and Basin Migration in the Dead Sea Rift. Tectonophysics 80:27–38

Zayadine F (1985) Caravan Routes Between Egypt and Nabataea and the Voyage of Sultan Baibars. In: Hadidi A (ed) Studies in the History and Archaeology of Jordan II. Dept Antiquities Jordan, Amman, pp 159–174

Zschocke K, Preuschen E (1932) Das urzeitliche Bergbaugebiet von Mühlbach-Bischofshofen. Mat. Urgeschichte Öster., Heft 6

Zwicker U (1980) Investigation of the Extractive Metallurgy of Cu/Sb/As Ore and Smelting Products from Norsuntepe (Keban) on the Upper Euphrates (3500–2800 B.C.). In: Oddy WA (ed) Aspects of Early Metallurgy. Brit Museum Occ Papers 17:13–26

Zwicker U (1981) Ancient Metallurgical Methods for Copper Production in Cyprus, part 1: Natural Copper-, Oxide-, Sulphate- and Silicate Ore. Bull Cyprus Ass Geol & Mining Eng 3:79–91

Zwicker U (1984) Metallographische und analytische Untersuchungen an Proben aus der Grabung der Bronzegießerei in der Phidias-Werkstatt von Olympia und Versuche zum Schmelzen von Bronze in flachen Tiegeln. Berliner Beitr. Archäometrie 9:61–94

Zwicker U (1985) Investigations from the Metallurgical Workshop at Kition. In: Karageorghis V, Demas M (eds) Excavations at Kition, vol V: The Pre-Phoenician Levels. Nicosia, pp 403–429

Zwicker U (1989) Untersuchungen zur Herstellung von Kupfer und Kupferlegierungen im Bereich des östlichen Mittelmeeres. In: Hauptmann A, Pernicka E, Wagner GA (eds) Archäometallurgie der Alten Welt / Old World Archaeometallurgy, Der Anschnitt, Beiheft 7, Deutsches Bergbau-Museum Bochum, pp 191–200

Zwicker U, Rollig H, Grembler E (1975) Untersuchung an Kupferschlackenproben von zwei frühgeschichtlichen Verhüttungsplätzen aus Timna (Negev). Metallwiss u. Technik 29(12): 1193–1197

Zwicker U, Greiner H, Hofmann KH, Reithinger M (1985) Smelting, Refining and Alloying of Copper and Copper Alloys in Crucible-Furnaces during Prehistoric up to Roman Times. In: Craddock PT, Hughes MJ (eds) Furnaces and Smelting Technology in Antiquity. Brit Museum Occ Papers 48:103–115

Appendix: Analytical Results

Table A.1a. Chemical composition of Cu- and Cu-Mn-ores from the Faynan district and from Wadi Khusheibah
Table A.1b. Description and geological assignement of samples from Table A.1a
Table A.2. Lead isotope abundance ratios of copper ores from Faynan
Table A.3. Lead isotope abundance ratios of copper ores from the "ancient" ore deposit of Timna
Table A.4. Bulk analyses of smelting slags of various ages from the Faynan-District
Table A.5. Semiquantitative plane scans of slags from Wadi Fidan 4
Table A.6. Wadi Fidan 4. Sample JD-8/13a. EMS-analyses of clinopyroxene in a Late Chalcolithic/EBA I smelting slag
Table A.7. Wadi Fidan 4. Sample JD-8/29a. EMS-analyses of åkermanite in an Early Bronze Age I slag
Table A.8. Wadi Fidan 4. Sample JD-8/24. EMS-analyses of merwinite in a Late Chalcolithic/Early Bronze Age I slag
Table A.9. Wadi Fidan 4. Sample JD-8/29a. EMS-analyses of diopside in a Late Chalcolithic/Early Bronze Age I slag
Table A.10. EMS-analyses of olivine (tephroite) in various slags from the Faynan-District
Table A.11. EMS-analyses of olivine in a slag from Ras en-Naqab, Early Bronze Age II/III
Table A.12. EMS-analyses of olivine (tephroite, knebelite) in various slags from the Faynan-District
Table A.13. EMS-analyses of pyroxenoides in Roman slags from Faynan 1
Table A.14. EMS-analyses of pyroxenoides in Roman slags from Faynan 1
Table A.15. Glass analyses of Early Bronze Age II/III and Roman slags
Table A.16. EMS-analyses of crednerite and of hausmannite
Table A.17. Chemical composition of copper and copper alloys from smelting sites in the Faynan-District
Table A.18. Chemical composition of copper ingots from the surroundings of Barqa el-Hetiye
Table A.19. Trace element concentrations in copper and slags from Faynan
Table A.20. Distribution coefficients $D_{Cu/S}$ of some trace elements between copper and slag from Table A.19
Table A.21. Lead isotope abundance ratios of metal drops, slags and an ore sample from Faynan

Table A.1a. Chemical composition of Cu- and Cu-Mn-ores from the Faynan district and from Wadi Kusheiba (AAS-analyses, data after Hauptmann et al. 1992). Not indicated in the table are: JD-24/1a = 1.64 µg g^{-1} U; JD-24/1b = 14.3 µg g^{-1} U. For stratigraphic details see Fig. 4.1. All values are given in µg g^{-1}, if not wt.-%; *na*: not analyzed

Sample no. DBM, Locality	Cu (%)	Mn (%)	Fe (%)	Pb (%)	Zn	Sn	As	Sb	Bi	Co	Ni	Ag	S (%)
Wadi Khalid													
JD-3/1b	16.8	1.11	0.39	0.30	240	<10	15	<1	11	14	30	<0.5	na
JD-3/3a	36	10.6	0.33	0.57	1100	<10	83	<1	3	150	100	<0.5	na
JD-3/5	35	4.10	0.99	0.76	900	40	44	<1	1	110	100	0.9	na
JD-3/14	1.63	21.6	1.90	0.85	800	45	74	<1	5	440	140	2	na
JD-3/15	55	0.22	0.08	0.05	1200	160	100	<1	<1	110	160	<0.5	na
JD-3/16	11.5	12.7	3.3	2.20	1300	<10	62	<1	1	180	66	2	na
JD-3/22	31	0.32	0.11	0.25	110	40	57	<1	3	16	6	<0.5	na
JD-3/29	33	0.13	0.02	0.001	210	20	<1	<1	<1	16	12	<0.5	na
JD-II/3	44	2.35	1.22	0.23	1400	37	150	4	8	130	100	<0.5	0.06
JD-II/5	20	0.02	1.96	0.17	240	51	8	1	3	18	25	0.3	2.37
Wadi Dana													
JD-I/22	6.4	28.9	0.91	2.09	1320	43	40	3	4	690	110	1	na
JD-I/26	6.7	31.9	2.21	0.65	2860	59	140	3	3	460	190	1	0.08
JD-I/26a	5.3	24.0	1.62	0.71	2060	na	na	na	na	300	210	na	na
JD-I/27	5.2	24.4	1.03	0.90	665	55	50	5	1	430	73	1	0.04
JD-I/28	3.7	33.9	6.20	0.58	1130	69	120	<1	3	460	210	1	0.04
JD-13/9	36	1.05	0.42	0.06	210	20	15	<1	1	120	50	<0.5	na
JD-13/14a	24	22.3	0.65	1.14	800	<10	57	<1	2	280	80	0.8	na

Table A.1a. Continued

Sample no. DBM, Locality	Cu (%)	Mn (%)	Fe (%)	Pb (%)	Zn	Sn	As	Sb	Bi	Co	Ni	Ag	S (%)
Wadi Dana (continued)													
JD-13/14b	48	0.84	0.18	0.06	620	20	33	2	1	34	46	<0.5	na
JD-13/14c	1.99	34.5	1.21	1.33	1620	50	20	<1	<1	620	100	<0.5	na
JD-III/12b	17.6	0.08	0.86	0.07	180	51	87	4	8	30	11	<0.5	0.47
JD-III/16	3.5	33.3	1.06	6.10	1250	51	65	2	5	990	35	1	na
JD-III/18	31	<0.01	0.41	0.006	na	na	3	<1	4	14	4	3	0.03
Wadi Ratiye, Roman mines													
JD-12/1	41	0.02	9.00	0.037	210	25	10	8	5	29	50	15	na
JD-GR/1	44	0.007	1.05	0.025	140	45	52	<1	5	16	34	12	3.16
JD-GR/2	27	0.002	1.16	0.016	115	53	<5	4	9	<5	15	3	1.29
JD-GR/5b	45	<0.002	2.38	0.018	110	na	8	1	3	97	12	7	na
JD-GR/5c	36	<0.002	1.20	0.03	80	na	6	2	2	5	6	3	1.76
JD-GR/6	43	0.01	4.9	0.016	170	na	21	4	7	14	15	29	3.58
JD-GR/7	48	0.002	0.06	0.018	60	na	3	<1	7	8	23	25	3.08
Wadi Abiad													
JD-WA/1	18.7	0.004	0.61	0.007	180	56	17	<1	<1	15	37	8	0.34
JD-WA/4	15.2	0.003	0.64	0.013	120	77	6	<1	<1	<5	15	1	0.13
Wadi Abu Kusheiba													
JD-24/1a	15.6	<0.002	0.17	0.02	110	40	21	<1	2	2	3	1	na
JD-24/1b	22.8	<0.002	0.17	0.009	100	20	31	<1	2	3	5	2	na

Table A.1b. Description and geological assignement of samples from Table A.1a

Sample no. DBM, Locality	Description	Stratigraphy
Wadi Khalid		
JD-3/1b	"Matrix-mineralization" in conglomerate	footwall cb3
JD-3/3a	Cu-Mn-ore	Top wall cb2
JD-3/5	Cu-bearing dolomite	Top wall cb2
JD-3/14	Cu-bearing siltstone	Top wall cb2
JD-3/15	Chrysocolla in Cu-Mn-bearing siltstone	Top wall cb2
JD-3/16	Chrysocolla in Cu-Mn- bearing siltstone	Top wall cb2
JD-3/22	Cu-silicates, vein mineralization in sandstone	Top wall cb2
JD-3/29	Cu-silicate, vein mineralization in sandstone	Footwall cb3
JD-II/3	Malachite in P-Mn-bearing shist	Top wall cb2
JD-II/5	Cu-ore in finegrained conglomerate	Footwall cb3
Wadi Dana		
JD-I/22	Mn-ore with Cu-silikacates in shist	Top wall cb2
JD-I/26	Mn-Cu- bearing shale	Top wall cb2
JD-I/26a	Mn-Cu- bearing shale	Top wall cb2
JD-I/27	Mn-Cu- bearing shale	Top wall cb2
JD-I/28	Mn-Cu- bearing shale	Top wall cb2
JD-13/9	Cu-bearing shale containing Mn-ore	Top wall cb2
JD-13/14a	Cu-Mn-ore in siltstone	Top wall cb2
JD-13/14b	Cu-ore from Cu-Mn-bearing siltstone	Top wall cb2
JD-13/14c	Mn-ore in siltstone	Top wall cb2
JD-III/12b	Siltstone containing Cu-ore	Top wall cb2
JD-III/16	Mn-ore containing Cu-mineralization	Footwall cb3
JD-III/18	"Matrix-mineralization" in conglomerate	Footwall cb3
Wadi Ratiye, Roman mines		
JD-12/1	"Tile ore" (Cu-Fe-ore) in sandstone	cb4
JD-GR/1	Cu-mineralization in veins, sandstone	cb4
JD-GR/2	Cu-mineralization in veins, sandstone	cb4
JD-GR/5b	Cu-mineralization in veins, sandstone	cb4
JD-GR/5c	Cu-mineralization in veins, sandstone	cb4
JD-GR/6	Cu-mineralization in veins, sandstone	cb4
JD-GR/7	Cu-mineralization in veins, sandstone	cb4
Wadi Abiad		
JD-WA/1	Cu-mineralization in veins, sandstone Cu	cb4
JD-WA/4	Cu-mineralization in veins, sandstone Cu	cb4
Wadi Abu Kusheiba		
JD-24/1a	Cu-concretion in sandstone	cb3 / cb4
JD-24/1b	Cu-concretion in sandstone	cb3 / cb4

Table A.2. Lead isotope abundance ratios of copper ores from Faynan. Indicated are also model ages following the the two-stage model of Stacey and Kramer (1975). For stratigraphic details see Fig. 4.1. Chemical analyses of these samples are given in the Table A.1

Sample no. DBM	Description		Locality	208Pb/ 206Pb	207Pb/ 206Pb	204Pb/ 206Pb	208Pb/ 204Pb	207Pb/ 204Pb	206Pb/ 204Pb	Model age (Ma)
JD-1/18a	Top wall cb2 (DLS)	Cu + Mn	Wadi Khalid	2.1206	0.8691	0.05553	38.189	15.652	18.009	560
JD-1/18b	Top wall cb2 (DLS)	Cu	Wadi Khalid	2.1170	0.8683	0.05553	38.120	15.636	18.007	540
JD-1/18c	Footwall cb3	Cu	Wadi Khalid	2.0629	0.8491	0.05421	38.055	15.663	18.447	260
JD-1/18d	Footwall cb3	Cu	Wadi Khalid	2.1019	0.8639	0.05534	37.979	15.610	18.069	440
JD-3/3a	Top wall cb2 (DLS)	Cu + Mn	Wadi Khalid	2.1166	0.8681	0.05551	38.132	15.638	18.015	530
JD-3/15	Top wall cb2 (DLS)	Cu	Wadi Khalid	2.1157	0.8683	0.05560	38.056	15.618	17.987	520
JD-3/16	Top wall cb2 (DLS)	Cu + Mn	Wadi Khalid	2.1190	0.8704	0.05585	37.944	15.585	17.906	510
JD-3/22	Top wall cb2 (DLS)	Cu	Wadi Khalid	2.1168	0.8678	0.05550	38.143	15.637	18.019	530
JD-GR/5a	cb4 (MBS)	Cu	Wadi Ratiye	2.0758	0.8387	0.05350	38.804	15.677	18.693	110
JD-GR/5b	cb4 (MBS)	Cu	Wadi Ratiye	2.0759	0.8393	0.05356	38.757	15.669	18.670	110
JD-GR/7a	cb4 (MBS)	Cu	Wadi Ratiye	2.0893	0.8474	0.05403	38.667	15.684	18.508	260
JD-13/14a	Top wall cb2 (DLS)	Cu + Mn	Wadi Dana	2.1167	0.8682	0.05560	38.071	15.615	17.986	510
JD-13/14b	Top wall cb2 (DLS)	Cu	Wadi Dana	2.1165	0.8677	0.05555	38.103	15.622	18.003	510
JD-24/1a	cb3/cb4 (MBS)	Cu	Wadi Abu Kusheibah	2.0457	0.8202	0.05225	39.155	15.698	19.140	−180
JD-24/1b	cb3/cb4 (MBS)	Cu	Wadi Abu Kusheibah	2.0470	0.8215	0.05233	39.114	15.698	19.108	−150

Table A.3. Lead isotope abundance ratios of copper ores from the "ancient" ore deposit of Timna. Measurements were performed in Oxford except of the last sample which was analyzed in Mainz. Except of the last 10 nos samples were taken by Leese et al. (1985/1986); here, also all chemical analyses are published. For definition of areas see Rothenberg (1980, p. 29). All localities are mentioned in Rothenberg (1988). For stratigraphic details see Fig. 4.1. Indicated are model ages following the two stage model of Stacey and Kramer (1975)

Sample no.	Oxford	Description		Locality	$^{208}Pb/^{206}Pb$	$^{207}Pb/^{206}Pb$	$^{204}Pb/^{206}Pb$	$^{208}Pb/^{204}Pb$	$^{207}Pb/^{204}Pb$	$^{206}Pb/^{204}Pb$	Model age (Ma)
23054P	D6	Amir-/Avrona-Fm.	Cu	Timna, –	2.1067	0.8628	0.05521	38.157	15.627	18.112	440
23057U	A1	Amir-/Avrona-Fm.	Cu/Fe	Timna, area A	2.1041	0.8578	0.05473	38.449	15.674	18.273	410
23060T	D17	Amir-/Avrona-Fm.	Cu	Timna, area D	2.0983	0.8552	0.05468	38.375	15.640	18.289	340
23061R	D31	Amir-/Avrona-Fm.	Cu	Timna, area DI	2.1066	0.8605	0.05499	38.311	15.650	18.186	430
23065U	T1	Amir-/Avrona-Fm.	Cu	Timna, area I	2.1106	0.8640	0.05528	38.183	15.630	18.091	460
23069X	N3	Amir-/Avrona-Fm.	Cu	Timna, area M	2.1103	0.8640	0.05527	38.179	15.632	18.092	460
23070P	P1	Amir-/Avrona-Fm.	Cu	Timna, area P	2.1106	0.8620	0.05502	38.363	15.667	18.176	470
23071Y	T2	Amir-/Avrona-Fm.	Cu	Timna, site 2	2.1172	0.8681	0.05562	38.068	15.608	17.980	500
23072W	T23	Amir-/Avrona-Fm.	Cu	Timna, area T	2.1245	0.8726	0.05575	38.110	15.652	17.938	620
23074S	T18	Amir-/Avrona-Fm.	Cu	Timna, area T	2.1017	0.8583	0.05490	38.283	15.634	18.215	380
23078V	U6	Amir-/Avrona-Fm.	Cu	Timna, area U	2.1270	0.8728	0.05576	38.144	15.652	17.933	620
23079T	U8	Amir-/Avrona-Fm.	Cu/Fe	Timna, area U	2.1183	0.8685	0.05555	38.136	15.635	18.003	540
23082S	D33a	Amir-/Avrona-Fm.?	Cu	Wadi Amram, site 33A	2.1057	0.8611	0.05499	38.293	15.658	18.185	450
23083Q	ARB64	archaeol. object?	Cu/Fe	Beer Ora, site 64	2.1081	0.8606	0.05497	38.353	15.656	18.193	440
23088R	D18	Amir-/Avrona-Fm.	Cu	Timna, site 18	2.1084	0.8618	0.05511	38.258	15.637	18.146	440
23089P	26A	archäolog. finding?	Cu	Timna, site 26A	2.1157	0.8657	0.05533	38.242	15.648	18.075	510
23090S	ARB9	Amir-/Avrona-Fm.	Cu	Timna, site 9	2.1243	0.8726	0.05588	38.018	15.616	17.897	580

Table A.3. *Continued*

Sample no.	Oxford	Description		Locality	$^{208}Pb/^{206}Pb$	$^{207}Pb/^{206}Pb$	$^{204}Pb/^{206}Pb$	$^{208}Pb/^{204}Pb$	$^{207}Pb/^{204}Pb$	$^{206}Pb/^{204}Pb$	Model age (Ma)
23094V	T11	Amir-/Avrona-Fm.	Cu	Timna, area S/T	2.1256	0.8731	0.05583	38.071	15.637	17.911	610
23098Y	T5	Amir-/Avrona-Fm.	Cu	Timna, area S/T	2.1225	0.8731	0.05584	38.007	15.634	17.907	600
23101Q	T8	Amir-/Avrona-Fm.	Cu	Timna, area S/T	2.1228	0.8716	0.05576	38.068	15.630	17.933	580
23102Z	T9	Amir-/Avrona-Fm.	Cu	Timna, area S/T	2.1243	0.8726	0.05588	38.018	15.616	17.897	580
23107P	T16	Amir-/Avrona-Fm.	Cu	Timna, area S/T	2.1153	0.8666	0.05547	38.135	15.624	18.028	500
23108Y	W18	Amir-/Avrona-Fm.	Cu	Timna, area S/T	2.1017	0.8583	0.05490	38.283	15.634	18.215	380
23113T	T23	Amir-/Avrona-Fm.	Cu	Timna, area S/T	2.1264	0.8736	0.05587	38.059	15.636	17.898	610
23114R	T24	Amir-/Avrona-Fm.	Cu	Timna, area S/T	2.1280	0.8737	0.05583	38.112	15.649	17.910	630
23116Y	T26	Amir-/Avrona-Fm.	Cu	Timna, area S/T	2.1157	0.8657	0.05533	38.242	15.648	18.075	510
23118U	C3	Amir-/Avrona-Fm.	Cu	Timna, area S/T	2.1145	0.8646	0.05518	38.323	15.671	18.124	520
6491		Amir-/Avrona-Fm.	Cu	Timna. –	2.1240	0.8739	0.05602	37.918	15.601	17.852	580
6597		Amir-/Avrona-Fm.	Cu	Timna. –	2.1038	0.8577	0.05487	38.343	15.632	18.226	370
T1		Amir-/Avrona-Fm.	Cu	Timna, mine T1	2.1103	0.8649	0.05544	38.066	15.602	18.038	450
T1329		Amir-/Avrona-Fm.	Cu	Timna, site 33b	2.1099	0.8675	0.05552	38.001	15.625	18.011	510
T1330		Amir-/Avrona-Fm.	Cu	Timna, site 88a	2.1136	0.8659	0.05535	38.185	15.642	18.066	500
T1337		Amir-/Avrona-Fm.	Cu	Timna, area T	2.1218	0.8718	0.05573	38.071	15.643	17.943	590
T1501		Amir-/Avrona-Fm.	Cu	Timna, shaft 14	2.1134	0.8666	0.05530	38.214	15.669	18.082	540
T1505		Amir-/Avrona-Fm.	Fe	Timna, near site 25	2.1125	0.8691	0.05568	37.943	15.610	17.961	520
TIM C		Amir-/Avrona-Fm.	Cu	Timna —	2.1231	0.8723	0.05591	37.971	15.602	17.885	560
TI/1		Amir-/Avrona-Fm.		Timna, area T	2.1260	0.8730	0.05586	38.058	15.628	17.901	600

Table A.4. Bulk analyses of smelting slags of various ages from the Faynan-District. Analyses were performed by AAS; all data are given in wt.-%; *na*: not analyzed

Sample no. DBM/ Period	Locality	SiO$_2$	TiO$_2$	Al$_2$O$_3$	FeO	MnO	MgO	CaO	BaO	K$_2$O	Na$_2$O	P$_2$O$_5$	S	Ni	Cu	Zn	Pb	Total
Late Chalcolithic/EBA I																		
JD-8/10a	Wadi Fidan 4	35.8	na	4.40	4.51	0.02	0.07	4.80	na	0.22	0.20	na	na	na	34.6	0.03	0.02	84.66
JD-8/10b	Wadi Fidan 4	39.5	na	1.10	1.90	0.02	0.60	5.50	na	0.35	0.17	na	na	na	32.6	0.01	0.01	81.75
JD-8/10c	Wadi Fidan 4	26.6	na	1.80	10.6	0.01	0.40	4.10	na	0.08	0.14	na	na	na	38.7	0.02	0.02	82.52
JD-8/10d	Wadi Fidan 4	32.3	na	2.10	3.90	0.02	0.50	4.80	na	0.19	0.17	na	na	na	36.4	0.03	0.01	80.37
JD-8/13a	Wadi Fidan 4	51.7	0.48	4.61	13.8	0.15	2.74	6.80	0.03	0.47	0.21	0.49	na	na	9.8	<0.01	0.07	91.38
JD-8/17a	Wadi Fidan 4	24.9	0.13	3.03	1.19	0.79	6.25	14.2	0.82	1.22	1.05	2.47	na	na	29.7	<0.01	0.13	85.90
JD-8/17b	Wadi Fidan 4	29.8	1.00	2.94	0.94	0.89	7.48	20.7	0.89	0.74	2.29	2.29	na	na	20.9	<0.01	0.11	90.92
JD-8/24	Wadi Fidan 4	19.4	0.82	1.79	0.72	1.35	9.08	21.3	1.06	0.43	0.42	1.50	na	na	23.6	<0.01	0.11	81.56
JD-8/29	Wadi Fidan 4	35.7	1.18	3.25	1.27	3.20	7.10	16.7	2.96	0.58	0.78	1.17	na	na	18.1	<0.01	0.30	92.29
Lz-276/2a	Wadi Fidan 4	26.2	0.12	1.56	4.31	0.14	0.69	1.11	0.02	0.21	<0.14	0.16	0.17	0.07	53.3	0.12	0.05	88.23
Lz-276/2c	Wadi Fidan 4	25.5	0.14	2.10	19.6	0.26	1.59	7.72	0.06	0.40	<0.14	0.23	0.06	0.07	28.5	0.06	0.02	86.32
Early Bronze Age II/III																		
JD-5/1a	Ras en-Naqab	32.3	0.22	4.55	2.65	23.4	4.96	13.0	1.93	1.01	0.38	2.12	<0.01	0.07	3.67	0.14	0.93	91.24
JD-5/1b	Ras en-Naqab	33.0	0.24	4.93	2.73	34.3	0.82	7.31	1.76	1.84	0.41	2.25	<0.01	0.08	3.58	0.11	0.98	94.30
JD-5/1c	Ras en-Naqab	31.7	0.17	4.06	2.05	30.4	1.88	13.0	1.78	1.40	0.33	3.67	<0.01	0.07	1.70	0.11	1.00	93.34
JD-5/1d	Ras en-Naqab	31.5	0.18	4.10	2.11	39.7	0.85	8.32	2.19	1.68	<0.14	1.58	0.04	0.08	1.77	0.10	0.99	95.14
JD-5/1e	Ras en-Naqab	33.5	0.27	5.90	3.68	19.4	8.14	16.0	1.18	1.91	<0.14	0.62	0.09	0.07	5.01	0.11	0.40	96.30
JD-5/1f	Ras en-Naqab	28.0	0.25	5.37	1.86	43.2	0.86	5.90	2.27	2.90	<0.14	0.77	<0.01	0.08	2.39	0.13	1.18	95.19
JD-5/1g	Ras en-Naqab	33.3	0.28	5.47	2.15	25.4	4.79	14.6	2.56	1.23	<0.14	0.63	0.14	0.08	2.65	0.14	0.64	94.05
JD-5/1h	Ras en-Naqab	28.8	0.19	4.62	5.24	26.5	2.82	15.7	1.81	1.42	<0.14	6.95	0.07	0.07	1.67	0.11	0.71	96.70
JD-5a/2a	Ras en Naqab	30.8	0.22	5.21	2.35	25.5	2.46	15.8	1.90	2.26	<0.14	6.21	0.01	0.07	1.49	0.09	0.88	95.30
JD-5a/2b	Ras en Naqab	29.0	0.23	4.96	2.32	35.5	3.29	11.0	2.03	2.19	<0.14	1.87	0.01	0.07	1.76	0.12	0.71	95.01
JD-5a/2c	Ras en Naqab	32.4	0.28	5.98	1.58	26.9	2.77	7.43	2.14	2.54	<0.14	0.42	0.01	0.07	2.50	0.12	0.56	85.73

Table A.4. Continued

Sample no. DBM/Period	Locality	SiO$_2$	TiO$_2$	Al$_2$O$_3$	FeO	MnO	MgO	CaO	BaO	K$_2$O	Na$_2$O	P$_2$O$_5$	S	Ni	Cu	Zn	Pb	Total
Early Bronze Age II/III (continued)																		
JD-5a/2d	Ras en Naqab	33.9	0.25	5.65	2.32	36.3	1.66	7.24	1.85	2.42	<0.14	1.68	<0.01	0.08	2.77	0.14	1.00	97.23
JD-5a/2e	Ras en Naqab	27.1	0.23	4.66	2.76	34.3	4.31	9.63	2.56	2.19	<0.14	0.55	<0.01	0.07	2.64	0.16	0.74	91.88
JD-9/1	Wadi Fidan 10	27.2	0.25	5.41	1.72	41.1	0.86	8.63	0.81	3.84	1.17	3.23	0.91	0.07	0.86	0.06	0.42	96.56
JD-20/1b	Wadi Faynan 4	28.4	0.22	4.92	6.97	36.4	0.75	9.24	2.01	1.64	<0.14	2.31	<0.01	0.08	1.94	0.16	1.06	96.14
JD-20/1c	Wadi Faynan 4	28.2	0.24	4.86	6.90	36.4	0.73	9.24	2.00	1.65	<0.14	2.30	0.03	0.07	1.94	0.16	1.07	95.74
JD-21/1a	Wadi Faynan 5	39.8	0.27	5.34	3.67	24.3	2.60	11.7	1.59	2.62	<0.14	1.45	0.02	0.07	1.30	0.07	0.61	95.46
JD-23/1a	Faynan 9	23.1	0.21	4.56	2.05	40.3	0.54	3.84	3.48	1.39	<0.14	0.81	1.22	0.08	8.07	0.14	1.54	91.33
JD-23/1b	Faynan 9	28.9	0.36	5.17	2.59	23.9	2.73	26.5	0.42	1.39	<0.14	1.07	0.04	0.08	2.35	0.23	1.08	96.80
JD-23/1c	Faynan 9	31.8	0.27	6.44	19.45	17.3	3.35	8.06	0.75	2.81	<0.14	0.92	<0.01	0.07	1.64	0.17	0.69	93.66
JD-23/1d	Faynan 9	28.1	0.21	4.40	3.25	36.3	1.59	10.9	5.45	1.50	0.43	3.05	0.05	0.07	1.49	0.13	0.95	97.87
JD-23/1e	Faynan 9	32.9	0.24	5.62	3.47	26.4	3.94	13.0	6.85	1.76	0.56	1.93	0.07	0.07	2.47	0.12	0.69	100.16
JD-23/1f	Faynan 9	27.5	0.26	5.63	3.07	28.2	3.60	13.3	2.71	2.80	0.31	5.88	0.10	0.07	2.90	0.11	0.69	97.11
Middle Bronze Age																		
JD-13/10a	Wadi Dana	35.5	0.19	2.31	6.96	45.4	0.81	4.55	0.78	1.59	<0.14	0.21	0.60	0.07	0.85	0.05	0.04	99.91
JD-13/10b	Wadi Dana	33.0	0.18	2.15	6.16	47.4	0.95	3.45	0.55	1.04	<0.14	0.16	0.94	0.07	1.68	0.05	0.03	97.80
JD-13/10c	Wadi Dana	36.0	0.22	2.71	4.91	34.5	1.46	10.5	0.65	1.41	<0.14	0.23	0.79	0.06	0.78	0.02	0.02	94.23
Iron Age II/III																		
JD-1/40	Faynan 5	26.4	0.25	5.01	2.17	37.4	1.37	15.3	1.28	3.06	<0.14	3.47	2.33	0.07	0.54	0.03	0.12	98.72
JD-2/2a	Kh. en-Nahas	29.8	0.23	5.05	3.32	41.0	1.38	7.55	0.56	2.75	0.50	3.40	0.76	0.07	0.36	0.04	0.02	96.80
JD-2/2b	Kh. en-Nahas	32.1	0.20	4.60	3.61	37.5	1.28	9.18	0.48	2.19	1.03	2.48	1.80	0.08	0.40	0.02	0.02	97.03
JD-2/2c	Kh. en-Nahas	34.3	0.22	4.93	3.34	39.5	1.11	5.76	0.72	1.78	0.44	1.45	0.83	0.09	0.31	0.02	0.02	94.83
JD-2/3	Kh. en-Nahas	51.3	0.24	5.30	2.49	20.5	1.09	4.28	0.56	2.67	0.59	1.82	0.45	0.06	1.37	0.03	0.16	92.85
JD-2/5a	Kh. en-Nahas	32.7	0.20	4.27	3.90	40.7	1.60	7.63	0.36	2.45	0.55	2.58	0.56	0.07	0.37	0.03	0.03	98.04

Table A.4. *Continued*

Sample no. DBM/Period	Locality	SiO$_2$	TiO$_2$	Al$_2$O$_3$	FeO	MnO	MgO	CaO	BaO	K$_2$O	Na$_2$O	P$_2$O$_5$	S	Ni	Cu	Zn	Pb	Total
Iron Age II/III *(continued)*																		
JD-2/5b	Kh. en-Nahas	30.4	0.21	3.78	3.34	43.7	1.40	7.63	0.21	2.30	0.55	2.04	1.07	0.07	0.35	0.03	0.01	97.13
JD-2/6	Kh. en-Nahas	33.5	0.26	5.53	3.73	36.8	1.47	7.48	0.43	2.95	0.85	2.17	0.98	0.07	0.59	0.03	0.09	96.97
JD-2/20b	Kh. en-Nahas	30.4	0.20	4.40	5.35	40.2	0.77	7.29	0.95	2.22	<0.14	3.45	0.53	0.07	0.56	0.06	0.08	96.76
JD-2/20c	Kh. en-Nahas	28.3	0.22	4.47	2.69	40.8	1.36	9.19	1.24	2.50	0.82	3.36	1.05	0.07	0.43	0.06	0.07	96.60
JD-5/3	Ras en-Naqab	33.3	0.25	5.59	3.38	28.6	3.00	13.4	1.64	1.97	0.42	2.01	<0.01	0.06	0.70	0.08	0.36	94.89
JD-5/3a	Ras en-Naqab	37.8	0.28	5.11	4.35	25.0	3.14	13.9	1.56	1.73	<0.14	2.33	0.03	0.07	0.83	0.08	0.39	96.65
JD-5/3b	Ras en-Naqab	33.7	0.25	5.01	3.53	24.4	2.75	12.4	1.63	1.65	<0.14	1.79	0.02	0.07	1.48	0.09	0.60	89.31
JD-5/3c	Ras en-Naqab	31.8	0.25	4.96	2.82	35.0	2.45	12.0	1.90	1.90	<0.14	1.74	0.10	0.07	0.38	0.06	0.39	95.82
JD-8/1	Wadi Fidan 4	25.3	0.32	6.43	2.10	34.0	1.46	16.0	0.68	3.59	<0.14	6.76	1.19	0.06	0.48	0.06	0.19	98.61
JD-11/2	Kh. el-Jariye	33.6	0.21	4.66	6.99	35.3	0.98	7.97	0.52	3.69	<0.14	3.46	0.61	0.07	0.61	0.05	0.04	98.73
JD-13/1b	Wadi Dana 1/6	29.9	0.16	2.43	4.35	49.2	0.92	5.26	0.39	0.61	0.61	0.28	0.58	0.08	1.47	0.07	0.02	96.31
Roman																		
JD-1/4	Faynan 1	32.6	0.20	4.89	3.52	36.7	1.33	7.92	0.65	3.26	0.72	2.31	0.82	0.07	0.61	0.06	0.27	95.95
JD-1/9	Faynan 1	40.5	0.22	2.55	1.67	43.2	1.00	2.20	0.98	1.57	<0.14	0.05	0.26	0.07	0.82	0.02	0.01	95.12
JD-1/10	Faynan 1	36.2	0.25	2.90	2.51	40.4	0.83	1.14	0.60	1.84	0.11	0.18	0.36	0.07	1.97	0.02	<0.01	89.40
JD-1/11	Faynan 1	47.8	0.22	2.63	2.35	29.8	1.22	6.03	0.53	2.63	0.40	0.07	1.63	0.07	0.51	0.00	<0.01	95.86
JD-1/21a	Faynan 1	40.9	0.34	3.42	2.66	39.8	0.20	0.97	1.50	3.17	0.29	0.17	0.06	0.07	0.22	0.01	0.14	93.89
JD-1/21b	Faynan 1	38.0	0.23	2.70	1.17	41.0	0.29	2.68	0.84	2.04	0.69	0.18	0.69	0.06	0.45	0.02	0.01	91.02
JD-1/23a	Faynan 1	42.6	0.30	4.86	5.49	33.9	0.49	2.29	0.65	1.32	0.42	0.16	0.16	0.07	0.49	0.01	0.01	93.16
JD-1/23b	Faynan 1	43.0	0.21	3.59	1.67	39.8	0.46	1.28	0.64	3.79	<0.14	0.12	0.14	0.07	0.35	0.00	<0.01	95.14
JD-1/23c	Faynan 1	43.8	0.31	5.03	5.40	35.9	0.60	2.38	0.67	1.43	0.42	0.14	0.18	0.07	0.46	0.16	0.01	96.92
JD-1/24a	Faynan 1	42.4	0.29	3.75	1.30	41.5	0.25	1.47	0.62	3.22	0.36	0.17	0.20	0.07	0.51	0.03	0.04	96.21
JD-1/24b	Faynan 1	41.2	0.20	3.32	5.80	37.9	0.73	1.93	0.70	2.42	0.41	0.23	0.31	0.06	0.31	0.01	0.01	95.56

Table A.4. *Continued*

Sample no. DBM/Period	Locality	SiO$_2$	TiO$_2$	Al$_2$O$_3$	FeO	MnO	MgO	CaO	BaO	K$_2$O	Na$_2$O	P$_2$O$_5$	S	Ni	Cu	Zn	Pb	Total
Roman (*continued*)																		
JD-1/24c	Faynan 1	41.4	0.26	3.35	5.72	38.1	0.46	1.76	0.82	2.24	0.32	0.15	0.54	0.07	0.52	0.02	0.02	95.69
JD-1/27a	Faynan 1	41.7	0.30	4.81	7.31	37.0	0.64	1.99	0.69	1.37	0.42	0.24	0.19	0.07	0.39	0.02	0.01	97.15
JD-1/27b	Faynan 1	43.5	0.21	3.97	1.93	38.1	0.77	2.42	0.72	3.88	0.58	0.13	0.13	0.07	0.27	0.01	0.01	96.68
JD-1/32	Faynan 1	43.8	0.41	4.09	1.92	34.4	0.94	3.85	0.15	3.31	<0.14	0.04	0.37	0.07	0.23	0.01	<0.01	93.56
JD-1/33d	Faynan 1	41.9	0.25	2.45	0.97	36.3	0.99	5.10	0.62	1.42	<0.14	0.09	0.22	0.07	0.61	0.01	<0.01	91.04
JD-1/38a	Faynan 1	38.7	0.29	3.71	10.2	41.5	0.15	0.09	1.76	0.63	<0.14	0.10	0.17	0.07	0.47	0.02	0.01	97.90
JD-1/38b	Faynan 1	43.0	0.35	3.07	1.87	38.3	0.64	3.24	1.19	2.83	<0.14	0.05	0.10	0.07	0.38	0.00	0.01	95.11
JD-12/3a	Wadi Ratiye	25.2	0.21	2.21	32.1	5.0	4.53	17.2	0.10	0.85	0.31	0.98	0.05	0.07	4.39	0.01	0.02	93.28
JD-12/3b	Wadi Ratiye	38.9	0.35	3.73	4.27	41.5	0.71	2.86	0.71	1.58	0.23	0.16	0.19	0.07	0.59	0.01	<0.01	95.85
Mamlukk period																		
JD-1/2a	Faynan 6	35.5	0.32	6.52	20.0	17.9	1.44	10.60	2.79	1.79	0.55	2.72	0.31	0.08	0.33	0.04	0.03	100.91
JD-1/2b	Faynan 6	30.0	0.26	5.58	29.5	16.7	1.34	5.82	2.56	1.63	0.35	2.42	0.26	0.07	0.60	0.11	0.10	97.32
JD-1/19	Faynan 1	36.0	0.30	6.22	36.0	13.4	0.54	3.47	0.57	1.53	0.52	0.77	0.14	0.07	0.46	0.05	0.03	100.03
Jd-6/2a	El-Furn	34.0	2.20	5.16	37.7	15.4	0.47	2.48	0.61	1.01	0.28	0.44	0.19	0.07	0.94	0.05	0.05	101.05
Jd-6/2b	El-Furn	67.7	0.34	12.49	4.03	1.3	0.85	0.50	0.11	4.71	3.22	0.19	0.06	0.06	0.37	0.01	0.01	95.96
Undated																		
JD-1/1a	Faynan 1	27.8	0.15	1.90	5.38	53.3	0.47	3.59	1.60	0.60	0.80	0.14	0.42	0.07	1.56	0.09	0.01	97.84
JD-1/1b	Faynan 1	36.2	0.25	5.82	3.66	31.2	1.34	8.83	0.35	2.54	2.50	2.63	1.15	0.06	0.38	0.02	0.04	96.95
JD-1/1c	Faynan 1	34.3	0.18	4.24	3.10	35.7	1.02	7.61	2.05	3.01	0.36	3.58	0.36	0.07	0.62	0.03	0.26	96.45
JD-7/2a	Ain el-Fidan	26.5	0.27	5.61	0.86	42.4	0.57	9.62	0.82	2.66	0.73	5.80	0.64	0.06	0.57	0.03	0.18	97.37
JD-7/2a	Ain el-Fidan	27.9	0.21	4.56	0.84	44.3	0.57	7.35	1.02	2.96	0.66	3.81	0.54	0.06	0.52	0.04	0.24	95.52
JD-7/3	Ain el-Fidan	34.9	0.23	4.77	1.08	35.1	0.49	8.16	0.71	2.40	0.90	4.05	0.60	0.07	1.03	0.04	0.34	94.90

Table A.5. Semiquantitative plane scans (ca. 4 mm^2) of slags from Wadi Fidan 4. They were analyzed by SEM/EDS. All data are given in wt.-%; *nd:* not detected

Sample no. DBM	SiO$_2$	TiO$_2$	Al$_2$O$_3$	FeO	MnO	MgO	CaO	K$_2$O	Na$_2$O	P$_2$O$_5$	Pb	Cu
JD-8/13a-1	50	0.3	5.1	10	0.2	3.8	11	0.5	nd	0.7	nd	16
2	57	0.4	6.1	15	0.2	1.7	7.2	0.6	nd	nd	nd	11
3	72	nd	5.0	11	0.8	1.8	6.1	0.2	nd	1.3	nd	2.1
4	70	0.4	3.3	10	0.1	1.4	4.6	0.4	nd	0.6	nd	8
5	79	0.2	2.2	6.8	nd	0.7	2.8	0.1	nd	0.5	nd	6.9
6	60	1.0	6.6	18	nd	1.8	9.2	0.9	nd	0.9	nd	1.7
7	59	0.7	6.6	18	0.1	3.0	10	1.0	0.7	0.3	nd	0.4
8	40	0.3	3.6	26	0.2	2.3	13	0.6	nd	nd	0.4	12
9	53	0.4	4.6	20	nd	2.9	11	0.5	nd	0.7	0.4	6
10	49	0.7	4.1	11	0.1	3.4	10	0.3	nd	0.2	nd	18
11	48	0.4	5.0	10	nd	3.9	11	0.5	nd	1.0	nd	18
12	52	0.4	4.7	10	0.1	3.8	11	0.5	nd	nd	nd	16
13	50	0.3	4.8	9.0	0.2	4.4	12	0.5	nd	nd	nd	17
14	85	0.6	3.6	3.0	0.2	0.5	0.8	0.3	nd	0.4	nd	2.8
15	47	0.4	5.5	10	0.1	3.9	11	0.4	nd	0.7	nd	19
16	46	0.4	4.8	12	0.2	3.1	10	0.5	nd	0.4	nd	19
17	46	0.6	5.8	10	0.2	4.0	11	0.5	nd	0.9	nd	20
18	51	0.4	6.0	12	nd	3.1	8.5	0.5	nd	0.5	nd	16
19	57	0.5	4.8	18	0.1	1.9	11	0.7	nd	0.6	n.d.	4.5
JD-8/24- 1	31	0.6	1.5	0.8	1.4	19	30	0.6	nd	1.9	nd	11
2	34	0.9	1.9	0.6	1.0	21	24	0.6	nd	1.4	nd	14
3	34	1.4	7.0	1.2	1.3	10	31	0.6	nd	0.9	nd	11
4	26	0.9	2.7	0.9	0.6	11	33	0.2	nd	2.0	nd	20
5	31	0.9	3.1	0.8	1.7	13	43	0.2	1.0	1.2	nd	4.6
6	32	1.4	3.6	1.6	2.6	11	39	0.2	nd	0.8	nd	7.1
7	32	1.1	5.9	1.4	1.6	11	35	0.3	nd	1.0	nd	8.9
8	24	0.2	1.1	0.8	1.6	13	41	0.2	nd	0.7	nd	16
9	34	1.1	4.7	1.3	1.7	9.3	35	0.7	nd	0.8	nd	10
10	30	1.1	3.0	1.4	2.1	11	42	0.3	nd	1.8	nd	6.7

Table A.5. *Continued*

Sample no. DBM	SiO$_2$	TiO$_2$	Al$_2$O$_3$	FeO	MnO	MgO	CaO	K$_2$O	Na$_2$O	P$_2$O$_5$	Pb	Cu
JD-8/24-11	32	1.5	4.8	1.9	3.2	9.1	37	0.1	nd	0.8	nd	8.5
12	32	1.1	3.2	0.9	2.0	11	39	0.3	nd	0.7	nd	9
13	32	0.9	2.5	0.9	2.1	12	41	0.2	nd	1.2	nd	6.3
14	31	0.6	2.1	1.5	2.0	15	39	0.1	nd	0.6	nd	7.4
JD-8/29- 1	44	2.5	4.5	1.8	4.4	8.1	25	0.5	nd	0.8	nd	7.7
2	44	2.3	4.6	1.4	3.9	8.5	23	0.4	nd	0.9	nd	10
3	43	3.0	4.5	1.5	4.1	8	24	0.4	1.2	0.4	nd	8.9
4	45	3.1	4.5	1.7	4.0	8.8	27	0.2	nd	0.3	nd	4.7
5	44	3.9	4.5	1.6	4.7	8.0	26	0.3	1.0	1.3	0.3	5.2
6	38	0.8	4.2	0.5	3.8	15	24	0.2	nd	0.2	nd	11
7	48	3.3	6.6	2.0	6.0	5.9	19	1.3	nd	1.3	nd	6
8	39	1.0	4.0	0.9	2.6	12	25	0.1	nd	0.3	nd	14
9	45	2.2	4.8	1.6	4.4	8.3	26	0.4	nd	0.6	nd	5.3
10	31	nd	4.4	0.8	nd	3.6	48	1.4	2.1	1.0	nd	7.3
11	45	2.1	3.7	2.2	4.3	8.5	22	0.2	nd	nd	nd	11
JD-8/17a-1	37	1.8	3.7	2.3	1.8	8.3	43	1.2	0.4	nd	0.60	1.6
2	37	1.6	2.8	1.2	1.8	8.3	43	0.9	nd	nd	nd	3.1
3	36	1.3	2.7	2.1	1.6	8.6	44	0.9	0.7	nd	nd	1.9
4	35	2.0	2.7	1.6	2.0	7.8	44	1.1	nd	nd	nd	3.7
5	28	1.0	2.4	0.8	1.1	8.9	32	0.7	nd	nd	nd	23
6	33	1.6	3.1	1.0	2.0	12	41	1.0	nd	nd	0.7	4.1
7	29	1.7	3.9	1.4	1.1	7.5	34	1.5	nd	nd	4.8	14
8	35	1.4	2.7	1.1	1.7	9.5	45	1.1	nd	nd	0.6	2.6
9	32	1.4	2.9	0.9	1.5	8	36	0.7	nd	nd	0.9	14
JD-8/17b-1	33	1.6	3.4	2.4	1.4	8.7	30	1.2	nd	nd	nd	17
2	29	2.2	3.1	1.5	2.1	9.1	38	0.5	nd	nd	nd	13
3	32	1.2	3.0	2.7	1.1	6.8	28	0.8	nd	nd	nd	22
4	32	0.9	2.7	2.7	1.1	6.6	31	0.9	nd	nd	nd	20
5	31	1.3	3.0	1.9	1.6	8.3	32	0.5	nd	nd	nd	19
6	31	0.9	3.2	2.4	1.90	8.2	32	0.5	nd	nd	nd	17

Table A.6. Wadi Fidan 4. Sample JD-8/13a. EMS-analyses of clinopyroxene in a Late Chalcolithic/EBA I smelting slag. *nd:* Not detected

	1	2	3	4	5	6	7	8	9	10
SiO_2	45.40	45.25	45.53	45.71	45.27	45.49	47.45	44.67	45.96	47.17
P_2O_5	0.36	0.15	0.27	0.35	0.09	0.28	0.31	0.29	0.12	0.11
TiO_2	0.07	0.04	0.07	0.08	0.03	0.06	0.04	0.07	0.04	0.03
Al_2O_3	1.96	2.25	2.10	2.06	2.26	2.10	1.63	2.23	2.29	1.97
FeO	13.25	13.24	12.88	13.47	12.61	13.27	11.16	13.99	11.91	9.98
MnO	0.07	0.07	0.06	0.04	0.07	0.05	0.11	0.08	0.06	0.08
MgO	10.40	9.73	10.21	10.63	9.51	10.30	11.62	10.01	10.00	10.76
CaO	22.73	22.81	22.65	22.87	22.86	23.04	23.32	22.77	22.91	23.32
CuO	4.30	5.09	4.42	3.29	6.06	3.84	2.73	3.56	5.02	5.00
K_2O	nd	0.02	nd	nd	0.01	0.01	nd	0.01	0.01	nd
Na_2O	0.12	0.13	0.13	0.10	0.16	0.11	0.08	0.10	0.12	0.12
Total	98.66	98.77	98.31	98.60	98.93	98.56	98.45	97.78	98.43	98.54
FeO	10.73	10.98	10.79	11.23	10.37	10.97	9.54	11.56	10.34	8.74
Fe_2O_3	2.80	2.51	2.31	2.48	2.49	2.57	1.80	2.70	1.75	1.38
Total	98.94	99.02	98.55	98.85	99.18	98.82	98.63	98.05	98.61	98.68
Mol related to 6 oxygens										
Si	1.810	1.812	1.820	1.815	1.814	1.813	1.860	1.800	1.835	1.863
Al	0.092	0.106	0.100	0.097	0.107	0.099	0.075	0.106	0.108	0.092
Ti	0.002	0.001	0.002	0.002	0.001	0.002	0.001	0.002	0.001	0.001
P	0.012	0.005	0.009	0.012	0.003	0.010	0.010	0.010	0.004	0.004
Fe^{3+}	0.084	0.076	0.070	0.074	0.075	0.077	0.053	0.082	0.053	0.041
Total	2.000	2.000	2.000	2.000	2.000	2.000	2.000	2.000	2.000	2.000
Cu	0.129	0.154	0.133	0.099	0.184	0.116	0.081	0.108	0.151	0.149
Fe^{2+}	0.358	0.368	0.361	0.373	0.347	0.366	0.313	0.389	0.345	0.289
Mn	0.002	0.003	0.002	0.002	0.003	0.002	0.004	0.003	0.002	0.003
Mg	0.618	0.580	0.608	0.629	0.568	0.612	0.679	0.601	0.595	0.633
CaO	0.970	0.979	0.970	0.973	0.982	0.984	0.980	0.983	0.980	0.987
K	0.000	0.001	0.000	0.000	0.001	0.000	0.000	0.000	0.000	0.000
Na	0.009	0.010	0.010	0.008	0.012	0.009	0.006	0.008	0.009	0.009
Total	2.086	2.095	2.084	2.084	2.097	2.089	2.063	2.092	2.082	2.070

Table A.7. Wadi Fidan 4. Sample JD-8/29a. EMS-analyses of åkermanite in an Early Bronze Age I slag. *nd:* Not detected; *na:* not analyzed

	1	2	3	4	5	6	7	8	9
SiO_2	43.93	44.67	43.26	43.24	42.21	42.44	42.34	42.51	42.41
P_2O_5	na	na	na	na	0.03	0.03	0.04	0.01	0.03
TiO_2	nd	nd	0.23	nd	nd	nd	nd	nd	nd
Al_2O_3	2.00	2.01	1.66	1.58	1.72	1.59	1.69	1.63	1.35
FeO	1.56	1.43	0.27	0.30	0.30	0.30	0.36	0.34	0.19
MgO	11.58	11.32	12.07	12.04	11.86	11.78	11.81	11.68	12.11
MnO	1.75	1.79	1.27	1.24	1.29	1.13	1.28	1.29	1.04
CaO	37.20	37.29	38.80	38.83	39.14	38.84	38.61	38.65	39.41
CuO	na	na	na	na	0.97	1.01	1.00	0.89	0.97
K_2O	na	na	na	na	0.04	0.05	0.04	0.04	0.05
Na_2O	nd	nd	0.40	0.83	0.68	0.65	0.66	0.67	0.50
Total	98.02	98.51	97.96	98.01	98.24	97.82	97.83	97.71	98.06
Fe_2O_3	1.73	1.59	0.30	0.33	0.33	0.33	0.40	0.37	0.21
$Total_K$	99.19	98.67	98.03	98.04	98.57	98.15	98.23	98.08	98.27
Mol related to 6 oxygens									
Si	2.024	2.044	2.003	2.004	1.972	1.987	1.982	1.991	1.982
Ti	–	–	0.008	–	–	–	–	–	–
Al	0.109	0.108	0.091	0.092	0.095	0.087	0.093	0.090	0.074
Fe^{3+}	0.060	0.055	0.011	0.012	0.012	0.012	0.014	0.013	0.008
Mg	0.795	0.772	0.833	0.832	0.825	0.822	0.824	0.815	0.843
Mn	0.068	0.069	0.050	0.049	0.051	0.045	0.051	0.051	0.041
Cu	–	–	–	–	0.043	0.036	0.035	0.032	0.034
P	–	–	–	–	0.001	0.001	0.001	–	0.001
Total	3.056	3.048	2.996	2.989	2.999	2.990	3.000	2.992	2.983
Ca	1.836	1.828	1.925	1.928	1.958	1.948	1.937	1.940	1.973
K	nd	nd	0.002	0.003	0.002	0.003	0.003	0.002	0.003
Na	nd	nd	0.036	0.057	0.061	0.059	0.060	0.061	0.046
Total	1.836	1.828	1.963	1.988	2.021	2.010	2.000	2.003	2.022

Table A.8. Wadi Fidan 4. Sample JD-8/24. EMS-analyses of merwinite in a Late Chalcolithic/Early Bronze Age I slag. *nd:* Not detected

	1	2	3	4	5	6	7	8	9
SiO_2	35.03	34.80	34.79	35.16	34.61	34.24	34.81	34.73	35.15
TiO_2	nd	nd	nd	nd	nd	nd	nd	nd	nd
P_2O_5	0.34	0.66	0.72	0.48	0.84	1.20	0.71	0.73	0.42
FeO	0.05	0.02	0.05	0.00	0.08	0.06	0.04	0.08	0.02
MnO	0.50	0.33	0.39	0.38	0.76	0.27	0.43	0.44	0.41
MgO	11.63	11.77	11.72	11.61	11.57	11.70	11.52	11.36	11.58
CaO	50.22	50.00	50.31	50.77	49.55	49.75	50.51	50.01	50.50
CuO	0.28	0.27	0.54	0.13	0.66	0.71	0.30	0.39	0.22
K_2O	0.02	0.06	0.05	0.04	0.02	0.03	0.06	0.07	0.04
Na_2O	0.08	0.19	0.16	0.15	0.16	0.16	0.16	0.16	0.11
Total	98.17	98.10	98.73	98.73	98.25	98.12	98.54	97.97	98.45
Mol related to 6 oxygens									
Si	1.963	1.947	1.940	1.958	1.939	1.918	1.943	1.949	1.962
Ti	0.000	0.000	0.000	0.000	0.000	0.000	0.000	0.000	0.000
P	0.016	0.031	0.034	0.023	0.040	0.057	0.033	0.035	0.020
Al	0.003	0.009	0.003	0.002	0.003	0.003	0.004	0.005	0.004
Total	1.982	1.987	1.977	1.983	1.982	1.987	1.980	1.989	1.986
Al	0.000	0.000	0.000	0.000	0.000	0.000	0.000	0.000	0.000
Fe	0.002	0.001	0.002	0.000	0.004	0.003	0.002	0.004	0.001
Mn	0.024	0.016	0.018	0.018	0.036	0.013	0.020	0.021	0.020
Cu	0.012	0.012	0.023	0.005	0.028	0.030	0.013	0.016	0.009
Mg	0.972	0.981	0.974	0.964	0.967	0.977	0.959	0.950	0.963
K	0.001	0.004	0.004	0.003	0.001	0.002	0.004	0.005	0.003
Na	0.009	0.020	0.018	0.016	0.017	0.018	0.018	0.018	0.012
Total	1.020	1.033	1.039	1.006	1.053	1.043	1.016	1.014	1.008
Ca	3.016	2.996	3.005	3.029	2.975	2.986	3.021	3.007	3.021

Table A.9. Wadi Fidan 4. Sample JD-8/29a. EMS-analyses of diopside in a Late Chalcolithic/Early Bronze Age I slag. *nd:* Not detected

	1	2	3	4	5	6	7	8	9	10
SiO_2	53.24	53.55	56.63	53.55	52.94	52.08	52.93	54.41	52.98	52.43
TiO_2	nd	nd	nd	nd	0.12	0.21	nd	0.06	nd	nd
Al_2O_3	2.35	2.37	0.89	2.55	2.64	3.07	1.78	3.63	3.51	4.31
FeO	1.64	0.81	0.23	1.45	1.51	1.83	1.09	1.94	1.98	2.44
MgO	16.17	16.23	17.30	16.02	15.90	15.21	16.69	15.24	15.41	14.33
MnO	1.95	1.74	1.11	2.04	1.79	2.19	1.45	2.60	2.22	2.72
CaO	24.29	24.39	24.82	24.39	24.67	24.55	24.31	22.80	25.00	24.67
K_2O	nd	nd	nd	nd	nd	nd	nd	nd	nd	nd
Na_2O	nd	nd	nd	0.04	0.08	nd	0.06	0.05	0.05	0.07
Total	99.64	99.09	100.98	100.04	99.65	99.14	98.29	100.73	101.15	100.97
Mol related to 6 oxygens										
Si	1.951	1.962	2.019	1.952	1.941	1.926	1.960	1.963	1.921	1.911
Al	0.049	0.038	nd	0.048	0.059	0.074	0.040	0.637	0.079	0.089
Total	2.000	2.000	2.019	2.000	2.000	2.000	2.000	2.000	2.000	2.000
Al	0.052	0.065	0.037	0.062	0.055	0.060	0.036	0.117	0.071	0.096
Ti	nd	nd	nd	nd	0.003	0.006	nd	0.002	nd	nd
Fe	0.050	0.025	0.007	0.044	0.046	0.057	0.034	0.059	0.06	0.074
Mn	0.061	0.054	0.034	0.063	0.056	0.069	0.046	0.079	0.068	0.084
Mg	0.883	0.887	0.919	0.871	0.869	0.838	0.921	0.820	0.833	0.778
Ca	0.954	0.958	0.948	0.953	0.969	0.973	0.964	0.881	0.971	0.963
K	nd	nd	nd	nd	nd	nd	nd	nd	nd	nd
Na	nd	nd	nd	0.003	0.006	nd	0.004	0.004	0.004	0.005
Total	2.000	1.989	1.945	1.996	2.004	2.003	2.005	1.962	2.007	2.000

Table A.10. EMS-analyses of olivine (tephroite) in various slags from the Faynan-District. *JD-1/7*: Faynan 1, Roman; *JD-2/2*: Khirbet en-Nahas, Iron Age II/III; *nd*: not detected; *na*: not analyzed

	JD-1/7					JD-2/2								
	1	2	3	4	5	1	2	3	4	5	6	7	8	9
SiO_2	31.59	31.32	31.43	32.61	31.04	32.31	29.82	30.83	32.08	27.89	28.65	26.19	29.06	30.03
TiO_2	0.04	0.04	0.02	0.08	nd	0.18	0.23	0.30	0.19	0.29	0.30	0.18	0.26	0.28
Al_2O_3	0.10	0.04	0.01	1.19	0.04	2.29	1.68	1.89	2.14	0.28	0.28	0.38	0.47	0.53
FeO	2.25	2.57	2.67	2.45	2.16	3.27	3.47	3.25	3.11	3.63	3.66	3.47	3.67	3.27
MnO	60.01	60.33	59.41	56.08	60.39	51.63	53.96	53.97	51.49	58.58	60.14	60.74	58.36	54.97
MgO	2.28	2.35	2.16	1.72	2.49	1.93	2.01	2.06	1.78	2.01	2.03	1.81	2.01	2.07
CaO	2.34	2.35	2.45	3.28	2.47	6.00	5.49	5.38	5.69	4.48	4.52	4.59	4.75	5.73
BaO	nd	nd	0.16	0.49	0.08	0.60	0.33	0.33	0.60	0.39	0.39	0.12	0.43	0.39
CuO	na	na	na	na	na	nd	0.01	nd	nd	0.01	0.01	0.03	0.01	0.02
K_2O	nd	nd	0.02	0.69	0.03	1.85	1.17	1.21	1.23	0.36	0.37	0.43	0.67	0.39
Na_2O	0.02	0.05	0.02	0.20	0.01	0.19	0.18	0.21	0.12	0.13	0.13	0.08	0.18	0.21
P_2O_5	na	na	na	na	na	nd	0.66	0.50	0.23	0.44	0.45	0.61	0.61	0.65
Total	98.63	99.05	98.35	98.79	98.71	100.25	99.00	99.93	98.66	98.49	100.93	98.63	100.48	98.54
Mol related to 4 oxygens														
Si	1.035	1.026	1.036	1.053	1.022	1.023	0.970	0.987	1.018	0.940	0.942	0.945	0.952	0.984
Ti	0.001	0.001	0.005	0.002	–	0.004	0.007	0.007	0.004	0.007	0.007	0.005	0.007	0.007
Al	–	–	–	–	–	–	0.023	0.006	–	0.053	0.051	0.055	0.028	0.016

Table A.10. Continued

	JD-1/7					JD-2/2								
	1	2	3	4	5	1	2	3	4	5	6	7	8	9
Mol related to 4 oxygens (continued)														
Fe^{3+}	–	–	–	–	–	–	–	–	–	–	–	–	–	–
Total	1.036	1.027	1.041	1.055	1.022	1.027	1.000	1.000	1.022	1.000	1.000	1.000	0.987	1.000
Al	0.004	0.002	0.000	0.045	0.002	0.085	0.042	0.055	0.071	0.001	–	–	–	0.005
Fe	0.062	0.070	0.074	0.066	0.059	0.087	0.095	0.087	0.084	0.102	0.101	0.099	0.101	0.090
Mn	1.666	1.674	1.658	1.534	1.684	1.395	1.500	1.484	1.498	1.632	1.636	1.629	1.601	1.537
Mg	0.111	0.115	0.106	0.083	0.122	0.091	0.097	0.098	0.085	0.101	0.099	0.092	0.098	0.101
Ca	0.082	0.082	0.087	0.114	0.087	0.214	0.197	0.185	0.196	0.162	0.159	0.168	0.167	0.202
Ba	–	–	0.002	0.006	0.001	0.008	0.004	0.004	0.008	0.005	0.005	0.002	0.006	0.005
Cu	–	–	–	–	–	–	–	–	–	–	–	–	–	–
K	–	–	0.001	0.029	0.001	0.075	0.059	0.060	0.050	0.016	0.016	0.019	0.028	0.001
Na	0.001	0.003	0.002	0.012	0.000	0.012	0.011	0.013	0.008	0.009	0.009	0.005	0.011	0.016
P	–	–	–	–	–	–	0.018	0.014	0.006	0.012	0.013	0.018	0.017	0.014
Total	1.926	1.946	1.929	1.889	1.956	1.967	2.023	2.000	2.006	2.040	2.001	2.009	2.029	1.989
Teph	89.6	89.5	89.6	88.6	88.9	82.1	83.6	84.0	84.2	96.2	86.3	86.2	85.8	83.5
Fo	6.0	6.1	5.7	4.8	6.4	5.3	5.4	5.5	4.8	5.3	5.2	4.9	5.2	5.5
La	4.4	4.4	4.7	6.6	4.6	12.6	11.0	10.5	11.0	8.5	8.4	8.9	9.0	11.0

Table A.11. EMS-analyses of olivine in a slag from Ras en-Naqab, Early Bronze Age II/III. *nd:* Not detected

JD-5/1i	1	2	3	4	5	6	7	8	9	10
SiO_2	29.44	29.06	29.58	29.36	29.81	31.17	29.87	29.05	31.24	31.11
TiO_2	0.25	0.25	0.30	0.28	0.30	0.57	0.36	0.26	0.62	0.52
Al_2O_3	0.82	0.90	1.12	0.90	1.10	0.88	0.98	0.91	3.23	3.12
FeO	1.68	1.26	1.55	1.26	1.55	0.93	1.53	1.24	2.23	2.13
MnO	57.24	56.10	55.21	55.87	54.87	53.59	55.21	55.95	48.04	47.90
MgO	4.78	5.70	5.33	5.28	5.30	5.92	5.43	5.69	2.96	2.85
CaO	2.47	2.43	2.30	2.37	2.06	0.25	2.03	2.45	4.59	4.49
BaO	0.42	0.41	0.43	0.50	0.23	2.08	1.42	0.46	1.69	1.70
CuO	2.29	2.26	2.51	2.35	2.34	2.28	2.52	2.26	4.02	3.91
K_2O	0.26	0.25	0.26	0.31	0.39	0.07	0.41	0.26	0.87	0.77
Na_2O	nd	nd	0.11	0.13	0.17	0.21	0.13	0.13	0.27	0.19
P_2O_5	0.68	0.67	nd	nd	0.80	0.56	nd	0.68	0.33	0.24
Total	100.33	99.29	98.70	98.61	98.92	98.51	99.89	99.34	100.09	98.93
Mol related to 4 oxygens										
Si	0.951	0.953	0.967	0.964	0.962	1.011	0.970	0.950	0.980	0.988
Ti	0.006	0.006	0.007	0.007	0.007	0.014	0.009	0.006	0.015	0.013
Al	0.031	0.035	0.026	0.029	0.031	–	0.021	0.035	0.005	–
Fe^{3+}	0.012	0.006	–	–	–	–	–	0.004	–	–
Total	1.000	1.000	1.000	1.000	1.000	1.025	1.000	0.995	1.000	1.001
Al	–	–	0.017	0.006	0.011	0.024	0.017	–	0.116	0.117
Fe	0.034	0.029	0.043	0.035	0.042	0.002	0.042	–	0.059	0.057
Mn	1.566	1.543	1.528	1.553	1.501	1.503	1.519	1.550	1.430	1.401
Mg	0.230	0.276	0.259	0.258	0.255	0.286	0.263	0.277	0.140	0.136
Ca	0.086	0.085	0.080	0.083	0.071	0.009	0.071	0.086	0.156	0.154
Ba	0.005	0.005	0.006	0.007	0.003	0.026	0.018	0.006	0.021	0.021
Cu	0.056	0.056	0.062	0.058	0.057	0.056	0.062	0.056	0.096	0.095
K	0.011	0.011	0.011	0.014	0.016	0.003	0.017	0.011	0.035	0.032
Na	–	–	0.007	0.009	0.011	0.014	0.009	0.009	0.017	0.012
P	0.019	0.018	–	–	0.022	0.015	–	0.019	0.009	0.007
Total	2.007	2.023	2.013	2.023	1.989	1.938	2.018	2.014	2.079	2.032
Teph	83.3	81.1	81.8	81.9	82.1	83.6	82.0	81.0	82.9	82.9
Fo	12.2	14.4	13.9	13.6	14.0	15.9	14.2	14.5	8.1	8.0
La	4.5	4.5	4.3	4.3	3.9	0.5	3.8	4.5	9.0	9.1

Table A.12. EMS-analyses of olivine (tephroite, knebelite) in various slags from the Faynan-District. *JD-1/21b*: Faynan 1, Roman; *JD-6/3*: El-Furn, Mamlukk period; *nd*: not detected

	JD-1/21b					JD-6/3												
	1	2	3	4	5	1	2	3	4	5	6	7	8	9	10	11	12	13
SiO_2	29.38	29.17	29.48	29.51	29.57	29.01	28.79	28.82	29.03	29.31	29.21	29.02	29.16	29.25	29.44	29.46	29.24	29.16
TiO_2	0.06	0.07	0.05	0.05	0.08	0.05	0.04	0.04	0.02	0.03	nd	nd	0.08	0.03	0.03	0.04	0.04	0.08
Al_2O_3	0.26	0.01	0.07	0.04	0.11	0.09	0.14	0.13	0.11	0.08	0.12	0.11	0.09	0.08	0.11	0.11	0.10	0.09
FeO	2.40	2.46	2.48	2.33	2.11	54.01	54.72	53.92	51.58	49.77	48.39	48.80	53.32	48.75	49.17	48.40	49.20	54.14
MnO	63.15	64.47	64.42	63.88	63.70	13.80	13.89	15.04	16.87	18.50	19.10	19.13	13.03	19.12	18.78	18.85	18.53	13.64
MgO	1.19	0.96	0.86	1.02	1.00	0.28	0.24	0.31	0.69	1.24	1.45	1.58	0.27	1.71	1.79	1.84	1.43	0.33
CaO	1.10	1.43	1.39	1.43	1.48	1.86	1.73	1.36	0.96	0.76	0.78	0.71	3.21	0.72	0.72	0.71	0.77	2.42
BaO	0.39	0.19	0.17	0.09	0.18	0.06	nd	0.02	nd	0.05	0.05	0.06	0.17	0.06	0.03	0.12	0.07	0.05
CuO	0.13	0.14	nd	nd	nd	0.02	nd	nd	0.10	0.08	nd	0.03	nd	nd	0.02	0.03	0.03	0.03
K_2O	nd	nd	nd	nd	0.04	nd	nd	nd	nd	0.01	nd	nd	nd	nd	nd	nd	nd	0.02
Na_2O	0.07	nd	0.01	nd	0.05	0.03	0.01	nd	0.02	0.01	0.06	0.01	0.01	0.02	nd	0.03	0.03	0.02
P_2O_5	0.14	0.06	0.14	0.14	0.05	0.09	0.05	0.08	0.07	0.10	0.19	0.19	0.10	0.16	0.11	0.06	0.12	0.09
Total	98.27	98.96	99.07	98.49	98.37	99.30	99.61	99.72	99.45	99.93	99.35	99.64	99.44	99.90	100.20	99.65	99.56	100.07
Fe_2O_3	0.00	0.48	0.17	0.06	0.00	0.37	0.59	0.61	0.43	0.44	0.45	0.74	0.33	0.62	0.48	0.28	0.42	0.49
FeO	2.40	2.03	2.33	2.27	2.11	53.67	54.19	53.36	51.19	49.38	47.98	48.14	53.02	48.19	48.74	48.15	48.82	53.70
$Total_k$	98.27	99.02	99.09	98.49	98.37	99.34	99.68	99.79	99.48	99.96	99.39	99.70	99.48	99.97	100.25	99.66	99.6	100.12
Mol related to 4 oxygens																		
Si	0.993	0.986	0.992	0.996	0.999	0.986	0.978	0.978	0.984	0.985	0.984	0.977	0.986	0.980	0.983	0.988	0.984	0.982
Ti	0.002	0.002	0.001	0.001	0.002	–	0.001	0.001	–	0.001	–	–	0.002	0.001	0.001	0.001	0.001	0.002

Table A.12. *Continued*

	JD-1/21b					JD-6/3												
	1	2	3	4	5	1	2	3	4	5	6	7	8	9	10	11	12	13
Mol related to 4 oxygens (continued)
Al	0.005	–	0.003	0.001	–	–	0.005	0.001	0.004	0.003	0.005	0.004	0.004	0.003	0.004	0.004	0.004	0.004
Fe³⁺	–	0.012	0.004	0.002	–	0.010	0.015	0.016	0.011	0.011	0.011	0.019	0.009	0.016	0.012	0.007	0.011	0.012
Total	1.000	1.000	1.000	1.000	1.001	0.995	1.000	0.995	1.000	1.000	0.989	1.000	1.000	1.000	1.000	0.993	1.000	1.000
Al	0.005	–	–	–	0.004	–	–	–	–	–	–	–	–	–	–	–	–	–
Fe²⁺	0.068	0.057	0.065	0.064	0.060	1.525	1.540	1.515	1.452	1.388	1.352	1.356	1.499	1.351	1.361	1.350	1.375	1.513
Mn	1.808	1.845	1.835	1.825	1.821	0.397	0.400	0.432	0.485	0.527	0.545	0.546	0.373	0.543	0.531	0.535	0.528	0.389
Cu	0.003	0.004	–	–	–	0.001	–	0.003	0.002	–	0.001	–	–	0.001	0.001	0.009	0.009	0.009
Mg	0.060	0.049	0.043	0.051	0.051	0.014	0.012	0.016	0.035	0.062	0.073	0.079	0.014	0.086	0.089	0.092	0.072	0.017
Ca	0.040	0.052	0.050	0.052	0.053	0.068	0.063	0.050	0.035	0.027	0.028	0.026	0.116	0.026	0.026	0.025	0.028	0.088
Ba	0.005	0.003	0.002	0.001	0.002	0.001	–	–	–	0.001	0.001	0.001	0.002	0.001	–	0.002	0.001	0.001
P	0.004	0.002	0.004	0.004	0.001	0.003	0.001	0.002	–	0.003	0.006	0.006	0.003	0.005	0.003	0.002	0.003	0.003
Na	0.005	–	0.001	–	0.003	0.002	0.001	–	0.001	–	0.004	–	0.001	0.001	–	0.002	0.002	0.001
K	–	–	–	–	0.002	–	–	–	–	–	–	–	–	–	–	–	–	0.001
Total	1.997	2.011	2.001	1.997	1.997	2.010	2.017	2.018	2.009	2.007	2.009	2.012	2.007	2.012	2.010	2.016	2.017	2.020
Fa	94.8	94.9	95.2	94.6	94.6	76.6	76.9	75.9	73.6	71.5	70.2	70.4	75.4	70.4	71.0	70.7	71.2	76.0
Teph	3.1	2.5	2.2	2.7	2.6	20.0	20.0	21.7	24.6	27.1	28.3	28.3	18.7	28.8	27.7	28.0	27.4	19.6
Fo	2.1	2.6	2.6	2.7	2.8	3.4	3.1	2.4	1.8	1.4	1.5	1.3	5.9	1.4	1.3	1.3	1.4	4.4
La																		

Table A.13. EMS-analyses of pyroxenoides in Roman slags from Faynan 1. *nd*: Not detected

	JD-1/7				JD-1/21b									
	1	2	3	4	1	2	3	4	5	6	7	8	9	10
SiO_2	47.59	42.17	47	48.13	46.00	46.09	44.91	45.37	45.6	45.53	45.55	45.53	45.60	45.69
TiO_2	0.16	0.34	0.18	0.04	0.07	0.07	0.1	0.07	0.09	0.07	0.08	0.1	0.07	0.11
Al_2O_3	3.24	6.03	1.25	0.63	0.11	0.36	0.36	0.43	0.21	0.4	0.27	0.24	0.28	0.3
FeO	1.49	2.67	1.32	1.15	0.7	0.69	0.9	1.08	0.02	0.8	0.8	0.91	0.65	0.82
MnO	26.99	22.88	31.59	34.46	46.92	43.57	48.63	45.67	48.01	47.27	46.83	48.17	47.76	47.22
MgO	0.74	0.56	0.49	1.44	1.04	0.79	0.97	0.55	1.03	0.9	0.95	1.04	1.24	1.24
CaO	15.91	14.77	16.58	13.82	3.9	7.24	2.5	5.21	2.96	3.81	3.76	2.81	3.02	3.18
BaO	1.58	5.29	0.22	0.31	0.12	0.34	0.14	0.18	0.17	0.12	0.18	0.09	0.13	0.03
CuO	nd	nd	nd	nd	nd	0.01	nd	nd	0.02	nd	0.06	nd	0.09	nd
K_2O	0.65	2.36	0.13	0.06	nd	0.24	nd	0.12	nd	nd	nd	nd	nd	nd
Na_2O	0.04	0.12	0.01	nd	nd	0.04	nd	0.03	0.04	nd	0.01	0.01	nd	0.01
P_2O_5	nd	nd	nd	nd	nd	0.06	0.03	0.03	nd	0.02	0.01	0.02	nd	0.04
Total	98.39	97.19	98.77	100	98.85	99.5	98.54	98.74	99.04	98.9	98.49	98.92	98.85	98.63
Mol related to 6 oxygens														
Si	1.976	1.840	1.962	1.996	1.995	1.979	1.970	1.975	1.984	1.979	1.987	1.982	1.983	1.985
Al	0.024	0.160	0.038	0.004	0.005	0.018	0.019	0.022	0.011	0.020	0.014	0.012	0.014	0.015
Total	2.000	2.000	2.000	2.000	2.000	1.997	1.989	1.997	1.994	1.999	2.001	1.994	1.997	2.000

Table A.13. *Continued*

	JD-1/7				JD-1/21b									
	1	2	3	4	1	2	3	4	5	6	7	8	9	10
Mol related to 6 oxygens *(continued)*														
Al	0.135	0.150	0.024	0.027	–	–	–	–	–	–	0.001	–	–	0.001
Ti	0.005	0.011	0.006	0.001	0.020	0.002	0.003	0.002	0.003	0.002	0.003	0.003	0.002	0.004
Fe	0.052	0.098	0.046	0.040	0.025	0.025	0.033	0.039	0.034	0.029	0.029	0.033	0.024	0.030
Mn	0.949	0.846	1.117	1.210	1.723	1.584	1.807	1.684	1.769	1.740	1.730	1.776	1.759	1.738
Mg	0.046	0.037	0.058	0.089	0.067	0.050	0.064	0.036	0.067	0.059	0.061	0.067	0.081	0.080
Ca	0.708	0.691	0.742	0.614	0.181	0.333	0.117	0.243	0.138	0.177	0.176	0.131	0.141	0.148
Ba	0.026	0.091	0.004	0.005	0.002	0.006	0.002	0.003	0.003	0.002	0.003	0.002	0.002	0.001
Cu	–	–	–	–	–	–	–	–	0.001	–	0.002	–	0.003	–
K	0.035	0.131	0.007	0.003	–	0.129	–	0.007	–	–	–	–	–	–
Na	0.003	0.010	0.001	0.000	–	0.037	–	0.003	0.003	–	0.001	0.001	–	0.001
P	–	–	–	–	–	0.002	0.001	0.001	–	0.001	–	0.001	–	0.002
Total	1.958	2.064	2.004	1.989	2.001	2.019	2.028	2.018	2.018	2.010	2.005	2.014	2.012	2.005

Table A.14. EMS-analyses of pyroxenoides in Roman slags from Faynan 1. *nd:* Not detected

	JD-1/33d										
	1	2	3	4	5	6	7	8	9	10	11
SiO_2	46.93	46.99	47.12	46.18	47.11	43.06	46.83	46.84	46.81	46.16	47.26
TiO_2	0.06	0.05	0.05	0.06	0.03	0.41	0.12	0.07	0.06	0.05	0.05
Al_2O_3	0.24	0.18	0.62	0.21	0.17	0.21	0.11	0.14	0.18	0.24	0.17
FeO	0.54	0.58	0.52	0.64	0.55	0.83	0.56	0.41	0.59	0.56	0.50
MnO	38.51	39.65	37.50	39.37	39.22	43.83	39.33	38.74	39.83	39.96	39.49
MgO	2.13	2.11	2.23	2.13	2.30	1.52	2.09	2.21	2.07	2.11	2.29
CaO	9.77	9.23	9.33	9.15	9.36	5.52	9.27	9.36	8.74	8.95	9.23
BaO	0.00	0.04	0.08	0.03	0.06	0.10	0.07	0.00	0.07	0.11	0.03
CuO	0.13	0.19	0.19	0.23	0.21	1.42	0.44	0.17	0.24	0.13	0.11
K_2O	0.00	0.00	0.38	0.00	0.00	0.17	0.00	0.00	0.00	0.00	0.00
Na_2O	0.03	0.05	0.11	0.06	0.03	0.01	0.09	0.04	0.09	0.05	0.02
P_2O_5	0.02	0.00	0.03	0.01	0.03	0.05	0.03	0.02	0.00	0.02	0.00
Total	98.36	99.07	98.16	98.07	99.07	97.13	98.94	98.00	98.68	98.34	99.15
Mol related to 6 oxygens											
Si	1.994	1.991	1.999	1.980	1.991	1.922	1.988	1.998	1.992	1.977	1.995
Al	0.006	0.009	0.001	0.011	0.009	0.011	0.006	0.002	0.008	0.001	0.005
Total	2.000	1.999	2.000	1.991	2.000	1.934	1.993	2.000	2.000	1.979	2.000
Al	0.006	0.000	0.029	0.000	0.000	0.000	0.000	0.005	0.001	0.000	0.003
Ti	0.002	0.001	0.002	0.002	0.001	0.014	0.004	0.002	0.002	0.002	0.002
Fe	0.019	0.021	0.019	0.023	0.020	0.031	0.020	0.015	0.021	0.020	0.018
Mn	1.386	1.422	1.347	1.430	1.404	1.657	1.414	1.399	1.436	1.450	1.412
Mg	0.135	0.133	0.141	0.136	0.145	0.101	0.132	0.141	0.131	0.135	0.144
Ca	0.445	0.419	0.424	0.420	0.424	0.264	0.421	0.428	0.399	0.411	0.417
Ba	0.000	0.001	0.001	0.005	0.001	0.002	0.001	0.000	0.001	0.002	0.000
Cu	0.004	0.006	0.006	0.008	0.007	0.048	0.014	0.005	0.008	0.004	0.004
K	0.000	0.000	0.021	0.000	0.000	0.010	0.000	0.000	0.000	0.000	0.000
Na	0.002	0.004	0.009	0.005	0.002	0.001	0.007	0.003	0.007	0.004	0.002
P	0.001	0.000	0.001	0.001	0.001	0.002	0.001	0.001	0.000	0.001	0.000
Total	1.999	2.007	1.999	2.029	2.005	2.129	2.015	1.998	2.006	2.028	2.001

Table A.15. Glass analyses (EMS) of Early Bronze Age II/III (JD-5, 21) and Roman slags (JD-1). Indicated are colours visible polarising microscopy. *na*: Not analyzed

Sample no.	Color	SiO$_2$	TiO$_2$	Al$_2$O$_3$	FeO	MnO	MgO	BaO	CaO	CuO	Na$_2$O	K$_2$O	P$_2$O$_5$	PbO	Total
JD-1/7, Faynan 1, Roman
1		40.3	0.22	3.79	2.55	36.3	1.19	3.31	6.80	na	0.47	2.59	na	na	97.52
2		43.0	0.32	6.84	2.42	20.8	0.40	6.22	10.1	na	0.79	3.98	na	na	94.87
3		41.8	0.34	6.23	2.63	20.0	0.36	6.06	11.7	na	0.65	3.43	na	na	93.20
4		43.0	0.37	7.09	2.55	20.7	0.34	5.46	8.95	na	0.83	4.53	na	na	93.82
5		38.9	0.20	4.23	2.55	37.0	1.18	2.48	6.29	na	0.43	2.72	na	na	95.98
6		41.7	0.21	5.84	2.06	25.1	0.65	2.41	7.36	na	0.82	3.80	na	na	89.95
JD-5/1i, Ras en-Naqab 1, Early Bronze Age II/III
1	Yellow	36.9	0.65	6.79	4.30	22.3	0.90	5.56	4.56	9.66	0.63	1.33	1.24	na	94.82
2	Yellow	36.2	0.46	6.84	3.52	20.9	0.65	5.38	7.56	9.11	0.64	1.69	1.07	na	94.02
3	Yellow	37.6	0.66	8.93	3.10	18.6	0.54	5.50	5.93	9.38	0.71	1.94	1.34	na	94.23
4	Yellow	36.9	nd	7.44	4.45	21.0	0.66	5.21	5.46	9.72	na	1.51	1.23	na	93.58
5	Yellow	37.7	na	8.10	4.22	19.7	0.66	5.21	5.86	9.81	na	1.73	1.27	na	94.26
6	Yellow	39.0	na	8.94	3.01	16.7	0.53	5.05	6.03	9.33	na	2.34	1.34	na	92.27
7	Yellow	36.9	na	6.82	4.27	22.3	0.92	5.56	4.65	9.66	na	1.32	1.24	na	93.64
8	Red	37.0	0.63	7.69	4.12	20.8	0.52	5.73	8.40	7.14	0.61	1.51	1.26	na	95.41
9	Red	39.2	0.44	7.81	4.05	23.0	0.57	4.15	7.80	4.07	0.70	1.85	1.11	na	94.75
10	Red	36.2	0.40	6.95	3.86	20.6	0.65	4.29	7.26	9.30	0.72	1.72	1.21	na	93.16
11	Red	37.9	0.53	7.44	4.34	20.0	0.66	5.32	5.44	9.71	0.61	1.31	1.13	na	94.39
12	Red	37.1	0.63	7.09	4.12	21.4	0.83	5.60	4.43	9.46	0.69	1.57	1.28	na	94.20
13	Red	39.5	na	8.34	3.03	16.8	0.58	5.69	6.84	8.74	na	2.66	1.67	na	93.85
14	Red	36.1	na	6.77	2.97	18.4	0.59	7.66	5.67	8.66	na	2.25	1.35	na	90.42

Table A.15. Continued

Sample no.	Color	SiO$_2$	TiO$_2$	Al$_2$O$_3$	FeO	MnO	MgO	BaO	CaO	CuO	Na$_2$O	K$_2$O	P$_2$O$_5$	PbO	Total
JD-5/1i, Ras en-Naqab 1, Early Bronze Age II/III (continued)
15	White	36.9	0.59	7.78	4.23	21.8	0.51	5.54	7.49	7.56	0.63	1.56	1.24	na	95.83
16	White	36.3	0.56	6.62	3.81	20.7	0.59	5.44	7.12	9.11	0.74	1.88	1.23	na	94.10
JD-21/1a, Wadi Faynan 5, Early Bronze Age II/III
1	Yellow	39.5	na	6.67	3.84	15.3	2.93	2.57	17.8	0.90	0.26	2.42	1.66	0.75	94.60
2	Yellow	40.5	na	7.01	3.80	15.8	3.38	2.50	19.0	0.66	0.25	2.62	1.62	0.75	97.89
3	Yellow	41.1	na	7.09	3.53	17.8	3.08	3.61	15.8	0.63	0.36	3.67	1.98	0.84	99.49
4	Yellow	40.4	na	7.04	3.53	17.1	3.22	2.93	17.1	0.66	0.34	2.93	2.15	1.00	98.40
5	Yellow	40.8	na	8.85	3.43	18.5	3.26	2.91	16.1	0.40	0.32	2.89	2.13	0.94	100.53
6	Yellow	41.4	na	6.31	3.38	18.5	3.12	3.02	15.2	0.94	0.31	3.16	2.02	1.01	98.37
7	White	41.8	na	7.23	3.38	16.0	3.15	2.84	17.7	1.12	0.28	2.80	2.02	0.69	99.01
8	White	42.0	na	7.26	2.87	13.6	3.26	2.79	19.8	1.35	0.33	3.17	2.11	0.38	98.92
9	White	43.0	na	7.41	2.80	15.1	3.05	3.11	17.8	0.45	0.26	2.94	2.83	0.49	99.24
10	White	41.9	na	7.23	2.87	15.5	3.20	2.69	18.8	0.60	0.31	2.84	2.73	0.64	99.31
11	Blue	38.3	na	6.69	4.06	15.6	3.90	3.98	17.4	4.96	0.31	2.64	1.53	1.13	100.50
12	Blue	39.0	na	7.00	3.22	12.1	3.98	2.60	25.3	1.26	0.29	2.34	1.51	0.48	99.08
13	Blue	41.8	na	7.13	3.26	16.6	3.25	2.55	17.4	0.73	0.30	2.07	2.06	0.61	97.71
14	Blue	40.7	na	7.22	3.19	13.8	3.24	2.52	21.2	0.53	0.26	2.50	2.32	0.37	97.85
15	Blue	41.2	na	7.2	3.13	13.7	3.33	2.52	21.4	0.50	0.23	2.49	2.01	0.38	98.09
16	Blue	40.5	na	7.01	3.58	17.5	3.14	2.80	17.3	0.71	0.29	2.62	2.09	0.69	98.23
17	Blue	39.5	na	7.04	3.83	20.9	3.06	3.66	13.5	0.50	0.28	2.56	2.37	0.90	98.10
18	Blue	33.4	na	7.40	3.44	19.6	1.94	4.47	12.9	0.59	0.31	2.63	2.39	2.04	91.11

Table A.16. EMS-analyses of crednerite (JD-1/33d) and of hausmannite (JD-5A/22b). Calculations of lamellae and rims in JD-1/33d based upon the formula of bixbyite (Mn,Fe,Cu)$_2$O$_3$. Calculations of fillings in between lamellae based upon hausmannite (Mn,Cu)O × (Mn,Fe)$_2$O$_3$. *JD-1/33d*: Faynan 1, Roman; *JD-5A/2b*: Ras en-Naqab, Early Bronze Age II/III; *nd*: not detected

JD-1/33d	Rim/Lamellae						Fillings					JD-5A/2b	1	2	3	4	5	6
SiO$_2$	0.47	0.54	0.53	0.53	0.71	0.61	0.43	0.55	0.43	0.53		SiO$_2$	0.94	0.36	0.36	0.40	0.51	3.95
TiO$_2$	0.31	0.20	0.40	0.51	0.71	0.72	0.67	0.42	0.36	0.39		TiO$_2$	0.89	0.11	0.22	0.14	0.17	0.80
Fe$_2$O$_3$	5.82	5.50	6.26	6.45	11.1	11.4	7.76	7.21	7.23	6.54		Al$_2$O$_3$	5.99	2.98	4.29	3.29	3.30	5.48
Mn$_2$O$_3$	79.2	74.8	78.9	78.5	78.7	79.8	55.0	55.7	56.4	55.3		Fe$_2$O$_3$	26.98	2.35	3.01	2.27	2.61	16.21
MnO	–	–	–	–	–	–	9.9	10.3	7.37	10.6		MnO	55.08	85.33	81.96	82.96	84.51	60.98
CaO	0.04	nd	nd	0.05	0.05	0.09	0.05	0.02	0.02	nd		MgO	2.30	1.16	1.51	1.55	1.46	2.14
CuO	11.2	13.2	12.2	11.3	8.45	4.59	23.3	22.3	25.5	21.3		CaO	0.43	0.18	0.18	0.32	0.33	1.45
Total	97.04	94.24	98.29	97.34	99.72	97.21	97.11	96.5	97.31	94.66		CuO	1.36	0.52	0.20	0.33	0.18	2.75
	Cations related to 3 oxygens						Cations related to 4 oxygens					Total	93.97	92.99	91.73	91.26	93.07	93.76
Si	0.01	0.02	0.01	0.01	0.02	0.02	0.02	0.02	0.01	0.02		MnO	26.10	28.94	28.44	27.83	28.71	24.68
Ti	0.01	–	0.01	0.01	0.01	0.01	0.02	0.01	0.01	0.01		Mn$_2$O$_3$	32.25	62.74	59.55	61.35	62.10	40.39
Fe^{3+}	0.12	0.11	0.12	0.12	0.22	0.23	0.24	0.22	0.22	0.20		Total	97.24	99.34	97.75	97.48	99.37	97.85
Mn^{3+}	1.69	1.67	1.68	1.68	1.61	1.66	1.69	1.72	1.73	1.73		Mol related to 4 oxygens						
Mn^{2+}	–	–	–	–	–	–	0.33	0.35	0.25	0.37		Mn^{2+}	0.818	0.914	0.903	0.890	0.901	0.754
Ca	–	–	–	–	–	–	–	–	–	–		Mg	0.127	0.064	0.084	0.087	0.081	0.115
Cu^{2+}	0.24	0.29	0.26	0.24	0.17	0.09	0.71	0.68	0.77	0.66		Ca	0.017	0.007	0.007	0.013	0.013	0.056
Total	2.07	2.09	2.08	2.06	2.03	2.01	3.01	3.00	2.99	2.99		Cu	0.038	0.015	0.006	0.009	0.005	0.075
												Total	1.000	1.000	1.000	1.000	1.000	1.000
												Mn^{3+}	0.908	1.781	1.699	1.764	1.751	1.109
												Fe^{3+}	0.751	0.066	0.085	0.065	0.073	0.440
												Al	0.261	0.131	0.190	0.147	0.144	0.233
												Ti	0.025	0.003	0.006	0.004	0.005	0.022
												Si	0.035	0.014	0.014	0.015	0.019	0.142
												Total	1.980	1.994	1.994	1.994	1.992	1.945

Table A.17. Chemical composition of copper and copper alloys from smelting sites in the Faynan-District (AAS-analyses). Samples analyzed by neutron-activation are marked by asterisks. Samples JD-8/2d, 28/1c, 38/1a,b are Cu-rich slags. Values are given in μg g^{-1}, if not wt.-% is indicated; *na*: not analyzed; *nd*: not detected; *WF4*: Wadi Fidan 4; *WFA*: Wadi Fidan A; *Naqab*: Ras en-Naqab; *Barqa*: Barqa el-Hetiye; *Dana*: Wadi Dana; *Nahas*: Khirbet en-Nahas; *Jariye*: Khirbet el-Jariye; *Dhubb*: Abu Dhubbaneh; *TWF*: Tell Wadi Faynan (from Hauptmann et al. 1992)

Sample no. DBM	Locality	Cu (%)	Fe (%)	Pb (%)	Zn	Sn	As	Sb	Bi	Co	Ni	Ag	Au	Se	Ir	S (%)
Chalcolithic/Early Bronze Age I																
JD-8/2c	WF 4	89.4	2.14	4.53	1900	<50	620	15	<5	1530	380	<5	na	na	na	na
JD-8/2d	WF 4	48.9	10.4	0.00	40	130	25	15	<5	10	30	20	na	na	na	na
JD-8/3a*	WF 4	90.3	2.72	2.91	1300	<40	540	5	5	1030	510	33	0.06	<1.5	<0.006	0.38
JD-8/3b	WF 4	83.6	0.9	0.06	120	<50	20	<10	<5	<30	300	<5	nd	na	na	na
JD-8/4h	WF 4	92.7	<0.025	0.10	30	<50	190	<10	<5	<30	300	<5	nd	na	na	na
JD-8/5	WF 4	98.3	0.03	0.05	40	<50	160	<10	<5	<30	140	<5	nd	na	na	0.67
JD-8/5a	WF 4	75.8	0.19	0.05	50	<50	120	<10	<5	40	190	<5	na	na	na	na
276/3a*	WF 4	98.2	0.01	0.05	6	<10	<5	1	<4	5	58	0.8	<0.01	0.5	<0.001	na
276/3b*	WF 4	82.2	0.66	0.01	78	70	60	1	<2	29	195	<0.5	0.01	1	<0.02	na
276/3c	WF 4	94.7	0.33	0.02	50	125	280	<10	<5	240	230	<5	nd	nd	nd	nd
276/3d	WF 4	94.1	0.01	0.03	<30	125	<10	<10	<5	<30	70	<5	nd	nd	nd	nd
JD-28/1c	Faynan 17	38.3	14.6	0.01	40	140	250	26	<5	40	110	180	na	na	na	na
JD-38/1a	WFA	37.2	0.5	0.14	270	<50	130	<10	<5	<30	45	20	na	na	na	na
JD-38/1b	WFA	22.4	51.6	0.02	70	<50	<10	<10	<5	<30	45	20	na	na	na	na
JD-38/1c	WFA	76.8	0.05	0.01	40	<50	60.0	<10	<5	40	185	<5	na	na	na	na
Early Bronze Age II/III																
JD-5/1	Naqab	99.0	0.01	0.04	<7	<20	210	13	<6	7	63	62	0.05	15	<0.002	0.79

Appendix: Analytical Results 369

Table A.17. Continued

Sample no. DBM	Locality	Cu (%)	Fe (%)	Pb (%)	Zn	Sn	As	Sb	Bi	Co	Ni	Ag	Au	Se	Ir	S (%)
Early Bronze Age II/III (continued)																
JD-5/1b	Naqab	84.6	0.49	2.64	130	<50	390	<10	<5	230	450	<5	nd	nd	nd	na
JD-5/1c	Naqab	83.6	0.25	0.15	30	130	70	<10	<5	<30	210	60	nd	nd	nd	na
JD-5/3d	Naqab	91.1	2.56	3.52	4800	120	420	<10	<5	980	915	<5	nd	nd	nd	na
JD-5/4a*	Naqab	85.8	1.58	13.10	4100	<30	540	10	1	425	650	13	<0.02	<1	<0.004	0.07
JD-5/8a	Naqab	96.2	0.03	1.25	<30	<50	220	<10	<5	<30	265	<5	nd	nd	nd	0.28
JD-5/8b	Naqab	90.7	0.03	0.43	50	<50	270	15	<5	<30	235	<5	nd	nd	nd	0.71
JD-5/8c	Naqab	92.9	1.82	2.71	1100	<50	500	<10	<5	1700	655	34	nd	nd	nd	0.28
JD-31/1a	Barqa	96.4	0.56	0.002	<30	<50	270	<10	<5	<30	150	80	nd	nd	nd	0.72
JD-31/1b	Barqa	98.3	0.03	0.002	<30	<50	175	<10	<5	<30	110	<5	nd	nd	nd	0.54
JD-31/1c	Barqa	93.5	1.35	0.07	750	<50	1400	<10	<5	430	350	<5	nd	nd	nd	0.46
Middle Bronze Age I																
JD-13/13	Dana	86.4	0.09	0.14	95	<35	680	3	<4	19	150	660	0.06	13	<0.005	0.5
Iron Age II																
JD-1/17*	Faynan 5	92.8	1.82	2.43	980	<30	300	3	<2	720	540	9.2	<0.02	<1.5	0.01	0.29
JD-2/19a*	Nahas	90.7	2.42	2.96	1690	<35	460	7	<2	870	370	6.8	<0.02	<1.5	0.01	0.53
JD-2/19b*	Nahas	95.3	2.19	0.71	2430	<55	500	4	<3	2040	580	8.8	<0.02	<2	<0.008	na
JD-2/19c*	Nahas	64.0	2.34	2.43	700	76	730	7	3	310	360	8.1	0.03	<1	<0.003	na
JD-2/19d*	Nahas	92.9	2.61	3.11	1650	<40	2060	12	<5	460	1350	39	<0.15	<2	<0.006	na
JD-2/19e*	Nahas	92.8	2.28	0.73	2430	<55	520	4	<5	2150	610	6.4	<0.04	<3	<0.009	na

Table A.17. Continued

Sample no. DBM	Locality	Cu (%)	Fe (%)	Pb (%)	Zn	Sn	As	Sb	Bi	Co	Ni	Ag	Au	Se	Ir	S (%)
Iron Age II (continued)																
JD-2/23a	Nahas	83.1	0.82	6.38	2700	<50	410	<10	<5	610	235	<5	nd	na	na	0.68
JD-2/23b	Nahas	89.7	2.06	1.49	2800	<50	1500	15	<5	350	670	<5	nd	na	na	0.49
Iron Age III																
JD-11/1b*	Jariye	92.7	2.41	0.66	690	<100	430	5	<4	950	520	<3	0.03	<2	<0.008	0.42
JD-1/3	Jariye	77.9	1.67	0.27	490	<25	310	2	<3	790	430	<1.5	<0.03	<1	<0.004	na
Roman																
JD-37/1a	Dhubb	84.7	0.096	2.0	50	57.5	240	700	40	500	1100	575	na	na	na	na
JD-37/1b	Dhubb	71.0	0.093	1.38	50	8400	480	160	15	180	410	200	na	na	na	na
Mamlukk																
JD-1/2c*	Faynan 6	84.7	2.55	4.19	1940	<60	520	6	<5	1550	400	7.8	<0.03	<2.5	<0.009	nd
JD-1/30	Faynan 1	19.3	76.50	0.23	1750	<90	1190	24	<2	8750	4250	<11	<0.06	<10	<0.03	0.08
JD-6/4a*	El-Furn	89.0	4.79	3.26	1570	<60	5370	34	<3	1740	2750	8.4	0.05	<2	<0.008	0.32
JD-6/4b*	El-Furn	41.8	52.90	1.09	4130	<400	2130	7	<3	23	8130	14	<0.09	<15	<0.01	0.68
JD-6/4c*	El-Furn	87.8	6.86	2.04	1690	<50	4430	26	<5	2620	3210	63	0.04	<3	<0.009	nd
Undated																
JD-22/1a	TWF	92.5	2.43	7.83	2000	<50	850	<10	30	2600	1200	140	na	na	na	na
397a	TWF	87.9	3.30	0.15	320	65	520	<10	<5	320	330	40	na	na	na	na
397b	TWF	87.2	3.21	3.18	2000	<50	630	12	11	1900	1000	30	na	na	na	na
397c	TWF	89.7	<0.025	1.50	80	<50	660	<10	13	45	1000	320	na	na	na	na

Table A.18. Chemical composition of copper ingots from the surroundings of Barqa el-Hetiye. Values given in µg g^{-1}, if not wt.-% are indicated

Sample no. DBM	Fe (%)	Pb (%)	Zn	Co	Ni	Ag	As	Bi	Sb
JD-31/22a	0.79	0.80	1050	1700	470	<5	330	<5	<10
22b	1.2	3.7	970	920	310	11	525	11	<10
22c	3.0	1.5	1230	1350	350	<5	445	<5	<10
22d	1.8	5.0	1530	1300	360	<5	620	<5	<10
22e	0.43	0.7	130	165	320	<5	570	<5	<10
22f	3.1	1.2	2400	1450	420	<5	740	<5	<10
22g	1.1	3.3	560	410	380	21	370	<5	<10
22h	3.0	1.1	1400	650	280	36	580	<5	<10
22i	2.7	3.1	1080	450	310	<5	400	28	<10
22l	1.7	4.7	2600	1000	480	14	680	10	12
22n	7.3	0.3	730	230	240	210	580	<5	68
22o	2.0	4.6	3800	2500	505	<5	760	<5	<10

Appendix: Analytical Results 373

Table A.19. Trace element concentrations in copper and slags from Faynan. Sb is always <10 mg g^{-1}. Values given in μg g^{-1}, if not wt.-% are indicated

Sample no. DBM	Slag								Copper						
	Fe (%)	Pb (%)	Zn	Co	Ni	As	Ag		Fe (%)	Pb (%)	Zn	Co	Ni	As	Ag
Early Bronze Age II/III: Ras en Naqab															
JD-5/11a	2.4	0.66	100	445	60	20	<5		0.006	0.38	<20	10	340	700	38
JD-5/11b	2.1	0.57	20	435	85	15	<5		0.006	0.32	<20	<10	335	770	40
JD-5/11c	3.0	0.98	1900	450	175	20	<5		0.03	0.077	60	<10	140	225	15
JD-5/11d	3.0	1.16	2400	440	160	20	<5		<0.002	0.05	20	<10	180	200	15
JD-5/11e	2.4	1.18	1600	360	140	25	<5		0.16	0.70	110	35	240	350	15
JD-5/11f	6.5	0.33	500	235	55	15	<5		1.24	0.47	165	690	455	640	65
Early Bronze Age II/III: Faynan 9															
JD-23/30a	2.4	1.6	1950	430	100	20	5		0.11	0.39	90	15	85	60	10
JD-23/30b	2.4	1.1	1200	275	75	40	8		0.023	0.54	70	15	155	170	5
JD-23/30c	3.0	1.1	1040	490	80	30	<		0.4	1.22	125	55	245	315	5
JD-23/30d	3.2	0.5	1220	185	85	30	<5		0.018	0.84	45	10	225	240	10
JD-23/30e	2.4	1.2	2000	560	160	55	<5		0.023	0.084	30	<10	20	150	65
JD-23/30f	3.4	1.3	1450	430	95	20	<5		0.003	1.78	<20	<10	400	465	35
JD-23/30g	1.5	1.2	1350	525	110	45	<5		0.03	2.4	35	18	540	590	50
JD-23/30h	2.3	1.2	1500	410	80	25	<5		0.007	0.34	<20	<10	100	135	10

Table A.19. *Continued*

Sample no. DBM	Slag Fe (%)	Pb (%)	Zn	Co	Ni	As	Ag	Copper Fe (%)	Pb (%)	Zn	Co	Ni	As	Ag
Early Bronze Age II/III: Barqa el-Hetiye															
JD-31/20a	2.8	0.54	1100	530	70	40	<5	0.58	1.6	710	860	530	630	17
JD-31/20b	1.2	0.077	330	190	30	30	<5	1.35	1.4	540	400	510	1400	20
JD-31/20c	1.7	1.1	1390	525	95	50	<5	0.78	1.1	40	25	<20	120	35
JD-31/20d	4.3	0.072	500	140	75	20	8	1.72	0.09	230	235	280	175	65
JD-31/20e	3.1	0.021	250	75	45	10	<5	0.005	0.03	<20	<10	205	235	100
Iron Age: Khirbet en Nahas															
JD-2/27a	3.1	1.3	1200	475	265	220	8	4.8	5.9	2760	1100	955	1100	35
JD-2/27b	4.8	0.45	610	440	100	45	<5	1.7	4.8	3350	430	560	1700	10
JD-2/27c	2.2	0.19	410	160	110	80		2.7	1.5	1710	520	940	1010	<5
JD-2/27d	2.5	0.5	100	570	80	110	5	2.0	8.3	570	805	460	2400	12
JD-2/27e	1.5	0.98	170	200	90	60	<5	1.7	9.7	955	640	400	925	5
JD-2/27f	2.2	0.067	175	245	50	55	<5	2.4	0.63	1800	730	1000	815	12
Roman: Faynan 1															
JD-1/30	3.2	1.8	515	665	155	80	7	1.4	11.4	2800	3020	1800	895	80
JD-1/33	8.8	0.08	640	230	65	15	<5	37.0	0.33	110	2500	565	230	25

Table A.20. Distribution coefficients $D_{Cu/S}$ of some trace elements between copper and slag from Table A.19

Sample no. DBM	Fe	Pb	Zn	Co	Ni	As	Ag
Roman: Faynan 1							
JD-1/30	0.44	6.3	5.4	4.5	11.6	11.2	11.4
33	4.2	4.2	0.17	10.9	8.7	15.3	>5.0
Iron Age: Khirbet en Nahas							
JD-2/27a	1.5	4.5	2.3	2.3	3.6	4.8	4.4
27b	0.35	10.7	5.5	0.97	5.6	38.4	>2.0
27c	1.2	7.9	4.2	3.3	8.6	12.6	<0.83
27d	0.80	16.6	5.7	1.4	5.8	21.7	>2.4
27e	1.1	9.9	5.6	3.2	4.4	15.4	>1.0
27f	1.1	9.4	10.3	3.0	20.3	14.8	>2.4
Early Bronze Age II/III: Ras en Naqab							
JD-5/11a	0.0025	0.57	0.21	0.022	5.7	34.7	>7.6
11b	0.0028	0.64	0.028	0.023	3.9	51.3	>8
11c	0.01	0.079	0.032	0.022	0.8	11.2	>3
11d	<0.0007	0.043	0.0082	0.023	1.3	9.8	>3
11e	0.067	1.1	1.2	0.078	4.0	17.5	>3
11f	0.19	1.4	0.33	2.9	8.3	42.6	>13
Early Bronze Age II/III: Faynan 9							
JD-23/30a	0.046	0.24	0.046	0.035	0.85	8.0	2
30b	0.0096	0.49	0.059	0.055	2.1	4.3	0.62
30c	0.133	1.016	0.12	0.11	3.1	10.5	>7
30d	0.0056	1.68	0.037	0.054	2.7	8.0	>2
30e	0.0096	0.07	0.015	0.018	0.75	2.7	>13
30f	0.0009	1.37	0.014	0.023	4.2	23.2	>7
30g	0.020	2.0	0.026	0.034	4.9	13.1	>10
30h	0.003	0.28	0.013	0.024	1.3	5.4	>2
Early Bronze Age II/III: Barqa el-Hetiye							
JD-31/20a	0.21	3.0	0.65	1.6	7.5	15.6	>3.4
20b	1.1	18.2	1.6	2.2	16.8	55.6	>4
20c	0.46	1.0	0.029	0.048	0.21	2.4	>7
20d	0.40	1.3	0.46	1.7	3.7	8.8	13
20e	0.0016	1.4	0.08	0.13	4.6	23.5	>20

Table A.21. Lead isotope abundance ratios of metal drops, slags and an ore sample from Faynan

Sample no. DBM	Description	Fundort	^{208}Pb/^{206}Pb	^{207}Pb/^{206}Pb	^{204}Pb/^{206}Pb	^{208}Pb/^{204}Pb	^{207}Pb/^{204}Pb	^{206}Pb/^{204}Pb
Chalcolithic/Early Bronze Age I								
JD-8/2c	Copper	Wadi Fidan 4	2.1193	0.8699	0.05570	38.046	15.617	17.952
JD-8/2d	Slag	Wadi Fidan 4	2.0674	0.8389	0.05354	38.615	15.668	18.678
JD-8/3a	Copper	Wadi Fidan 4	2.1207	0.8706	0.05571	38.068	15.628	17.951
JD-8/5a	Copper	Wadi Fidan 4	2.1156	0.8690	0.05564	38.024	15.618	17.973
276/3a	Copper	Wadi Fidan 4	2.1188	0.8700	0.05571	38.031	15.617	17.950
276/3b	Copper	Wadi Fidan 4	2.1173	0.8695	0.05566	38.037	15.620	17.965
JD-28/1c	Slag	Faynan 17	2.0859	0.8472	0.05413	38.536	15.652	18.474
JD-38/1a	Slag	Wadi Fidan A	2.1157	0.8691	0.05565	38.020	15.619	17.970
JD-38/1b	Slag	Wadi Fidan A	2.0830	0.8443	0.05388	38.659	15.670	18.560
JD-38/1c	Copper	Wadi Fidan A	2.1163	0.8676	0.05562	38.052	15.600	17.980
Early Bronze Age II/III								
JD-1/1	Slag	Faynan 1	2.1219	0.8719	0.05583	38.005	15.617	17.911
JD-5/1	Copper	Ras en-Naqb	2.1140	0.8669	0.05543	38.141	15.641	18.042
JD-5/4a	Copper	Ras en-Naqb	2.1200	0.8704	0.05573	38.042	15.619	17.945
Middle Bronze Age/Iron Age I								
JD-13/13	Copper	Wadi Dana	2.1186	0.8696	0.05566	38.066	15.624	17.968
Iron Age II								
JD-1/17	Copper	Faynan 5	2.1185	0.8693	0.05569	38.044	15.611	17.958
JD-2/19c	Copper	Kh. en-Nahas	2.1211	0.8708	0.05572	38.064	15.627	17.945
JD-2/19d	Copper	Kh. en-Nahas	2.1183	0.8696	0.05560	38.098	15.639	17.98
JD-2/19e	Copper	Kh. en-Nahas	2.1177	0.8698	0.05567	38.043	15.625	17.964
JD-2/21	Copper ore	Kh. en-Nahas	2.0830	0.8536	0.05460	38.148	15.634	18.314
Iron Age III								
JD-11/3	Copper	Kh. el-Jariye	2.1197	0.8699	0.05560	38.125	15.646	17.985
Roman								
JD-1/9	Slag	Faynan 1	2.1189	0.8689	0.05558	38.121	15.632	17.991
Mamlukk								
JD-1/2c	Copper	Faynan 6	2.1208	0.8702	0.05565	38.108	15.635	17.968
JD-1/30	Copper	Faynan 1	2.1162	0.8687	0.05560	38.063	15.625	17.986
JD-6/4	Copper	El Furn	2.1165	0.8688	0.05561	38.060	15.624	17.983

Index

Geographical Index

A

Aarja (Oman) 246
Abu Dhubbaneh 369
Abu Matar (see Tell Abu Matar)
Abu Qurdiyah 1
Ad-Dawadimi 63
Aegean 13, 17, 35, 37, 83, 152, 180, 202, 229, 232, 282, 289, 295
Afghanistan 156, 260, 282
African iron smelting 252
Afunfun (Niger) 161, 163
Ain el-Fidan 133
Ain Ghazal 147, 258–260, 289
Ain Ziq 152, 285, 286
Al-Jadida 63
Al-Qurayqira 48, 49, 64, 134, 141
Al-Wajh (Saudi Arabia) 63
Almyras (Cyprus) 15
Alps 14, 25, 168, 235, 243, 244, 246, 248, 253
Anatolia 2, 11–14, 29, 36, 37, 55, 61, 68, 83, 111, 148, 154, 157, 160, 179, 201, 223, 229, 239, 246, 248, 255, 257, 270, 274, 281, 282, 288, 289, 293–299, 300, 302, 307, 308
Aqaba 39–41, 48, 49, 55, 57, 63, 137, 148, 158, 228, 231, 261, 290, 291, 293
Arabah 284
Arad 6, 142, 151, 218, 272, 274, 280, 285, 287, 293, 296
Arisman 232, 239, 286
Arslantepe (Anatolia) 11, 14, 38, 217, 229, 239, 248, 270, 296, 299, 300
Ashqelon Afridar 262
Az-Zuwaydidiyah 63

B

Bab edh-Dhra 273, 274, 280
Bahrain 153, 298
Baja 257, 260
Barqa el-Hetiye 44, 88, 89, 91, 141, 143, 151, 152, 204
Batan Grande 107, 220
Beer Ora 96, 155, 156, 208, 364
Beer Resisim 152, 285, 286
Beersheba 42, 44, 48, 148–150, 152, 218, 227, 261, 268, 271, 289–291, 302, 306
Beidha 63, 257–260, 289, 304
Beth Yerah 274
Bilad el-Sham 40, 308
Bir Nasib (Sinai) 14, 62, 154, 231
Bir Safadi 5, 148, 261, 265–267, 269, 271, 291, 293, 294
Birkat Ram 51
Black Sea 11, 296
British Isles 17, 149, 218, 222

C

Cabezo Juré (Huelva, Spain) 228
Canaan 1
Çatal Hüyük (Anatolia) 110, 157, 177
Caucasus 29, 295, 299, 308
Çayönü Tepesi (Anatolia) 111, 160, 179, 227, 248, 255
Central Europe 15
Chrysokamino (Crete) 29, 180, 231, 246, 250
Cudeyde 239
Cyprus 3, 10, 12, 14, 15, 19, 36, 53, 55, 60, 61, 68, 69, 79, 82, 83, 87, 153, 154, 157, 181, 201, 211, 234, 239, 243, 244, 246, 250, 251, 254, 288, 291, 296, 299

D

Dead Sea 1, 2, 40–42, 46, 49, 55, 57, 59, 65, 260, 273, 274, 285, 288
Değirmentepe (Anatolia) 217, 218, 227, 239
Deir Alla 210
Dilmun 153, 297

Geographical Index

E

Ebla 241, 298
Edomite kingdom 1, 3, 6, 41, 132, 154, 306
Egypt 62, 83, 126, 153, 154, 215, 220, 229, 231, 260, 262, 274, 289, 292, 299, 302, 303
El-Furn (see also Ngeib Asiemer) 64, 91, 126, 127
En Yahav 142, 210, 230
Enkomi 243
Ergani Maden 56, 60–62, 82, 201, 243, 250, 270, 271, 282, 285, 296, 299, 300
Erzgebirge 17, 177
Euphrates 56

F

Fatmalı Kalecik 62
Faynan 48, 69, 70, 73, 74, 76, 77, 79, 88, 91, 185, 342, 345, 348–351, 358, 361, 363, 365–376
Fertile Crescent 40, 275

G

Georgia (Caucasus) 296
Gharandal 41, 64
Gilat 261, 262
Givat Sasgon (Timna area) 67, 257, 285, 303
Golan Heights 50, 51, 261
Göltepe 223, 239
Gözlu Kule 239
Gulf of Aqaba 39, 40, 55, 57, 63, 148, 261
Gulf of Suez 55

H

Hallan Çemi Tepesi 255
Har Arif 63
Har Kahal 63
Har Yeruham 152, 285, 286
Har Zayyad 285
Hassek Höyük 36, 38, 201, 270, 296, 299, 300
Hazeva 230
Hebron 285
Hijaz 2, 42, 51, 57, 60, 63, 152
Hisarlik 239

I

Iberian Peninsula 218
Ikawaten (Niger) 161, 163
Imsayea 63

Iran 26, 55, 63, 69, 154, 156, 160, 179, 184, 197, 208, 217, 232, 239–241, 243, 255, 256, 260, 281, 286, 296, 299, 308
Israel 1
Isthmia 157

J

Jericho 39, 47, 257, 258, 261, 272, 274–280, 282, 284, 285, 289, 296–299, 301
Jordan Valley 1, 45, 49, 56, 59, 138, 257, 261, 262, 265, 271, 273, 274, 289, 296, 298
Jordanian Plateau 41–43, 45, 46, 52, 144, 288
Judean Desert, Israel 263

K

Kaman-Kalehöyük 296
Karanova 223
Kazakhstan 282
Kea Island (Cyclades) 231
Keban 12, 60, 62, 250
Kephala 231
Kfar Monash 201, 274, 297, 299, 300
Khirbet el-Ghuwebe 64, 91, 132
Khirbet el-Jariye 53, 64, 69, 89, 91, 131, 132, 147, 245, 253, 306, 369
Khirbet el-Kerak ware 274
Khirbet el-Msas 142
Khirbet en-Nahas 17, 53, 64, 86, 89, 91, 97, 103, 108, 120, 123, 126–128, 130, 132, 153, 192, 209, 245, 253, 306, 350, 358, 369
Khirbet Faynan 43, 44, 48, 91, 94, 97, 109
Khirbet Hamra Ifdan 18, 36, 88, 130, 134–136, 141, 142, 150, 151, 153, 199, 203, 239, 241, 245, 286, 288, 293, 306
Kisabekir 299
Kition (Cyprus) 78, 157, 243
Kornos 251
Kura-Araxes 274, 296, 299, 302, 308
Kythnos 14, 29, 30, 180, 231, 236

L

Lachish 152, 218, 285
Lahn-Dill-area 184
Lapithos 201
Laurion 12, 17, 35
Lebanon 2, 60, 274
Limassol 61, 299
Lythrodonda 251

M

Maadi 289, 292, 299, 300, 303
Maan 42, 210
Madsus 130
Magan 297
Magdalensberg 243
Makhtesh Ramon 63
Maysar (Oman) 179, 239
Mediterranean Sea 36, 45, 55, 294, 297, 302
Merhgarh III (Pakistan) 217
Meser 139, 218
Mesopotamia 5, 40, 61, 153, 239, 272, 293, 297, 302
Midian 2, 282
Mitterberg (Austria) 13, 178, 227, 232, 243
Moab 2, 3, 39
Monte Romero 167
Murgul (Turkey) 160, 167, 179, 226, 296

N

Nahal Besor 262, 264, 265, 268, 272, 294
Nahal Hemar 260, 289
Nahal Issaron 147, 257, 261
Nahal Mishmar 5, 30, 76, 201, 261, 263, 264, 269, 291, 292, 294, 295, 297, 308
Nahal Qanah 295
Nahal Shehoret 64
Nahal Tuweiba 64, 155
Nahal Zeelim 295
Natufian 3
Negev 58
Nevali Çori 160, 179, 225–227, 229, 248
Neve Noy 261
Ngeib Asiemer 103, 126
Nichoria 157
Nile Delta 243, 292
Norsuntepe (Anatolia) 68, 158, 161, 173, 177, 217, 218, 226, 227, 296, 299, 300
Numeira 63, 272–285, 296

O

Olympia 157
Oman 12–14, 55, 82, 126, 127, 156, 168, 179, 181, 191, 192, 194, 234, 235, 239, 246, 248, 297, 298, 300, 301
Ottoman Empire 3

P

Palestine 3, 255, 273
Paltental 243
Pella 296, 301
Peqi'in Cave 261
Petra 4, 41, 42, 51, 53, 63
Pfyn 223
Phunon 39
Pinon 39
Poliochni 202, 282
Polis 251
Politiko Phorades (Cyprus) 15, 154, 243
Punon 39

Q

Qalb Ratiye 67, 75, 78, 79, 81, 85, 91, 96, 112–115, 137, 138, 144, 145, 149, 155, 213
Qantir-Piramesse 243
Qurayyah (NW-Arabia) 282

R

Ras en-Naqab 88, 91, 104, 123–125, 187, 193, 350, 369
Red Sea 1, 10, 40, 41, 55, 303, 305
Rift Valley 40, 41, 55, 56, 59, 60, 63, 64, 68
Rio Tinto (Spain) 14, 17, 53, 87

S

Samanalawewa 236–239
Saqqara 220
Sardinia 34, 36, 82, 210, 235
Saudi Arabia 57, 59, 63, 83
seir 39
Selevac 110, 158
Serabit el-Khadim 62, 83, 260, 304
Seriphos 231
Shahdad (Iran) 239–241
Shahr-i Sokhta (Iran) 26, 160, 179, 97, 239, 243
Shanidar (Iraq) 255
Shehoret 64, 67, 230
Shim at-Tasa 63
Shiqmim 5, 14, 148, 158, 161, 178, 218, 223, 261, 264, 290, 291, 295
Siegerland 235
Sinai Peninsula (see also Serabit el-Khadim, Bir Nasib) 2, 83, 256
Skouriotissa (Cyprus) 251
Southern Levant 2, 256, 257, 262
Susa 297
Syria 2, 41, 55, 59, 60, 174, 241, 271, 274, 283, 297, 298, 302

T

Tal-i Iblis (Iran) 217
Taurides 55, 298

Geographical Index

Tell Abu Hamid 261, 265–267, 271, 291, 296
Tell Abu Matar 5, 14, 29, 70, 138, 161, 163, 177, 218–220, 224, 227, 261, 264–269, 289, 290, 296
Tell esh-Shuna 201, 262, 296–298
Tell ez-Zeiraqun 242, 283
Tell Fidan 148
Tell Halaf 271
Tell Halif 262
Tell Hujayrat al-Ghuzlan 14, 48, 90, 139, 148, 150, 152, 158, 218, 228, 262, 263, 164, 290, 291, 293, 303, 304
Tell Khuera (Syria) 174
Tell Magass 48, 152, 290, 291, 293
Tell Ramad 256, 288
Tell Wadi Faynan 109
Tepe Ghabristan 217
Tepe Sialk 177, 232
Tepecik (Anatolia) 14, 217, 239
Thasos 12
Thermi 202, 284
Timna 1, 2, 43, 56, 57, 63, 67–69, 77–82, 89, 90, 99, 102, 103, 110, 122, 138, 146, 147, 148, 150, 152, 154, 158–160, 170, 173, 174, 180, 182, 183, 194, 210–214, 228–231, 244, 248, 250–254, 261, 262, 268, 270, 272, 274, 282, 286, 290, 292, 293, 302, 307, 346, 347
Transcaucasia 274, 302
Trentino 107, 168, 178, 243
Troodos mountains 299
Troy 239, 282
Tuleilat Ghassul 261, 266, 270, 271, 283, 296
Tülintepe (Anatolia) 14, 30, 217
Tyrol 243

U

Umm an-Nar 297, 298, 301
Umm Bogma 56, 62
Umm el-Amad 91, 144, 145, 306
Umm ez-Zuhur 64, 75, 91, 130, 153
Umm Qatafa 271
Uruk 301, 302
Uvda Valley 257, 261, 273, 274, 303
Uzbekistan 282

W

Wadi Abiad 67, 71, 75, 96, 112, 115, 126, 137, 138, 146, 155, 343, 344
Wadi Abu Barqa 63
Wadi Abu Dubbana 142
Wadi Abu Khusheibah 1, 4, 64, 77, 81, 82, 85, 96, 155, 156, 271, 343–345

Wadi Abu Qurdiyah 1, 4
Wadi Amram (Israel) 64, 67, 96, 155, 214, 230, 346
Wadi Arabah 1, 2, 5, 12, 20, 36, 37, 39–46, 49, 53, 55–65, 68, 69, 83, 121, 128, 131–133, 136, 140, 148, 156, 163, 201, 232, 243, 256, 273, 291, 293, 296, 297, 303, 305
Wadi Ba'Ba (Sinai) 231
Wadi Dana 41, 42, 47, 52, 56, 64, 69, 70, 75, 89, 91, 103, 108, 122, 123, 132, 151–153, 186, 211, 342–345, 349, 350, 369, 376
Wadi Dara (Egypt) 231, 304
Wadi el-Ghuwebe 132
Wadi Faynan 4, 19, 40, 43, 47, 48, 53, 88, 94, 109–111, 147, 149, 150, 289, 349, 367, 369
Wadi Fidan 4, 14, 43, 45, 47, 52, 53, 70, 85, 88, 91, 109, 114, 130–134, 136, 138–150, 152, 153, 158, 164, 173, 174, 185, 189, 191, 194, 211–213, 218, 225, 226, 228, 229, 257, 263, 264, 289–291, 306, 341, 348–350, 352, 354–357, 369, 376
Wadi Fidan 4 86–88, 139, 140, 218, 226, 228, 264, 289, 290, 304, 348, 350, 376
Wadi Ghazzeh 70, 262, 264–268, 272, 294
Wadi Ghuwebe 41, 42, 46, 49, 69, 120, 124, 126–130, 132, 133, 153, 154
Wadi Ghwair 42, 46, 47, 49, 88, 89, 91, 94, 95, 107, 111, 112, 140, 141, 147, 188, 229, 257
Wadi Hasa 46
Wadi Huwar 63
Wadi Jariye 131, 132, 153
Wadi Khalid 64, 69, 70, 75, 85, 89, 91, 96, 112, 115–123, 126, 132, 147, 150–153, 155, 211, 286, 342, 344, 345
Wadi Khuneizira 46
Wadi Maghara (Sinai) 62, 260
Wadi Musa 46, 63
Wadi Nehushtan 67
Wadi Nimra (Timna) 214, 215
Wadi Ratiye 71, 75, 81, 112, 115, 343–345, 351
Wadi Riqueita 62
Wadi Sheger 42, 43, 48, 94
Wadi Tar (Sinai) 56, 62, 201, 281
Wadi Tawahin 156
Wadi Yitim 48

Y

Yarmukian 109
Yotvata 174, 194, 230, 261, 303

Z

Zagros 40, 55, 60, 61, 255, 298, 299
Zeytindağ (Anatolia) 12

Subject Index

A

Acacia spec. 53
African Plate 56
åkermanite 22, 164, 165, 174–176, 225, 306, 341, 355
albite 22
alite 165, 177
alloying 29, 150, 157, 219, 242, 301
Alter Mann 9
alteration of host rock 12, 24, 30, 61
alumosilicates 176
amazonite 260, 261, 288
Amir Formation 66, 73
analysis
 –, chemical 27, 264, 266, 275
 –, isotope 31, 79, 264, 275
andesite 42, 57, 111
annealing 291, 292
anorthite 22
antimony 5, 294, 299
anvil stone 104, 130, 137, 142, 246
apatite 247, 261
Arabah-sites 273
Arabian Plate 55–57, 60, 298
archaeometallurgy 7
argillitization 250
arsenical copper 28–30, 63, 149, 280, 281, 291, 295–297, 301
Arsenkupferzeit 296
arsenopyrite 301
artifact 8, 264, 270, 283
 –, analysis 10, 154
 –, arsenic concentration 280, 281
 –, Chalcolithic 11, 132
 –, composition 277, 280, 281
 –, greenstone 257
 –, lead isotope analysis 31
 –, lead isotope ratios 268, 282, 287
 –, metal 13, 17, 27–32, 35, 36, 76, 149, 201, 203, 207, 208
assaying 219
atacamite 114, 257, 261
atomic absorption spectrometry 16
Avrona Formation 73
Ayyubid-Mameluk 'industrial complex' 127
azurite 10, 12, 62

B

Ballengefüge 170
baryte 25, 66, 165, 189, 260
bellows 94, 95, 99, 154, 221, 236, 239, 243, 245
bioclimatic zones (of Jordan) 45
bisbeeite 69
bixbyite 165, 190, 197–199, 368
bloomery process 15, 232, 233, 253
blowpipe 221
blow-tube 100
borax 249
bornite 60, 61, 70, 179
botanical investigations 8, 16, 51, 120
Boudouard-equilibrium 221, 234
brass 205, 219, 284
braunite 165, 189, 193, 197
buffer equilibria 22
bustamite 165, 185, 194, 195
Byzantine 3
 –, period 4
 –, ruins 92, 93

C

calcio-olivin 247
calcite 63, 71, 72, 161, 165, 176, 258, 261
 –, in rocks 63, 71, 72, 258
 –, in slags 161, 165, 176
camp fire 217
carbon-14 dates 16, 17, 32, 88, 90
cassiterite 282
casting mould 9, 110, 135, 142, 240, 263, 286, 290, 303
celsian 165, 174
cementation 10, 12, 34, 60, 61, 205, 219
cementite 210
Central Europe, smelting sites 15
chain, metallurgical 7
chalcocite 12, 60, 70–72, 149, 165, 178
Chalcolithic 3
 –, copper from Faynan 261
chalcopyrite 12, 61, 179
chalk 58, 259, 289, 295
chaine d'operatoire 7, 9, 240
charcoal 15, 16, 17, 23, 51–53, 87, 94, 96–99, 102, 103, 107, 109, 123, 128, 140, 145, 155, 163, 164, 179, 189, 192, 209, 217, 219, 221–223, 227, 234, 247–249, 252, 253
chemical analysis 16, 19, 21, 22, 27, 31, 35, 73, 81, 128, 182, 251, 256, 264–266, 275, 277, 280, 284, 345, 346
 –, bulk analyses 19, 31, 159–162, 175, 176, 181–183, 186, 193, 341, 348
 –, micro-plane scans 159, 175, 176

Subject Index

–, of rocks 27, 73, 128, 345, 346
–, of slags 19, 21, 22, 163, 183, 251
–, spot analyses 16, 19
Chenopodiaceae spec. 53
chlorite 250
chromite 25
chrysocolla 69, 114, 240, 257, 261, 288, 344
classification of modern copper smelting processes 207
clay rod 105, 231
climate 45
clinopyroxene 22, 162, 169, 170, 172–175, 195, 247, 248, 341, 354
–, composition 173
co-smelting of copper ores 179
coloured sand- and claystones 255, 257
coin 4, 85, 94, 103, 121, 126, 144, 155
converter-process 174
copper
 –, arsenical 295
 –, arsenic-antimony-alloy 294
 –, arsenic-nickel-alloy 297
 –, chemical analyses 200, 264
 –, chlorides 25, 68, 151, 240, 257, 260, 305
 –, composition 199, 200
 –, deposit 63
  –, genesis 63
  –, geochemistry 73
  –, geological framework 63
  –, lead isotope composition 37
  –, ore 68
 –, export 255
 –, flaw casting 241
 –, homogeneity 202
 –, iron in 207
 –, lead isotope
  –, abundance ratio 212, 214
  –, analyses 211, 264
 –, lump 241
 –, native, impurities 27, 29, 203, 211, 281
 –, ore 77
 –, prill 241
 –, processing 239
 –, pure 291
 –, silicates 69, 70, 149, 163, 173, 257, 258, 260, 261, 272, 305, 344
 –, smelters 50
 –, smelting 21
  –, technology 217
 –, sponge 223
 –, stability ranges 222

–, sulphides 11, 26, 62, 71, 159, 179, 257, 305
–, trace element 199, 202, 203, 205
–, partitioning 204
coronadite 71
covellite 12, 26, 60, 61, 71, 165, 178
crednerite 165, 189, 190, 198, 199, 341, 368
crescent-shaped ingots 135, 142, 203, 273, 283–286, 288, 303
cristobalite 22, 25, 164–171
crucible 14, 18, 19, 138, 139, 142, 148, 150, 157, 158, 163, 179, 208, 217–220, 223, 225–229, 239, 240, 249, 270, 275, 290, 294, 306
–, slags 157–159, 163, 177
–, smelting 205, 217, 221–223, 243
–, thermodynamics 219
cuneiform texts 297
crushed slags 110
cup cake 241
cuprite 12, 60, 62, 71, 72, 105, 114, 158, 159, 162, 164–174, 177, 179, 187, 188, 193, 196, 200, 220–225, 227, 228, 232, 241, 248, 255, 305–307

D

Daba-Siwaqa-Marble 258, 260
dating 3–5, 8, 9, 14–18, 85, 87, 90, 96, 97, 99, 108, 109, 112
–, carbon-14 17, 32, 85, 90, 128
–, physical 16, 128
–, thermoluminescence (TL) dating 18, 90, 91, 123, 148, 290
deforestation 8
delafossite 18, 157, 164, 165, 169–174, 177, 179, 224–227, 248, 306
dendrites 11, 178, 188, 189, 193, 196, 209, 225
diopside 22, 164, 165, 170, 172–176, 306, 341, 357
dioptase 257, 259, 260, 288
dolomite 43, 53, 58, 62, 65, 66, 70, 75–78, 97, 115, 117, 124, 126, 161–163, 166, 175–177, 184, 200, 211–213, 224, 233, 250, 258, 275, 282, 305, 344
–, -Limestone-Shale Unit 43, 65, 66, 76, 77, 97, 115, 117, 163, 184, 200, 211–213, 258, 275, 305
–, sandy 75, 78, 161, 163, 224, 233, 250
domeykite 62
domestic mode production 4, 151, 264

Subject Index

E

Early Bronze Age 3
 –, copper from Faynan 261
 –, metal trade 272
 –, slag 180
 –, smelting 228
Early Islamic 3
Edomite 39, 128, 132, 153, 154, 306
enstatite 173
ethnographic studies 9, 251
eutectic 21, 25, 26, 162, 168, 170–172, 178, 189, 210, 251, 252

F

fahlore 29
fayalite 21–23, 157, 161, 162, 171, 173, 177, 185, 191–193, 195, 229, 233, 247–249, 294, 306
Faynan
 –, agriculture 47
 –, climate 45
 –, copper
  –, deposit 63
  –, production 147
 –, fuel supply 50
 –, geography 40, 86, 87
 –, geology 39
 –, importance 255
 –, irrigation 47
 –, lithostratigraphy 66
 –, mining district 85
  –, development 145
 –, nature 39
 –, ore 257
  –, trade 255
 –, palaeogeography 65
 –, slag 158, 182
  –, production 147
  –, phases 165
 –, smelting 85
  –, furnace 50
 –, topography 42
 –, toponymy 39
 –, vegetation 50
 –, vicinity 94
 –, water 47
 –, wind 49
feldspar 22
fenestrated axe 275, 283
ferrifayalite 178, 193, 248
ferroåkermanite 22
ferrobustamite 194, 195
ferrosilite 173, 195
Ferrum Noricum 235, 243
Fertile Crescent 5
fieldwork 9
firing temperatures 160, 163, 164, 184, 226, 248
flowing point temperature 185
fluorapatite 258, 261, 288
fluorite 249
fluxing agent 21, 24, 35, 72, 79, 101, 102, 148, 163, 207, 211, 227, 228, 234, 249–251, 305, 307
forsterite 161, 162, 191, 192
free enthalpy 23
166, 167
fuel supply 50
furnace 102
 –, conglomerate 189, 240
 –, construction 101
 –, reconstruction 244
 –, shaft 252
 –, wind-powered 229, 237, 238

G

galena 63, 148
gehlenite 22, 175, 247
geography 40
glass 165, 195
 –, lead-silicate 165
glaucochroite 191, 192
glauconitic chalk 295
gold 3, 10, 12, 62–64, 150, 156, 246, 261, 299
gossan 11, 229
granite 53, 111, 141
green colour 255
greenstones 257, 260, 288

H

Haloxylon persicum 44, 53
Hamada 41
hammerstones 16, 104, 130, 131, 137, 138, 153, 240, 246
Hashemite 40
hausmannite 165, 188–190, 193, 198, 233, 249, 341, 368
heating stage microscopy 185
hedenbergite 22, 172, 175, 195, 247
Hellenistic Period 3, 4
hematite 61, 71, 72, 162, 172, 226, 250
hercynite 22
hilgenstockite 165, 189
hollandite 71
hydrothermal ore deposits 11, 59, 69, 179, 250

I

ignimbrites 57
imprinting phase 4, 6
inductively coupled plasma with optical
  emission spectrometry 16
industrial stage 4, 293
ingot 135, 242
  –, crescent-shaped 283, 285
  –, trace elements 204
iron
  –, carburised iron 235
  –, Cu-Fe-alloys 253
  –, Fe-phoshate 209
  –, oxides, stability ranges 198
  –, phosphide 208
Iron Age 3
  –, smelting 242
irrigation 47
  –, Roman-Nabatean 48
iscorite 248
Islamic Middle Ages 3
Israelites 39

J

jacobsite 165, 190, 198
jarosite 12, 250
johannsenite 165, 194
Juniperus phoenicea 51, 52

K

kalifeldspar 22
kaliophilite 22
kalsilite 22
kaolinite 250, 264
karstic ore deposits 12, 62
kirschsteinite 22
knebelite 165, 186, 189, 192, 193, 195, 307
koutekite 62

L

ladyfingers 105, 231
  –, at Timna 231
laihunite 178, 193, 248
larnite 175, 176, 191, 192
Late Antique 3
Late Bronze Age 3
lateral zoning 11, 12, 70, 183
lead
  –, abundance ratios 33
  –, isotope analysis 16, 24, 31, 33–35,
    59, 79, 202, 264, 275

lead-silicate 165
ledeburite 210
leucite 22, 247
Levant Platform 55
lime plaster 259, 260
limestone 40, 53, 60, 65, 141, 147, 249, 250,
  258, 289
limonite 11, 21, 75, 171, 250, 305
Lisan Lake 59
listwaenite 299
litharge 62, 240
loellingite 29

M

maceheads 263, 268, 295, 297, 299, 300
Madsus mining district 130
maghemite 15
magmatite 62
magnetite 15, 23, 61, 162, 164, 165,
  169–172, 174, 179, 193, 195, 224–227,
  233, 235, 247–249, 306, 307
malachite 10, 12, 62, 69–72, 114, 131,
  132, 149, 151, 179, 219–222, 228,
  240, 255, 257, 258, 261, 288, 290,
  305, 344
Mameluk Period 180
  –, slag 180
manganese
  –, for fluxing 234
  –, ores in rocks 62, 68, 71, 97
  –, oxides, stability range 198
  –, phases in slag 165
  –, sulphide 208
manganite 71
marble 59, 258, 289
Massive Brown Sandstone 66, 71, 75–77,
  79, 96, 115, 270
massive sulphide deposits 12, 60, 61
material sources 55
matte smelting 25, 154, 207, 243
melilite 22, 174, 175
mercury 60, 219
merwinite 164, 165, 174–177, 306, 341,
  356
metal
  –, chemical analysis 27
  –, distribution pattern 288
  –, hoard 263
  –, production 263
  –, trade 272
    –, archaeological aspects 272
metallurgical chain 7, 8
metallurgy, domestic 263
meteorite 209

Subject Index

method, analytical 16
microlite 175
Middle Bronze Age 3
Midianite 128, 142, 152, 282, 303
migmatite 57
mine 112–121
mining
 –, dating 85
 –, district 112, 115, 116, 130, 144
 –, double shaft 133
 –, sites, carbon-14 dates 90
modeling, crucible smelting 227
monticellite 176, 191, 192
Mössbauer-spectroscopy 178
Moringa peregrina 53
mushistonite 28

N

Nabatean period 4
Najd fault system 56
native copper 11, 12, 14, 30, 61, 62, 69, 75, 111, 148, 163, 201, 255, 256, 289
Near East
 –, geology 55
 –, ore deposits 59
Neolithic
 –, Pottery 3
 –, Pre-Pottery, 3
nepheline 22, 247
Nerium oleander 52, 53
natural draught furnaces 229–233, 236, 238, 239, 243, 306
neutron activation analysis 16, 275
nickel
 –, -arsenide minerals 295, 297, 298
 –, metal 12, 241
nickeline 298
Noranda-process 207
Nubian Sandstone 58, 260

O

obsidian 288, 295, 296
Olea europaea 52
olivines 22
ophiolite 59–61, 82, 298–300
 –, Baer-Bassit, Hatay 61, 299
 –, -obduction 55
 –, Samail 298–300
 –, Troodos 55, 60, 299
ore
 –, deposit 1, 2, 9–14, 17, 27, 29, 31–36, 38, 47, 55–64, 67, 68, 73, 75, 78–83, 141, 148, 150, 179, 180, 199, 210, 217, 229, 243, 250, 251, 255, 256, 263, 273–275, 281, 282, 289, 293, 296–300, 302–305, 307
 –, trade 255
Ottoman period 4
oxhide ingot 36, 37, 78, 79, 211, 242, 254, 287, 291
oxidation zone 11, 12, 29, 250
oxides, stability range 235
oxygen partial pressure 22, 190, 193, 195, 197, 198, 206, 222, 226, 233, 234, 248, 249, 307

P

paleobotany 51
paratacamite 69, 71
partial melting 25
partitioning coefficient 204–207
partridgeite 165, 189, 190, 197
pegmatite 260
periclas 161, 165, 176
peridotite 61, 298
perlite 210
Phoenix dactylifera 53
phosphate 165
phosphorite 25, 66, 75, 78, 79, 258
pig iron 210, 235
Pistacia cf. *atlantica* 52
phase diagram (see also *system*) 20, 162, 163, 185
planchéite 70, 257
plastered skulls 289
porphyrite 64
phosphorous
 –, belt (Jordan) 210
 –, Fe-phosphate 209–211
 –, in host rocks 344
 –, in slags 78
Precambrian rock formation 41, 45, 55–57, 60, 62, 63, 83, 133, 260
precursory stage 4, 288
prestige items 263, 292
Procavia capensis 50
processes, metallurgical 232
propylitization 250
PIXE-analysis 31, 280, 284
pyrite 60, 61, 70
pyrolusite 71, 165, 189, 190
pyroxene 19, 21, 168, 172, 173, 194, 195, 224
pyroxenoid 195, 247, 307, 341, 363, 365
pyroxferroite 195
pyroxmangite 165, 194

Q

Qalb Ratiye mining district 112
quartz 249
 –, blue 70
 –, inclusions in slags 25, 161, 167, 169, 170
quartzite 78
Quercus callipronis 51, 52

R

radiocarbon 16, 17, 32, 90
 –, data
  –, Feinan 88
  –, Timna 87, 148, 262, 290
 –, dates 16, 17, 32, 88, 90
radiolites 61
rain fall (in Jordan) 46
ramsdellite 71
recycling 9, 14, 37, 153, 240, 241
redox-conditions 20, 22, 219, 248, 307
reduction 233
 –, process 191
refining slag 26, 157, 174
refractory 15, 16, 78, 100, 162, 168, 223
re-melting 204, 208
Rennfeuerverfahren (see slagging process, bloomery smelting) 232
restite 26, 98, 253
Retama raetam 52, 53
rhabdite 209
rhodonite 165, 185, 193–195, 233, 249
rhyolites 57, 60, 64, 168
roasting of ores 51, 69, 243, 244
Roman Period 3, 4

S

Salawwan Fault 64
salite 165, 172, 174
satellite 85
scanning electron microscopy 16
Scheibenreißen 167
schreibersite 165, 209
scorodite 29
secondary zoning 11, 12
self-fluxing ore 70, 72, 78, 233, 250, 251, 307
sericite 250
serpentinite 61, 255
settlement, Pre-Pottery Neolithic 256
shaft furnace 127, 162, 170, 195, 227, 233, 238, 244, 248, 251, 252, 254
shannonite 22
Shehoret Formation 66

silica bricks 164, 170
silicates, stability range 235
silver production at Fatmalı Kalecik 62
sintering 164, 176, 177
site catalogue 91
slag 157
 –, ancient 157
 –, archaeometallurgical 157
 –, CaO-MgO-rich silicate slag 160
 –, chemical bulk analysis 183
 –, chemistry 159, 182
 –, composition 21, 160
 –, copper content 182
 –, crushed 130
 –, crushing 245
 –, dating 184
 –, eutectic formation 252
 –, Faynan 158, 181
 –, Fe-rich silicate 164, 166, 247
 –, flowing point 186
  –, temperature 185
 –, 'free silica' 167
 –, furnace slag 97–99, 175, 189, 208, 247, 252, 253
 –, heap 104, 122
 –, hemispherical temperature 186
 –, investigation 18
 –, lead isotope abundance ratios 212, 214
 –, manganese-rich 180
 –, melting behavior 159, 182
 –, mineralogical phases 169, 190
 –, Mn-rich silicate slag 78, 307
 –, natural dosage rate 91
 –, petrography 164, 186
 –, phase content 247
 –, processing of slags 6, 8, 124, 129
 –, site 246
 –, softening 185, 186
 –, sulphide inclusions 178
 –, tapped 98, 99
 –, tap slag 26, 97–99, 101, 129, 132, 153, 154, 189, 246, 247, 252, 253
 –, TL-age 91
 –, trace elements 205
slagless metallurgy 14, 149, 150, 228
smelting
 –, controllable 242
 –, crucible 217
  –, modeling 227
  –, reconstruction 223
 –, dating 85
 –, experimental studies 219, 236, 251
 –, furnace 95, 106–108
 –, in wind-powered furnaces 228

Subject Index

–, Iron Age 242
–, Mameluk Period 242
–, processes 246
–, site 13, 16, 134
  –, carbon-14 dates 90
  –, documentation 13
  –, Roman 96
–, slag 105
–, technology 217
sphalerite 60, 61
spherolites 187
stannite 28
steel 154, 162, 195, 219, 232, 234, 235, 243
stockwork-mineralisation 12, 59, 61
stone
  –, anvil 130
  –, hammer 130
stratabound ore deposit 68
study, archaeometallurgical 157
sulphidic ore deposit 12
system
  –, CaO–FeO–Al$_2$O$_3$–SiO$_2$ 22
  –, CaO–FeO–SiO$_2$ 21, 22, 160, 163
  –, CaO–MgO–SiO$_2$ 162, 163, 175, 176
  –, CaO–MgO–SiO$_2$–Al$_2$O$_3$ 162, 233
  –, CaO–SiO$_2$–FeO–Al$_2$O$_3$ 21, 22
  –, CaO–SiO$_2$–MgO(+MnO+FeO) 161
  –, Cu$_2$O–f–S,A 160
  –, Cu–Fe 207
  –, Cu–Fe–O 172, 226
  –, Cu–Fe–P 209
  –, Cu–Fe–P–O 210
  –, Cu–Fe–S 179
  –, Cu–O 23, 222, 233
  –, Cu–O–SiO$_2$ 162
  –, CuO–Cu$_2$O–SiO$_2$ 173
  –, Cu–S–O 178
  –, enstatite–ferrosilite–wollastonite 173
  –, fayalite–tephroite–larnite 192
  –, Fe-C-P 210
  –, FeO–Al$_2$O$_3$–SiO$_2$ 21
  –, Fe-oxide–SiO$_2$ 25
  –, Fe–Si–O 22, 193, 235, 249
  –, forsterite–tephroite–larnite 192
  –, MnO–(FeO+MgO)–CaO 195
  –, Mn–Si–O 23, 233, 235, 249
  –, SiO$_2$–CaO–FeO 185
  –, SiO$_2$–CaO–MnO 183–185
  –, SiO$_2$–CaO–MnO(+FeO+MgO) 184, 196
  –, SiO$_2$–CaO–MnO–FeO 183
  –, SiO$_2$–Fe oxide 168
  –, SiO$_2$–FeO–Al$_2$O$_3$ 22
  –, SiO$_2$–FeO–MnO–CaO 183

–, SiO$_2$–MnO$_2$–MnO 191
–, SiO$_2$–PbO 168

T

Tamad Fault 56
Tamarix jordanis 52, 53
tannurs 138
tap slag (see *slag*)
tempering 292, 293
tenorite 177
tephroite 165, 184–193, 234, 247, 249, 253, 307
Tethyan Eurasian metallogenic belt 55, 298
texture of slags 19, 70, 159, 162–164, 169, 177, 186, 187, 189, 190, 195
Themed Fault 56
Theophilus Presbyter 253
thorium decay 32
tin 3, 17, 28, 60, 150, 154, 157, 223, 271, 274, 275, 280, 282, 283, 294, 296, 301
  –, bronze 150, 271, 294, 301
topography 42
toponymy 39
trace element 79
  –, analyses 16, 19, 38
  –, in copper 27, 62, 74–76, 199, 202–204, 251, 307, 373, 375
  –, in slags 10, 24
  –, partitioning between metal and slag 31, 199, 200, 204–207, 375
Transjordanian Block 57
tridymite 22, 25, 164–171
turquoise 62, 260, 288, 289
tuyère 95, 100, 219, 220

U

ultrabasic rocks 59, 61, 299
ulvite 247
Umm el-Amad mining district 144
Umm ez-Zuhur mining district 130
uranium decay 32, 34
urudu, urudu-luh-ha 241

V

vaisselle blanche 289
vegetation 14, 39, 40, 43, 44, 49–53, 87
vogtite 195
vredenburgite 190, 198

W

Wadi Abiad mining district 115

Wadi Fidan, slag composition 160, 161
Wadi Ghwair find spots 111
Wadi Khalid mining district 116, 117
water 47
white ware 111, 148, 289
wind-powered furnaces 228–230, 232, 236–239
wind 49, 50, 95, 106, 107, 123, 124, 228–239, 243
wollastonite 22, 162, 173, 176, 185
wuestite 22, 161, 195, 233, 247–249

X

X-ray fluorescence spectrometry 16

Z

Zagros tectonic line 298
Zakimat el-Hassa Fault 56
zinc 63, 219, 233, 284
Ziziphus spina-christi 44
zones, bioclimatic 45